INTERNATIONAL DIRECTORY

OF

AGRICULTURAL ENGINEERING INSTITUTIONS

REPERTOIRE INTERNATIONAL D'INSTITUTIONS DE GENIE RURAL

REPERTORIO INTERNACIONAL DE INSTITUCIONES DE INGENIERIA RURAL

Agricultural Engineering Service
Agricultural Services Division

Service du génie agricole
Division des Services Agricoles

Servicio de Ingeniería Agrícola
Dirección de Servicios Agrícolas

Food and Agriculture Organization of the United Nations
July 1983   Rome, Italy

The designations employed and the presentation of material in this publication do not imply the expression of any opinion whatsoever on the part of the Food and Agriculture Organization of the United Nations concerning the legal status of any country, territory, city or area or of its authorities, or concerning the delimitation of its frontiers or boundaries.

Les appellations employées dans cette publication et la présentation des données qui y figurent n'impliquent de la part de l'Organisation des Nations Unies pour l'alimentation et l'agriculture aucune prise de position quant au statut juridique des pays, territoires, villes ou zones, ou de leurs autorités, ni quant au tracé de leurs frontières ou limites.

Las denominaciones empleadas en esta publicación y la forma en que aparecen presentados los datos que contiene no implican, de parte de la Organización de las Naciones Unidas para la Agricultura y la Alimentación, juicio alguno sobre la condición jurídica de países, territorios, ciudades o zonas, o de sus autoridades, ni respecto de la delimitación de sus fronteras o límites.

M-90

ISBN 92-5-001397-3

The copyright in this book is vested in the Food and Agriculture Organization of the United Nations. The book may not be reproduced, in whole or in part, by any method or process, without written permission from the copyright holder. Applications for such permission, with a statement of the purpose and extent of the reproduction desired, should be addressed to the Director, Publications Division, Food and Agriculture Organization of the United Nations, Via delle Terme di Caracalla, 00100 Rome, Italy.

Reproduction interdite, en tout ou en partie, par quelque procédé que ce soit, sans l'autorisation écrite de l'Organisation des Nations Unies pour l'alimentation et l'agriculture, seule détentrice des droits. Adresser une demande motivée au Directeur de la Division des publications, Organisation des Nations Unies pour l'alimentation et l'agriculture, Via delle Terme di Caracalla, 00100 Rome, Italie, en indiquant les passages ou illustrations en cause.

Este libro es propiedad de la Organización de las Naciones Unidas para la Agricultura y la Alimentación, y no podrá ser reproducido, ni en su totalidad ni en parte, por cualquier método o procedimiento, sin una autorización por escrito del titular de los derechos de autor. Las peticiones para tal autorización especificando la extensión de lo que se desea reproducir y el propósito que con ello se persigue, deberán enviarse al Director de Publicaciones, Organización de las Naciones Unidas para la Agricultura y la Alimentación, Via delle Terme di Caracalla, 00100 Roma, Italia.

© FAO 1983

INTRODUCTION

This Directory is an up-to-date issue of the last edition published in 1973.

Names and addresses of central government services are listed, as well as international and national institutes dealing with land and water development, farm power and machinery, rural electrification, farm buildings and farm work organization. For each institution information is provided on scientific staff, training and research, recent publications and languages of correspondence. The institutions are arranged by countries in the alphabetical order of their _English geographical_ names; within the countries the institutions are listed by states, provinces or cities. The information on each institute is presented in the language chosen by that institute for correspondence with FAO.

H.J. von Hülst
Chief
Agricultural Engineering Service
Agricultural Services Division

Rome, 26 July 1983

INTRODUCTION

Le présent Répertoire est l'édition mise à jour de l'ouvrage publié en 1973.

Noms et adresses des services officiels sont donnés, ainsi que ceux des instituts nationaux et internationaux s'occupant de la mise en valeur des terres et des eaux, de l'énergie et du machinisme agricoles, de l' électrification des campagnes, des constructions rurales et de l'organisation du travail en agriculture. Pour chacun de ces organismes, des renseignements sont fournis sur le personnel scientifique, les activités de formation et de recherche, les publications récentes et les langues utilisées pour la correspondance. Ils sont groupés par pays, classés selon l'ordre alphabétique de leur désignation géographique en anglais; à l'intérieur des pays, les institutions sont énumérées par états, provinces ou villes. Les renseignements sur chacun des instituts sont donnés dans la langue choisie par l'institut même pour sa correspondance avec la FAO.

H.J. von Hülst
Chef du Service du génie agricole
Division des Services Agricoles

Rome, le 26 Juillet 1983

INTRODUCTION

El presente Repertorio es una versión actualizada de la última edición aparecida en 1973.

Los nombres y direcciones de los servicios oficiales centrales están aquí elencados, así como los de los institutos nacionales e internacionales que se ocupan en el fomento de tierras y aguas, fuerza motriz y maquinaria agrícola, electrificación rural, edificaciones rurales y organización del trabajo agrícola. Para cada institución se dan datos sobre su personal científico, capacitación e investigación, publicaciones recientes e idiomas en que mantiene correspondencia. Las instituciones van ordenadas por orden alfabético de sus nombres geográficos en inglés; dentro de cada país se enumeran las instituciones por estados, provincias o ciudades. La información relativa a cada instituto se dan en el idioma elegido por dicho instituto para su correspondencia con la FAO.

H.J. von Hülst
Jefe del Servicio de Ingeniería Rural
Dirección de los Servicios Agrícolas

Roma, el 26 de Julio 1983

GOVERNMENT SERVICES AND INSTITUTES BY COUNTRIES

SERVICES GOUVERNEMENTAUX ET INSTITUTS PAR PAYS

SERVICIOS GUBERNAMENTALES E INSTITUTOS POR PAISES

Countries	Page	Pays	Page	Países	Página
Afghanistan	1	Afghanistan	1	Afganistán	1
Algeria	1	Afrique du Sud	329	Alemania occidental (R.F.)	123
Argentina	2	Algérie	1	Alemania oriental (R.D.)	118
Australia	6	Allemagne occ.(R.F.)	123	Alto Volta	462
Austria	19	Allemagne oriental (R.D.)	118	Arabia Saudita	326
Bahrain	21	Arabie Saoudite	326	Argelia	1
Bangladesh	22	Argentine	2	Argentina	2
Belgium	23	Australie	6	Australia	6
Benin	28	Autriche	19	Austria	19
Bolivia	29	Bangladesh	22	Bangladesh	22
Botswana	30	Bahrain	21	Bahrain	21
Brazil	31	Belgique	23	Bélgica	23
Bulgaria	38	Bénin	28	Benín	28
Burma	39	Birmanie	39	Birmania	39
Burundi	39	Bolivie	29	Bolivia	29
Cameroun	40	Botswana	30	Botswana	30
Canada	41	Brésil	31	Brasil	31
Central African Republic	63	Bulgarie	38	Bulgaria	38
Chad	64	Burundi	39	Burundi	39
Chile	64	Cameroun	40	Camerún	40
China (People's Rep. of)	69	Canada	41	Canadá	41
Colombia	72	Chili	64	Chad	64
Congo	76	Chine (Rép. Populaire)	69	Checoslovaquia	78
Costa Rica	76	Chypre	77	Chile	64
Cuba	76	Colombie	72	China (Rep.Pop.)	69
Cyprus	77	Congo	76	Chipre	77
Czechoslovakia	78	Corée	240	Colombia	72
Denmark	85	Costa Rica	76	Congo	76
Dominican Republic	92	Côte d'Ivoire	221	Corea	240
Ecuador	93	Cuba	76	Costa de Marfil	221
Egypt, Arab Republic of	98	Danemark	85	Costa Rica	76
El Salvador	101	Egypte, Rép. Arabe d'	98	Cuba	76
Ethiopia	102	El Salvador	101	Dinamarca	85
Finland	102	Equateur	93	Ecuador	93
France	105	Espagne	335	Egipto, Rep. Arabe de	98
Gambia	118	Etats Unis d'Amerique	411	El Salvador	101
German Democratic Rep.	118	Ethiopie	102	España	335
German Fed. Republic	123	Finlande	102	Estados Unidos de América	411
Ghana	146	France	105	Etiopía	102
Greece	148	Gambie	118	Filipinas	293
Guatemala	153	Ghana	146	Finlandia	102
Guinea	154	Grèce	148	Francia	105
Guyana	154	Guatemala	153	Gambia	118
Hungary	154	Guinée	154	Ghana	146
Iceland	160	Guyane	154	Grecia	148
India	161	Haute Volta	462	Guatemala	153
Indonesia	188	Hongrie	154	Guinea	154
Iran	193	Ile Maurice	251	Guayana	154
Iraq	194	Inde	161	Hungría	154
Ireland	195	Indonésie	188	India	161
Israel	200	Irak	194	Indonesia	188
Italy	204	Iran	193	Irak	194
Ivory Coast	221	Irlande	195	Irán	193
Jamaica	221	Islande	160	Irlanda	195
Japan	222	Israël	200	Isla Mauricio	251
Kenya	238	Italie	204	Islanda	160
Kampuchea	240	Jamaïque	221	Israel	200
Korea 240	240	Japon	222	Italia	204
Lebanon	242	Kampuchea	240	Jamaica	221
Lesotho	244	Kenya	238	Japón	222

Countries (cont'd)

English		French		Spanish	
Liberia	245	Lesotho	244	Kampuchea	240
Libya	245	Liban	242	Kenia	238
Luxembourg	246	Liberia	245	Lesotho	244
Madagascar	246	Libye	245	Libano	242
Malawi	247	Luxembourg	246	Liberia	245
Malaysia	248	Madagascar	246	Libia	245
Mali	250	Malaisie	248	Luxemburgo	246
Mauritania	250	Malawi	247	Madagascar	246
Mauritius	251	Mali	250	Malasia	248
Mexico	251	Maroc	254	Malawi	247
Morocco	254	Mauritanie	250	Mali	250
Nepal	255	Mexique	251	Marruecos	254
Netherlands	256	Népal	255	Mauritania	250
New Zealand	275	Niger	280	México	251
Niger	280	Nigéria	280	Nepal	255
Nigeria	280	Norvège	284	Níger	280
Norway	284	Nouvelle Zélande	275	Nigeria	280
Pakistan	288	Ouganda	374	Noruega	284
Panama	291	Pakistan	288	Nueva Zelandia	275
Peru	292	Panama	291	Países Bajos	256
Philippines	293	Pays-Bas	256	Pakistán	288
Poland	300	Pérou	292	Panamá	291
Portugal	316	Philippines	293	Perú	292
Puerto Rico	321	Pologne	300	Polonia	300
Romania	322	Porto Rico	321	Portugal	316
Rwanda	326	Portugal	316	Puerto Rico	321
Saudi Arabia	326	République Centrafricaine	63	Reino Unido	393
Senegal	327	République Dominicaine	92	República Centroafricana	63
Sierra Leone	329	Roumanie	322	República Dominicana	92
Somalia	329	Royaume Uni	393	Rumanía	322
South Africa	329	Rwanda	326	Rwanda	326
Spain	335	Samoa de l'Ouest	469	Samoa Occidental	469
Sri Lanka	339	Sénégal	327	Senegal	327
Sudan	344	Sierra Leone	329	Sierra Leone	329
Surinam	345	Somalie	329	Siria	355
Swaziland	345	Souaziland	345	Somalia	329
Sweden	346	Soudan	344	Sri Lanka	339
Switzerland	350	Sri Lanka	339	Sudáfrica	329
Syria	355	Suède	346	Sudán	344
Tanzania	357	Suisse	350	Suecia	346
Thailand	360	Surinam	345	Suiza	350
Togo	361	Syrie	355	Surinam	345
Trinidad and Tobago	362	Tanzanie	357	Swazilandia	345
Tunisia	363	Tchad	64	Tailandia	360
Turkey	365	Tchécoslovaquie	78	Tanzania	357
Uganda	374	Thaïlande	360	Togo	361
Union of Soviet Socialist Republics	375	Togo	361	Trinidad y Tobago	362
United Kingdom	393	Trinité et Tobago	362	Túnez	363
United States of America	411	Tunisie	363	Turquía	365
Upper Volta	462	Turquie	365	Uganda	374
Uruguay	463	Union des Républiques Socialistes Soviétiques	375	Unión de Repúblicas Socialistas Soviéticas	375
Venezuela	465	Uruguay	463	Uruguay	463
Viet-Nam	468	Venezuela	465	Venezuela	465
Western Samoa	469	Viet-Nam	468	Viet-Nam	468
Yugoslavia	469	Yougoslavie	469	Yugoslavia	469
Zaire	474	Zaire	474	Zaire	474
Zambia	475	Zambie	475	Zambia	475
Zimbabwe	476	Zimbabwe	476	Zimbabwe	476
International Institutions and Organizations	479	Institutions et Organisations Internationales	479	Instituciones y Organizaciones Internacionales	476

NOTE/NOTA

Organizations which have no staff listed are those which have not replied to the FAO questionnaire. They are, therefore, included for the record only and the information on their activities has been repeated from the previous edition of the Directory.

Les organisations dont le personnel n'apparaît pas sur la liste sont celles qui n'ont pas répondu au questionnaire envoyé par la FAO. Elles sont donc mentionnées pour mémoire et les renseignements fournis sur leurs différentes activités proviennent de l'édition précédente du répertoire.

Las organizaciones para las cuales no existe una lista de personal son aquellas que no contestaron al cuestionario de la FAO. Por lo tanto, éstas organizaciones se han incluido solo pro memoria y la información sobre sus actividades se ha tomado de las ediciones previas del repertorio.

AFGHANISTAN

Kabul

Agricultural Engineering Department, Faculty of Agriculture, Kabul University
Kabul

Status: Institution established in 1967 and financed by the Ministry of Education.

Activities: 5 years study courses; first 2 years at the Faculty of Engineering on general engineering, last 3 years on agricultural engineering (irrigation, drainage, farm power and machinery) leading to a degree in engineering - Service courses to agricultural students on farm mechanics and irrigation - Investigation on native basin and furrow irrigation.

ALGERIA

Alger

Direction du Génie Rural et de l'Hydraulique Agricole, Ministère de l'Agriculture et de la Réforme Agraire (Directorate of Agricultural Engineering, Ministry of Agriculture and Agrarian Reform)
12, Bd. Colonel Amirouche, Alger   Tél.: 647900

Activité: Grande hydraulique: Etudes régionales et de programmation conduites en commun avec le Service des Grands Travaux d'Hydrauliques(S.E.G.G.T.H.), le Service des Etudes Scientifiques (S.E.S.) relevant tous deux du Ministère des Travaux publics et le Plan, ce dernier jouant le rôle de coordinateur - Projets d'équipement pour l'irrigation jusqu'aux bornes desservant les ilots d'irrigation ou les prises d'arrosage - Réalisation des équipements collectifs sur le terrain - Appui technique pour l'équipement interne des domaines et l'assistance technique aux irrigants - Exploitation et entretien des équipements dans les périmètres - Vente de l'eau. - Moyenne hydraulique: Tous aménagements hydro-agricoles basés sur l'irrigation autres que les grands périmètres et intéressant une collectivité - Etudes de projets - Mise en valeur et gestion des équipements - Création d'aire d'irrigation. - Petite hydraulique: Irrigations individuelles, à partir des sources, puits, forages, pompes ou dérivations d'Oued - Equipements internes et assistance aux irrigants - Hydraulique pastorale pour l'alimentation des hommes et des animaux dans la zone de steppe et la zone saharienne.

Correspondance: arabe, français.

El Harrach

Laboratoire de Génie Rural, Institut Agricole d'Algérie (Agricultural Engineering Laboratory, Agricultural Institute of Algeria)
El Harrach (anciennement Maison Carré)

Activité: Enseignement et vulgarisation - Machinisme agricole - Hydraulique agricole et constructions rurales.

Correspondance: arabe, français, anglais, allemand.

ARGENTINA

Buenos Aires

Departamento de Ingeniería Rural - Facultad de Agronomía - Universidad de Buenos Aires
(Agricultural Engineering Department, School of Agronomy, University of Buenos Aires)
Avenida San Martín 4453, Buenos Aires, República Argentina - Código Postal: 1417

Colombino, A.A.	Ing.Agr.	Director
Piscitelli, M.A.	Ing.Agr.	Director Sustituto

Cátedra de Maquinaria Agrícola:

Colombino, A.A.	Ing.Agr.	Profesor Titular
Pollacino, J.C.	Lic.Mec.Agr.	Profesor Adjunto
Siffredi, J.G.	Ing.Agr.	Profesor Adjunto
Araolaza, A.M.	Ing.Ind.	Jefe de Trabajos Prácticos
Arena, I.B.	Ing.Agr.	Jefe de Trabajos Prácticos
Barzizza, R.A.	Ing.Agr.	" " "
Noacco, N.E.	Ing.Agr.	" " "
del Olmo, F.L.	Lic.Mec.Agr.	" " "
Peraldo, E.M.	Ing.Agr.	" " "
Raggio, J.B.	Ing.Agr.	" " "
Scapola, J.C.	Ing.Agr.	" " "
Sosa, R.O.	Ing.Agr.	" " "
Colombo, C.A.	Lic.Mec.Agr.	Ayudante de 1º
Eisler, C.R.M.	Ing.Agr.	" "
Gallo, E.	Ing.Agr.	" "

Cátedra de Hidrología Agrícola:

Luque, J.A.	Ing.Agr.	Profesor Honorario
Araujo, E.	Ing.Agr.	Jefe de Trabajos Prác.
Pariani, S.	Ing.Agr.	" " "
Burghi, O.	Ing.Agr.	Ayudante de 1º
Franke, R.	Ing.Agr.	" "
Genova, L.	Ing.Agr.	" "
Grillo, M.	Ing.Agr.	" "

Cátedra de Planificación de Espacios Verdes:

Piscitelli, M.A.	Ing.Agr.	Profesor Asociado
Nizzero, G.R.	Ing.Agr.	Jefe de Trabajos Prác.
Tchalidy de Outeriño, E.	Ing.Agr.	Ayudante de 1º

Cátedra de Topografía:

Firmenich, V.	Ing.Civ.	Profesor Adjunto
Vicente, C.	Ing.Agr.	Jefe de Trabajos Prác.

Cátedra de Construcciones Rurales:

García, J.J.	Ing.Agr.	Profesor Titular

Carácter y estructura: La Facultad de Agronomía puso en práctica el 10 de diciembre de 1969 la estructura departamental. - El Departamento de Ingeniería Rural es pues, una unidad docente y de investigación dependiente de las autoridades de la Facultad y por lo tanto de caractér gubernamental.

Actividades en ingeniería rural: (a) Enseñanza: En el Departamento se realiza la enseñanza teórica y práctica de cada una de las asignaturas que lo componen, que integran el Plan de Estudios de la Facultad de Agronomía. La concurrencia es de 200 a 350 alumnos por año por cada materia. El título que otorga la Facultad de Agronomía es el de Ingeniero

Agrónomo. (b) Investigación: Maquinaria Agrícola: Ensayos de maquinaria agrícola
Hidrología Agrícola: Estudios sobre cálculo y aplicación de riego por aspersión.
Topografía: Estudios de aplicación de técnicas topográficas a problemas de ingeniería
rural. Planeamiento de Espacios Verdes: Estudios sobre cespedes, árboles ornamentales
latifoliados, transplante de ejemplares adultos y arbolado de calles y avenidas.
Planificación de espacios para instituciones oficiales.

Publicaciones desde 1973: en "Facultad de Agronomía" y "Mecánica Aplicada a la Maquinaria Agrícola".

Correspondencia: Español, Inglés, Francés, Italiano.

Instituto Nacional de Tecnología Agropecuaria (INTA) Centro Nacional de Investigaciones Agropecuarias - Departamento de Ingeniería Rural
C.C. 25 - 1712 - Castelar - Argentina

Casares, J.M.	Ing.Agr.	Jefe Departamento
Delafosse, R.M.	Ing.Agr.	Maquinas sembradoras
Ferrando, J.C.	Ing.Agr.	Maquinas de labranza
Gil Espinosa, E.	Ing.Maq.	Fuentes de energía no convencionales
Larragueta, O.	Lic.Mec.Agr.	Máquinas pulverizadoras
Lostri, A.	Ing.Agr.	Ensayos de tractores
Pensotti, G.A.J.	Ing.Agr.	Máquinas de labranza
Smith, J.	Ing.Agr.	Máquinas para forrajes y heno

Carácter y estructura: Creado en 1958 - Gubernamental.

Actividades: Ensayo de tractores y máquinas agrícolas - Investigación y extensión en maquinaria agrícola.

Publicaciones desde 1973: Ensayo acople rápido ACCORD - Evaluación económica de un nuevo sistema de cosecha mecánica de batata. Evaluación de seis años de experimentación con labranza cero en soja de segunda sobre cosecha de trigo. - Interpretación de los ensayos de tractores. - Boletines de ensayo de tractores de acuerdo a la Norma IRAM 8005 - Ensayo n. 320 - Tractor Fahr - Mod D 181 F - Ensayo n. 321 - Tractor Massey Ferguson - Mod MF 235 - Ensayo n. 322 - Tractor Fiat 400 DT + Ensayo n. 323 - Tractor Fiat 600 DT

Correspondencia: Español, Inglés, Italiano, Francés.

Facultad de Agronomía - Departamento de Ingeniería Rural - Cátedra de Hidrología Agrícola
Universidad de Buenos Aires

Araujo, E.M.	Jefe de T. Prácticos
Franke, Roberto	Ayudante Primero
Genova, Leopoldo J.	Ayudante Primero
Grillo, Mario	Ayudante Primero
Pariani, Susana	Jefe de T. Prácticos

Estructura: Una cátedra: Hidrología Agrícola - Una asignatura: "Riego y Drenaje".

Actividades docentes: Dictado de la Asignatura: Riego y Drenaje - 2° Semestre - El dictados ocupa aproximadamente 4 meses. Numero promedio alumnos: 250. Actividades de investigación: a) Determinación de parámetros para el riego gravitacional en el area del Gran Buenos Aires - b) Recopilación bibliográfica de "Riego por goteo" para posteriores aplicaciones - c) Review de los equipos mecanizados de aspersion y autopropulsados de existencia en plaza, para riego complementario.

Publicaciones: en "Facultad de Agronomía. UNBA".

Correspondencia: Español, Inglés, Francés.

La Plata

Facultad de Agronomía, Universidad Nacional de la Plata, Cátedra de Maquinaria Agrícola, República Argentina
Calles 60 y 118, La Plata (C.P. 1900)

Chiesa, C.A.	Ing.Agr.	Profesor Titular (por concurso de antecedentes)
Foti, R.D.	Ing.Agr.	Profesor Adjunto
Jorajuria Collazo, D.	Ing.Agr.	Jefe de Trabajos Prácticos
Bilo, A.O.	Ing.Agr.	Jefe de Trabajos Prácticos
Balbuena, R.	Ing.Agr.	Ayudante Diplomado
Foster, H.	Ing.Agr.	Ayudante Diplomado
Arraras, E.		Ayudante Alumno
Etchevest, D.		Ayudante Alumno
Marcenat, A.		Ayudante Alumno
Remorini, R.		Ayudante Alumno
Spiga, S.		Ayudante Alumno

Nota: el Ing.Agr. Carlos Alberto Chiesa, se ha especializado en Maquinaria Agrícola y Tractores, con especial énfasis a la Ingeniería Agrícola.

Status: Esta Facultad de Agronomía es la de más antigua data en el país. Comenzó su actividad en el actual Instituto Fitotécnico de Santa Catalina en los albores de 1883, en la localidad de Llavallol, Provincia de Buenos Aires. Fecha en que se implantaron los estudios agronómicos superiores a nivel universitario en la República Argentina. Posee carácter gubernamental nacional.

Actividades: Enseñanza - Maquinaria agrícola y tractores. Materia de 5° año de la carrera de Ingeniería Agronómica (nó Agrícola). Promedio anual de alumnos: 150. - Investigación y/o Experimentación: Actualmente: desarrollo de banco para pulverizaciones de herbicidas. Ensayo y experimentación de máquinas de intersiembra de forrajeras.

Publicaciones: Planes 122 y 220 de C.A.F.P.T.A. sobre "subsolado" y "cosecha de forrajeras" realizados en el ex-C.E.E.M.A. (Primer y único centro de enseñanza y experimentación de Maquinaria Agrícola). Otras publicaciones sobre "Higiene laboral y seguridad en trabajos agrícolas mecanizados"."Ingeniería Agrícola, una profesion con futuro". Código Normalizado de la O.C.D.E. - "Maquinaria Agrícola" Facultad de Agronomía de la Plata año 1979.

Correspondencia: Castellano, Inglés, Francés, Italiano y Portugués.

Morón

Instituto de Ingeniería Rural, Instituto Nacional de Tecnología Agropecuaria (INTA)
(Agricultural Engineering Institute)
Villa Tesei, Morón, Provincia de Buenos Aires

Carácter: Institución dependiente del inta y fiscalizada por la Secretaría de Agricultura y Ganadería de la Nación.

Actividades: Investigación, experimentación, ensayo y extensión - Se desarrollan materiales tales como: tractores y maquinaria agrícola - Investigaciones y experiencias relacionadas con la mecanización y motorización agrícola, desarrollando los recursos técnicos necesarios para aumentar la eficiencia de la producción agropecuaria - Estudio de eficiencia mecánica, agronoómica y proyectos de máquinas y motores agrícolas para la labranza, cultivo, tratamientos sanitarios, cosecha de cereales, maíz, papas, etc. y para huerta, frutales y forestales - Investigación de procedimientos mecánicos y otros para el secado, limpieza y desinfección, almacenamiento y transportes de los productos agropecuarios, y cuanto conviene a las construcciones rurales - Estudio de las condiciones técnicas y eficiencia de las máquinas herramientas, motores y tractores en razón de sus destinos - Estudios, proyectos, modificaciones y construcciones de máquinas agrícolas experimentales.

Correspondencia: Español, inglés, francés, italiano, portugués.

Santa Rosa - La Pampa

Facultad de Agronomía, Universidad de La Pampa (Faculty of Agriculture, University of La Pampa)
Avenida Mitre No. 63, Santa Rosa, La Pampa

Tucumán

Facultad de Agronomía y Zootecnica, Universidad Nacional de Tucumán (Faculty of Agronomy and Zootechnics, University of Tucumán)
Avenida Roca 1900 - Casilla de Correos 125 - 4.000 - San Miguel de Tucumán - Argentina
Tel.: 30139 - 34155 - Interno 324

Budeguer, M.E.	Ingeniero Mecánico
Diaz Botta, C.A.	Ingeniero Agrónomo, Director de Departamento
Martinez Ribo, R.C.	Ingeniero Agrónomo
Rosales, F.A.	Ongeniero Agrónomo

Carácter y Estructura: La Facultad fue creada en Diciembre de 1947, siendo una institución oficial, financiada por el Gobierno Nacional.

Actividades: a) Enseñanza: Curso de "Elementos de Mecánica y Maquinarias Agrícolas". Es un curso normal que se dicta cuatrimestralmente una vez por año y que integra el plan de estudio de la carrera de Ingeniero Agrónomo. b) Investigación: La cátedra "Elementos de Mecánica y Maquinarias Agrícolas" desarrolla el plan de investigación: "Estudio de equipos para preparación de Suelo y Cultivo de la Caña de Azúcar".

Publicaciones: en Rev.Agron. del Noroeste Argentino; Miscelánea; Serie Didáctica.

Correspondencia: Español.

AUSTRALIA

New South Wales

Snowy Mountains Engineering Corporation
P.O. Box 356, Cooma North, N.S.W. 2630   Tel.: COOMA (0648) 21777

Price, D.G.   BE, FTS, FIE Aust., M.ASCE   Director

Status: Semi-governmental

Activities in Agricultural Engineering: Planning, design and implementation of projects to obtain optimum control and utilization of water resources, involving development of potential for land use, irrigation, water supply, flood control, and hydro-power production.

Publications since 1960: Some 200 papers and reports on engineering and economic aspects of rural and water resources development.

Correspondence: English.

Division of Irrigation Research, Institute of Biological Resources, CSIRO,
Private Bag, Griffith, NSW, 2680, Australia

Kriedemann, P.E.	B.Agr.Sc.(Hons), Ph.D.
Garzoli, K.V.	B.Mech.E., ME, M.I.E.
Blackwell, J.	NDA, NDAgr.E. MAIAS
Speed, R.E.W.	M.Sc.

Status: Established 1961. Part of a statutory Government authority.

Activities: Energy conservation in greenhouses, development of environmental control facilities for greenhouses.

Publications: in Farmers Newsletters, J.Agric.Engng. Res., Sol.Energy Prog. Aust. N.Z.

The School of Civil Engineering, University of New South Wales
Box 1, Kensington 2033, N.S.W.   Tel.: Sydney 633 0351

Branch: Water Research Laboratory at King Street, Manly Vale.

Status: Semi governmental, run by Council of the University.

Activities: Undergraduate courses in irrigation of ten hours lectures attended by undergraduate fourth year civil engineering students, approximately 70 per year - Post graduate course in irrigation and drainage comprising 60 hours lectures and 30 hours tutorial, attended by graduates proceeding to Master of Engineering Science degrees, approximately five students per year - Undergraduate courses leading to degree in civil engineering - Post graduate courses leading to degree of Master of Engineering Science - Research in water requirements of crops - Water balance studies on catchments - Design and construction of small earth dams - Sprinkler irrigation design research - Agricultural engineering economics.

Publications: Engineering aspects of water harvesting and the Keyline Plan - Distribution patterns for irrigation sprinklers - Tests on irrigation sprinklers - Tests on piles and fittings used in irrigation - Agricultural engineering economics.

Correspondence: English.

Department of Agricultural Engineering, School of Mechanical and Industrial Engineering, University of New South Wales
Kensington, N.S.W. Australia, 2033   Tel.: Sydney 663 0351

Department developed within parent institution. First officially listed in 1968. Financed by Federal and State Governments. - Short courses only for other institutions - Experimental work on pasture band sowing - mechanical processing of ginger - mechanical harvesting of apples and peaches - automatic sorting of tomatoes by colour - mechanization of cocoa and coconut plantations.

Publications: Agricultural Engineering in Australia; Journal of Agricultural Engineering Research.

Correspondence: English.

The Water Research Foundation of Australia
Post Office Box 47, Kingsford, N.S.W. 2032   Tel.: Sydney 663 4257

Branches: State Branches in each of the five Australian States.

Status: Established in 1956 - Institution directed by a Board of Trustees - Financed by contributions from private individuals, industrial and commercial organizations and grants from various State governments.

Activities: Hydraulics of surface irrigation - Small earth dams - Reduction of evaporation from stock water tanks - Irrigation.

Publications: Research bulletins and reports issued periodically.

Correspondence: English.

Hawkesbury Agricultural College
Richmond, N.S.W.

Status: Governmental institution, administered by State Department of Agriculture, NSW.

Activities: Engineering is treated as a subject in a three year agriculture diploma course. The course includes principles, adjustment and maintenance of farm machinery, thermodynamics of tractors and other power units; irrigation and drainage; soil and water engineering; farm structures and environmental engineering - Extensive trials with spray irrigation - Some testing of machines and implements.

Correspondence: English.

Agricultural Engineering Section, New South Wales Department of Agriculture
McKell Building, Rawson Place, Sydney, N.S.W.

Holligan, P.J.	B.E.	Head of Section
Finlayson, W.H.	G.D.E.	Senior Mechanization Officer

Agricultural Engineering Centre, Glenfield:

Wingate-Hill, R.	B.Sc.(Agric.), M.Sc. (Agric.Eng.), M.A., M.Sc., (Mech.Eng.)	Leader of Research, Tillage
Haggett, K.	B.App.Sc. (Chemistry)	Alternative Fuel Production
Sanders, K.F.	B.E.	Equipment Design
Brown, G.A.	B.E.(Agric.)	Power and Machinery
Andrews, A.S.	B.E.(Agric.)	Alternative Fuels
Danh, H.Q.	B.E.(Hons.), M.E.Sc., Ph.D.	Systems Analyst, Machinery Management

Agricultural Research Station, Condobolin:

Palmer	B.E.	Tillage Research
Kruger	B.E.(Agric.)	Tillage Research

Regionalized Agricultural Mechanization Officers:

Carlin, C.T.	Wagga Wagga	Mechanization Extension
Hill, A.J.	Leeton	Mechanization Extension
Whiston, W.D.	Orange	Mechanization Extension
Murphy, A.P.	Dubbo	Mechanization Extension
Lund, R.D.	Yass	Mechanization Extension
De Silva, S.	Gunnedah	Mechanization Extension
Wedd, S.	Lismore	Mechanization Extension

Status: Agricultural Engineering Section was established 1960. Agricultural Engineering Research Centre was established 1970. Finance: Government Funds supplemented by Industry Grants.

Activities: Research, development and field evaluation of farm machinery, alternative fuel production and utilization, development of specialized equipment for field and laboratory research workers. Decentralized engineering extension service.

Publications: Permanent electric fencing, on-farm fuel production, dynamometer performance of a diesel engine fueled with rapeseed oil, numerous mechanization publications and reports.

Correspondence: English.

Soil Conservation Service of N.S.W.
P.O. Box R201, Royal Exchange, Sydney, N.S.W. 2000.

Allen, J.R.	B.E., M.I.E. Aust.	Senior Engineer

Thirteen (13) Area Offices: Condobolin, Cooma, Cowra, Goulburn, Gunnedah, Inverell, Kempsey, Orange, Scone, Tamworth, Wagga, Wellington, Yass.

Six (6) Research Centres: Cowra, Gunnedah, Invernell, Scone, Wagga, Wellington.

Status: Established under the Soil Conservation Act, 1938. Government Institution.

Agricultural Engineering activities: (a) Teaching: nil. (b) Research: Conservation tillage: stubble management, reduced cultivation, direct drilling; Hydrological research: Effect of land use on erosion (arable, grazing, forestry); Use of hydraulic structures for gully erosion control, and runoff disposal.

Publications: "The Soils of New South Wales, their Characterization, Classification and Conservation". "Urban Erosion and Sediment Control". "Rainfall in New South Wales, with particular reference to Soil Conservation". "Journal of the Soil Conservation Service of N.S.W."; quarterly issue.

Correspondeence: English.

Water Resources Commission
Ibis House, 201/211 Miller Street - Box 952, P.O. North Sydney, N.S.W. 2060 - Tel.922 0121

Name	Qualifications	Position	Area
Williamson, W.H.	M.Sc.	Senior Hydrogeologist -	Groundwater Investigation
Hind, M.C.	B.Sc.	Asst. Hydrogeologist	" "
Roberts, D.J.	B.Sc.	"	" "
Woolley, D.R.	M.Sc., GRAD DIP ADMIN	"	" "
Sundararamayya, T.	M.Sc., M.A.	"	" "
Merrick, N.P.	M.Sc., GRAD DIP DATA PROC.	"	" "
McKibbin, D.B.	B.Sc.	"	" "
Krumins, I.	B.Sc.	"	" "
Drury, L.W.	B.Sc.	"	" "
Gates, G.W.	M.Sc., B.A.	"	" "
Williams, R.M.	B Sc.	"	" "
Ross, J.B.	B.Sc.	"	" "
Wooldridge, D.	B.Sc.	"	" "
Harwood, R.C.	B.Sc.	"	" "
Martinez, M.	B.Sc.	"	" "
Odins, J.	M.Sc., GRAD DIP DATA PROC.	Geophysicist	" "
O'Neill, D.	B.Sc., GRAD DIP APP. GEOPH.	"	" "
Beckham, J.	B.Sc.	"	" "

Farm Water Supplies in Sydney:
Name	Qualifications	Position	Area
Stapleton, R.C.	A.S.T.C.	Principal Engineer -	Farm Water Supplies
Wardle, A.O.	A.S.T.C.	Senior "	" " "
Moor, D.W.	B.E.	"	" " "
Tuite, J.B.	B.E.	"	" " "
Clement, K.D.	A.S.T.C.	District "	" " "
Sawkins, G.C.	B.E.	"	" " "
Singh, M.	B.Sc.(Eng.)	"	" " "
Irvine, D.O.	B.E.	Engineer, Boring	Bore Construct.

Farm Water Supplies Branch in Dubbo:
Name	Qualifications	Position	Area
Johnston, D.I.	B.E.(Min.)	Senior Engineer Boring	Bore Constr.
Thiem, N.D.	M.Eng.Sc.	District Engineer	Farm Water Supplies

Farm Water Supplies Branch in Albury:
Name	Qualifications	Position	Area
Brett, J.R.	A.S.T.C.	District Engineer	Farm Water Supplies

Farm Water Supplies Branch in Bega:
Name	Qualifications	Position	Area
Exell, G.A.	A.B.T.C.	District Engineer	Farm Water Supplies

Farm Water Supplies Branch in Cootamundra:
Holden, G.W.	M.Eng.Sc.	District Engineer	Farm Water Supplies

Farm Water Supplies Branch in Forbes:
Mann, K.   B.E.   District Engineer   Farm Water Supplies

Farm Water Supplies Branch in Grafton:
Kingman, J.J.   B.E.   District Engineer   Farm Water Supplies

Farm Water Supplies Branch in Maitland:
McLennon, G.I.   A.S.T.C.   District Engineer   Farm Water Supplies

Farm Water Supplies in Moree:
Hogwood, B.G.   B.E.   District Engineer   Farm Water Supplies

Farm Water Supplies Branch in Tamworth:
Ying, P.   M.Eng.Sc.   District Engineer   Farm Water Supplies

Activities in Agricultural Engineering: The Farm Water Supplies Branch undertakes: a) Construction of bores for supply of water to farms and towns and for investigation of groundwater resources. b) The investigation and design of water supply projects for stock, domestic and irrigation purposes on individual farms and for groups of farms. The works involved comprise bores, wells, earth dams and ring tanks, excavated tanks concrete weirs and diversion structures, cattle and concrete supply and drainage channels, pumping plants, pipe reticulation networks, and flood and spray irrigation systems.

Research Branch in Sydney:
Flint, S.E.   B.Sc.Agr.   Principal Research Officer
Pels, S.   M.Sc.   Groundwater hydrology,
                     Senior Research Officer drainage and reclamation.

Research Branch in Leeton:
Stannard, M.E.   B.Sc.Agr.   Field supervision of research from Leeton and
                     Senior Research Officer Griffith.
Kelly, I.D.   B.Sc.Agr.   Land evaluation, soil salinity
                     Research Officer
Beecher, H.G.   B.Rur.Sc.   Land evaluation, irrigation problems

Research Branch in Griffith:
van der Lelij, A.   Ir.Ag.   Sub-surface drainage, irrigation hydrology
                     Research Officer
Hoey, D.   B.Nat.Res.   Sub-surface drainage, irrigation hydrology
                     Research Officer

Research Branch in Deniliquin:
Schroo, H.   Ir.Ag.   Drainage and reclamation, irrigation hydrology

Activities in Agricultural Engineering: The Research Branch of the Commission undertakes soil geomorphological and stratigraphic investigations relating to land-use in irrigated areas and as a prerequisite for evaluating the potential and effects of irrigation developments. Research and investigations are carried out into ground water hydrology in irrigated areas, saline sub-surface drainage design and operation, land reclamation, surface water quality.

Correspondence: English.

The Electricity Authority of New South Wales
P.O. Box 456, North Sydney, N.S.W. 2060

The Electricity Authority of N.S.W. is an instrumentality of the New South Wales State Government. It was established in 1946, but agricultural engineering interests did not actively commence until 1956.

Activities: Research, Experiment, Testing, Extension.   **Publications:** bulletin.

Correspondence: English.

Queensland

Department of Primary Industries
G.P.O. Box 46, Brisbane, Qld. 4001   Tel.: 2240414

Engineering Services Sections:

Comalco House: G.P.O. Box 46, Brisbane, Qld. 4001   Tel.: 2244056
Hay, V.T.   Dip.Mech.Eng.,   Officer in Charge, also working in
                     Executive Engineer   greenhouse design, construction and
                                               facilities.

Toowoomba, Qld. 4350:
Grevis-James, I.W.   B.E. Agr., M.Eng.Sc.,   Engineer Division I: Tractor/
                     M.Eng.Agr.   Implement Management
Norris, C.P.   B.E. Agr.   Engineer Division I: Tillage and
                                               Planting Machinery
Collinge, M.   B.E. Agr.   Engineer Division II: Spray
                                               Application Technology
Walsh, P.A.   B.E. Agr.   Engineer Division II: Agricultural
                                               machinery extension
Ryan, I.A.   B.E. Agr.   Engineer Division II: Agricultural
                                               machinery extension
Casey, K.D.   B.E. Agr.   Engineer, Division II: Intensive
                                               animal industry housing and en-
                                               vironmental control and waste
                                               management.

Ormiston, Qld. 4163:
Franklin, T.G.   B.E. Tech.   Engineer Division I: Horticultural
                                               machinery

Biloela, Qld. 4715:
Robotham, B.G.   B.E. Agr.   Engineer Division II: Agricultural
                                               machinery extension

Atherton, Qld. 4883
McPhee, J.E.   B.E. Agr.   Engineer Division II: Agricultural
                                               (emphasis on peanuts) machinery
                                               extension.

Status: Directed and financed by State Government.

Activities: Research and extension to mechanization of field experiments, crop establishment, harvesting, grain drying, stubble mulching – Research and extension related to intensive animal housing including strcutures, controlled environment, treatment of animal wastes, preparation and distribution of feed.

Queensland Agricultural College, Lawes, Q 4345, Australia

The College was established in 1897, and tertiary courses in Agricultural Engineering have been given since 1930. Since 1967, it has been a College of Advanced Education, and is administered by the Queensland Agricultural College Council. It is financed jointly by the College Council, the Queensland Government and the Commonwealth Government.

Activities: (a) Tertiary courses of four years duration are offered in Rural Technology, Horticultural Technology, Rural Management, Food Technology, Food Service Management, Poultry Technology and Property Valuation. Agricultural engineering subjects including basic engineering, farm power and machinery, agricultural processing, environmental control, and soil and water engineering, comprise approximately one-fifth of the total time allocated to courses in the first three years. Fourth year students may specialize in Agricultural Engineering. Average student numbers in the four years are 90, 60, 40 and 25. - (b) Research: Hay drying, field evaluation of tractors and machinery, rural safety and irrigation.

Publications: Queensland Agricultural College Annual.

Correspondence: English.

Darling Down Institute of Advanced Education
c/o Post Office, Darling Heights, Toowoomba, Qld. 4350, Australia   Tel.: 301300 (076)
                                                                     Telex: 40010

O'Shea, J.A.	B.E. (Civil)	Head of Programme (Hydrology)
James, K.R.	B.E. (Agr.)	Lecturer (Agr. Machinery)
Rixon, A.J.	B.Agr.Sc., Ph.D.	Senior Lecturer (Agric.Science)
Smith, R.J.	M.Eng.Sci.	Lecturer (Hydrology)

Status: College of Advanced Education. Established in 1967. Funding from Commonwealth and State funds with grants from industrial and private sources.

Activities: (a) Teaching: Degree course over four years. Average of 35 undergraduate students. Basic engineering subjects in the first year. Specialized Agricultural Engineering course over the next three years. - Research: Sunflower oil for diesel fuels, Air Pressure drop in grain dryers, Tillage Forces, Tillage Dynamometers, Stump Jump Design, Wind Energy, Crop Drying, Tractor Cab Noise, Farm Fencing, Microprocessor Control, Strip Cropping, Soil Conservation, Catchment Modelling, Flood Detention, Channel Seepage, Groundwater Modelling, Micro-meteorology.

Publications: Strip Cropping, Farm Fencing, Vegetable Oil and Diesel Fuel Substitutes, Kenaf Fibre Separation, Soil and Water Salinity.

Correspondence: English.

S o u t h   A u s t r a l i a

Roseworthy Agricultural College
Roseworthy, South Australia 5371

Thistlethwayte, Dr. B.	Director
Tunks, R.B.	Head, Department
Rowland, P.W.	Lecturer
Harris, P.L.	Lecturer

Status: College of Advanced Education. Semi Government Institution.

Activities: Machinery development dry land farming.

Publications: Rural Safety, Tractor Hydraulics.

Correspondenve: English.

Victoria

Irrigation Services Division, State Rivers and Water Supply Commission
590 Orrong Road, Armadale, Vic. 3143   Tel.: (03) 509 9511

Robinson, E.P.	M.Agr.Sc., M.Eng.Sc.	Chief Irrigation Officer (efficient water use in irrigation)
Webster, A.	B.Sc.(Agr.)	Senior Research Officer (drainage of irrigated land)
Jones, L.D.	B.Agr.Sc.	Senior Irrigation Officer (efficient water use in irrigation)
Patto, P.M.	B.Sc.(Hons.)	Senior Irrigation Officer (efficient water use in irrigation)
Lavis, A.R.	B.Agr.Sc.	Research Officer (efficient water use in irrigation)
Poulton, D.C.	B.Sc.(Hons.)	Research Officer (drainage of irrigated land)
Dickinson, P.J.	B.Agr.Sc.	Research Officer (drainage of irrigated land)
Rendell, R.J.	B.E.(Agr.)	Research Officer (efficient water use in irrigation)
Aughton, D.J.	B.E.(Agr.)	Research Engineer (efficient water use in irrigation)
Benkhauser, B.J.	B.E.(Agr.)	Research Engineer (efficient water use in irrigation)

District offices: There are five district offices of the Irrigation Services Division.

Status: The Irrigation Services Division is a part of a government body supervised by a board of Commissioners.

Agricultural Engineering activities: a) Research and extension work: Reclamation of saline and waterlogged soil by tile drainage and ground-water pumping - Improvement in the efficiency of water use for irrigation and its distribution, including automatic irrigation - Environmental studies - Investigation of seepage from channels. Advisory service on land layout for irrigation, drainage, and irrigation practices.

Publications: Annual Reports - Water Talk, irrigation advisory journal, published three times a year - Publications on automatic irrigation, drainage, salinity, farm water supplies, methods of irrigation and land layout. - Principles of Irrigation, State Rivers and Water Supply Commission, 1974, pp. 56, illustrated with photographs and diagrams, out of print, will be reprinted 1980.

Correspondence: English.

Dookie Agricultural College (1886)
Victoria 3647   Tel.: (058) 286371

Grant, S.G.	Dip.Mech.Eng., Dip.El.Eng.	Lecturer in Ag. Engineering
Lawrie, W.G.	B. Eng.(agric.), Dip.Mech.Eng., Trained Technical Teachers Cert.	Instr. in Ag.Eng. Hydraulics
Fell, R.I.	Dip.Build.Constr.	Building Instructor

Activities: Diploma of Applied Science - Agriculture (120) - Diploma of Applied Science - Food Production Hort. (25).

Correspondence: English.

Longerenong Agricultural College
Dooen 3401 Victoria, Australia

Pyke, W.J.	B.E.(Ag), Dip.Mech.Eng.	Agricultural Engineering - Soils, water ag. machinery
Riddell, J.R.	Dip.Ag.	Instructor in Agricultural Engineering

Status: Est. 1889. Governed by Ag.Ed. Div., Department of Agriculture, Victoria.

Activities: Teaching - Course in Applied Science (Agriculture) Diploma - 100 Students.

Correspondence: English.

Division of Mechanical Engineering, Commonwealth Scientific and Industrial Research Organization
P.O. Box 26, Highett, Victoria 3190

Rawlings, B.	Ph.D., C.Eng.	Chief of the Division
Taylor, P.A.	B.Sc., C.Eng.	Group Leader
Czech, J.	Ing.	Forestry Machinery
Elder, W.B.	A.G.Inst.Tech., C.Eng.(Aust)	Crop preservation
Ghaly, T.F.	M.Sc.	Crop preservation
Hunter, A.J.	Ph.D.	Crop preservation
Quick, G.R.	Ph.D.	Crop production
Shanks, D.H.	Ph.D.	Forestry machinery
Sutherland, J.W.	M.Eng.Sc.	Crop preservation
Thorpe, G.R.	Ph.D.	Crop preservation

Status: Government institution with some financial support from the Wheat Industry Research Council and other industry bodies. It is a Division in the Institute of Industrial Technology of the Commonwealth Scientific and Industrial Research Organization.

Activities: Research in agricultural and forestry engineering in the areas of crop production (tillage and seeding), crop preservation (grain and seed drying; physical methods of insect control in bulk grain including refrigerated aeration of bulk storages and disinfestation by heat treatment) and tree harvesting (clear felling of large pine trees; continuous harvesting of smallwood).

Publications since 1973: Refrigerated aeration of grain storages; high temperature disinfestation of bulk wheat; grain drying; oilseed drying; aeration of peanuts; tree felling dynamics; aerial spraying; grape vine spraying; aerodynamic spreader for agricultural aircraft; narrow-row cotton harvester.

Correspondence: English.

Soil Conservation Authority - Engineering Section
378 Cotham Road, Kew. 3101 Victoria, Australia

Findlay, G.H.	NE (Agr)	Chief Engineer
Garrett, B.K.	Dip.CE, Dip. Geol.	Regional Engineer (North Eastern Region)
Thomas, P.R.	Dip.CE	Regional Engineer (North Western Region)

Nichol, E.   BCE, Dip.CE   Regional Engineer
                                     (South Eastern Region)
Cox, T.W.   Dip.CE   Regional Engineer
                                     (South Western Region)

Branches: The Soil Conservation Authority has offices in 25 country centres.

Status: Government Department.

Agricultural Engineering Activities: Advice to landholders and public bodies on soil conservation. Advice to landholders on farm water supply schemes. Investigations into reduction of seepage losses from farm dams.

Publications: "Soils Aid" pamphlets on various aspects of soil conservation. "Construction Aids" and "Maintenance Aids" for erosion control works. Manual on Guidelines for Minimising Soil Erosion and Sedimentation on Construction Sites in Victoria.

Correspondence: English.

State Electricity Commission of Victoria
Monash House, 15 William Street, Melbourne - Box 2765Y GPO Melbourne, Vic., Australia 3001
                                                        Tel.: 6150433
                                                        Telex: 31153
White, R.G.   M.I.E. Aust.   Industrial and
                                     Rural Services Engineer

Status: Semi-governmental institution.

Activities: Rural utilization of electricity - development of farm-electric methods and equipment - research on: Potato Cool Stores - horticultural applications of electricity - dairy water heating investigations - feed-lot engineering problems - crop drying.

Publications: Series of technical literature on "How Electricity Aids the Pig, Dairy, Poultry, Potato Farmer and Horticulturist".

Correspondence: English.

National Safety Council of Australia, Federal Secretariat
377 Little Collins Street, Melbourne, Victoria, 3000   Tel.: 67 7278

Lewis, G.C.   Head of Publications Department

Branches: each capital city of Australia.

Status: Not established as such.

Publications: Quarterly magazine "Australian Family Safety", includes rural safety articles. Circulation 22,500.

Correspondence: English.

Agricultural Engineering Section, Department of Civil Engineering, University of Melbourne
Parkville, Victoria 3052, Australia

Schmidt, L.C.	B.E.(Civ.), M.Eng.Sc., Ph.D.	Chairman of Department
McMahon, T.A.	B.E.(Agr.), Ph.D.(NSW)	Professor. Hydrology, water resources
Turner, A.K.	B.C.E., M.E.	Reader. Hydrology, irrigation, waste treatment
Angus, D.E.	B.Sc., Ph.D.(Calif.)	Reader. Micrometeorology, heat transfer
MacMillan, R.H.	B.E.(Agr.), M.E.(Canter.)	Senior Lecturer. Mechanization, physical properties of agricultural materials
Pamment, N.B.	B.Sc.(Syd.), Ph.D.(NSW)	Lecturer (part-time). Waste treatment
Watts, P.J.	B.E.(Agr.)	Tutor. Evapotranspiration
Field, B.W.	B.E.(Mech.), M.Eng.Sc., Ph.D., Dip.Ed.	Research Engineer. Sheep handling
Burrow, R.P.	B.E.(Agr.), M.Eng.Sc.	Research Engineer. Sheep handling
Bryson, J.A.	B.E.(Mech.)(Adel.)	Research Engineer. Sheep handling
Freeman, R.B.	B.E.(Agr.)	Engineer, Technology Transfer. Sheep handling
Redding, G.J.	B.E.(Civ.), M.S.(Calif.)	Research Engineer. Farm buildings

Branches: Mt. Derrimut Field Station of the University — Agricultural Engineering Centre of the Victorian Department of Agriculture, Werribee (sheep handling laboratory).

Status: University of Melbourne: An independent body supported by Commonwealth and State finances and grants from industrial and private funds. Agricultural Engineering established as a separate entity in 1956.

Activities: Teaching elementary engineering to 1st and 2nd years of B.Agr.Sc. course (Faculty of Agriculture and Forestry); 60 students. — Teaching in 3rd and final year (4th) of B.E.(Agr.) course (Faculty of Engineering); 12 students. — Post-graduate training for M.Eng.Sc., M.AGr.Sc., M.Env.Eng., Ph.D. — Current research projects on: hydrology, irrigation, soil physics, traction, energy conversion, buildings, sheep handling, waste treatment.

Publications: Many papers in technical journals and in our Technical Report series, relating to irrigation, drainage, soil physics, grain separation, hydrology, environmental impact, sheep handling, groundwater, mechanization.

Correspondence: English.

Agricultural Engineering Centre
Werribee, Vic. 3030

Brown, W.T.	B.E., MIE Aust.	Principal Engineer
Redding, G.J.	B.E., M.S., MIE Aust.	Senior Research Engineer
Hutchings, M.J.	B.E.	Senior Services Engineer
Bakker, A.C.	B.Sc., M.Sc.	Senior Research Officer
Gould, I.V.	Dip.Ag. Sci., M.Sc.(Eng), MIE Aust.	Research Engineer
Foster, M.P.	B.E. MIE Aust.	Agricultural Engineer
Simmons, E.K.	B.Ag.Sci.	Scientific Officer
Reid, P.J.	B.E.	Agricultural Engineer
Huzzey, J.	B.E.	Agricultural Engineer
Martin, M.	B.E.	Agricultural Engineer
Murphy, J.P.	B.Ag.Sci.	Scientific Officer

Simpfendorfer, H.	Dip.Ag.Sci.	Agricultural Officer
Miller, M.	Dip.Ag.Sci.	Agricultural Officer
Bartels, L.F.	B.Ag.Sci.	Scientific Officer
Larkin, M.E.		Agricultural Engineer
Rigby, A.	B.E.	Agricultural Engineer
Wollin, A.	B.E.	Research Engineer
Lyle, C.	B.E.	Agricultural Engineer
Makin, A.	Dip.Ag.Sci.	Agricultural Officer
Young, G.	B.E.	Agricultural Engineer
Pyke, W.	B.E.	Senior Agricultural Engineer
Connellan, G.	B.E.	Senior Agricultural Engineer
Cox, G.	B.E.	Chemical Engineer

Status: A Branch of the Department of Agriculture, Victoria, supplementary funding from Commonwealth and industry.

Activities: Research and extension in all aspects of agricultural engineering. Current work includes fundamental draft and tillage studies, minimum tillage cultivation, solar energy applications in greenhouses, energy conservation in greenhouses, solar energy application and energy conservation in farm dairies, automatic control of natural ventilation for animal housing, drainage of soils, tractor performance, mechanical fruit harvesting, energy audit in dairy factories, computer control of fruit and vegetable storage environment, packing and handling of export fruit, performance of diesel engines on vegetable oils, mole drainage, farm building plan service, solid fuels for tobacco curing, pesticide application, fodder conservation, cattle transport, deer housing, windrowing of oilseeds, irrigation equipment and practices.

Publications: A number of research reports in the areas of greenhouse design, energy use in agriculture, drainage of soils, tractor performance, mechanical fruit harvesting, fruit and vegetable packaging, performance of diesel engines on vegetable oils. In addition a large number of tractor test reports and a comprehensive set of extension pamphlets on all aspects of agricultural engineering.

Correspondence: English.

Agricultural Engineering Society (Australia)
191 Royal Parade, Parkville 3052, Vic., Australia

Foster, M.	B.Eng.(Agr.)	Secretary

Victorian Branch: Agricultural Engineering Centre, Department of Agriculture, Werribee 3030.

<u>Queensland Branch</u>: P.O. Box 49, Darling Heights 4350

<u>Status</u>: Formed in Melbourne on 4 May 1950. Incorporated in 1953 under the Companies Act in Victoria. Governed by a General Committee. Financed by members' subscriptions.

<u>Principal Objects and Activities</u>: (a) To promote, encourage and co-ordinate the study, development and profession of Agricultural Engineering and farm mechanization in all its branches; (b) To encourage, promote or undertake investigations and research in Agricultural Engineering or allied fields of activity; (c) To hold Conferences, Lectures, Field Days: to discuss subjects affecting Agricultural Engineering generally, and to publish proceedings of the Society or any other publications which may seem to further the objects of the Society.

<u>Publications</u>: Branch Newsletters sent to members in Australia. National Journal (Agricultural Engineering Australia) published twice yearly and is available on separate subscription.

<u>Correspondence</u>: English.

Western Australia

<u>Muresk Agricultural College</u>
Northam. Western Australia 6401

Edghill, G.   M.Sc.(Agric. Man), B.Sc.   Course Co-ordinator
                     (Ag.Eng.)

<u>Status</u>: Department of the Western Australian Institute of Technology. Para-Statal.

<u>Agricultural Engineering activities</u>: Teaching: Current - Diploma course in Farm Management. Proposed - a) Option in Bachelor of Business (Agriculture) Degree; b) Option in Graduate Diploma in Engineering. - Research: Farm Machinery reliability in Western Australia.

<u>Correspondence</u>: English.

AUSTRIA

Petzenkirchen

Bundesanstalt für Kulturtechnik und Bodenwasserhaushalt (Federal Institute for Land and Water Development)
A-3252 Petzenkirchen, Pollnbergstr. 1, tel.no. 0043-07416-2108

Blümel, F. Dipl.Ing.Dr. Director of the Institute and Head of the Division "Physikalische, chemische und bodenmechanische Untersuchungen"
Leder, N. Dipl.Ing. Head of Division "Landeskultureller Wasserbau und Hydrogeologie"
Klaghofer, E. Dipl.Ing.Dr. Head of Division "Bodenwasserwirtschaft und Bodenwasserschutz".

Status: Subsidiary institute of the Federal Ministry of Agriculture and Forestry, Vienna.

Activities: Basis of assessment and planning of agricultural water engineering, in particular drainage, irrigation, river regulation, land slides, erosion, agricultural sewage and mud disposal; changes of ground water regime by power plants, sewage disposal plants, dumping grounds and road construction; research on measuring techniques of capillarity in connection with the water supply of plants, improvement of soil water conditions, ferrous compounds in the soil, obstruction of drainpipes and purification, drain filter, use of machinery, weed control in ditches, eutrophication.

Publications: Evaporation of differently used surfaces, soil erosion and losses of nutrients, reduction and oxidation in the soil, drain filter, weed control in ditches, mechanization of ditch maintenance, soil loosening, eutrophication, water preservation in the soil.

Correspondence: German, English.

Wien

Division IIB7 Bundesministerium für Land- und Forstwirtschaft (Division IIB7 Federal Ministry of Agriculture and Forestry)
A-1010 Wien, Stubenring 1

Schröfl, J. Dr.Dipl.Ing. Ministerialrat.

Mechanization, farm building construction, energy utilization.

Österreichisches Kuratorium für Landtechnik (Agricultural Engineering Council for Austria)
A-1040 Wien, Schwindgasse 5.

Kandler, K. Ing. Managing Director.

Mechanization, farm building construction, application of energy in agriculture.

Institut für Landtechnik und Energiewirtschaft, Universität für Bodenkultur Wien
(Institute of Agricultural Engineering and Power Economy, University of Agriculture, Vienna)
A-1190 Wien, Peter Jordan-Strasse 82.

Rossbrucker, H. Dr.Ing.Prof. Director
Rehrl, K. Dr.Ing.Prof. Emeritus
Häusler, F. Dr.Ing. Assistant
Weingartmann, H. Dipl.Ing. Assistant

Activities: Teaching: Agricultural engineering, unit operations; Research: Drying of cereals and forage crops; efficient use of farm machinery, alternative-energy, machinery of the animal-husbandry.

Publications: Articles in the journals 'Die Bodenkultur', 'Praktische Landtechnik', 'Der Förderungsdienst'.

Correspondence: German, English.

Abteilung IV 5 Bundesministerium für Land- und Forstwirtschaft (Dept. IV 5 Federal Ministry of Agriculture and Forestry)
A-1010 Wien, Stubenring 1.

Drainage, irrigation, regulation of small waters.

Wieselburg

Bundesversuchs- und Prüfungsanstalt für landwirtschaftliche Maschinen und Geräte
(Federal Experimenting and Testing Station for Farm Machinery and Implements)
A-3250 Wieselburg, Rottenhauserstrasse 1   tel. 07416/2175, 2176

Scientific staff:

Reichmann, E. Dr.Dipl.Ing. Director

Department 1 - Machine Testing:
Zehetner, J. Dipl.Ing. Head of Department
Lehner, F. Dipl.Ing.

Department 2 - Agricultural Engineering Research:
Pernkopf, J. Dipl.Ing. Head of Department
Lampel, H. Dipl.Ing.
Wörgetter, M. Dipl.Ing.

Department 3 - Agricultural Engineering Development and Metrology
Schrottmaier, J. Dr.Dipl.Ing. Head of Department
Hütl, G. Dipl.Ing.

Department 4 - Farm Work Management
Wernisch, A. Ing. Head of Department

Department 5 - Information, Documentation and Library
Hammerschmid, W. Dipl.Ing. Head of Department

Department 6 - Mechanization of Mountain Farms, Soil Tillage Methods
Sieg, R. Ing. Head of Department

Department 7 - Electronic Data Processing Service
Nestelberger, J. Ing. Head of Department

Branches: do not exist.

Status: Federal institution, established in 1947 and supervised by the Ministry of Agriculture and Forestry, Vienna, Austria.

Activities: a) Testing of agricultural machinery and implements: testing according to national and international test codes; development of new test procedures.
b) **Research** (basical and practical research in the field of Agricultural Engineering): energy problems in agriculture; alternative fuel sources (vegetable oils, ethanol); alternative paths for heating and drying; usability of waste biomasses for energy production (production of biogas, combustion of crop residues); new ways in wood use (combustion, gas generation). Possibilities of improvement of agricultural vehicles; stability of vehicles on a slope; soil compaction by agricultural machines. Ergonomic **problems** on agricultural vehicles, especially comfort on **driver** seat; protection of implements by electronically programmable overload clutches; **safety problems** in silos (carbon dioxide). Improvement of work efficiency in the field, farm and house; economics and premises for farm work and farm management; data **collection and** preparation for agricultural advisory service. c) Teaching: teaching of technical subjects to the students of the Agricultural Engineering Department of the "Höhere Landwirtschaftliche Bundeslehranstalt-Francisco Josephinum" at Weinzierl near Wieselburg.

Publications: a) General publications: annual reports of the Institute. b) Research publications: Body adequate shape of seats of tractors and implements - Protection of Man against vertical vibrations of tractors and motor cars (theory and practice), numerous illustrations - Electronical procedure of measuring the distribution of grains of single seed drills, numerous illustrations - Study concerning the usability of wood gas generators for fuelling diesel engines in agriculture, with illustrations - Knowledge concerning motor cars, with illustrations - Measuring inertia and location of centre of gravity of agricultural vehicles by oscillation methods, numerous illustrations - Austrian contribution to Ceneca Colloquium 1980 - Vegetable oils for diesel engines, a guarantee for the supply with food in the case of a fuel supply crisis. c) Other publications: braking of agricultural tractors and trailers; preparation of fruit juice in the farm-household; examples for teaching working procedures; modern soil tillage methods; mechanization of mountain farms.

Correspondence: English, German.

B A H R A I N

Ministry of Commerce and Agriculture, Agriculture Directorate
P.O. Box 251

Technical Section: Land and Water

Technical Section: Farm Machinery Maintenance

Activities: Irrigation, drainage, soil and farm machinery.

BANGLADESH

Dacca

Agricultural Engineering Division - Bangladesh Agricultural Research Institute
Joydebpur, Dacca

Scientific Staff:
Mazed, M.A.	B.Sc.Eng.(Bangladesh) M.Agr.Eng.(U.K.) MIE (Bangladesh) MISAE (India) Head of Division
Satter, M.A.	B.Sc.Agric.Eng.(Bangladesh), Senior Scientific Officer
Rashid, M.H.	B.Tech.(India) AMIE (Bangladesh), Senior Scientific Officer
Roy, K.C.	B.Sc.Agric.Eng. (Bangladesh) M.Sc.Eng.(AIT) AMIE (Bangladesh) Senior Scientific Officer
Rahman, M.M.	B.Sc.Agric.Eng.(Bangladesh)
Islam, M.S.	B.Sc.Agric.Eng.(Bangladesh)

Branches with other location: There are four other branches of this division located at: (i) Regional Agricultural Research Station, Jamalpur, Bangladesh; (ii) Regional Agricultural Research Station, Jessore, Bangladesh; (iii) Regional Agricultural Research Station, Hathazari, Chittagong, Bangladesh; (iv) Hill Agricultural Research Station, Ramgarh, Chittagong Hill Tracts, Bangladesh.

Status: Autonomous Institution, established in 1974, under the governing body of the Ministry of Agriculture.

Agricultural Engineering activities: Research for design and development of appropriate agricultural machinery and implements suitable for Bangladesh conditions. Test and modification of the existing traditional agricultural implements. Extension/transfer of technology to the farmers' field through the regional stations. Testing and standardization of agricultural machinery and implements imported from other countries.

Publications: Bangladesh Journal of Agriculture 1978, 1979; Agricultural Mechanization in Asia 1978, 1979; Proceedings of 2nd Workshop of Potato Research Workers, BARI, 1979.

Correspondence: English and Bengali.

Mymensingh

Faculty of Agricultural Engineering and Technology, Bangladesh Agricultural University
Mymensingh, Bangladesh

Department of Irrigation and Water Management

Department of Farm Machinery and Power

Department of Agricultural Engineering and Basic Engineering

Department of Food Technology and Rural Industries

Status: Date of establishment of the Faculty: August 1963. Financed by the Government of the People's Republic of Bangladesh.

Activities: Teaching: The courses offered are for Bachelor's and Master's degree. The type of degree awarded is: B.Sc.Agri.Eng. with majors in Farm Power and Machinery; Irrigation and Water Management; Agricultural Production Eng.; Food Technology. - M.Sc.Agr.Eng. in Farm Power and Machinery; Irrigation and Water Management; Agricultural Production Engineering; Food Technology. - Research.

BELGIUM

Bruxelles

Direction du Génie rural, Ministère de l'Agriculture (Department of Agricultural Engineering, Ministry of Agriculture)
17a, Avenue de la Toison d'Or, 1060 Bruxelles

Cornet d'Elzius, Cl.	Ingénieur agronome, région tropicale
	Ingénieur en Chef-Directeur
Van der Voorde, A.	Ingénieur agronome Génie rural
	Ingénieur Principal – Chef de service
Dackweiler, P.	Ingénieur agronome, zones tempérées et zones tropicales
Timmermans, J.	Ingénieur agronome, Ingénieur principal
Reyns, L.	Ingénieur agronome Génie rural,
	Ingénieur principal
de Donnea de Haloir, R.	Architecte
Van de Velde, E.	Architecte

Services provinciaux:

Anvers: 166, Mechelsesteenweg, 2000 Antwerpen (Anvers), ir. Blommaert R., Ing.Agron.
Brabant: 29-31 Steenweg op Elsene, 1050 Brussel, ir. Vaerewijk, Ingénieur agronome Génie rural
Flandre occidentale: 6, Twijnstraat, 8000 Brugge, ir. De Paepe, J., Ingénieur agronome
Flandre orientale: 68 Gebroeders Van De Veldestraat, 9000 Gent, ir. Baele, H., Ing.agron.
Hainaut: 1 rue des Echelles, 7000 Mons, ir. Coulon, P., Ingénieur agronome Génie rural
Liege: 26 rue des Guillemins, Bte 51, 4000 Liege, ir. Piron, F., Ingénieur agronome
Limbourg: 73, Demerstraat, 3500 Hasselt, ir. Moens de Hase Ph., Ing.Agronome
Luxembourg: Clos des Seigneurs, 1 (cité Administrative), 6620 Neufchâteau
ir. Galloy, A., Ing.Agr. Génie rural
Namur: 29, avenue Gouverneur Bovesse, Bte 5, 5100 Jambes, ir. Cavalier, H., Ing.Agronome

Statut: Le "Service du Génie rural" de l'Etat a été créé par arrêté royal du 6 novembre 1919. L'arrêté royal du 13 mars 1972 l'a fait passer au rang de "Direction du Génie rural".

Activités: Ce service doit rechercher les mesures propres à permettre une amélioration de l'habitat rural, de l'alimentation en eau potable des fermes isolées, de l'électrification des écarts ruraux, de l'utilisation du froid, ainsi que du développement de la mécanisation et de l'organisation scientifique du travail en agriculture et en horticulture.

Publications: Tracts, brochures et plans-types relatifs aux matières citées plus haut.

Correspondance: français, néerlandais.

Direction de l'Hydraulique agricole, Ministère de l'Agriculture (Department of Agricultural Hydraulics, Ministry of Agriculture)
Administration centrale: Avenue de la Toison d'Or, 17A - 1060 Bruxelles

Bouttefeux, F.	Ingénieur agronome	Ingénieur en Chef-Directeur
Strubbe, J.	Ingénieur agronome	Ingénieur principal-Chef de service
Sohet, N.	Ingénieur agronome (génie rural); Ingénieur principal	
Van Hecke, W.	Ingénieur agronome	

Services extérieurs:

Flandre: correspondance néerlandais.

Lambrecht, W.	Ingénieur agronome (génie rural), Ingénieur Twijnstraat 6 - 8000 Bruges
Goderis, W.	Ingénieur agronome (génie rural); Ingénieur principal - Chef de Service, Gebroeders Van de Veldestraat 68 - 9000 Gand
O, K.	Ingénieur agronome (génie rural); Ingénieur Mechelsesteenweg 166 - 2000 Anvers
Francois, J.	Ingénieur agronome; Ingénieur Demerstraat 73 - 3500 Hasselt

Wallonie: correspondance français.

—	Rue des Quatre Fils Aymon 9 - 7000 Mons
Van Damme, P.	Ingénieur agronome; Ingénieur Avenue Gouverneur Bovesse 29 - 5100 Namur
Daxhelet, C.	Ingénieur agronome; Ingénieur Rue du Luxembourg 22 - 5400 Marche-en-Famenne

Activités: Cette direction est chargée de l'amélioration du régime hydrologique des terres agricoles et de la voirie rurale. Sa compétence s'étend également aux cours d'eau non navigables, aux polders et aux wateringues.

Direction de l'Aménagement de l'Espace rural, Ministère de l'Agriculture
17a, Avenue de la Toison d'Or, 1060 Bruxelles

Vanwelden, M.	Ing.Agron. génie rural - Ingénieur en Chef-Directeur de la Direction de l'Aménagement de l'Espace rural
Bouvigne, R.	Ing.Agron. - Ingénieur
Charue, B.	Ing.Agron. génie rural - Ingénieur
Demuynck, J.	Ing.Agron. - Ingénieur principal - chef de service
Vankerckvoorde, D.	Ing.Agron. - Ingénieur

Services Provinciaux:

Flandre occidentale: 4, Kelkstraat - 8000 Brugge

Dewaerle, R.	Ing.Agron.

Flandre orientale: 69, Gebr. Vande Veldestraat - 9000 Gent

Gijselinck, R.	Ing. principal

Anvers et Brabant: 166, Mechelsesteenweg - 2000 Antwerpen

Dingenen, R.	Ing.Agron.

Limbourg et Brabant: 73, Demerstraat - 3500 Hasselt

Lobel, J.	Ing.Agron.

Namur et Brabant: 29, Bte 8 Avenue Gouverneur Bovesse - 5100 Jambes

Jacquemart, R.	Ing.principal, chef de service

Hainaut et Brabant: 1, rue des Echelles - 7000 Mons

Gonze, P.	Ing. principal

Liège: 26, Bte 051 Rue des Guillemins - 4000 Liège

Mesureur, F.	Ing. principal

Luxembourg: 23, rue Busleyden - 6700 Arlon
Stuyvaert, J.   Ing. principal

Structure et activités: Ce service se charge
a) Pour compte des Exécutifs Régionaux, de régler les formalités administratives découlant de l'application des lois sur le remembrement légal de biens ruraux. b) Pour compte du Département de l'Agriculture, des avis à donner sur base de l'arrête royal du 17 juin 1965 relatif à l'intervention du Ministère de l'Agriculture en ce qui concerne l'élaboration des plans d'aménagement et la désignation de terrains industriels.

Correspondance: français, néerlandais.

Gent

Faculteit van de Landbouwwetenschappen (Faculty of Agricultural Sciences - State University - Ghent - Belgium)
Rijksuniversiteit - Gent (R.U.G.) - Coupure links, 533 - 9000 Gent.   Tel. 09/23.69.61

- Onderzoekcentrum voor Boerderijbouwkunde (Centre de Recherches des Constructions Rurales)
Debruyckere ir. Dr.   Professeur associé
Nicolaus, A. ir.   Maître de Conférences
Peerlinck P. ing.techn.

- Centrum voor de Studie van het Stalklimaat (Centre pour l'Etude de la Climatisation des Etables)
Christiaens, J.   Assistant ir.
Neuckermans, G.   Assistant ir.
Vanderbiest, W.   ing.techn.
Vyncke, L.   ing.techn.
Velghe, K.   ing.techn.

- Studiecentrum voor Toegepaste Electriciteit in Land-en Tuinbouw (Centre de Recherches pour les Applications Rurales de l'Electricité)
De Groote, A.   Assistant ir.

- Onderzoekcentrum voor Grote Productie - Eenheden in de Veeteelt 9Centre de Recherches des Grosses Unités de Production bovine et porcine)
Ghijsens, A.   ing.techn.

- Landbouwmechanica (Laboratoire de Mécanique Agricole)
Moerman, J.J.   ir. Professeur ordinaire Titulaire
Van Loocke, E.   ir. Premier Assistant

- Laboratorium voor Land en Tuinbouwmachines (Laboratoire de Machines Agricoles et Horticoles)
Van Lancker, J.   ir. Professeur ordinaire Titulaire
Kermis, L.   ir. Dr. Premier Assistant

- Laboratorium voor Hydraulica, Topografie en Cultuurtechniek (Amélioration foncières)
Heyndrickx, G.   ir. Professeur ordinaire Titulaire
Hesters, M.   Assistant ir.

Station affiliée: Ferme expérimentale à Melle

Statut: Institution gouvernamentale, placée sous le contrôle du Ministre de l'Education Nationale et de la Culture

Activités concernant le Génie Rural: Formation: L'institution donne des cours d'instruction durant cinq ans et sanctionnés par le diplôme d'Ingénieur Agronome. - Recherches et publications sur: les constructions rurales - la climatisation des étables - désodorisation et épuration du lisier - les applications de l'électricité dans l'agriculture et l'horticulture - les exigences et les problèmes techniques de la mécanisation de l'agriculture - la mécanisation de la production animale: l'équipement de traite, d'alimentation et de nettoyage - la dynamiqye du sol - systèmes de récolte.

Correspondance: Néerlandais, Français, Anglais, Allemand.

Gembloux

Station de Génie Rural de l'Etat (National Station of Agricultural Engineering)
Chaussée de Namur, 146 - B 5800 Gembloux

Dufey, V.	Directeur
Pletinckx, A.	Chef de Travaux
Vitlox, O.	Assistant

Structure: Fondation: en 1912. Organisme gouvernemental - Ministère de l'Agriculture - Doté de la personnalité juridique.

Activités: Recherche et expérimentation - a) Mécanisation des productions végétales: céréales, plantes industrielles, fourrages. b) Mécanisation des productions animales. c) Interactions sol-machines. d) Economie d'énergie et de matières. - Essais officiels de machines - Normalisation.

Publications: C.R.A., Gembloux; Bulletin de Recherches Agronomiques, Gembloux.

Correspondance: Français, néerlandais, anglais, allemand, espagnol, italien.

Centre d'Etude de la Mécanisation en Agriculture C.E.M.A.G. (Center for the Study of Mechanization in Agriculture)
Chaussée de Namur, 146 - B 5800 Gembloux

Dufey, V.	Président
Depoorter, J.	)
Legrand, E.	) Ingénieurs agronomes
Prade, J.	)
Van Camp, R.	)

Structure: Fondation: 1960 - Organisme contractuel - Ministère de l'Agriculture.

Activités: Recherche, expérimentation et information sur la mecanisation et la gestion de la ferme mécanisée. - Détermination des performances pratiques des machines - Etude des besoins en main-d'oeuvre - Charges et économie de la mécanisation.

Publications: C.R.A., Gembloux - Indicateur des performances et des coûts d'utilisation de l'équipement agricole, CEMAG, tome I et II.

Correspondance: Français, néerlandais, anglais, allemand, espagnol, italien.

Merelbeke

Rijksstation voor Landnouwtechniek (Station de Recherches du Génie Rural de l'Etat
(National Institute of Agricultural Engineering)
B9220 Merelbeke, van Gansberghelaan 115   Tel.: 91/52.18 21

Maton, A.M.	Docteur ès sciences, Ing.Agron.	Directeur
Daelemans, J.	Docteur en scienc.agron., Ing.Agron.	Sous-directeur
Taveirne, W.	Ing. Agron.	Chef de travaux
Lips, J.	Ing.Agron.	Chef de travaux
Dierickx, W.	Docteur en scienc.agron., Ing.Agron.	Chef de travaux
Priem, R.	Ing.Agron., Ing. Environnem., M.Sc. Envir. Eng.	Chef de travaux
Desmet, J.	Ing.Agron.	Assistant

Statut: L'institution dépend de l'Administration de la Recherche Agronomique du Ministère de l'Agriculture de l'Etat belge.

Activités: Drainage - Technologie des matériaux de drainage - Nettoyage mécanique des fossés - La technique de séchage du houblon, du foin, de la chicorée, de graines de betterave, des céréales - Elaboration de plans-type pour séchoirs - Construction et chauffage des serres - Méthodes rationnelles d'exécution des travaux en agriculture, élevage et horticulture, y comprise la mise en oeuvre de l'ordinateur - Utilisation rationnelle des machines et outils agricoles - Aménagement des étables, porcheries et poulaillers - Problèmes d'environnement en rapport avec l'élevage (traitement du lisier, contrôle des mauvaises odeurs).

Publications: dans des revues scientifiques belges et étrangères et depuis novembre 1964, aussi dans des propres "Communications" dont, jusqu'à présent 80 numéros ont paru. Trois livres ont été publiés, dont un en trois pays.

Correspondance: Neérlandais, anglais, français, allemand, espagnol.

Louvain

Département du Génie Rural, Faculté des Sciences Agronomiques (Agricultural Engineering Department, Faculty of Agronomical Sciences)
Kardinaal Mercierlaan, 92. B-3030 Heverlee (Louvain)

Statut: Département fondé en 1952 par l'Université Catholique de Louvain.

Activitées: Enseignement et formation: 5 années - 2 candidatures + 3 années de cours à orientation Génie Rural; Etudes et recherches: Mécanisation Rurale - Conditionnement des Produits Agricoles - Besoins en eau des plantes - Etudes de Bassins Hydrographiques - Automatisation des Mesures et de leur Récolte à courte et à longue distance - Projets de Génie Forestier - Locomotion hors route - Aménagements Ruraux

Publications: Buletins et comunications.

Correspondance: français, anglais, néerlandais, allemand.

Laboratorium voor Boerderijbouw, Cultuurtechniek en grondhydraulica (Laboratoire de Recherche de constructions Rurales, d'Améliorations Foncières et d'Hydraulique Agricole) (Research Station for Farm Buildings, Land Reclamation and Agricultural Hydraulics)
Kardinaal Mercierlaan, 92, Heverlée, Leuven (Louvain)

Statut: Institution sous le contrôle de la Faculté des Sciences Agronomiques de l'Université de Louvain et du Ministère Belge d'Agriculture.

Activité: Cours de génie rural avec cycle de 5 années d'études supérieures pour le titre d'ingénieur du génie rural - Recherches et expérimentations sur les constructions rurales - La climatisation d'étables - Irrigation par aspersion - Courants d'eau souterrains - Lutte contre les inondations.

Publications: Travaux d'amélioration dans le cadre du remembrement des terres - Etude des irrigations en général et des arrosages en particulier - Description du régime d'eau de bassins hydrographiques naturels et artificiels pour des travaux de génie rural et civil - La lutte économique et technique contre les inondations par l'installation de bassins de réserve à digues élevées - L'assèchement des polders et les stations de pompage - Méthode nouvelle du calcul de rebattement de la nappe aquifère - Bilans thermiques et hydriques dans les porcheries d'élevage.

Correspondance: Néerlandais, français, allemand, anglais.

BENIN

Porto Novo

Direction du Génie Rural et des Améliorations Foncières
B.P. 268, Porto Novo

Salouffou, Z.	Ingénieur de l'Equipement Rural	Directeur

Service Machinisme Agricole:
Damien, M.	Ingénieur du Génie Rural	Chef Service

Service Hydraulique Agricole:
Agbossou, E.	Ingénieur Agronome, Specialité Génie Rural	Chef Service

Service Habitat et Desserte Ruraux:
Ahoussinou, R.	Ingénieur Adjoint des Services	Chef Service

Service Topographie:
Agonvi, E.	Adjoint Technique du Génie Rural	Chef Service

Annexes situées dans d'autres localités: Divisions Génie Rural aux Centres d'Action Régionale pour le Développement Rural (CARDER)

a) *Province de l'Atlantique:*
Hessavi, P.	Ingénieur de l'Equipement Rural	Chef Division

b) *Province de l'Ouémé:*
Zanfognon, N.	Ingénieur Agronome Spécialité Génie Rural	Chef Division

c) *Province du Zou:*
N'Da N'Po, J-P.	Ingénieur du Génie Rural	Chef Division

d) *Province du Borgou:*
Dotche, M.	Ingénieur de l'Equipement Rural	Chef Division

e) <u>Province de l'Atacora</u>:
   Tchougourou, A.   Ingénieur de l'Equipement Rural   Chef Division

f) <u>Province du Mono</u>:
   Ahonon, B.   Adjoint Technique du Génie Rural   Chef Division

<u>Structure</u>: Date de fondation: 20 Octobre 1966 par decret n. 544/PR. Organe de Contrôle: Ministère du Développement Rural et de l'Action Coopérative. Organe Financier: Budget National (Ministère des Finances).

<u>Activités concernant le Génie Rural</u>: Formation - La Direction du Génie Rural dispense un certain nombre de cours au Lycée Agricole de Sékou en vue de la formation des Adjoints Techniques du Génie Rural et au Complexe Polytechnique Agricole Niveau I de Porto-Novo pour la formation des Agents Techniques du Génie Rural. Elle assure en outre les stages pratiques de ces Etudiants.

<u>Correspondance</u>: Français.

BOLIVIA

Cochabamba

<u>Facultad de Ciencias Agronómicas, Universidad Mayor de San Simón</u> (Faculty of Agricultural Science, University of San Simón)
Casilla 658, Cochabamba

<u>Dirección Nacional de Cuencas Hidrográficas, Ministerio de Asuntos Campesinos y Agropecuarios</u> (National Direction of Hydrographic Basins)
Av. Camacho No. 1471, La Paz, Bolivia

<u>Personal científico</u>:
Ing.Agr. M.C. Jaime Sejas Albornoz   Director
Ing.Agr. José Herrera   Departamento de Riegos
Ing.Agr. Ramiro Iriarte   Departamento de Suelos
Ing.Agr. M.C. Marcial Marcías   Departamento de Suelos
Ing. Civil M.C. José Ortíz   Departamento de Ingeniería
Ing. Civil Jorge Quino   Departamento de Ingeniería
Ing.Agr. Antonio Sainz   Departamento de Suelos
Ing.Agr. Arnaldo Sanjinéz   Departamento de Mecanización

<u>Dependencias en otras localidades</u>:

Existen 5 oficinas regionales y son:

Regional Suelos - Riegos en Cochabamba
Regional Suelos - Riegos en Oruro
Regional Suelos - Riegos en Tarija
Regional Suelos en Santa Cruz
Regional Suelos en Chuquisaca

<u>Carácter y Estructura</u>: La Dirección Nacional de Cuencas Hidrográficas con sus Departamentos Técnicos fué creada mediante Resolución Ministerial No. 85/80 de fecha 8 de febrero del año en curso, tiene carácter gubernamental y la dirige y financia el Ministerio de Asuntos Campesinos y Agropecuarios.

Actividades en ingeniería rural: No es una institución de enseñanza; por tanto, no ofrece cursos de estudios - Se tiene trabajos de investigación y experimentación en los Sistemas de riegos "La Angostura" en Cochabamba, "Tacagua" en Oruro y en el Programa Desarrollo Agrícola por Riego BOL/78/005 en Cochabamba y Tarija en aspectos de uso y manejo de tierras y aguas. - Los Departamentos técnicos prestan asistencia técnica sobre: Clasificación, uso y manejo de suelos y aguas, Ingeniería Agrícola y mecanización agrícola a todo el sector agropecuario.

Publicaciones: Desde 1973 a la fecha, solamente se publican informes técnicos de: suelos, Riegos, Ingeniería y Mecanización Agrícola, mecanografiados y a veces publicados por la imprenta del MACA, cuyas copias son distribuidas a la Biblioteca, a las Instituciones de-centralizadas y desconcentradas del Ministerio e Instituciones privadas que solicitan asistencia técnica.

Correspondencia: En español.

Instituciones de Ingeniería Rural: Otras instituciones que tienen relación con ingeniería rural, son: (a) Servicio Nacional de Desarrollo de la Comunidad (S.N.D.C.) Av. Ecuador esquina Fernando Guachalla; (b) Instituto Nacional de Fomento Lanero (INFOL) Edificio Avenida - Av. 16 de Julio No. 1490; c) Instituto de Desarrollo Rural del Altiplano (IDRA) Edificio Presencia - Av. Mariscal Santa Cruz; d) Instituto Nacional de Colonización (INC) Calle Junín esquina Indaburo; e) Centro de Desarrollo Forestal (CDF) MACA, Av. Camacho No. 1471 (6. Piso); (f) Instituto Boliviano de Tecnología Agropecuaria (IBTA) MACA - Av. Camacho No. 1471 (5. Piso)

BOTSWANA

Gaberones

Botswana Agricultural College
Gaberones

Status: Institution financed by Government.

Activities: Teaching courses in general agriculture to about 50 students per year, with agricultural machinery as part of the syllabus and in-service training courses up to 10 days.

Gaborone

Agricultural Research Station, Ministry of Agriculture, Department of Agricultural Research
Private Bag 0033, Gaborone, Botswana   Tel. 2841 ext 49 65 75

Oland, K.   Director of Agriculture Research

Evaluation of Farming Systems and Agricultural Implements:
Middlemiss, C.P.	B.Sc.(Agric.) Leeds, D.T.A. Trinidad	Team Leader
Horspool, G.D.	C.D.A. N.D.A. T.Eng(CEI), M.I. Agr.E.	Agricultural Engineer
Riches, C.R.	B.Sc.(Agric.Bot.) Reading M.Sc.(Agric.Ext.) Reading	Extension Training Officer
Lightfoot, C.W.	B.Sc.(Agric.Bot.) Reading. M.Sc.(Trop. Agric.Dev.) Reading	Systems Agronomist

Activities: Design. Modification and evaluation of farming systems and agricultural implements for Botswana.

Dryland Farming Research Scheme:
Jones, M.J.	B.Sc.(Chem)	Team Leader, Tropical Agronomist
St. Andrews	Ph.D.(Soil Science) Aberdeen	
Rees, D.	B.Sc.(Envir.Sci.) East Anglia	
	Ph.D.(Crop Phys.) Edinburgh	
Sinclair, J.	M.A.(Nat.Sci.) Aberdeen, M.Sc.(Physics) Nairobi	

Activities: Basic research into the environmental constraints (meteorological, edaphic) on dryland arable crop production in semi-arid regions - Operations include: Investigation of different types and depths of tillage and their effect on soil properties and crop root development. Physiological and microenvironmental studies of crop growth, with particular reference to the efficiency of water utilization.

Correspondence: English.

B R A Z I L

Universidade de Brasília - UnB, Departamento de Engenharia Agrícola
907 Norte, Cidade Universitaria, Brasilia, D.F.

Departamento de Engenharia, Instituto de Tecnologia, Universidade Federal Rural do Rio de Janeiro (Agricultural Engineering Department, Technology Institute, Federal Rural University of Rio de Janeiro)
Seropédica 23460, Rio de Janeiro

Corrêa, A.	A.M., A.E.	Head of Department, Prof.	Farm machinery and tractors
Bruno, B.	A.E.	Vice-Head of Department, Prof.	Farm Power

Sub-division of Mechanics:
Castanheira, R.G.	M.E.	Farm mechanics
Dalcolmo, E.L.	A.E., M.S.	Farm machinery
Jacintho, J.L.	A.E.	Farm mechanics
Montalvo, M.	A.E., Ph.D..	Farm machinery and tractors
Santana, G.C.	A.E.	Farm machinery and tractors
Silva, N.N.	A.E.	Farm machinery and power

Sub-division of Hydrology:
Duarte, E.F.	A.E.	Irrigation and drainage
Couto, J.L.V.	A.E., M.S.	Hydrology
Lôu, W.C.	C.E., Ph.D.	Hydrology
Mello, J. L.P.	A.E.	Irrigation and drainage
Silva, A.T.	A.E., M.S.	Irrigation and drainage

Sub-division of Surveying:
Almeida, R.V.	A.E.	Photometry
Eharaldt, E.	C.E.	Surveying
Lima, A.A.	C.E.	Surveying
Mueller, S.K.	A.E.	Surveying
Silva, M.S.	A.E.	Surveying

Status: This is a governmental institution, since 1913.

Activities: Agricultural Engineering is taught as a part of the college's general agricultural courses, taken in a four year period and leading to the degree of "Engenheiro Agrônomo" (Agriculturist Engineering). There is an average of 800 students attending this college.

Correspondence: Portuguese, Spanish, English, French.

Serviço Público Federal - Superintendência de Documentação e Informação
SAS Quadra 02 Ed. INPI Lotes 1/2, 70 070 Brasilia, D.F.

Universidade Federal da Bahia, Faculdade de Agronomia
44.380 Cruz das Almas, BA

Universidade Federal do Ceará - UFC, Centro de Ciências Agrarias, Departamento de Engenharia Agrícola
Cx. Postal 354, 60.000 Fortaleza, CE

Universidade Federal de Goiás, Departamento de Engenharia Rural
Cx. Postal 697, 76.600 Goiânia, GO.

Instituto de Pesquisa Agropecuária do Centro Sul (South Centre Research Institute in Agriculture and Farmstock (IPEACS)
Km 47 Estrada Rio, Sao Paulo, Rio de Janeiro ZC 26

Agricultural Engineering Section

Status: Date of establishment, 1965. It is a Governmental body financed by the Federal Government.

Activities: It has only specialized agricultural engineering courses. Work in irrigation, drainage and soil management is being carried out.

Correspondence: Portuguese, English.

Escola de Agronomia do Maranhão
Av. Lourenço Vieira da Silva, s/n - Cx. Postal 9, 65.000 São Luis, MA

Minas Gerais

Ministério da Educação e Cultura, Universidade Federal de Viçosa, Centro de Ciências Agrárias, Departamento de Engenharia Agrícola (Ministry of Education and Culture, Federal University of Viçosa, Center of Agrarian Science, Agricultural Engineering Department)

Hara, T.		Head of Department
Agricultural Climatology:		
Alves, A.R.	M.Sci.	Instructor
Coelho, D.T.	M.Sci., Ph.D.	Associate professor
Costa, J.M.N.	M.Sci.	Assistant professor
Sediyama, G.C.	M.Sci., Ph.D.	Associate professor
Vianello, R.L.	M.Sci., Ph.D.	Associate professor
Vieira, H.A.	M.Sci.	Assistant professor
Agricultural Mechanization:		
Cruz, J.M.	M.Sci.	Instructor
Dias, G.P.		Instructor
Mah, M.M.	M.Sci.	Instructor
Mantovani, A.		Full professor
Martyn, P.J.	Ph.D.	Associate professor
Mewes, B.O.		Instructor
Vieira, L.B.		Instructor
Agricultural Process Engineering:		
Carvalho, G.R.		Technical staff
Correa, P.C.	M.Sci.	Instructor
Dalpasquale, V.A.	M.Sci.	Instructor
Giudice, P.M. del	M.Sci.	Full professor
Hara, T.	M.Sci.	Full professor
Pinheiro Filho, J.B.	M.Sci., Ph.D.	Associate professor
Silva, J.S.	M.Sci., Ph.D.	Assistant professor
Farm Structure:		
Baeta, F.C.	M.Sci.	Instructor
Homem, A.C.F.		Instructor
Oliveira, J.L.	M.Sci., Ph.D.	Assistant professor
Peloso, E.M. del		Full professor
Hydraulic, Irrigation and Drainage:		
Bernardo, S.	M.Sci., Ph.D.	Full professor
Decículi, W.	M.Sci.	Assistant professor
Ferreira, P.A.	M.Sci., Ph.D.	Associate professor
Loureiro, B.T.	M.Sci., Ph.D.	Associate professor
Martinez, M.A.		Instructor
Ramos, M.M.	M.Sci.	Assistant professor
Soares, A.A.		Instructor

Branches with other location: CEPET, CEP – 38360 – Capinópolis, MG, Brasil; EMAF, CEP – 35 663 – Florestal, MG, Brasil.

Status: The Ag. Engineering Dep. was founded in 1972.

Agricultural Engineering activities: (a) Teaching: Under-graduate courses: Agricultural climatology, applied mechanics, Agricultural machinery, Agricultural meteorology, Power and Agricultural mechanization, Agricultural machinery design, Agricultural hydraulics, Hydraulics, Irrigation and drainage, Drainage, Rural hydraulic structure, Farm structure I, Farm structure II, Rural Electrification, Storage and processing of vegetable products I, Storage and processing of vegetable products II.

Graduate courses: Climatology physics, Evapotranspiration, Meteorology dynamics I, Agricultural machinery, Water movement in soil; sprinkle and drip irrigation, Agricultural land drainage, Applied hydrology, Surface irrigation, Instrumentation and electrical measurement system, Rural electrification; Storage and Processing of vegetable products II, Physical properties of Biological materials, Special problems, Seminar, and Research.

Publications: Revista CERES; Boletim de Extensão; Revista Informe Agropecuário; Revista Experientiae; Revista Brasileira de Armazenamento; Pesquisas em Andamento.

Correspondence: Portuguese, Spanish, English.

Departamento de Recursos Naturais Renováveis, Secretaria da Agricultura do Estado de Minas Gerais (Department of Natural Resources, Secretariat for Agriculture)
Rua Curitiba 656, 4º andar, 30.000 B. Horizonte, Minas Gerais

Service of Soil Conservation Planning
Service of Soil Conservation Execution

Motorized work teams in: Belo Horizonte, Coromandel, Curvelo, Felixlândia, Medina, Muriaé, Sao Domingos do Prata.

Status: Governmental institution, established in 1964. It absorbed the former "Soil Conservation, Irrigation and Drainage Service.

Activities: Study of natural resources of watersheds and river basins - Planning of irrigation, drainage and small earth dam projects - Mechanical execution of soil conservation work, terracing, dam building, canal digging and road construction.

Publications: Rain maps for Minas Gerais - Principles of irrigation - Overhead and sprinkler irrigation in Central Minas Gerais - programme of small watershed development with motorized teams.

Correspondence: Portuguese, Spanish, English, French.

Faculdade de Ciências Agrárias do Pará, Departamento de Engenharia Agrícola
Cx. Postal 917, 66.000 Belem, PA

Universidade Federal do Paraná, Setor de Ciências Agrárias, Departamento de Engenharia Agrícola
Cx. Postal 672, 80.000 Curitiba, PR

P e r n a m b u c o

Directoria de Defesa do Solo, Secretaria da Agricultura do Estado de Pernambuco
(Direction of Soil Conservation, Secretariat for Agriculture of the State of Pernambuco)
Recife, Pernambuco

Activities: Extension work on soil conservation and irrigation. Research work.

Correspondence: Portuguese, English, Spanish.

Universidade Federal Rural de Pernambuco, Departamento de Engenharia Agrícola
Cx. Postal 2.071, 50.000 Recife, PE

Rio Grande do Sul

Departamento de Engenharia Rural, Faculdade de Agronomia e Veterinária, Universidade Federal do Rio Grande do Sul (Department of Agricultural Engineering, College of Agriculture, University of Rio Grande do Sul)
Caixa Postal 776, 90.000 Porto Alegre, RS

Status: Governmental institution, supervised by Federal Ministry of Education. Founded in 1910.

Activities: Agricultural engineering training as part of the general agricultural courses of the college over 4 years, leading to the agricultural degree of BS, "Engenheiro Agronômo". The average number of students attending this college is 350. The average of students on agricultural engineering courses is 150. - Research on: Forage production mechanization; Mechanical cultivation; Irrigation of maize and soybean; Irrigation, drainage and water management.

Publications: Text books on: Farm power and agricultural machinery; Farm buildings; Irrigation and drainage.

Correspondence: Portuguese, Spanish, English and French.

Universidade Federal de Pelotas, Escola Superior de Agronomia "Eliseu Maciel",
Campus Universitário, Bairro Baroneza, Cx. Postal 767, 96.100 Pelotas, RS.

Universidade Federal de Sta. Maria, Centro de Ciencias Rurais
97.100 Santa Maria, RS.

Instituto Riograndese de Arroz (IRGA)
Cx. Postal 1.927, 94.000 Gravataí, RS

São Paulo

Escola Superior de Agricultura "Luiz de Queiroz", Departamento de Engenharia, Universidade de Sao Paulo (Agricultural College Luis de Queiroz, Department of Agricultural Engineering, Agricultural University of Sao Paulo)
Cx. Postal 9, 13.400 Piracicaba, SP

Status: Governmental institution.

Activities: Agricultural Engineering is taught as part of the general agricultural courses of the college over 5 years, leading to the agricultural degree of "Engenheiro Agrônomo". The agricultural engineering courses concern mechanics, engines and farm machinery. - Research on Tractor testing; Minimum tillage of sugar cane fields; Experiments with herbicides in several cultures; Study of wear of piston rings with the application of darioisotopes; Mechanization of the harvest of cotton, maize, sugar cane, peanuts, beans, rice, potatoes.

Publications: Problems of tractors and farm machinery; Experiments with agricultural implements; Mechanization of crops; Application of herbicides; Dynamometry; Development of a Brazilian tractor industry; Technology of wood and iron.

Correspondence: Portuguese, English, Spanish.

Instituto Agronômico de Campinas
Avenida Barão de Itapura 1481 - C. Postal 28 - Campinas - SP - Brazil

Divisão de Solos - Soil Division - C. Postal 28 - Campinas - SP - Brazil
Verdade, F.C.	Director General
Amaral, A.Z.	Director of Division

Seção de Conservação do Solo - Soil Conservation Section
Lombardi Neto, F.	Head of Section - Ph.D. in USA
Dechen, S.C.F.	Dr. in Agronomy in Brazil
Benatti Jr., R.	M.S. in Brazil
Silva, I.R.	Eng.Agr.
Castro, O.M.	Eng.Agr.

Seção de Irrigação e drenagem - Irrigation and Drainage Section
Arruda, F.B.	Ph.D. in USA
Vieira, S.R.	Ph.D. in USA
Donzelli, J.L.	Eng.Agr.

Seção de Fotointerpretação - Photointerpretation Section
Chiarini, J.V.	Eng.Agr. Head of Section
Donzelli, P.L.	Dr. in Agronomy in Brazil
Nogueira, F.P.	M.S. in Brazil
Peres Filho, A.	M.S. in Brazil
Barbieri, J.L.	Eng.Agr.

Status: Governmental Institution of São Paulo State.

Agricultural Engineering activities: Research in: Soil erosion control, Economic methods of irrigation, Effects of different systems of irrigation in plants, Land use using aerial photos, and Agricultural Planning.

Publications: The Instituto Agronômico publishes a scientific Journal "Bragantia" and Circulars and Technical Bulletins (Boletim Técnico).

Correspondence: Portuguese, Spanish and English.

Divisão de Engenharia Agrícola (IAC) - (Agriculture Engineering Division (IAC)
13.200 - Jundiaí, SP - P.O. Box 26, Brazil

Corrêa, H.G.	Eng.Agr.	Head of Division with 13 technical officers concerned with testing, development and research of farm machinery.

Soil Working Machinery Section:
Silveira, G.M. da	Eng.Agr.	Head

Planting, Tillage and Pesticides Machinery Section:
Corrêa, H.G.	Eng.Agr.	Head
Costa, J.A.S.	Eng.Agr.	
Maziero, J.V.G.	Eng.Agr.	
Mello Moraes, R.A.D. de	Eng.Agr.	

Harvesting and Agricultural Products Processing Machinery Section:
Moreira, C.A.	Eng.Agr.	Head
Testa, A.	Eng.Agr.	

Projects and Materials Section:
Kurachi, S.A.H.	Eng.Agr.	Head
Bernardi, J.A.	Eng.Agr.	

Traction and Power Machinery Section:
Menezes, J.F.	Eng.Agr.	Head
Rochelle, E.	Eng.Agr.	

Mechanization Sector:
Armond, G.	Eng.Agr.	Head
Braga, C.A.S.	Eng.Agr.	

Status: Government institution, subject to Instituto Agronômico.

Activities: This is one of Instituto Agronomico's Divisions for S. Paulo's State Agriculture Secretariat — It tests, develops and researches on tractors and agricultural equipment. It has a track of concrete, land and grass.

Research on: farm traction and power machinery, farm tools — agricultural construction materials; farm buildings; experiments, tests, improves, and develops farm machinery.

Publications: Seed distributors; Field plot seeder; Abrasion impact; Off-center rotary cutter; Lime spreader; Pesticides applicator; Disc harrow; Traction and power for disc plows; Soil moving; Weeds control; Soil structure; Sub-soilers; Pasture mechanization; Cotton fibers and yarn; Erosion losses with use of agricultural machinery; Farm machines test; Traction tests; Coffee sprayed leaves; Seed metering devices; Sprayers performances.

Correspondence: Portuguese, Spanish, English.

Divisão de Mecanizaçao Agrícola, Departamento de Engenharia e Mecânica da Agricultura, Secretaria da Agricultura (Agricultural Mechanization Division, Department of Agricultural Engineering, Secretariat for Agriculture)
Avenida Francisco Matarazzo 455, C.P. 8366, São Paulo   Tel.: 628287
                                                                   626235

Subdivision of Analyses and Farm Machinery Testing

Farm Mechanization Branch

Regional Farm Mechanization Branches

Schools for tractor operators in: Jabuticabal, Pirassununga, São Manuel.

Status: Government institution established in 1949 and financed by the Secretariat of Agriculture, São Paulo State.

Activities: Training courses for tractor operators; Experimental work on farm machinery and mechanization in general; Testing of tractors, farm implements and machinery, and processing equipment; Elaboration of farm building plans.

Publications: "Boletim da Divisão de Mecanizaçao Agrícola", published irregularly.

Correspondence: Portuguese, Spanish, English.

BULGARIA

Sofia

<u>Nauchnoizsledovatelski Institut po Mechanizatziya i Electrificatziya na Selskoto Stopanstvo (NIMESS)</u> (Research Institute for Mechanization and Electrification of Agriculture)
Chaussée Bankja 3, 1731 Sofia, 101.

Stoyanov, Chr. N.	M.Sc. Professor	Director
Grancharov, Vl. G.	Eng. M.T.Sc. R.A.	Deputy-Director

<u>Status</u>: Governmental, supervised by the National Agro-Industrial Union (NAPS) which performs the functions of the Ministry of Agriculture. It was established in 1954.

<u>Activities</u>: Scientific and research work for establishing of new industrial technologies for mechanized performance of the agricultural works, system of machines and complex mechanization of the agricultural production, plant-growing and animal husbandry. Research and design-construction work for creating of complex technological lines (KTL) for vine-growing, fruit-growing, vegetable-growing, forage-production, plant-growing and animal husbandry. - Development of machines and equipment for loading and unloading works in the agricultural transport, energetics, hydraulics, equipment for drying and storage of agricultural products. Development of the problems on electrification and automation in agriculture. Development of the problems on the technical maintenance and repair of the agricultural machines. Developing prognosis for development of mechanization.

<u>Publications</u>: "Farm Mechanization" and "Agricultural Machinery".

<u>Correspondence</u>: Russian, English, French, German.

<u>Institut po Selskostopansko Mashinostroene (ISSM) Russe</u> (Institute of Agricultural Engineering (ISSM) Russe)

<u>Mashinoizpitatelna Stantziya (MIS) Plovdiv.</u> (Machine Testing Station (MIS) Plovdiv.)

<u>Mashinoizputatelna Stantziya (MIS) Russe.</u> (Machine Testing Station (MIS) Russe)

<u>Centralna Experimentalna Baza, Botevgrad</u> (Central Experimental Basis, Botevgrad)

<u>Status</u>: Governmental, supervised by the National Agro-Industrial Union (NAPS).

<u>Activities</u>: Carries out scientific research studies and design-constructors and testing works in the field of the mechanization, electrification and automation of agriculture, tractors and agricultural machinery construction. Develops and introduces complex line machines for different agricultural works. Treats the questions of technology organization standardization in agricultural engineering.

<u>Publications</u>: "Farm Mechanization" and "Agricultural Machinery".

<u>Correspondence</u>: Russian, English, French, German.

BURMA

Rangoon

Department of Mechanized Agriculture, Ministry of Agriculture and Forests
Rangoon

Activities: Coordination and Supervision of tractor services and farm mechanization operations.

Division of Research, Institute of Economics
University Estate, Rangoon

Activities include investigations of problems of modernization and development of agriculture.

Pyinmana

Institute of Agriculture
Yezin, Pyinmana

BURUNDI

Bujumbura

Département du génie rural, Ministère de l'Agriculture et de l'Elevage (Agricultural Engineering Department, Ministry of Agriculture and Animal Husbandry)
B.P. 1850, Bujumbura   Tél. 2869

Activité: Exécution de travaux de génie rural - Contrôle de travaux de génie rural exécutés par des tiers - Etudes de projets de drainage et d'irrigation.

Correspondance: Français.

Institut des Sciences Agronomiques du Burundi (ISABU)
B.P. 795, Bujumbura

Activities: Applied agronomic research and farm management.

Université du Burundi
B.P. 1550, Bujumbura   Tél.: 3288

CAMEROUN

Yaoundé

Direction du Génie Rural et de l'Hydraulique Agricole (Ministère de l'Agriculture)
(Directorate of Agricultural Engineering and Hydraulics)
Yaoundé, Cameroun

Personnel scientifique:

Ngassam, J.	Ingénieur Agronome dipl. de l'Institut National Agronomique (INA) de Paris. Ingénieur Civil du Génie Rural dipl. de l'Ecole Nationale du Génie Rural (ENGREF) de Paris	Directeur de l'Institut
Mbepi, H.	Ingénieur des Techniques de l'Equipement Rural dipl. de l'Ecole Nationale des Ingénieurs des Travaux Ruraux et des Techniques Sanitaires de Strassbourg.	Directeur-Adjoint du Génie Rural
Ngwessitcheu, V.	Ingénieur de l'Equipement Rural, dipl. de l'Ecole Inter-Etats d'Ingénieurs de l'Equipement Rural (EIER) de Ouagadougou (Haute-Volta)	Chef de Service de l'Hydraulique Rurale et des Aménagements Ruraux à la Direction du Génie Rural
Ngankou, J-M.	Ingénieur de l'Equipement Rural dipl. de l'EIER de Ouagadougou (Haute-Volta)	Chef de Service Provincial du Génie Rural pour la Province du Centre-Sud à Yaoundé.
Medjo, S.	Ingénieur de l'Equipement Rural dipl. de l'EIER de Ouagadougou (Haute-Volta)	
Bemo, N.	Ingénieur de l'Equipement Rural dipl. de l'EIER de Ouagadougou	Chef de Service Prov. du Génie Rural pour la Province du Sud-Ouest à Buea.
Tangwan, E. Isari	B.Sc. of Engineering diplomé de d'Hale Sélassié (Ethiopia)	Chef de Service Prov. du Génie Rural pour la Province du Littoral à Douala.
Momo, J.	Ingénieur des Travaux du Génie Rural	Chef de Service Prov. du Génie Rural pour la Province de l'Ouest à Bafoussam.
Owono, P.	Ingénieur des Travaux du Génie Rural	Chef de Service Prov. pour la Province de l'Est à Bertoua.
Achu-Ngu, C.	Technicien principal des Techniques Industrielles	Chef de Service Prov. pour la Prov. du Nord-Ouest à Bamenda.

Annexes du Génie Rural situées dans d'autres localités:

- 7 services provinciaux dans les 7 provinces du pays.
- 40 sections départementales situées dans les 40 départements du pays.

Structure: Le Service du Génie Rural au Cameroun date des années 1953/1954.

Activités concernant le Génie Rural: a) Formation: Beaucoup d'ingénieurs actuellement en poste ont été formés à l'étranger en dehors de l'Afrique: France, Bélgique, etc. Actuellement des Ecoles de formation existent en Afrique et au Cameroun. - Ecole Inter-Etats d'Ingénieurs de l'Equipement Rural de Ouagadougou (Haute Volta) pour la formation d'Ingénieurs en moyenne par an pour le Cameroun: 5 Ingénieurs. - Ecole Inter-Etats de l'Hydraulique et de l'Equipement Rural de Kamboinse (Haute Volta) pour la formation des Techniciens du Génie Rural: 5 Techniciens en moyénne par an pour le Cameroun. - Ecole Nationale de Technologie de Yaoundé (Cameroun) pour la formation des techniciens du Génie Rural: 25 à 30 techniciens en moyénne par an pour l'option génie rural.

Recherches: néant. Mais activités variées dans le domaine de l'Hydraulique Villageoise, l'Hydraulique Agricole, constructions rurales; aménagement et équipement rural, etc.

Correspondance: En Français de préférence, mais aussi en Anglais.

Centre National d'Etudes et d'Expérimentations du Machinisme Agricole (CNEEMA)
C'est un centre de recherches rattaché directement au Ministère de l'Agriculture (Yaoundé).

CANADA

Alberta

Vermilion College
Vermilion, Alberta, Canada
Agricultural Engineering Division

Department of Agricultural Engineering, University of Alberta
Edmonton, Alberta

Anderson, A.W.	M.Sc. Assoc.Prof.	Operation Research
Domier, K.W.	Ph.D. Prof.	Farm Power
Grimmer, E.G.A.	B.Sc. Assoc. Prof.	Forest Engineering
Harrison, H.P.	Ph.D. Prof.	Farm Machinery
MacHardy, F.V.	Ph.D. Prof.	Economics of Agricultural Mechanization
McQuitty, J.B.	M.Sc. Prof.	Farm Buildings
Micko, M.M.	Ph.D. Assoc. Prof.	Wood Science
Rapp, E.	M. Sc. Prof.	Soil and Water
Stephanson, B.T.	M.Sc. Prof. Chairman of Department	Materials Handling

Branches: Research Farm, Ellerslie, Alberta.

Status: Semi-governmental institution.

Activities: Teaching: Four year course for B.Sc.(Agr.Eng.) in Engineering and four year course for B.Sc.(Eng.Agrol.) in Agriculture degree. Ten to fifteen students graduate each year. Master of Science degree with eight to ten students graduating each year. Research Activities: Tractive characteristics of Alberta soils - Determination of the optimum amount of tractor ballast - Cultivator chassis mechanics - Minimum tillage - Field machinery requirements - Methods of solving agricultural systems engineering problems - Cost analysis of machinery systems - Farmstead planning - Draft, torque and power requirements of rod weeders - Soil forces for a simple inclined blade - Treatment of forage with sulphur dioxide - Low pressure dehydration - Water table behaviour of glacial till soils - Using weather data for irrigation planning in Alberta - Hydraulics of drain materials - Seepage reduction in dugouts - Atmospheric contaminants in total confinement livestock buildings - Heat and moisture loads in confinement livestock buildings - Thermal characteristics of livestock buildings - Control of H_2S evolution from anaerobic swine manure - Systems engineering analysis of environmental control alternatives in confinement livestock housing - Fertilizer placement - Power utilization of large tractors - Analysis of fertilizer flow from factory to field - Potato planters - Variation of wood properties, physical, chemical and anatomical - Softwood and hardwood, aspen in particular - Suitability for reconstituted wood products, aspen wafers, composition boards and boned panels - Wood adhesive and curing parameters.

Publications: Pumpability and viscosity tests of crankcase oils - Traction performance characteristics of 13-6-28 tires on agricultural soils - Comparison of predicted and measured drawbar performance of tractors on agricultural soils - Performance parameters of tractors equipped with singles, duals and four wheel drive - Traction characteristics of two-wheel drive, four wheel drive and crawler tractors - Wheel traffic effects on soil compaction and growth of wheat - Comparisons of calculated weight transfer to measured weight transfer on an agricultural tractor - Viscosity changes in multigraded motor oils during service - Cultivator chassis mechanics - Methods of determining capacity of farm machinery - Estimating tractor drawbar soil from soil properties - Instrumentation of a multi component sensor - Effect of moisture content of wheat on threshing - Simulated particle dynamics with combine walkers - Friction in and energy required for extruding hay - Soil reacting forces from laboratory measurements with disks - Soil reacting forces for disks from field measurements - Soil reacting forces from field measurements with sweeps - Design of vertical rotary tiller blades for reforestation - Programming for minimum-cost machinery combinations - A method for sizing farm machines for weather dependent operations - An automatic guidance system for farm tractors - Economics of farm machinery utilization - The influence of the weather on field tractability in Alberta - Product terms in mathematical programming - Optimum combining time for minimum cost - The influence of machinery inventory size and price on farm income - Some physical factors affecting fungal population in stored wheat - The influence of ventilation on distribution and dispersal of atmospheric gaseous contaminants - Harvest simulation - Feedlot pollution - A literature review - Failure in concrete slats - Concentration-temperature relationships of atmospheric gaseous contaminants - An evaluation of harvest simulation as an aid to decision making - Moisture changes in dry and tough wheat in unventilated storage subjected to a cooling-warming cycle - Beef cattle performance in total confinement and open feedlot - Modelling the dynamic thermal characteristics of confinement livestock housing - The effects of beef housing systems on gaseous contaminants removed by ventilation - The effects of beef housing systems on moisture removed by ventilation - Aerobic and anaerobic storage of beef cattle wastes - Fortran IV program to predict the thermal environment within total confinement livestock housing - Hydrogen sulfide evolution from anaerobic swine manure - Pollution potential measurement of beef cattle wastes - Estimating plant nutrients in liquid manure from beef cattle - Chemical control of hydrogen sulfide from anaerobic swine manure. I. Oxidizing agents - Manure gases in the animal environment - a literature review - Dust in the animal environment - Modelling the thermal environment within total confinement livestock housing - Chemical control of hydrogen sulfide from anaerobic swine manure - II. Iron compounds - Data acquisition system for measuring environmental variables within confinement animal units - Some physical factors affecting dust concentrations in a pig facility - Solar heating natural air for grain drying - Design of an air-speed sensor system - Fuel and fertilizer production from dairy manure - An annotated bibliography of farm animal wastes - Feedlot pollution - nitrogen under some central Alberta feedlots - Response of calves to atmospheric hydrogen sulfide and ammonia - Some considerations in environmental control in pig housing - Warning - manure gases are dangerous - Feedlot pollution in Central Alberta - Maize storage under hot humid conditions - Heat and moisture loads in dairy barns - Heat and moisture loads in swine feeder barns - Comparison of methods for measuring conductance heat losses from livestock buildings - The method study of irrigation systems - Farm work study - Some applications of chronocyclegraphy in agricultural work study - Inverse square laws govern the operational scale of farm implements - The force platform, chronocyclegraph, and reaction timer as training aids for agricultural workers - A computer programme for the evaluation of alternative methods - Alberta harvesting problems gaps, leavings, damage - Ideal farm systems design, a guide to farm rationalization - Factors affecting plant stands in potato fields - 1969 potato harvester trials - Soil reclamation by deep-well pumping in the Maple Creek upper "V" area of Saskatchewan - Seepage control in concrete-lined irrigation ditches - Performance of some canal dugout linings on the Canadian prairies - Soil salinity and drainage problems: causes, effects, management - Performance of shallow subsurface drains in glacial till soils - Performance of tile drains under irrigation in southern Alberta - The problem of drainage - A hydrologic budget for a southern Alberta irrigation district - Drainage and salinity - Plastic film linings for dugouts - Land levelling on irrigated farms - Tractor operator roll over protection - Availability and usefulness of Alberta agricultural data - Work systems design in agriculture (an Ideals Concept) - Viscosity changes in multigraded motor oils during service - Analysis of tractor tests: a statistical approach - Performance of coulters for minimum tillage seeding - Correlation of cone index with traction characteristics - Thermal comfort of cab operators - Heating and ventilating of tractor cabs - Traction analysis of Nebraska Tractor Tests - Maximum of optimum tractive efficiency - No-tillage grain production in the Edmonton region - Prediction of drawbar power - several approaches - Performance characteristics of cup and pick type potato planters - Exotherm curves - a reproducible curing parameter - One-step method of making veneer overlaid plywood - Accelerated curing of wood panelling - Thermomechanical properties of crosslinked adhesives.

Correspondence: English.

International Development Research Centre, c/o Department of Agricultural Engineering, University of Alberta, Edmonton, Canada

Edwardson, W.	Ph.D.	Program Officer, Operation Research, Food Processing
Schmidt, O.G.A.	M.Sc.	Program Officer, Industrial Engineering Power and Machinery
Yaciuk, G.	Ph.D.	Program Officer, Solar Energy, Food Drying and Storage and Processing.

Engineering and Home Design Sector, Engineering and Rural Services Division, Alberta Department of Agriculture
Agriculture Building, 9718 - 107 Street, Edmonton, Alberta, Canada T5K 1E4

Calver, G.	Associate Director, Edmonton
Martens, J.	Coordinator, Water Pumping, Lethbridge
Chawla, K.	Processing Engineer, Edmonton
Kienholz, J.	Mechanical Engineer, Edmonton
West, B.	Animal Waste Mgmt. Engineer, Red Deer

Engineering Field Services

Ulrich, W.	Branch Head, Edmonton
Johnston, R.	Water & Sewage Engineer, Edmonton
Darby, D.	Farm Structures Engineer, Lethbridge
Green, M.	Farm Machinery Engineer, Airdrie
Borg, R.	Regional Engineer, Red Deer
Dill, G.	Regional Engineer, Lethbridge
Wendel, G.	Regional Engineer, Barrhead
Kennedy, B.	Regional Engineer, Vermilion
Wood, D.	Regional Engineer, Fairview
Winchell, W.	Regional Engineer, Airdrie
Wasylik, L.	Ag.Eng. Technologist, Vermilion
Cornwell, B.	Ag.Eng. Technologist, Fairview
Williamson, K.	Ag.Eng.Technologist, Red Deer
Kenzie, O.	Ag.Eng.Technologist, Lethbridge
Livingstone, A.	Ag.Eng.Technologist, Barrhead
Barlott, P.	Branch Head, Edmonton
Scott, D.	Systems Engineer, Edmonton
Thornton, E.	Ass't Systems Engineer, Edmonton

Status: Institution financed and sponsored by the Government.

Activities: Applied research, testing and extension work on home buildings, farm buildings, livestock production, materials handling, waste management, farm equipment care and maintenance, rural water supplies, plumbing and sewage disposal, land development, electrical appliances and equipment.

Publications: Plans and publications on Agricultural Engineering and Home Design for general distribution. Available upon request and free of charge.

Correspondence: English.

Canada Department of Agriculture, Research Station, Lethbridge, Alberta
Canada T1J 4B1

Scientific staff

Andrews, J.E.   Ph.D.   Station Director

Soil Science Section
Hobbs, E.H.   B.Sc.   Irrigation, soil moisture
Lindwall, C.W.   B.Sc.   Tillage
Oosterveld, M.   Ph.D.   Hydrology
Sommerfeldt, T.G. Ph.D.   Drainage, reclamation of saline soils

Plant Science Section
Mains, W.H.   M.Sc.   Forage handling

Status: Government institution.

Agricultural engineering activities: a) Teaching: nil. b) Research: Field machinery use, tillage, and soil conservation under semi-arid conditions - Irrigation techniques and evapotranspiration - Agrometeorology - Salinity and drainage - Hay and silage harvesting, storage, and retrieval.

Publications: Surface trash conservation with tillage machines - seeding into trash cover - Evapotranspiration for various crops as related to meteorological factors - Local climate and its effect on overwinter moisture storage - Irrigation application methods as related to water distribution and salinity - Drainage of soils as influenced by soil type, physical properties, irrigation methods, and water table proximity.

Correspondence: English.

Fairview College, Fairview, Alberta
Box 3000, T0H 1L0

Scharf, J.S.   Dean of Instruction
Iftody, B.   Agriculture Mechanics Instructor
McNaught, W.B.   Agriculture Mechanics Instructor
Wischoff, J.A.   Agricultural Mechanics Instructor

Status: Originated in 1951 as the Fairview School of Agriculture and Home Economics. Governed by local Board of Governors. Financed by Government of the Province of Alberta.

Activities: Teaching: Diploma in Agriculture.

Correspondence: English. French accepted, but is not teaching language.

Olds College, Olds, Alberta
Canada

Boettger, S.E.   B.Sc.   Electrical Engineering
Hamilton, R.S.   B.E.   Agricultural Engineering
Lipsit, M.L.   B.Sc.   Agriculture, Department Chairman
Taylor, M.E.   B.E.   Agricultural Engineering

Status: Established Agricultural Engineering, Technologist Program in 1976. This is a public college, funded partially by government grants.

Activities: Teaching: Students may receive an Agricultural Engineering Technologist Diploma.

Correspondence: English.

British Columbia

Bio-Resource Engineering Department, University of British Columbia
Vancouver, B.C. V6T 1W5, Canada

Staley, L.M.	B.A.Sc., M.Sc.(A.E.)	Acting Head of Department
Bulley, N.R.	B.A.Sc., Ph.D.	Water Quality and Waste Management
Coulthard, T.L.	B.Eng., M.Sc.	Professor Emeritus, Irrigation and Drainage
Lo, V.	B.S., M.Sc., Ph.D.	Food Process Engineering
Watson, E.L.	B.A.Sc., M.Sc.	Professor Emeritus, Food Engineering
Zahradnik, J.W.	B.S., M.Sc., Ph.D.	Aquaculture and Systems Engineering
Position Vacant		Soil and Water Engineering
Jackson, N.		Machine Shop Technology
Pehlke, J.		Instrumentation and Electronics Technology
Liao, P.		Analytical Chemistry

Status: Operates under provincial government university act and is supervised by the President and Board of Governors. Engineering program accredited by Canadian Council of Professional Engineers.

Teaching Activities: Study courses are provided leading to degree of B.A.Sc. in engineering, B.Sc. degree in agricultural mechanics, M.Sc. degree in agricultural engineering, and M.Sc. degree in agricultural mechanics. Ph.D. interdisciplinary.

Research Activities: Water quality, waste management and odor control from livestock and food processing enterprises; bio-mass conversion; mechanical means of weed control in fresh water lakes; coupled aquaculture and agriculture systems; hydrodynamic aspects of commercial production of aquatic organisms; physical properties of food materials; energy conservation in livestock and greenhouse structure; solar energy storage in greenhouses; tool and equipment development for fence building; drainage design and flow measurement; mathematical modelling of groundwater movement.

Publications: Thin layer drying; rehydration of freeze dried foods; thermal properties of food; flow properties of livestock waste; thermal environment influences on heritability traits and growth of poultry; moiré-patterns measure soil deformations produced by tillage tools; electronic color sorting of fruit; field chamber designs for air pollutant influence on plants; solar energy use by greenhouses; modelling hydrological and groundwater movement in relation to soils; electronic size and color grading of tomatoes; sensory evaluation of odors; aerobic stabilization of livestock wastes; nitrogen balances in fields involving waste disposal and cropping practices; thermal processes in food processing; kinetics of food utilization by oysters; computer simulation of retort control.

Correspondence: English.

Canadian Department of Agriculture Research Station
Summerland, B.C., Canada

Moyls, A.L.	Ph.D.	Head of Agricultural Engineering
Parchomchuk, P.	B.A.Sc., M.S.	Agricultural Engineering

Status: 1950 established Ag.Engineering. Financed by Federal Government.

Activities: Fruit drying, Calcium dip for apples, concentrating solar energy collector, trickle irrigation, fruit processing equipment, fruit handling.

Publications: Drying of fruit purées – Construction of commercial fruit dryer – Use of intermittant chlorine dosage to prevent emitter blockage – Report vent silencing – A simplified method for agitation processing of canned foods – Temperature effects on emitter discharge rates.

Correspondence: English.

Engineering Branch, British Columbia Ministry of Agriculture
33832 South Fraser Way, Abbortsford, British Columbia, Canada V2S 2C5

Windt, T.A.	Director
Baehr, B.E.	Assistant Director
Johnson, P.D.A.	Farm Structures Engineer
Kilpatrick, R.C.	Farm Mechanization Engineer
van der Gulik, T.W.	Water Management Engineer
May, K.E.	Regional Engineer, Vernon, B.C.
Metzger, J.F.	Regional Engineer, Dawson Creek, B.C.
Van Kleeck, R.J.	Regional Engineer, Summerland, B.C.

Status: Government Institution.

Agricultural Engineering Activities: The objectives of the Branch are to provide an Engineering Design and Advisory Service to farmers and ranchers throughout the Province of British Columbia. This service covers most aspects of on-farm engineering needs including land drainage, irrigation, farm structures, waste management, mechanization, and land and range development. Information is extended by on-farm visits, meetings, short courses and publications.

Publications: The primary type of publication issued by the Branch is a 2 to 4 page leaflet called "Engineering Notes". These cover a very wide range of agricultural engineering related topics and are continually updated and reissued as required.

Correspondence: English language.

Manitoba

Department of Agricultural Engineering, University of Manitoba
Winnipeg, Manitoba, R3T 2N2

Laliberte, G.E.	Ph.D.	Soil and water conservation
Britton, M.G.	Ph.D.	Farm structures
Buchanan, L.C.	BSAE	Farm structures, materials handling
Gauer, E.L.	M.Sc.	Irrigation and drainage
Johnson, D.J.	BSA	Farm power and machinery
Lapp, H.M.	M.Sc.	Soil and water conservation, farm structures
Muir, W.E.	Ph.D.	Farm structures, grain storage
Penkava, F.F.	Ph.D.	Soil and water conservation
Philp, J.D.	BSA	Farm power and machinery, machining, welding
Putnam, J.G.	Dipl.Agr.	Soil and water conservation, machining

Reitsma, S.Y.	Ph.D.	Instrumentation
Shlosser, H.J.	B.Eng.	Instrumentation
Tennenhouse, F.	BSA	Farm power and machinery
Terry, P.D.	B.Sc.(EE)	Grain storage
Townsend, J.D.	Ph.D.	Farm power and machinery
Wallace, H.A.H.	M.Sc.	Plant pathologist, grain storage
White, N.D.	Ph.D.	Grain storage
Wold, C.E.	Dipl.Agr.	Farm power and machinery, welding

Status: Established in 1906. An administrative subdivision of the Faculty of Agriculture, Governed by the Board of Governors, a 24-member body with 12 members appointed by the Lieutenant-Governor-In-Council of the Province of Manitoba and 12 members elected and/or appointed from among the academic staff and students of the University. Financed through the Board of Governors by the Universities Grants Commission, a provincial granting body appointed by the Government of Manitoba, and by direct research grants from federal, provincial and private granting agencies.

Activities: Teaching: Teaching activities include direction and instruction of MSc and BSc curricula in Agricultural Engineering and instruction only in Bachelor of Science and Diploma programmes in Agriculture. - Research activities span the broad areas of irrigation and drainage, animal waste management, grain storage and handling, farm structures and animal housing environment, instrumentation, farm power and machinery and greenhouse management.

Publications: Irrigation and drainage, grain storage and drying, farm structures, animal housing environment, instrumentation, greenhouse production, energy.

Correspondence: English.

Agricultural Engineering Section, Technical Services Branch - Production Division
Manitoba Agriculture
911 Norquay Building, 401 York Avenue, Winnipeg, Manitoba R3C 0V8, Canada

Campbell, B.	B.Sc.Eng.	Agricultural Engineer (Extension)
Friesen, O.H.	B.Sc.Eng., M.Sc.	Chief Agricultural Engineer (Power and Machinery)
Hodgkinson, D.	B.Sc.Eng.	Agricultural Engineer (Waste Management)
Jackson, D.	B.Sc.Eng.	Agricultural Engineer (Extension)
Plohman, G.	B.Sc.Eng.	Agricultural Engineer (Extension)
Rae, R.M.	B.Sc.Agr., M.Sc.	Agricultural Engineer (Extension)
Shwaluk, D.	B.Sc.Eng.	Agricultural Engineer (Extension)
Small, D.S.	B.Sc.Eng.	Agricultural Engineer (Special Crops)
Wilson, T.	B.Sc.Eng.	Agricultural Engineer (Extension)
(Vacant)		Agricultural Engineer (Farm Buildings)

Farm Machinery Board:
Chinn, R.	Secretary-Manager	

Agri-Water Branch:
Cousin, J.	B.Sc.Eng.	Agricultural Engineering (Water)
Griffin, E.	B.Sc.	Director
Klassen, G.	B.Sc.Eng.	Agricultural Engineering (Irrigation)
Lasuik, T.	Civil Tech.	Design Technologist
Pasquill, R.	Dip.Agr.	Agricultural Engineering (Drainage)
Tokarz, M.	B.Sc.Eng., M.Sc.	Agricultural Engineering (Design)

Activities: Farm Buildings — Develop and provide plans of farm buildings. Provide information to farmers on the selection, operation and construction of farm buildings. — Livestock waste management: Provide information on the handling and disposal of livestock wastes. Assist farmers to comply with environmental regulations. — Farm Power and Machinery: Evaluation of farm machinery for use in crop production and farm mechanization. Provision of information to farmers on selection, use adjustment, maintenance and economics of machinery. Assistance in the development of new machines and methods in cooperation with manufacturers and farmers. — Farm Machinery Act: Administered by a Board comprised of seven members including the Chairman. Responsible for processing complaints and repossessions filed with the Board under the Act. Also responsible for licencing and bonding dealers. — Agri-Water: Provide information to farmers and small communities on the location of underground water sources, well design, plan of water distribution systems, water treatment, pumps, well screening and equipment. — Irrigation and Drainage: Provision of advisory services and studies on farm irrigation and drainage.

Correspondence: English.

New Brunswick

Agricultural Engineering Branch, New Brunswick Department of Agriculture and Rural Development
P.O. Box 6000, Fredericton, New Brunswick, Canada, E3B 5H1

Gilchrist, E.D.	P.Eng., B.Sc.Agr.	Director
Durant, W.C.	P.Eng., B.Sc.Agr.	Assistant Director, Farm Structures
Allen, D.		Woodstock, District Engineer, Fruit and Vegetable Storage
Arsenault, L.		Grand Falls, District Engineer, Beef Structures
Banks, G.A.		Moncton, District Engineer, Soil and Water
Collette, L.		Fredericton, District Engineer
Daigle, J.L.		Grand Falls, Soil and Water Conservation
Everett, C.F.		Fredericton, Farm Machinery and Safety
Folkins, L.		Sussex, District Engineer, Dairy Structures
Gartley, C.		Moncton, Marshland Development
McKendy, K.		Bathurst, District Engineer, Greenhouses
Milburn, P.		Fredericton, Draianage
Trenholm, R.		Fredericton, Land Resource Planning

Branches with other location: The Branch has six regional offices strategically located throughout the Province.

Status: The Branch was established in 1943. It is a Government institute, managed by a director who is responsible through the Deputy Minister to the Minister of Agriculture.

Agricultural Engineering activities: Extension and service work in Farm Structures, Land and Water Conservation and Development, Land Resource Planning and Management, Farm Machinery and Safety.

Publications: Potato Mechanization Series — Planting, spraying, cultivation; Harvesting and Mechanical Injury; Controlling Soil losses; Guidelines for farmstead wiring; Forced

Air Drying for High Quality Hay, Bulk Potato Storage; Guide for manure handling and disposal. A number of circulars on a wide range of Agricultural Engineering subjects.

Correspondence: English and French.

Research Station, Research Branch, Department of Agriculture
Fredericton, New Brunswick

Status: Agricultural Engineering Facilities. Established in 1966 under the authority of the Canada Department of Agriculture.

Activities: Research: potato harvesting and storage, cereal harvesting, forage harvesting and storage, apple harvesting, water erosion control by parallel graded terraces, infiltration, evapotranspiration percolation studies. Potato harvesting, hay drying, silage unloading, apple harvesting.

Correspondence: English, French.

Nova Scotia

Nova Scotia Agricultural College
P.O. Box 550, Truro, NS

Adams, J.	M.Sc., Structures and Environment	Associate Professor
Cunningham, J.D.	B.Sc., Structures and Equipment	Assistant Professor
Havard, P.L.	M.Sc., Soil and Water, Solar	Lecturer
MacAulay, J.T.	M.Sc., Power and Machinery	Associate Professor
Townsend, G.E.	B.Sc., Structures and Environment	Assistant Professor

Status: Nova Scotia Department of Agriculture

Activities: a) Agricultural Engineering Technician, Diploma, 20 - Farm Equipment Technician, Diploma, 10 - Agricultural Engineering (leading to Degree), Diploma 12.

Correspondence: English.

Nova Scotia Department of Agriculture and Marketing - Extension Services Branch, Extension Agricultural Engineering Division

Milligan, D.C.	B.Sc.(Agr.Eng.)	Chief Engineer
Higgins, J.K.	M.Sc.(Agr.Eng.)	Supervisor
Bishop, G.	B.Sc.	Extension Engineer, Farm Structures
Browning, D.		Extension Engineer, Crop Processing Research
Cochrane, L.	B.Sc.	Extension Engineer, Farm Structures
Honey, L.		Extension Engineer, Farm Structures
Kolstee, H.		Extesnion Engineer, Land Protection
MacLennan, R.		Extension Engineer, Drainage
Swinkels, P.		Extension Engineer, Farm Machinery
Vermeulen, A.		Extension Engineer, Farm Structures

Status: Date of establishment – 1938

Activities: See Nova Scotia Agricultural College.

Correspondence: English.

Nova Scotia Technical College, Bio-Resources Engineering Department
P.O. Box 1000, Halifax, NS

O n t a r i o

School of Engineering, Ontario Agricultural College, University of Guelph
Guelph, Ontario, N1G 2W1

Status: Private institution, supervised by President and Board of Governors. Agricultural Engineering established in 1946.

Activites: Study courses are given at the bachelors, masters and doctoral level. Bachelor level B.Sc.(Eng.) in Agricultural, Biological and Water Resources Engineering, four year programme; Master level M.Sc., 1 to 2 years; Ph.D. level, 2 to 3 years.

Research activities: Power and Machinery; Mechanization of Fruit and Vegetable Planting, Harvesting and Processing; Combine Cleaning Systems; Transport of Agricultural Materials on Oscillating Conveyors; Terra-Mechanics and Traction; Efficiency of Systems of Power Supply; Development of Aerial Survey Techniques for Agriculture; Optimization of Machinery for Harvesting Corn in Ontario; – Soil and Water: Water Management; Hydrology. – Structures and Environment: Thermodynamics Applied to Ventilation; Stress Distribution in Non-Homogeneous Materials; Structural Components; Quantitative Analysis of Snow Accumulation; Wind Effect on Ventilation; Determination of Suitable Coatings for the Interior of Metal Bins used to Storage Propionic Acid Treated Corn; Animal Waste Treatment; Engineering Properties of Composite Materials; Drying of Poultry Manure; Gas Exchange of Silo Containers – Electric Power and Processing: Physical and Rheological Properties of Agricultural Materials; Thermal Properties of Soybeans and Rape-seed; Use of Waste Heat for Greenhouse Heating; Application of Low Cost Automation Technology to the Agriculture and Food Processing Industries; Reaction Kinetics in Fermentation Process – Miscellaneous: Computer Studies of the Wave Phenomenon; Rural Municipal Services; In Vitro Study of the Hemodynamics of Blood Flow in the Cardio-Pulmonary System.

Publications: There is a series of technical publications on a variety of subjects based on completed research.

Correspondence: English.

Contact Men for the Canada Plan Service (School of Engineering, University of Guelph)
Guelph, Ontario

Brubaker, J.E.	MSA, P.Eng.	Chairman, Manager, Energy Management Resource Centre, Ontario Ministry of Agr. and Food, 801 Bay Street, Toronto, Ont. M7A 2B2 (416) 965-9519.
Johnson, P.		Agricultural Engineering Branch, B.C. Department of Agriculture, 203-33780 Laurel Street, Abbotsford, British Columbia V2S 1X4 (604) 853-6451.

Darby, D.E.	P.Eng.	Engineering Field Services Branch, Alta Dept. of Agriculture, Agriculture Centre, Lethbridge, Alberta T1J 4C7 (403) 329-5114.
Wrubleski, E.M.	P.Eng.	Research Specialist, Research Branch, Saskatchewan Dept. of Agriculture, 1318 Winnipeg Street, Regina, Sask. S4P 3V7 (306) 565-6562.
Friesen, O.		(New man to be appointed). Technical Services Branch, Manitoba Dept. of Agriculture, 911 Norquay Building, Winnipeg, Manitoba R3C 0P8 (204) 946-7801.
Arnold, J.B.	P. Eng.	Ontario Ministry of Agr. and Food, School of Engineering, University of Guelph, Guelph, Ontario N1G 2W1 (519) 824-4120 ext. 2463.
Belzile, G.	Agr.	CP 269 St. Martine, Quebec, Canada J0S 1V0 (514) 427-2017.
Durant, W.C.		Asst. Director, Agr. Engineering Branch, N.B. Dept. of Agr. and Rural Development, P.O. Box 6000, Fredericton, New Brunswick, E3B 5H1 (506) 453-2691.
Higgins, J.K.	P.Eng.	Supervisor, Agricultural Engineering, Nova Scotia Dept. of Agr. and Marketing, Truro, Nova Scotia, B2N 5E3 (902) 895-1571.
Linkletter, G.		Farm Building Engineer, P.E.I. Dept. of Agr. and Forestry, P.O. Box 1600, Charlottetown, Prince Edward Island C1A 7N3 (902) 892-5465.
Pascua, M.	Ag.Eng.	Structures and Machinery, Nfld. Dept. of Forestry and Agriculture, Bldg. T-815, Pleasantville, St. John's, Newfoundland A1A 1R1 (709) 737-2639 or 2640.

Correspondence: English.

Agricultural Engineering Extension Service - Agricultural Engineers in Ontario

Brubaker, J.E.	MSA, P.Eng.	Supervisor, School of Engineering, University of Guelph, Guelph, Ontario N1G 2W1.

Special Projects:
Arnold, J.B.	M.Sc., B.Sc. P.Eng.	School of Engineering, University of Guelph, Guelph, Ontario N1G 2W1.

Silos:
Bellman, H.E.	MSA, P.Eng.	OMAF, Box 1330, Walkerton, Ont. N0G 2V0.

Poultry Buildings:
Bird, N.A.	M.Sc., P.Eng.	OMAF, Box 398, 413 Hibernia Street, Stratford, Ontario N5A 5W2.

Swine Buildings:
Boyd, K.G.	B.Sc., P.Eng.	OMAF, Box 730, Petrolia, Ontario N0N 1R0.

Fruit and Vegetable Storages:
Clarke, K.A. MSA, P.Eng. OMAF, Hort. Research Station, Vineland Station, Ontario. L0R 2E0

Grain Drying:
Clayton, R.E. BSA, B.Sc. P.Eng. OMAF, R.C.A.T. Campus, Ridgetown, Ontario N0P 2C0

Energy:
Cuthbertson, H. B.Sc., P.Eng. OMAF, Box 651, Kingston, Ontario. K7L 1H3
Fleming, R.J. B.Sc., P.Eng. OMAF, Box 159, Clinton, Ontario. N0M 1L0
Fraser, H.W. B.Sc. OMAF, 207 Greenwich Street, Brantford, Ontario. N3S 2X7

Milking Machines:
Garland, G.A. M.Sc., B.Sc., P.Eng. OMAF, Box 370, Alliston, Ontario. L0M 1A0

Drainage, Western Ontario:
Gregg, R.G. BSA, P.Eng. OMAF, Newmarket Plaza, Newmarket, Ontario. L3Y 2N1
Hilborn, D.R. B.Sc., P.Eng. OMAF, Box 129, Cayuga, Ontario. N0A 1E0

Ventilation:
Huffman, H.E. B.Sc., P.Eng. OMAF, 195 Dufferin Avenue, London, Ontario. N6A 1K7

Steel Design:
Johnson, J.W. B.E.Sc., P.Eng. OMAF, 594 Talbot Street, St. Thomas, Ontario. N5P 1C7

Beef Buildings:
Kains, F.A. B.Sc., P.Eng. OMAF, 279 Weber Street North, Waterloo, Ontario. N2J 3H8

Drainage, Eastern Ontario:
Kelly, R.D. BSA, P.Ag. OMAF, K.C.A.T. Campus, Kemptville, Ontario. K0G 1J0

Structures, Northern Ontario:
Kirik, M. BSA, B.Sc., P.Eng. OMAF, 222 McIntyre Street W., North Bay, Ontario. P1B 2Y8
MacPherson, T.W. B.Sc., P.Eng. OMAF, Box 540, 666 Notre Dame Street, Embrun, Ontario. K0A 1W0
Milne, R.J. BSA, P. Ag. OMAF, Box 666, Woodstock, Ontario. N4S 7Z5

Farm Machinery Management:
Moggach, G.S. BSA, P.Ag. OMAF, 322 Kent Street West, Lindsay, Ontario. K9V 2Z9

Reinforced Concrete Maple Syrup Equipment:
Mullen, K.W. B.Sc., P.Eng. OMAF, 181 Toronto Street South, Markdale, Ontario. N0C 1H0

Farm Ponds:
Myslik, J.P. B.Sc., P.Eng. OMAF, 3 Elizabeth Street South, Brampton, Ontario. L6Y 1P7
Plue, P.S. B.Sc., P.Eng. OMAF, 26 Thorncliff Place, Nepean, Ontario. K2H 6L2

Environment:
Presant, D.E. BSA, P.Ag. OMAF, Box 340, Stirling, Ontario. K0K 3E0

Surface Drainage:
Slater, G.M. B.Sc., P.Eng. OMAF, Box 579, Alexandria, Ontario. K0C 1A0

Greenhouses:
Stone, R.P. B.Sc., P.Eng. OMAF, Box 820, Brighton, Ontario. K0K 1H0

Horse Buildings Forage Machinery:
Weeden, J.K.   B.Sc., P.Eng.   OMAF, Box 159, Fergus, Ontario. N1M 2W7
Wood, W. H.   B.Sc.   OMAF, 46 Fox Street, Essex, Ontario. N8M 2S2

Status: Agricultural Engineering Extension was established in 1950 under the Extension Branch of the Ministry of Agriculture and Food. It is still located and financed there.

Activities: Provides an engineering consulting service to farmers and closely allied industry in drainage, farm water supply including ponds, farm structures, farm machinery, waste management, ventilation and refrigeration and pollution abatement.
- Provides a farm management service for clients reviewing costs, returns and benefits.
- Research activities are minimal being restricted to small applied projects on farms to evaluate new concepts or material.

Publications: in "Canadian Society of Agricultural Engineering" and "American Society of Agricultural Engineering".

Correspondence: English.

Agricultural Engineering Section, Kemptville College of Agricultural Technology
Kemptville, Ontario K0G 1J0

Status: Governmental institution supervised by the Ontario Ministry of Agriculture and Food.

Activities: Teaches agricultural engineering as part of a two-year diploma course in agriculture to student body of institution - Audio-visual support for the entire institution, e.g. still and motion picture photography, closed circuit television, audio and video recording - Applied agricultural engineering research on problems pertaining to Eastern Ontario agriculture - Currently under study are: Quick drying of agricultural products - Soil pressures encountered by tillage machines - Development of computer programmes related to selection and management of agricultural equipment.

Publications: A.S.A.E. Transactions and National Conference on Automatic Control, Edmonton - Automatic Draft Control of Self-propelled Wagons - Evolution of a Mechanized Harvesting Aid - Socio Economic Aspects of Automation.

Correspondence: English.

Engineering and Statistical Research Institute, Research Branch, Agriculture Canada
Bldg. No. 94, Ottawa, Ontario K1A 0C6

Voisey, P.W.	F.I., Mech.E.	Director
Mitchell, K.B.		Administration

Mechanization and Systems:
Feldman, M.	B.E., M.Sc.	Head of Section; Mechanization
Bishop, D.G.	B.Sc.	Systems analysis
Lievers, K.W.	B.Sc.(Agr.), M.Sc.	Systems analysis
Reid, W.S.	B.Sc.(Agr.), M.Sc.	Mechanization
Van Die, P.	B.Sc.(Eng.), M.Sc.	Energy

Research Service:
Brach, E.J.	D.E.E., Dip.Mil.Electronics	Head of Section; Electronics
Buckley, D.J.	B.E., M.Sc.	Electronics
Timbers, G.E.	B.S.A., M.S.A., Ph.D.	Food process engineering

Statistical Research:
Lefkovitch, L.P.	B.Sc.	Director, Statistical Res. Service
Poushinsky, G.P.	B.Sc., M.Sc.	Head of Section, Statistics (on leave)
Binns, M.R.	M.A., Dip.Stat.	Statistics
Jui, P.Y.	M.Sc., Ph.D.	Statistics
Lin, C.S.	B.Sc., M.Sc., Ph.D.	Statistics
Morse, P.M.	M.A.	Statistics
Price, K.R.	B.Sc., M.Math.	Statistics
Thompson, B.K.	B.Sc., M.Math., Ph.D.	Statistics
Williams, C.J.	B.S.A., M.SC., Ph.D.	Statistics
Wolynetz, M.S.	B.Math., M.Math., Ph.D.	Statistics
Hodgins, D.K.	B.Sc.	Head, Systems and programming
Francis, L.M.A.	B.A.	Systems and programming
Hobbs, J.D.	B.Sc.	Systems and programming
Wu, J.	B.Sc.	Systems and programming
Vanasse, T.M.		Head, Computing services

Structures and Environment:
Turnbull, J.E.	B.S.A., M.S.A.	Head of Section and Director, Canada Plan Service; Farm structures
Hore, F.R.	B.S.A., M.S.	Water resources
Jackson, H.A.	B.Sc.(Eng.), M.Sc.	Storages
Munkoe, J.A.	B.S.A., M.Sc., Ph.D.	Livestock housing
Phillips, P.A.	B.Sc.(Agr.), M.Sc.	Waste management

Technical and Scientific Information:
Montgomery, G.F.	B.Sc.(Agr.)	Head of Section

Technical Services:
Caron, J.G.		Head of Section

Status: In January 1978, the Engineering Research Service and the Statistical Research Service were amalgamated to form the Engineering and Statistical Research Institute (ESRI). This recognized the contribution of engineers and statisticians to Research Branch programs. The Institute is responsible for providing a broad and comprehensive service to the Research Branch, the Department of Agriculture, and elsewhere on all aspects of engineering and biometrics with collaborative association in all areas of the research program.

Activities: Provides an engineering research and development service to the Canada Department of Agriculture in areas where instrumentation apparatus or machines form a major part of the study - Devises new methods of measurement, control and processing in agricultural research - Supplies engineering consultation and advisory service to federal and provincial departments of agriculture, to universities and the agricultural industry in the areas of agricultural mechanization, automation, farm structures, environmental control and water resources – Collects, collates and disseminates information on all aspects of engineering related to agriculture – Carries out liaison and coordination of agricultural engineering research and development in Canada.

Publications: Engineering and Statistical Research Institute, Research Branch Report 1976-1978; Eng.Res.Serv., Agric.Can., Rep.; J. Agric.Eng.Res.; Can. Agric.Eng.

Correspondence: English, French.

Canadian Society of Agricultural Engineering
151 Slater St., Suite 907, Ottawa, Ontario K1P 5H4

Desilets, D.J.		President 1980-81
Rapp, E.		Secretary-Treasurer

Status: Technical Society, affiliated with the Agricultural Institute of Canada.

Activities: Annual technical meeting with papers presented in: Power and Machinery, Soil and Water, Structures and Environment, and Electric Power and Processing.

Publications: Canadian Agricultural Engineering, two issues per year.

Correspondence: English.

Vineland Research Station, Research Branch, Agriculture Canada
Vineland Station, Ontario L0R 2E0

Menzies, D.R.   Ph.D.   Agricultural Engineering (Mechanical)

Status: The research Station is primarily concerned with horticultural crop protection, has eighteen scientists, one of whom is concerned with engineering.

Activities: Research and development – pesticide application methods and equipment – integrated control – uniform droplets – efficacy – nozzle wear – abrasion – coverage – sprayer design – airflow – drift – droplet distribution.

Publications: Droplet generator, laboratory study of droplet size, number/unit area, concentration of material, deposit, on host pests – Nozzle abrasion, wear, hydraulic grape sprayer, wettable powder formulations, wear rates – Effect of droplet size, number/unit area, concentration on mortality of European Red Mite – Behaviour of mites exposed to varying distributions of droplets/unit area, concentration of pesticide, probability of contact – Strawberry plot sprayer – pesticide evaluation work, hydraulic, hooded. Modification of hydraulic nozzle droplet size spectrum, trajectory separation – Experimental orchard sprayer design, three fan air-blast – Mechanical insect trap design, orchard insect, population sampling – Mortality of Oriental fruit moth, effect of droplet density and exposure time – Instrumented test chamber, forage compaction – Friction coefficients of alfalfa, high pressure – Pesticide deposit evaluation, rapid method to handle test larvae – Mechanical fruit harvesting, physical-biological properties of peaches – Sprayer evaluation, comparison of high and low volume boom sprayers, tomatoes – Mechanical harvesting, catching surface performance – Optimization model, mechanical peach harvesting – Mechanical harvesting, growth regulation effect on physical properties of peaches – Insect mortality, toxicity response of codling moth to particulate residues, phosmet – Spray coverage uniformity, effect of volume median diameter, and application rate, apple trees – Insect mortality; correlation of spray coverage and phosmet residues, Oriental fruit moth.

Correspondence: English.

Ridgetown College of Agricultural Technology (R.C.A.T.)
Ridgetown, Ontario N0P 2C0

Jung, R.V.	B.Sc.(Eng.), M.Sc.	Farm Power and Machinery Processing
Sojak, M.	B.S.A., B.A.Sc.	Farm Water Management, Farm Structures
Spieser, H.	B.Sc.(Eng.)	Energy Management

Status: Established 1951; administered by Ontario Ministry of Agriculture and Food (Provincial Government)

Activities: Teaching: teaching agricultural engineering subjects as a part of various two-year agriculture-related diploma courses. Research: research in land drainage, livestock housing environment, planting and harvesting. Maintaining an Energy Management resource centre. Addressing meetings, conducting farmer short courses, and other extension activities.

Correspondence: English.

Quebec

Agricultural Engineering Department, Macdonald Campus of McGill University
Ste. Anne de Bellevue, Québec H9X 1C0

McKyes, E.	B.Eng., M.Sc., Ph.D.	Chairman – Soil and water, Agricultural machinery, Agricultural structures and environment
Broughton, R.S.	B.S.A., B.A.Sc., S.M., Ph.D.	Soil and water
Jutras, P.J.	B.S.A., M.S.A.E.	Agricultural structures and environment, Soil and water
Kok, R.	B.E.Sc., Ph.D.	Food processing, Computers
Norris, E.R.	B.S.A., M.Sc., Ph.D.	Agricultural machinery, Physical properties
Raghavan, G.S.V.	B.Eng., M.Sc., Ph.D.	Processing, Soil and water.

Status: Private Institution.

Activities: Courses leading to the bachelor's degree in agricultural engineering yielding professional engineering status, B.Sc.(Agr.Eng.). Courses and research leading to the degrees of M.SC. and Ph.D. in agricultural engineering. Research in the areas of hydrology, drainage, irrigation, environment control, waste management, structures, agricultural machinery systems, tillage, food processing, physical properties of biological materials.

Publications: Canadian Agricultural Engineering; Transactions of the American Society of Agricultural Engineering; Macdonald Journal; Journal of the Irrigation and Drainage Division, ASAE; Journal of Terramechanics.

Correspondence: English and French.

Brace Research Institute, Macdonald Campus of McGill University
Ste. Anne de Bellevue, P.O. Box 900, Quebec, Canada H9X 1C0

Lawand, T.S.	B.Eng.(Chem.Eng.), P. Eng. Dip.Eng. M.Sc.(Agr.Eng.)	Aspects of Arid Land Development. Director, Desalination, Renewable Energy
Alward, R.	B.Sc.(Eng.), M.Sc.(Eng.)	Renewable Energy, Research Associate
LeNormand, J.	B.Eng., M.Eng.(Metallurigcal)	Renewable Energy, Assist. Director
Papadopoli, N.	B.Eng. (Civil)	Renewable Energy, Research Assistant
Skelton, A.	B.Arch.	Renewable Energy, Research Assistant

Status: Established in 1961 as an integral part of the Faculty of Engineering, McGill University, Montreal, Quebec, Canada. Financed by private endowment. Undertakes contract research, missions.

Objectives: The Institute is charged with undertaking research and development programmes in the fields of engineering and applied agriculture for the specific purpose of assisting rural communities in the underdeveloped arid zones of the world with special reference to the supply of fresh water, desalination, efficiency of water use and local, natural energy utilization, i.e. the application of solar and wind energy. Particular emphasis is that of the Intermediate or Appropriate Technology Approach .

Research activities: The Institute undertakes laboratory research, development, tropical testing, field evaluation, consulting and extension work in the areas of conversion of saline water to fresh irrigation and other water pumping applications using local energy sources; electricity and shaft power production from wind machines; solar energy utilization for water distillation, water heating, cooking, drying and processing of agricultural produce, heating, the adaptation of greenhouses for arid zones using seawater as a feed source, the redesign of greenhouses for cold northern regions, low cost wood stoves, biogas, producer gas units, etc.

Training: The Institute provides "stage" training for persons interested in our activities (primarily from or for developing areas of the world). Stage prospectus available on request.

Publications since 1965: Annual reports – publications list – technical reports – technical reprints – miscellaneous reports– do-it-yourself leaflets – engineering reports. These publications are aimed at the dissemination of results of research in the tropics listed above at all levels from the academic to the rural village. Reports have been published in English, French, Spanish and Arabic.

Correspondence: English, French, Arabic and Spanish.

Département de Génie rural, Institut de Technologie Agricole (Agricultural Engineering Department, Agricultural Institute of Technology)
C.P. 70, St. Hyacinthe, P.Q., Canada

Guay, B.	B.A., L.S.A., M.Sc.	Head of Department

Teaching staff:

Bernard, F.	B.A., B.Sc.A.
Cayer, B.	B.A., B.Th., B.Sc.A.
Corbeil, J.	B.A., B.S.A.
Deslandes, R.	B.Sc.A.
Drouin, A.	B.A., B.Sc. App.ing
Hébert, J-P.	B.Sc.
Mourot, G.	B.A., ing.agr.

Status: Date of establishment: 1962. Governmental institution.

Activities: Three-year course in agricultural mechanization and drainage. Average number of students attending the courses: 50.

Correspondence: French, English.

Direction de l'hydraulique agricole, du machinisme et des constructions rurales, Ministère de l'Agriculture, des Pêcheries et de l'Alimentation du Quebec
1020 route de l'Eglise, 4e étage - Sainte-Foy (Québec) C1V 3V9

Robert, S.   P.Eng.   Head

Status: Established in April 1970 as a Governmental body.

Activities: Testing and extension work in farm machinery, farm construction, farm hydraulics.

Publications: Testing and extension bulletins.

Correspondence: French, English.

Saskatchewan

Conservation and Land Improvement Branch, Saskatchewan Department of Agriculture
Administration Building, 3085 Albert Street, Regina, Saskatchewan S4S 0B1, Canada

Danyluk, J.F.	B.Eng.	
Ireland, B.H.	B.Eng.	
Coghlan, D.O.	B.Eng.	
Veroba, A.M.	B.Eng.	
Turek, S.S.	B.Eng.	
Carnduff, R.D.	B.Eng.	Irrigation Engineer

Status: Government.

Activities: Extension service and development assistance on agricultural water development projects, primarily irrigation, drainage and flood control - Reclamation of abandoned farm lands.

Correspondence: English.

Family Farm Improvement Branch, Saskatchewan Department of Agriculture
1318 Winnipeg Street, Regina, Sas.

Moen, P.O.	B.S.A.	Director

Agricultural Engineering Services:

Wrubleski, E.M.	B.E., M.Sc.	Supervisor
Protz, A.	B.S.A., B.S.A.E.	
Gebhardt, P.	B.S.A.E.	
Bayne, G.	B.E.	
Padbury, G.E.	B.E.	

Farmstead Engineering Services:

Carlson, C.W.	B.S.A.	Supervisor
Harder, E.	B.S.A.	
Hill, L.	B.E.	
Henley, W.T.	B.Sc.E.	
McKnight, K.	B.E., M.Sc.	
Smith, R.	M.S.A.E.	
Vopni, R.	B.S.A.	

Outposted offices: The Branch has one head office in Regina, Saskatchewan, and eleven field offices throughout Saskatchewan where the 15 field technicians and 6 agricultural engineers provide professional and technical services.

Status: The Branch was established under the Family Farm Improvement Act (1960) and is a Branch of the Saskatchewan Department of Agriculture (provincial).

Activities: Extension Services are provided through short courses and published reports — Field calls are made by technicians and specialists to assist small communities in designing water and sewage systems, and to help farmers plan farmstead layouts and livestock production facilities (including pollution control structures) — Research: limited research is carried out within the Branch on subjects directly related to the Branch's activities: water research and development (includes testing water quality, supply development and water treatment system design) farm structures and systems research — research relating to pollution by livestock.

Publications: The Branch publishes bulletins which deal with specific agricultural engineering problems in Saskatchewan and through these suggest possible methods of solving the problems.

Correspondence: English.

Agricultural Engineering Department, University of Saskatchewan
Saskatoon, Sask.

Zoerb, G.C.	B.E., M.Sc., Ph.D.	Head of Department	Power and Machinery

Electric Power and Processing:
Lampman, W.B.   B.E., M.Sc.

Farm Structures:
Moysey, E.B.   B.E., M.Sc.
Sokhansanj, S.   B.Sc., M.Sc., Ph.D.

Power and Machinery:
Berg, J.E.   B.E.
Bickner, R.L.   B.Sc.
Bigsby, F.W.   B.E., M.Sc., Ph.D.
Campbell, R.J.   B.Sc.
Chopiuk, R.G.   B.E.
Docking, E.A.   B.Sc.
Dodds, R.A.
Fritzler, B.A.   B.Sc.
Grier, D.G.   B.E.
Kent, G.L.   B.E.
Lal, R.   B.E., M.Sc., Ph.D.
Leach, P.A.   B.E.
Milne, W.G.   B.Sc.
Norum, E.B.   B.Sc.
Reding, D.F.   B.E.
Reed, W.B.   B.E.
Smart, D.R.   B.Sc.
Strayer, R.C.   B.E., M.E.

Soil and Water:
GRay, D.M.   B.S.A., M.Sc., Ph.D.
Murray, J.M.   B.E., M.Sc.
Norum, D.I.   B.E., M.Sc., Ph.D.
Steppuhn, H.W.   B.E., Ph.D.

Status: Private institution, supervised by President and Board of Governors (1917).

Activities: **Teaching:** Study courses are provided leading to the following degrees: B.E. in agricultural engineering (attended by about 15 students each year) – B.Sc. in agriculture (about 6 students each year) – M.Sc. in agricultural engineering (about 5 students each year) – Ph.D. in agricultural engineering (about 3 students each year).
Research: Development and study of controlled droplet sprayer – Development of a mechanical feeder to deliver large round hay bales into a mobile feed mix-mill – The development of a discer for simultaneous shallow placement of seed, deep tillage, deep placement of fertilizer and incorporation of herbicides for minimum tillage practices – S-qther attachment for snow management – Evaluation of, and recommendation for, operation of low pressure center pivot sprinklers – Static and dynamic pressures in grain bins – Evaluation of rapeseed oil as a fuel for farm diesel engines – Hydrology of the Prairie Environment – Snow Management for Surface Mined Areas – Dynamic Pressure in Grain Bins – Automation of Agricultural Equipment – Energy conservation in livestock buildings –

The Impact of snow and snow management practices to agricultural production – Grain quality and energy efficiency of a recirculating-air grain dryer – The modelling of soil water resources – Aerodynamics and sublimation of wind transported snow – Study of heat and mass transfer in porous heat storage materials such as rock and concrete – Swather attachment for snow management – Development and evaluation of automatic surface irrigation – Development and performance of heat recovery equipment for use in livestock ventilation.

Publications: Textbook: Instrumentation and Measurement for Environmental Sciences, 212 pp. Publisher: American Society of Agricultural Engineers. Editor: Z.A. Henry. Associate Editors: G.C. Zoerb, G.S. Birth. Published in 1975. – Textbook: Oilseed and Pulse Crops in Western Canada. 1975. pp. 703. Publisher: Western Cooperative Fertilizers, Calgary. Editor: J.T. Harapiuk. Chapter 22 by Moysey, E.B. and E.R. Norum.
Research Papers (published and unpublished): A laboratory study of grain straw separation – Comparison of methods of direct-seeding wheat on stubble land in Saskatchewan – Development of the Roto-thresh combine – A cleaning system aspirator for combines – A review of monitoring devices for combines – The effect of temperature and moisture on the thermal properties of rapeseed – Rock bed system captures escaping heat – Winter ventilation of livestock barns with heat exchangers – Empirical estimates of furrow irrigation stream advance – Friction characteristics of rapeseed and flaxseed – The effect of temperature and moisture on the thermal properties of rapeseed – The effect of soil moisture on infiltration as related to runoff and recharge – Snow management on the Canadian Prairies – Automatic header-height control system for windrowers – Ramifications of mechanized and automated agriculture – A fluidic transducer for a tillage depth control system – A furrow-following tractor guidance system – Developments in automatic controls for agricultural equipment – Energy use in farm power – Liquid to air heat exchangers for livestock buildings – Hydro-thermal stresses in a rice kernel-a finite element approach.

Correspondence: English.

Canada Agricultural Research Station, Swift Current, Saskatchewan
P.O. Box 1030, Swift Current, Sask. S9H 3X2   Tel.: (306) 773-4621

Nicholaichuk, Dr. W.	B.E., M.Sc., Ph.D.	Hydrology Research
Dyck, F.B.	B.E., M.Sc.	Equipment Design
Jame, Y.-W.	B.E., M.Sc., Ph.D.	Irrigation Research
Stumborg, M.	B.Eng.	Energy Research
McLaughlin, D.N.	B.Eng., M.Sc., Ph.D.	Instrumentation

Status: Established 1938.

Agricultural Engineering activities: Research – Hydrology: Snow management, sewage and effluent irrigation, seepage and evaporation control, agricultural runoff and water quality, dryland salinity – Equipment Design: design, development and testing of laboratory and small plot equipment – Irrigation: irrigation scheduling, irrigation ditch maintenance, thermocouple psychrometry and soil water measurements – Energy: energy use in agriculture, excluding solar and wind – Instrumentation: farm equipment, laboratory and plot equipment.

Publications: Sewage effluent irrigation, evaporation and seepage control, agricultural runoff and water quality, plot seeding and harvest equipment, irrigation ditch construction and maintenance equipment, irrigation scheduling, irrigation ditch maintenance, salinity, agricultural research laboratory equipment (e.g. automatic pot waterers, etc.)

Correspondence: English.

CENTRAL AFRICAN REPUBLIC

Bangui

Direction du Génie Rural, Ministère du Développement Rural (Agricultural Engineering Directorate - Ministry of Rural Development)

Sakoma, M.	Ingénieur de l'Equipement Rural	Directeur

Service de l'Hydraulique et Aménagement Ruraux

Doutambaye, C.	Ingénieur de l'Equipement Rural	Chef de Service
M'Baye, J-P.	Ingénieur Hydraulicien	Chef du Bureau Hydraulique
N'Ditifei Boy-Sembe, M.	Ingénieur des Travaux	Chef du Bureau des Aménagements Ruraux

Service de la Pédologie et Conservation des Sols

N'Gouanze, F.	Ingénieur Agro-Pédologue	Chef du Service
Kawalec, A.	Docteur Pédologue (Expatrié)	Conseiller Technique FAO
Safa, M.	Ingénieur Agro-Pédologue	Chargé des prospections
Damegaza, D.	Ingénieur Agro-Pédologue	Chargé des prospections
Mabe-Boto, V.	Ingénieur Agro-Chimiste	Chargé du laboratoire

Service d'Entretien des Pistes Rurales

Komaria, S.	Ingénieur d'Application en construction	Chef de Service
Bodenane, G.	Technicien Supérieur du Génie Rural	Contrôle du matériel
Dakadouhon, B.	Ingénieur Génie Civil	Construction ouvrages

Service de la Mécanisation Agricole

M'Ballea-Mandaba, A.	Ingénieur Agronome Machiniste	Chef de Service
Petro, S.	Adjoint Technique du Génie Rural	Section machines

Bureau d'Etudes

Ore, P.	Ingénieur des Travaux du Génie Rural	Chef de Bureau
N'Gaissona, J.	Adjoint Technique du Génie Rural	
Guitermbi, F.	Technicien Supérieur en construction.	

Annexes situées dans d'autres localités

1° **Inspections du Génie Rural**

Centre-Sud	- Bangui	Angoure, G., Adj.Tech.Génie Rural
Ouest	- Bouar	Touazoumbona, E., Ing.T.Génie Rural
Nord-Ouest	- Bossangoa	Toussonekeya, J., Ing.T.Génie Rural
Centre	- Bambari	M'Boutou, M., Ing.Trav.Génie Rural
Sud-Est	- Bangassou	Bogny, N., Ing.Travaux Génie Rural
Nord-Ouest	- Bria	N'Goulaka, E., Ing.T. Génie Rural

2° **Secteurs du Génie Rural**

Bimbo M'Baiki	) Centre Sud	Bouar Berberati	)	Ouest
Bossangoa Bozoum	) Nord-Ouest	Bambari Sibut	)	Centre
Bangassou Mobaye	) Sud-Est	Bria Birao N'Délé	)	Nord-Est

Structure: L'organe directe du Génie Rural est gouvernemental. C'est une institution dépendant du Ministère du Développement Rural créée vers les années 1953 sous l'administration coloniale du Territoire de l'Oubangui-Chari.

Activités concernant le Génie Rural: Entretien des pistes rurales et ponts sur pistes rurales - Hydraulique agricole et humaine - Construction hangars de stockage des produits agricoles - Aménagements agricoles (photos aériennes, pédologie ...) - Aménagement du Territoire - Habitat rural.

Publications: Bibliographie - 1948-1979: Travaux concernant les sols de la R.C.A. par A. Kawalec (FAO); F. N'Gouanze (Senasol); D. Damegaza (Senasol). Potentiel Agroécologique de la R.C.A. - le Caféier par A. Kawalec (FAO); M. Safa (Senasol).

Correspondance: Français et Anglais.

CHAD

Fort-Lamy

Direction du Génie Rural et de l'Hydraulique Agricole, Ministère de l'Agriculture (Agricultural Engineering Service, Ministry of Agriculture)
B.P. 47, Fort-Lamy

Structure: Fondé Décembre 1951 sous Ministère de l'Aménagement du Territoire et de l'Habitat.

Activités: Recherches. Aménagements ruraux, digues, puits, constructions.

Correspondance: Français.

CHILE

Chillán

Instituto de Ingeniería Agrícola de la Universidad de Concepción (Agricultural Engineering Institute, University of Concepción)
Casilla 537, Chillán, Chile

Regadío y Drenaje:
Valenzuela, A.	Ing.Agr. M.Sc. Ph.D.	Regadío, Estructuras de Riego
Holzapfel, E.	Ing.Agr. M.Sc.	Regadiío, Métodos de Riego
Gonzales de la Fuente,A.	Técnico Topógrafo	Topografía, Dibujo
Jara, J.	Ing.Agr.	Regadío, Relación agua-planta
Ramos, E.	Ing.Civil	Hidráulica - hidrología
Salgado Seguel, L.	Ing.Agr. M.Sc.	Drenaje
Villarroel, E.	Lic. Matemático M.Sc.	Matemáticas, Computación
Matta, C.R.	Ing. Ejecución Agrícola	Técnico
Velazco, J.	Ing. Agrícola, M.Sc.	Hidrología

Mecanización:
Ibánez C., M.	Ing.Agr.	Mecanización Agrícola
Hetz, E.	Ing.Agr. M.Sc.	Mecanización Agrícola
Concha C., L.	Técnico Mecánico	Motores
Celis, R.	Ing.Civil Mecánico	Diseño de máquinas
Reyes, F.	Ing. Civil Metalúrgico	Prueba de equipos
Maenzono, L.	Ing. Agrícola M.Sc.	Diseño
Candia, B.	Ing.Ejec.Mec.	Técnico

Proceso de Productos Agrícolas:

Milanese, A.	Ing. Civil Químico	Plantas Agroindustriales
Fuentes, J.	Ing.Agr.	Proceso de granos
Melín, P.	Ing.Agr.	Proceso de frutas
Loyola, M-Cristina	Técnico Químico	Laboratorio
Phillips, A.L.	Ing.Agr. Ph.D.	Proceso de productos

Carácter y Estructura: Creado en 1978 por la Universidad de Concepción a partir del antiguo Departamento de Ingeniería Agrícola, y con el apoyo de PNUD y FAO. Es un organismo semigubernamental.

Actividades en Ingeniería Rural: a) Ofrece cursos de pregrado para las carreras de Agronomía, Ing. Forestal y Medicina Veterinaria. Atiende alrededor de 250 estudiantes por año. b) Dicta un curso de postgrado para el grado de Magister en Ing.Agrícola, con 10 vacantes anuales.

Cursos Pregrado: Dibujo - Topografía I - Mecanización - Maquinaria Agrícola I - Fundamentos de Maq. de uso ganadero - Unidades Motrices - Matemáticas para Ing.Agrícola - Hidráulica - Operación de Tractores e Implementos Agrícolas - Proceso de Productos - Fundamentos de Riego - Materiales de Construcción - Topografía II - Fundamentos de Riego - Estructuras de Riego - Evapotranspiración - Métodos de Riego - Hidrología - Maquinaria Agrícola II - Topografía (Ingeniería Forestal) Tractores Agrícolas - Electrificación Rural - Drenaje. - Cursos Postgrado: Matemáticas para Ingenieros - Computación y Métodos Cuantitativos - Dinámica de Fluidos - Mecánica - Termodinámica y Transferencia de calor - Técnicas de Riego y Drenaje - Maquinaria Agrícola Aplicada - Proceso de Productos Agrícolas - Dinámica del agua en el suelo - Diseño de Riego por aspersión y goteo - Hidrología avanzada - Drenaje avanzado - Recursos hídricos - Proyectos de Riego - Evapotranspiración Avanzada - Estructuras de riego avanzado - Formulación y Evaluac. de Proyectos - Programación Lineal y Dinámica - Análisis de Sistemas - Análisis numérico - Construcciones Agrícolas - Diseño y uso de taller - Instrumentación - Análisis de tractores - Mecanización Agrícola - Diseño y uso de taller - Diseño de máquinas Agrícolas - Selección de máquinas agrícolas - Prueba de equipos agrícolas - Resistencia de material - Electrificación Rural - Construcciones Agrícolas - Ingeniería de Alimentos - Industrialización de Prod. Agrícolas - Control de calidad - Diseño de Plantas Procesadoras - Procesos agrícolas especiales.

Publicaciones: Estación Experimental, Boletín Técnico - Circular Informativa.

Correspondencia: Español (preferencia), Inglés, Francés, Portugués, Alemán.

Santiago

División de Protección Agrícola - Servicio Agrícola y Ganadero - Ministerio de Agricultura (Plant Protection Department, Agricultural and Livestock Division, Ministry of Agriculture)
Casilla #4647, Santiago,Chile

Morales Valencia, O.   Director de la División

Depto. Defensa Agrícola:
Rodriguez Díaz, F.   Ingeniero Agrónomo

Depto. Diagnóstico y Vigilancia:
Herrera Autter, S.   Ingeniero Agrónomo

Depto. Laboratorios:
Fajardo Arce, L.   Ingeniero Agrónomo

Dependencias en otras localidades: Existen oficinas en las 13 regiones del país como Direcciones Regionales, además de varias oficinas por cada región para atender los distintos sectores.

Caractér y Estructura: Fecha de origen: 1896. Financiamiento: gubernamental.

Correspondencia: Español.

Departamento de Ingeniería y Suelos (Facultad de Agronomía, Universidad de Chile)
(Engineering and Soils Department (Faculty of Agronomy, University of Chile)
Casilla 1004 Santiago, Chile

Fritsch, N.	Director	
Acevedo, E.	Ing.Agr. Ph.D.	Relación Suelo-Agua-Planta
Benavides, C.	Ing.Agr.	Fertilidad de Suelos
Carrasco, A.	Químico M.S.	Química de Suelos y Aguas
Castillo, H.	Lic. en Biología	Agroclimatología
Estevez, H.	Const. Civil	Construcciones Rurales
Fajardo, H.	Ing.Agr. M.S.	Fotointerpretación
Fritsch, N.	Ing.Agr. M.S.	Riego y Drenaje
Fontaine, G.	Ing. Agr.	Maquinaria Agrícola
García de Cortazar, V.	Ing.Agr. M.S.	Relación Suelo-Agua-Planta
Gonzales, W.	Tec.Topógrafo	Topografía
Gurovich, L.	Ing.Agr. Ph.D.	Física de Suelos
Hernandez, J.	Ing. Mil.	Fotoaerometría
Jordan, E.	Ing.Agr. M.S.	Riego y Drenaje
Luzio, W.	Ing.Agr. M.S.	Pedología
Menis, M.	Ing.Agr.	Pedología
Mery, J.	Ing.Agr.	Maquinaria Agrícola
Moraga, J.	Ing.Agr.	Conservación de Recursos Naturales
Munita, J.	Ing.Agr. M.S.	Edafología
Munoz, M.		Fotoaerometría
Novoa, R.	Ing.Agr. Ph.D.	Agroclimatología
Opazo, J.	Ing.Agr.	Fertilidad de Suelos
Ossandon, E.	Ing.Agr.	Maquinaría Agrícola
Peña, H.	Ing.Civil	Hidrología
Peralta, M.	Ing.Agr.	Conservación Suelos y Aguas
Pino, I.	Ing.Agr. M.S.	Fertilidad de Suelos
Quezada, M.	Ing.Agr. M.S.	Riego y Drenaje
Santibanez, F.	Ing.Agr. Dr.	Agroclimatología
Selles, G.	Ing.Agr.	Riego y Drenaje
Stöckle, C.	Ing.Agr.	Riego y Drenaje
Torres, J.	Arquitecto	Construcciones Rurales
Trujillo, A.	Ing.Agr. M.S.	Conservación de Recursos Naturales
Urra, M.	Ing.Agr.	Dibujo Técnico
Varnero, M.T.	Químico	Microbiología de Suelos
Vera, W.	Ing.Agr.	Edafología
Villa, R.	Ing.Agr.	Maquinaria Agrícola
Villavicencio, J.	Ing.Civil	Mecánica de Fluídos

Caractér y Estructura: La Institución fué establecida en 1971 y es de carácter gubernamental.

Actividades en ingeniería rural: a) Enseñanza: Geología y Geomorfología, Edafología, Fertilidad de Suelos, Maquinaria Agrícola, Riego y Drenaje, Topografía y Conservación de Recursos Naturales. Estas asignaturas son parte del curriculum de la carrera de Ingeniero Agrónomo que concede la Universidad de Chile. Además se imparten las siguientes asignaturas, para la mención de Ingeniería y Suelos: Física de Suelos, Pedología, Relación Suelo-Agua-Planta, Fertilizantes, Topografía Agrícola, Metodos de Riego, Drenaje Agrícola, Química de Suelos y Aguas, Maquinaria Agrícola Especial, Fotointerpretación, Dibujo Técnico, Hidraulica Aplicada, Motores y Tractores, Agrometeorología, Cartografía de Suelos, Mecanica Agrícola, Mecanica de Materiales, Hidrología, Construcciones Rurales. - El numero de alumnos promedio por asignatura es de 35. - b) Investigación: Se realizan investigaciones en genesis y clasificación, fertilidad, química y microbiología de suelos. Relación Suelo-Agua-Planta y manejo de agua, agrometeorología y fisiología de cultivos. Evaluación de máquinaria y mecanización Agrícola.

Publicaciones: Se han realizado numerosas publicaciones sobre los temas que se están investigando.

Correspondencia: Idioma español, inglés y francés.

Escuela de Agronomía, Universidad Católica de Chile (School of Agronomy, Catholic University of Chile)
Campus San Joaquin, Av. Vicuña Mackenna 4860

Dominguez, J.I.　　　　　　　　Decano

Carácter: Privado, con aportes del Estado.

Actividades en Ingeniería Rural: Cursos con un promedio de 50 alumnos - Se concede título de ingeniero agrónomo a quien cumpla con todos los requisitos para el caso.

Correspondencia: Español, inglés, francés.

Corporación de Fomento de la Producción (CORFO), Gerencia de Desarrollo
(Chilean Development Corporation (CORFO), Development Management)
Calle Ramóm Nieto No.920, Santiago, Chile - One World Trade Center, Suite 5151, New York, N.Y. 10048, USA

Velásquez, M.H.　　　　Ing.　　　　　　　　Gerente

Subgerencia de Operaciones: Darrigrandi, L., J.A. Ing.　　Subgerente

Area Energía:
Hernández, E.　　　　　Ing.　　　　　　　　Jefe Area

Area Agrícola:
Zelada, G., L.　　　　Ing.Agrónomo　　　　Jefe Area

Instituciones dependientes: a) INTEC - Instituto de Investigaciones Tecnológicas Director Ejecutivo: Bartolomé Dezerega - b) IREN - Instituto Nacional de Investigación de Recursos Naturales. Director Delegado: Enrique Junemann.

Caracter: Institución semifiscal de administración autónoma, establecida en 1939.

Actividades: Información básica sobre recursos naturales renovables - Investigación aplicada sobre mecanización agrícola. Estudios y proyectos de desarrollo agrícola. Electrificación Rural.

Publicaciones: Informes Técnicos - Estudios - Investigaciones aplicadas - Dirigirse a Gerencia de Desarrollo por lista detallada.

Correspondencia: Español, Inglés.

Instituto de Ingeniería Agraria y Suelos, Facultad de Ciencias Agrarias, Universidad Austral de Chile (Institute of Agricultural Engineering, Faculty of Agricultural Sciences, Southern University of Chile)
Independencia 641, Casilla 567, Valdivia, Chile

Ellies Schmidt, A.	Dr.rer hort. Ingeniero Agrónomo	Director de la Unidad
Daroch Pérez, R.	Ingeniero Agrónomo, M.Sc.	Especialista Mecaniz. Agrícola
Ellies Schmidt, A.	Ingeniero Agrónomo, Dr rer. hort.	Especialista en Física de Suelos, Mecánica de Suelos, Conservación de Suelos, Riego y Drenaje.
Nissen Mutzenbecker, J.	Ingeniero Agrónomo Dr.rer. hort.	Especialista en Física de Suelo, Hidraulica Agrícola, Riego y Drenaje.
MacDonald Hadida, R.	Ingeniero Agrónomo M.Sc.	Especialista en Reconocimiento de Suelos, Clasificación de Suelos, Fotocorogrametría.
Silva Norambuena, B.	Ingeniero Agrónomo M.Sc.	Especialista en Nutrición Vegetal, Fertilidad.

Caracter y Estructura: El Instituto fue creado al iniciarse la Universidad Austral de Chile en Septiembre de 1954. Desde 1954-1980 los temas que principalmmente abordaba el Instituto fue Suelos y Nutrición Vegetal. Desde 1980 cambio la orientación del Instituto a Ingeniería Agraria y Suelos.

Actividades en Ingeniería Rural: a) Enseñanza: La Enseñanza que imparte el Instituto es parte de la formación de estudiantes que estudian la carrera de Ciencias Agrarias, para obtener el título profesional de Ingeniero Agronomo. Las Asignaturas que imparte la Unidad que se circunscriben al area de la Ingeniería Rural son: Maquinaria Agrícola, Mecanización Agrícola, Motores y Bombas, Física de Suelos, Conservación de Suelos, Hidráulica Agrícola, Riego y Drenaje, Reconocimiento y Clasificación de Suelos. b) Investigación: Las investigaciones que actualmente estan en marcha de proceso y que tienen relación con la Ingeniería Agraria son: capacidad de soporte, mecánica de Suelos; Balance hidrológico en suelos agrícolas y forestales; Caracterización Física Mecánica de Suelos como antecedentes para un uso apropiado e intensivo; Diseño de un dosificador para sembradoras en papel; Diseño de una línea de extracción de gases en una planta fumigadora; Diseño de un sistema de Riego por Goteo.

Publicaciones: Chile Agro Sur; Circular informativa, Dep. Ing.Agr., Universidad de Concepción.

Correspondencia: preferentemente en Español, Inglés y Aleman.

CHINA
(People's Republic of China)

Nanjing

Nanjing Research Institute of Agricultural Mecahnization
Liuying, Nanjing, People's Republic of China   Tel. (025) 41695   Cable: 6593

Zhu Jian-qun   General Director
Shui Xin-yuan   M.S. Agric.Engineering   Vice-Director

First Research Division (Mechanization Problems in General):
Ru-Jin-yu   Senior Engineer   Head of Division

Second Research Division (Technology of Rice Mechanization):
Jiang Yao   M.S. Agric.Engineering   Head of Division

Third Research Division (Exploitation of Machinery):
Feng Kuei-zhen   Engineer   Head of Division

Fourth Research Division (Plant Protection):
Chen Guo-fan   Engineer   Head of Division

Fifth Research Division (Technical Information and Library):

Sixth Research Division (Testing Methods and Equipment).

Status: The Institution was established in 1957 as an organization under the Chinese Academy of Agricultural Science, the Ministry of Agriculture. It ceased to function for several years during the 70s. Now, with all the staff, it is transferred to, and financed by the Ministry of Agricultural Machinery; and moved to a new site at Liuying, still in Nanjing.

Agricultural Engineering activities:

Main Research subjects:
(1) Rice Mechanization   Gu Qian-an   Senior Engineer
(2) Grain Drying   Chang Xu-you   Engineer
(3) Regional study of Mechanization and Machinery System:
                                      Ke Jie   Senior Engineer
(4) Rice Transplanter   Jiang Yao   M.S. of Agricultural Engineering
(5) Small Rice Combine   Chen Nan-yun   Engineer
(6) Plant Protection Machinery   Chen Guo-fan   Engineer
(7) Rotary Tilling Machinery   Wang Quan   Engineer

Publications:: Institution bulletins and Technical reports.

Correspondence: In English and Chinese.

Beijing

Chinese Academy of Agricultural Mechanization Sciences
No. 1 Beishatan, Deshengmen Wai, Beijing, China

(a) <u>Agricultural Engineering (Leader or Head of Department):</u>

Guo Dongcai	Director of CAAMS
Hua Guozhu	Vice Director of CAAMS
Wang Wanjun	Vice Director and Chief Engineer of CAAMS
Fu Lizhi	Vice Director of CAAMS
Wen Yi	Vice Director of CAAMS and Head of Livestock and Poultry Equipment Research Institute (LPERI)
Song Hongding	Vice Director of CAAMS and Head of Material and Technology Research Institute (MTRI)
Li Shouren	Deputy Chief Engineer of CAAMS

(b) <u>Scientific Staff in charge of Agricultural Engineering subjects:</u>

Bai Quingrong	Head of Farm Machinery Information Research Division
Bian Yaogang	Head of Hydraulic Technique Research Division
Dong Ghohua	Head of Harvesting Machinery Section Farm Implement Research Institute (FIRI)
Erwin Engst	Head of LPERI
Feng Bingyuan	Head of Scientific Research Work Managing Department
Gao Xiansheng	Head of Power Unit Testing Division
Hou Zhiming	Head of Tillage Equipment Section of FIRI
Hu Zhong	Head of Farm Machinery Information Research Division
Huan Qingyun	Head of Technology Research Section of MTRI
Joan Hinton	Head of LPERI
Li Huaizhen	Head of Livestock and Poultry Equipment Research Section
Li Zhenyu	Head of FIRI
Li Zhifang	Head of Material Research Section of MTRI
Liu Baige	Head of Earthworking Machinery used on Farmland Section of FIRI
Liu Dazheng	Head of Irrigation and Drainage Equipment Research Division
Ma Ji	Chief Engineer of FIRI
Niu Zhanbiao	Head of Farm Machinery Experiment Station
Shen Kerun	Head of Planting and Crop Protection Equipment Section of FIRI
Shi Yuwen	Head of Farm Machinery Information Research Division
Tu Bingheng	Head of Power Unit Testing Division
Wang Tieren	Head of Scientific Research Work Managing Department
Wang Yanzhang	Head of FIRI
Wang Zhipei	Head of Livestock and Poultry Equipment Research Section of LPERI
Yang Jiaiwai	Head of FIRI
Yang Yi	Head of Measuring Technique and Instrumentation Research Division
Yi Ruitang	Head of Scientific Research Work Managing Department
Yu Jiyun	Head of Harvesting Machinery Section of FIRI
Yue Zhongying	Head of MTRI
Zhan Jize	Head of Research Institute for Farm Mechanization (RIFM)
Zhang Dewen	Head of Planting and Crop Protection Research Section of FIRI

Zhang Tianming	Head of Farm Machinery Information Research Division
Zhang Yiping	Head of RIFM
Zheng Xiaoying	Head of Irrigation and Drainage Equipment Research Division
Zhu Shenyou	Head of Tillage Equipment Section of FIRI

Branches with other location: Only in Beijing.

Status: The Chinese Academy of Agricultural Mechanization Sciences was founded in 1962 under the leadership of the Ministry of Agricultural Machinery (governmental).

Agricultural Engineering activities: CAAMS is a comprehensive scientific research organ. Its main tasks cover: assisting in drawing up the state program of scientific research of agricultural machinery; carrying out research and development as well as test of the products, assemblies and parts of major farm machinery; researching the manufacture technology and materials of key assemblies and parts; formulating national standards and ministry standards of agricultural machinery; researching and exchanging information of farm machinery; testing domestic and foreign agricultural machinery.

Publications: Three nationwide periodicals are edited and published: (1) "Farm Machinery" mainly concerns the technique and experience of operation and repairing. (2)"Farm Mechanization" publishing mainly the articles of science and techniques of farm machinery. (3) "Farm Machinery Journal" publishing mainly the research papers in cooperation with Chinese Society of Agricultural Machinery.

Correspondence: Chinese or English.

Institute for Stored Grain Research *
Meinyang City, Szechuan Province

Tan Xianchang	Engineer
Ren Xihong	Engineer

Institute for Storage Facilities Research *
Zhenzhou, Henan Province

Yang Yanyou	Drying Engineer
Yang Shizhong	Civil Engineer

* These addresses were provided by Consultants of the Food and Agriculture Organization of the U.N.

COLOMBIA

Bogotá

Instituto de Investigaciones Tecnológicas (Institute for Technological Research)
Avenida 30 No. 52-A.77, Bogotá, D.E.   Tel.: 235 00 66

Ayala, J.	Director	Químico, Estudios complementarios de Ingeniería Química
Díaz, D.	Subdirector de Investigación	Químico Especialista en Manejo, Empaque y Conservación de Alimentos
Flechas, G.	Subdirector de Consultoría	M.S. Ingeniería Química. Diseño de Procesos y de Planta.
McCormick, A.	Subdirector de Servicios Industriales	M.S. en Química Agrícola
Cardeño, J.	Subdirector administrativo y Financiero	Ing. Químico y de la Industria de Fertilizantes.

Status: El Instituto de Investigaciones Tecnológicas se creó en 1958 como una entidad autónoma con personería jurídica. Cuenta con el patrocinio de la Caja de Crédito, la Federación Nacional de Cafeteros, el Banco de la República, la Empresa Colombiana de Petróleos y el Instituto de Fomento Industrial. Entidades que tienen el carácter de gubernamentales a excepción de la Federación Nacional de Cafeteros que es un Ente privado. Estas entidades contribuyen anualmente al sostenimiento del Instituto con sumas que en conjunto pueden llegar a ser del Orden del 20% del presupuesto de ingresos de la entidad. El 80% restante proviene de contratos celebrados entre el Instituto y Entidades Nacionales y Extranjeras (Gubernamentales privadas) y particulares, por el desarrollo de proyectos específicos.

Actividades de Ingeniería Rural: Investigaciones sobre: cadena de frío para exportación de perecederos - Desarrollo de tecnologías para procesar frutas tropicales (V. gr. puré de banano, concentrado de maracuyá, etc.) - Diseño de sistemas de almacenamiento de papa fresca - Desarrollo de sistemas de conservación de yuca fresca - Desarrollo de formulaciones y tecnologías para substituir el trigo en la erlaboración de pan y pastas - Labores : el proceso de fabricación de panela (azúcar en bruto) - Desarrollo de sistemas artesanales para la conservación de pescado - Desarrollo de concentrados proteícos y proteínas texturizadas a base de harinas de soya y semillas de algodón - Estudios sobre pérdidas postcosecha de alimentos en Colombia - Investigaciones sobre residuos de plaguicidas en productos agrícolas y pecuarios Colombianos.

Publicaciones: Informe anual de actividades - Revista bimestral "Tecnología" - Conservación de pulpas de algunas frutas tropicales mediante un método químico (Valor del ejemplar $60.00) - Elaboración de panela (Valor del ejemplar $80.00).

Correspondencia: Español, Inglés.

Instituto Colombiano Agropecuario (Colombian Agricultural Institute)
Apartado Aéreo 151123, Bogotá, Colombia

División de Ingeniería Agrícola:
Rodriguez-Amaya, C.   Jefe   Ingeniero Civil M.S. (Suelos y Aguas)
                                       Ph.D. (Sistema de Recursos Hídricos)

Programa de Recursos de Agua y Tierra:

Forero-Saavedra, J.A.	Jefe	
	Ingeniero Agrónomo	M.S. (Ingeniería de Riegos)
Suarez-Montes, J.G.	Agrólogo	M.S. (Física de Suelos)
Villaneda-Vivas, E.	Agrólogo	
Molano-Cogua, M.	Ingeniero Agrónomo	M.S. (Candidato) (Riegos y Drenaje)
Lopez-Jimenez, H.	Ingeniero Agrícola	
Lopez-Varon, J.V.	Ingeniero Agrónomo	
Caicedo, A.	Ingeniero Agrónomo	
Palacio.Saldarriaga, M.	Ingeniero Agrícola	M.S. (Candidato)(Riegos y Drenajes)

Programa de Procesos Agropecuarios:

Moreno-Pinzon, F.	Jefe	
	Ingeniero Mecánico	M.S. Ingeniería Agrícola (Procesos)
Pinto-Serrano, R.	Ingeniero Agrónomo	M.S.
García-Reynel, H.	Ingeniero Agrónomo	
Abarca-Pinzon, A.	Ingeniero Agrícola	

Programa de Maquinaria Agrícola:

Restrepo-Henao, L.A.	Jefe	
	Ingeniero Agrónomo	
Guerrero-Jimenez, L.	Ingeniero Agrónomo	
Camacho-Garcia, H.	Ingeniero Mecánico	
Rincon-Cardenas, C.J.	Ingeniero Mecánico	M.S. Diseño de Máquinas
Piedrahita-Velasquez, R.	Ingeniero Agrícola	
Rodriguez-Rodriguez, M.	Ingeniero Agrónomo	

Sede Principal: Centro Nacional de Investigaciones Agropecuarias "Tibaitatá", Apartado Aéreo 151123, Bogotá, Colombia

Sub-sedes: C.N.I.A. Palmira, Apartado Aéreo 4128, ICA, Palmira, Valle, Colombia - Centro Regional de Investigaciones Agropecuarias, Nataima ICA, Apartado Aéreo 527, Ibagué, Colombia - C.N.I.A. Turipaná, Apartado Aéreo 206, ICA, Montería, Córdoba, Colombia - Centro Regional de Investigaciones Agropecuarias, Obonuco, Apartado Aéreo 339, ICA, Pasto Nariño, Colombia.

Status: 1953 como: División de Investigaciones Agropecuarias (DIA)
1963 como: Instituto Colombiano Agropecuario(ICA) Entidad descentralizada del Ministerio de Agricultura. De caracter gubernamental con financiación del Gobierno Colombiano.

Actividades: Investigación: (1) Agua y Tierra - Estudios basicos para Riego y Drenaje - Registro y Descripción de las Propiedades Físicas de los Suelos de Colombia. - Riego por Superficie: Manejo de agua en Cocotero - Manejo de agua en el Cultivo del Algodón - Manejo de agua en Ajonjolí - Riego por aspersion: Evaluación de un sistema de Riego por aspersión tipo Cañon y Programación del Riego en el CNIA Tibaitatá - Evaluación de Aspersores de giro rápido y recomendaciones para su Manejo en Zonas de Minifundio. - Riego por Goteo: Determinación de tres Láminas de Riego en Tomate Chonto - Efecto del Riego por Goteo y la Fertilización Nitrogenada en la Producción de tomate Manapal - Efecto de la lámina de Riego por Goteo en la Producción de Fresa, Van Tioga Californiana - Lámina de Riego por Goteo en Melón - Lámina del Riego por Goteo en el Cultivo del Ají Pimentón - Evaluación Económica de los diferentes metodos de riego: Riego por Goteo y por aspersión en la producción de papa (Solanum tuberosum L.) - Drenaje de Campos Agrícola:

Espaciamiento de drenes superficiales en maíz y algodón con riego suplementario.
(2) Procesos Agropecuarios - Acondicionamiento de Granos: Secamiento: Evaluación de Secado de Maíz a diferentes contenidos de Humedad - Organización de Trapiches Paneleros: Evaluación del Proceso de fabricación de Panela - Molienda de Caña: a) Determinación de los parámetros que influyen en la extractión, capacidad y potencia de los molinospaneleros; (b) Diseno, construcción y evaluación de un sistema de preparación de caña para alimentar molinos paneleros - Energía utilizada como combustible en las hornillas - Precalentamiento de aire de combustión en hornillas paneleras - Procesos de Evaporación y concentración - Consideraciones para el diseno de hornillas paneleras - Acondicionamiento y Procesamiento de raíces y tubérculos - Comparación de trés sistemas de almacenamiento de papa - Cosecha de granos: Incidencia de la velocidad, el cilindro y la humedad en la trilla de arveja con máquinas estacionarias - Comportamiento de una trilladora estacionaria en cebada y fríjol - Evaluación de pérdidas de grano en cosecha de arroz, cebada, sorgo y soya realizada con combinada - Fuentes de Energía no convencionales en el Procesamiento de Productos Agrícolas - Diseno y construcción de un prototipo de secador de bagazo para trapiches paneleros - Evaluación de un quemador con cascarilla de arroz - Diseno, construcción y evaluación de colectores solares para secado de grano y de foraje. (3) Maquinaria Agrícola - Mecanización de zonas planas - Evaluación de Siembras Mecanizadas sin Labranza Previa - Comparación de tres tipos de cultivadoras en el control de Malezas y sus efectos sobre el Suelo en el Cultivo de Maíz (tea Maize) - Mecanización de Zonas de Ladera: Evaluación de algunos sistemas de Labranza por las pérdidas de Suelo y Agua - Reducción de Operaciones y Siembra sin Labranza de Maíz en Ladera - Mecanización en el Cultivo de la Cana Panelera - Diseño de Maquinaria: Diseno y construcción de una Cosechadora de Semilla de Pasto Brachiaria - Evaluación de los mecanismos dosificadores de las Sembradoras de Grano Grande con Semillas Nacionales.

Publicaciones: Revista ICA; Boletín Promocional.

C a l i

Universidad del Valle (Valle University)
Apartado Aéreo 21-88, Cali

Integrated programme with Nacional University, Faculty of Agronomy at Palmira, Valle.

Status: This programme was established in 1968 as a part of the Engineering Division, which also has programmes in Civil, Mechanical, Chemical, Electrical, Sanitary and Systems Engineering.

Activities: (a) A five-year programme which offers courses for 120 students (average) in: Soil and water - Processing of agricultural products - Agricultural machinery - Agricultural structures and rural electrification. (b) Research in the above mentioned areas.

Correspondence: Spanish, English, French, Italian.

M a n i z a l e s

Facultad de Agronomía, Universidad de Caldas (Faculty of Agriculture, Univ. of Caldas)
Apartado Aéreo 275, Manizales, Caldas

Carácter: La Facultad de Agronomía de Manizales inició labores en febrero de 1950 y es financiada con fondos del Gobierno Nacional.

Actividades en ingeniería rural: Cursos de topografía, agrimensura, hidráulica, riegos y drenajes, maquinaria agrícola y construcciones rurales.

Correspondencia: Español, Inglés.

Medellín

Carrera de Ingeniería Agrícola, Facultad de Agronomía, Universidad Nacional de Colombia, Seccional Medellín (Agricultural Engineering Program, Faculty of Agronomy, National University of Colombia, Medellín Sectional)

López, J.A.	Ingeniero Agrícola, Director Carrera de Ingeniería Agrícola

Sección de Riegos y Drenajes:

Bustamante, F.	Ingeniero Agrónomo, M.S. en Ingeniería de Riegos y Drenajes
Bustamante, I.D.	Ingeniero Agrónomo, M.S. en Manejo de Suelos y Aguas
Mercado, G.E.	Ingeniero Agrónomo
Salazar, C.A.	Ingeniero Agrícola, M.S. en Ingeniería de Riegos y Drenajes

Sección de Mecanización Agrícola:

Alvarez, A.	Ingeniero Agrónomo
Alvarez, F.	Ingeniero Agrícola
Cortés, E.	Ingeniero Agrícola
Gómez, J.M.	Ingeniero Agrícola
González, L.J.	Ingeniero Agrícola
Puyana, J.O.	Ingeniero Agrónomo, M.S. en Maquinaria Agrícola

Sección de Procesos Agrícolas y Sistemas:

López, J.A.	Ingeniero Agrícola
Castro, G.	Ingeniero Agrícola, M.S. en Procesos Agrícolas
Mora, M.M.	Ingeniero Agrícola
Pino, J.	Ingeniero Agrónomo, M.S. en Procesos Agrícolas
Porras, L.A.	Ingeniero Agrícola
Restrepo, G.	Ingeniero Agrónomo, Ph.D. en Sistemas Agrícolas

Carácter: Institución Gubernamental establecida en noviembre de 1964 y aprobada por la Universidad Nacional de Colombia el 2 de diciembre de 1965.

Actividades en Ingeniería Agrícola: a) Enseñanza: Se imparten principalmente cursos a la carrera de Ingeniería Agrícola, la cual tiene una duración de 10 semestres, durante los cuales se capacita al estudiante en Recursos de Agua y Tierra, Procesos Agrícolas, Maquinaria Agrícola y Estructuras Agrícolas. Al terminar sus estudios se le concede el título de Ingeniero Agrícola. - Se ofrecen cursos para otras carreras, tales como: Agronomía, Zootecnica, en el área de Riegos, Maquinaria Agrícola y Procesamiento de Productos Agrícolas. - El promedio de alumnos por curso es de quince (15). b) Investigación: Sección de Riegos y Drenajes: Riego por goteo - Estabilidad estructural de cuencas hidrográficas - Sección de Mecanización Agrícola: Máquinas hidráulicas - Evaluación de implementos - Sección de Procesos Agrícolas y Sistemas: Ambientación animal - Utilización de energía solar en secado de granos.

Publicaciones: Artículos de investigación: Métodos de riego - Captación de Agua lluvia - Estabilidad estructural de suelos - Diseño de maquinaria agrícola - Evaluación de maquinaria agrícola - Ambientación animal - Secado de granos - Procesamiento de frutas.

Correspondencia: Español.

CONGO

Brazzaville

Service Central du Génie Rural, Ministère de l'Economie Rurale
B.P. 13   -   Brazzaville

Liwanga-Vakazy, Z.	Ingénieur de l'Equipement Rural	Chef de Service Central du Génie Rural

Okemba, A.   Ingénieur Hydraulicien
Miere, P.   Ingénieur des Travaux Ruraux
Okogna, B.M.   Ingénieur des Travaux Ruraux
Ngabogo   Ingénieur des Travaux Ruraux
Opoma, S.P.   Ingénieur des Travaux Ruraux
Bayonne, R.   Ingénieur des Travaux Ruraux
Ondongo, P.   Ingénieur des Travaux Ruraux
Boyama, R.   Ingénieur des Travaux Ruraux
Mbitsi, P.   Technicien Hydraulicien
Paka-Paka   Technicien Hydraulicien

Annexes situées dans d'autres Localités: Six (6) Divisions Génie Rural Régionales
       Kouilou   (Pointe-Noire)
       Niari   (Loubomo)
       Pool   (Kinkala)
       Cuvette   (Owando)
       Sangha   (Ouesso)
       Likouala (Impfondo)

Structure: Les tous derniers textes qui réorganisent le Génie Rural au Congo datent de 1977. Ce Service est placé sous la tutelle du Ministère de l'Economie Rurale.

Activités concernant le Génie Rural: a) Formation: URSS, CUBA, Pays d'Afrique (Haute-Volta) - Les étudiants sortent avec des diplômes d'Ingénieur de l'Equipement Rural, Hydraulicien, en Machinisme Agricole ... après cinq ans d'étude. b) Les activités actuellement en cours au Génie Rural sont: l'Hydraulique Villageoise (fonçage des puits dans le milieu rural), la construction des hangars agricoles.

Correspondance: Française.

COSTA RICA

San Pedro

Facultad de Agronomía, Universidad de Costa Rica (Faculty of Agronomy, University of Costa Rica) - Apartado Postal 337, San Pedro de Montes de Oca

Actividades en Ingeniería Rural: Cursos de hidráulica agrícola - maquinaria agrícola y construcciones rurales.

CUBA

La Habana

Facultad de Ingeniería Agronómica, Universidad de la Habana (Faculty of Agricultural Engineering, University of La Habana) - La Habana

Actividades en ingeniería rural: Cursos de hidráulica agrícola, maquinaria agrícola y construcciones rurales.

CYPRUS

Nicosia

Department of Agriculture, Ministry of Agriculture and Natural Resources
Nicosia, Cyprus

Soil Conservation and Farm Machinery Subdivision:

Savvides, K.L.	B.Sc.(Agr.), M.Sc.	Officer-in-charge
Michaelides, Ph.	B.Sc.(Agr.Eng.), M.Sc.	Soil Water Engineering
Pissourios, D.	B.Sc.(Agr.), M.Sc.	Soils
Kalathas, Chr.	B.Sc.(Agr.Mech.), Cert.	Land Reclamation and Rural Development (Basi)
Nicolaou, P.	B.Sc.(Agr.Mech.)	
Savvides, A.	Cert.Agr., Cert.Mech.	

Activities: Planning and application of soil conservation and land development projects – Application of the World Food Programme Project No.418 "Soil Conservation and Tree Planting" – Implementation of the Soil Conservation Law – Operation and maintenance of crawler tractor fleet and ancillary equipment used for soil conservation and land reclamation works – Operation, use and maintenance of agricultural machinery and implements on agricultural farms and nurseries – Hire service of new agricultural machinery to farmers – Demonstration of tractors and equipment in collaboration with local importers and dealers of agricultural machinery – Training courses to farmers on the maintenance and operation of agricultural machinery – Advisory service to local manufacturers of agricultural machinery – Advisory service to Grain Commission, Olive Products Marketing Board and the Potato Marketing Board for related machinery – Training course and advice for cold stores of fruits.

Water Use Section:

Chimonides, S.J.	B.Sc.(Agr.), M.Sc. (Irrigation)	Officer in charge
Kalimeras, P.	B.Sc.(Agr.), Cert.(Irrigation)	Ass. Officer in charge
Motitis, Chr.	B.Sc.(Agr.)	Nicosia
Zemenides, S.A.	H.N.D.(Mech.Eng.) Member of the American Society of Mechanical Engineers.	Nicosia
Pentayitis, A.	B.Sc.(Agr.)	Nicosia
Oroundiotis, G.	B.Sc.(Agr.), M.Sc.(Irrigation)	Paphos
Tsappis, N.	B.Sc.(Agr.), M.Sc.(Irrigation)	Limassol
Savvides, A.	B.Sc.(Agr.), M.Sc.(Irrigation)	Nicosia
Herodotou, Chr.	B.Sc.(Agr.)	Nicosia
Photiou, Chr.	B.Sc.(Agr.)	Famagusta

Activities: a) Development planning. Designing and cost estimates of improved irrigation systems such as Sprinkler, Drip (Trickle), Mini sprinklers (micro - jets), hose - basin, pipe - basin and gated pipe - furrows irrigation systems. Active participation in all stages for each Irrigation Project (main and minor). – b) Advisory work. General follow-up of the above systems. Advice on installation, operation and maintenance, irrigation demonstration fields, salinity problems in relation to irrigation and preparation of irrigation schedules. – c) Water use publicity, education and training campaign. Issues on various subjects related to the field of irrigation. Contribution with articles to the "Country man" (Agricultural magazine). Radio and T.V. Agricultural Programmes. – In-service training courses for the technical staff and short training courses for farmers at Agricultural Training Centers, group field trips, lectures and talks, etc.

Correspondence: Greek, English, Turkish.

Water Development Department, Ministry of Agriculture and Natural Resources
Nicosia, Cyprus

Lytras, C.S.   M.Sc., D.I.C., B.Sc., Hydrogeologist,   Director
                      Engineering Geologist

Activities: Planning, investigation, design, operation and maintenance of: Dam projects – Irrigation distribution systems and other irrigation works – Domestic water supplies – Antiflood works – River training works – Aquifer recharge works – Drilling of boreholes and exploitation and control of ground water.

Publications: Dams of Cyprus. Published 1974 by Government of Cyprus.

Correspondence: English, Greek.

Agricultural Research Institute
Nicosia, Cyprus

Samios, Th.   B.Sc.(Agr.), Salonica, Greece. Specialization in Agricultural
                     Engineering Durham University, England
Photiades, J.   B.Sc.(Agr.Eng.), University of Newcastle, England

Status: Established in 1962 by the Government of the Republic of Cyprus.

Activities in Agricultural Engineering: Experimental work on the use of various types of machinery and implements – Modification on imported farm field equipment – Cultural practices – Operation, maintenance and repairs of tractors, field equipment and experimental machinery used by the Institute.

Publications: Annual Reports of the Agricultural Research Institute.

Correspondence: Greek, Turkish, English.

CZECHOSLOVAKIA

Bratislava

Polnohospodárske stavby, odborové riaditelstvo (Agricultural Building Industries, Mamagement Corporation)
Steinerova ul. 72, 883 37 Bratislava

Affiliated: Polnohospodárske stavby, np., Bratislava (Agricultural Construction Works)
            "   "   " Prešov
            "   "   " Zvolen
           Potravinostav, np., Bratislava (Foodstuff Industry Construction Works)
           Hydromeliorácie, np., Nitra (Amelioration and Water Use Works)

Status: The Corporation was established in 1963 and is directed by the Ministry of Agriculture and Food Industry. It is financially self-supporting.

Activities: Development of structural elements for application to the agricultural building industry and to agricultural reclamation, irrigation and drainage works – Development of power machinery for construction works – Development of structures and construction systems for agricultural buildings.

Publications: New construction technologies in drainage and irrigation works and in farm building construction – New building materials and structural elements for agricultural buildings.

Correspondence: Slovak, German, Russian.

Výskumný ústav závlahového hospodárstva (Research Institute of Irrigation Farming)
Vrakuňská cesta, 834-21 Bratislava – Podunajské Biskupice

Department of Irrigation techniques and mechanization

Department of water regime and nutrition of plants

Department of irrigation agrotechnics

Department of economics and organization

Department of scientific information

Status: State institution established in 1959 by the Slovak Agricultural Academy and financed by the Government.

Teaching activities: Courses for irrigation technicians.

Research activities: Plant production and protection under irrigation – Water requirement and nutrition of plants – Physiology of irrigated crops – Hydropedology and biology of irrigated soils – Irrigation techniques and irrigation works – Mechanization of operation – Use of Plastics – Economics and organization.

Publications: The results of investigations carried out are published in the annual "Vedecké práce" (with summaries in Russian, English and German).

Correspondence: Slovak, Russian, English, German, French.

Brno

Fakulta provozně ekonomická, obor mechanizační a Fakulta agronomická, obor zahradnický, Vysoká škola zemědělská v Brně (Faculty of Agricultural Economics, Line of Mechanization and Faculty of Agronomy, Line of Horticulture, University of Agriculture, Brno)
Zemědělská 1, 662 65 Brno

Department of Farm Mechanization and Electrification (Faculty of Agricultural Economics)

Department of Field Mechanization (Faculty of Agricultural Economics)

Department of General Engineering and Repairs (Faculty of Agricultural Economics)

Department of Applied Physics (Faculty of Agricultural Economics)

Department of Farm Buildings (Faculty of Agricultural Economics)

Department of Horticultural Engineering (Faculty of Agronomy)

Status: Governmental institution supervised by the Ministry of Education.

Activities in agricultural engineering: Courses for students of the Faculty of Agricultural Economics and of the Faculty of Agronomy, attended by an average number of 250 students for the general courses and by an average of 75 students for the specialized courses.

Research on: Tillage equipment – Vegetables growing mechanization – Weed control – Orchards and vineyards mechanization – Forage cutting and harvesting – Field crop harvesting – Sugar beet harvesting.

Publications: Textbooks on: Tractors – Handbook for tractor drivers – Agricultural machines – Implements and equipment in Agriculture – Farm electrification – Mechanization of plant production.

Fakulta lesnická, Vysoká škola zemědělská v Brně (Faculty of Forestry, University of Agriculture, Brno)
Zemědělská 3, 662 66 Brno

Department of Forest Mechanization and Automation

Status: Governmental institution supervised by the Ministry of Education.

Activities in forestry engineering: Courses attended by an average of 250 students in the general courses.

Research on: Mechanical devices for forestry operations – Development and testing of power chain saws – Mechanical means for wood logging.

Publications: Textbooks on Mechanical devices for forest operations – Mechanical means for logging – Mechanical means for felling and trimming – Industrial log conversion depots.

Correspondence: Czech, English, German, Russian.

Zetor Brno, Výzkumný ústav traktoru (Zetor Brno Tractor Research Institute)
Brno 32

Vrána, B.	Ing.	Director
Junek, J.	Ing.	Head of the Research and Development Section
Musil, J.	Ing.	Head of the Testing Station

Status: Established at ZKL Brno Tractor Factory in 1960.

Agricultural engineering activities: Research, development, design and testing of agricultural tractors, including tractor engines, gear boxes, hydraulic equipment, safety cabs, etc.

Publications: Reports on the results of both research and testing of tractors, tractor engines, gear boxes, hydraulic equipment, safety cabs, etc. in Czechoslovak and foreign journals and bulletins.

Correspondence: Czech, Russian, English, German, French.

Nitra

Odbor polnohospodárskej mechanizácie, Prevádzkovo - ekonomická fakulta, Vysoká škola polnohospodárská (Department of Agricultural Mechanization, Economic Faculty, College of Agriculture)
Nábrežie mládeže, Nitra, Slovensko

Affiliated institutes:

Institute of Technical Mechanics, ul.plk. Gagua 10, Nitra

Institute of Repair Technology, ul.plk. Gagua 10, Nitra

Institute of Farm Machinery Management, Nábrezie mladeze, Nitra

Status: Governmental institution, established in 1962.

Teaching activities: Courses in: Mathematics - Physics - Mechanics of solids - Strenght and elasticity of materials - Engineering design - Hydrotechnics - Heat technics - Farm tractors and motors - Repair technology - Organization of maintenance and service - Field crop mechanization - Theory of field machinery - Drainage machinery - Harvesting machinery - Management of field machinery - Technology of mechanized field operation - Farm buildings - Transport of agricultural materials. The duration of the studies is 10 semesters and the courses lead to a degree of engineer of agricultural mechanization. The average number of students is 700.

Research activities: Functional design of working elements in agricultural machines with regard to their agrophysical and technological functions.

Publications: Textbooks on: Maintenance and service of agricultural machines - New methods of cereal harvesting - Farm buildings - Acta technologica agriculturae, Universitatis Agriculture (Collection of Papers) edited by College of Agriculture, Nitra.

Praha

Oddělení mechanizace a elektrifikace zemědělství Ministerstva zemědělství a lesnictví (Department of Mechanization and Electrification of Agriculture, Ministry of Agriculture and Forestry) Těšnov 65, Praha 11

Activities: Coordination of all development, research and testing work for the mechanization of agriculture and forestry in Czechoslovakia.

Mechanizační fakulta, Vysoké školy zemědělské v Praze (Faculty of Mechanization, College of Agriculture of Prague)
Suchdol, Kamýcká ulice, Praha 6

Status: Governmental institution, established in 1952. It is financed by the Ministry of Education.

Teaching activities: Principal courses: Mathematics – Physics – Mechanics of solids, liquids and gases – Engineering design – Engineering materials – Electrotechnology – Experimental statistics – Automation – Field-crop production – Animal production – Farm tractors (three courses) – Farm machinery (three courses) – Mechanization of animal production – Irrigation and drainage machinery – Repair technology and organization – Farm buildings and structures – Sociology of rural life – Technology of mechanized field operations – Duration of the studies: 5 years – Degree awarded, after state examination and thesis, title of "inzenýr" (abbr. "Ing.") – Total average of students during the academic year: 650.

Research activities: Technology of tractors, machinery, equipment and mechanized operations with regard to the practical farm service and agrophysical properties of processed materials.

Publications: Collection of Papers, Faculty of Mechanization in Prague – Textbooks and books on: Farm machines – Wheel and track tractors, operation properties – Electrical equipment of tractors – General engineering in agriculture – Machines and equipment for animal production – Exercise on mechanics – Repeating the algebra, trigonometry and analytical geometry.

Correspondence: Czech, Russian, German, English, French.

Katedra zemědělských strojů, fakulta strojní, České vysoké učení technické (Department of Agricultural Machinery, Faculty of Mechanical Engineering, Czech Technical University) Suchbátarova 4, Praha 6 – Dejvice

Status: Governmental institution, established in 1952, directed by the Ministry of Education and financed by the Government.

Teaching activities: Study courses in engineering design of farm machinery including tractors, implements and equipment for soil preparation, irrigation, drilling, cultivation, crop protection, harvesting, drying, storage and processing – Food preparation and distribution – Dairy equipment – Materials handling – Courses in experimental work, testing and technology of production of agricultural machines and implements. Main degree awarded to candidates: M.Sc. (Eng.) after a full time study (five years) and final State examination – Postgraduate fellowships (three – five years) and candidates' dissertations in general mechanical engineering and theory of machinery – Degree awarded: Candidate of sciences in engineering, corresponding to the degree of doctor of philosophy (Eng.) – Average number of students attending the study courses: 20 to 25.
Research activities: Experimental and theoretical research in agriculture machines and their parts and testing of machines in cooperation with the State Testing Station and agricultural engineering institutions.

Publications: Textbooks on Agricultural Machines I (Technology of production) – Agricultural Machines II (Testing) – Machines for Farm Mechanization I – Machines for Farm Mechanization II – Machines for Farm Mechanization III – Machines for Soil Preparation – Machines for Crop Protection – Machines for Irrigation – Potato Harvesters.

Correspondence: Czech, English, French, Russian, German.

Výzkumný ústav meliorací (Research Institute for Land Reclamation and Improvement)
255 80 Praha 516-Zbraslav

Votruba, J.	Ing.Dr.	Director
Raděj, J.	Ing.Dr.	Scientific Secretary

Department of Drainage:
Švihla, V.   Ing.Dr.   Head of Department

Department of Soil Conservation:
Pasák, V.   Ing.Dr.Doc.   Head of Department

Department for Land Reclamation and Improvement in the Slovakian Socialist Republic
834 00 Bratislava

Raučina, Š.   Ing.Dr.   Head of Department

Land Drainage Research Station:
370 01 České Budějovice   (Southern Bohemia)
073 01 Sobrance   (Eastern Slovakia)

Peat Bog Research Station:
391 91 Borkovice   (Southern Bohemia)

Grassland Research Station:
353 21 Závišín   (Western Bohemia)

Recultivation Research Station:
435 06 Albrechtice   (Northern Bohemia)
702 00 Ostrava   (Northern Moravia)

Status: State institution established in 1954 and financed by the Government.

Activities - Training: Postgraduate courses for so-called "scientific aspirants". Degree awarded: "Candidate of Science" in Land Reclamation and Improvement. Average number of students: 10.
Research: Hydropedology - Hydraulics and Hydrology - Applied Physical Chemistry - Drainage - Water Regime Regulation - Recultivation of Soils in the Regions damaged by Mining and Industrial Activities - Soil Conservation Measures - Peat Bog Hydrology - Substances with Soil Improving Effects, etc.

Publications: The results of research work are published in the "Vědecké práce" (Scientific Works) with resumés in Russian and English - "Scientific Monographs", - "Agriculture and Forestry Amelioration Abstracts" and other papers concerning all activities of the Institute.

Correspondence: Czech, Russian, English, German, French.

Výzkumný ústav zemědělské techniky (Research Institute of Agricultural Engineering)
163 07 Praha 6, K šancím 50

Fiala, J.	Dr.habil.	Director
Masková, H.(Mrs.)	Ph.D.	Vice-Director
Pastorek, Z.	Ph.D.	Scientific Secretary

Section of Research Planning and Coordination (under Director)

Mareš, Z.	Ph.D.	Head, Dept. of Research Plan and Methodology
Pavlík, J.		Head, Dept. of International Cooperation
Višinský, J.	Ph.D.	Head, Dept. of National Coordination
Cempírek, B.	M.Sc.	Head, Dept. of Scientific and Technical Information

Section of Technological Systems:

Pick, E.	Ph.D.Assoc.Prof.	Head
Sedlák, J.	Ph.D.	Head, Dept. of Plant Production
Vegricht, J.	Ph.D.	Head, Dept. of Animal Production

Section of Energy and Machinery Utilization:

Haš, S.	Ph.D.	Head
Špelina, M.	Ph.D.	Head, Dept. of Utilization of Machinery
Fišer, Z.	M.Sc.	Head, Dept. of Transport and Material Handling
Kosek, J.	Ph.D.	Head, Dept. of Utilization Energy

Section of Principles of Agricultural Engineering:

Maskova, H. (Mrs.)	Ph.D.	Head
Havelík, J.	Dr.habil.	Head, Dept. of Agrophysical Systems
		Laboratory of Agrochemistry and Biology
		Laboratory of Computer Applications
		Laboratory of Measuring and Instrumentation

Status: Governmental institution, administered and financed by the Federal Ministry of Agriculture and Foods of Czechoslovakia. Founded in 1951.

Activities: Postgraduate training of Candidates of Science (CSc), a degree corresponding to Ph.D. Field: Agricultural Engineering.
Research Activities: Fundamentally in agricultural engineering covering both principles and production technologies with special regard to large-scale production methods.

Publications: Scientific papers (with summaries) are published in the journal "Zemědělská technika" (edited by ÚVTIZ Praha). Current year results are published in the Annual Report (with summaries) available free and on exchange basis. Recent book editions: VIELEBIL et al.: Mechanizace nových a modernizovaných stájí. SZN Praha, 1976 (Mechanization of new and modernized stabulation barns). Cz. 373 pp. SPELINA et al.: Stroje druhé generace v zemědělství. SZN Praha, 1976 (Agricultural machines of the second generation). Cz. 306 pp. BLAŽEK et al.: Progresívní technologie sklizně, konzervace a skladování krmiv. SZN Praha, 1978 (Progressive Technologies of harvesting, preserving and storaging feeds). Cz. 297 pp.

Correspondence: Czech, Russian, English, French, German, Spanish.

Výzkumný ústav zemědělských strojů (Research Institute of Agricultural Machinery)
149 43 Praha 4, Chodov - cables: VÚZS Praha Chodov, Telex 122255 vuzs c, tel. 763941

Homolka, J.	Ing.	Director
Novak, J.	Ing.	Head, Engineering Research Division
Souček, J.	Ing.	Research
Hutla, D.	Ing.	Prognostics
Sulek, V.		Information Service
Kolář, Z.	Ing.	Testing

Status: Established in 1953, incorporated in the group of associated enterprises of Zbrojovka Brno and under control of the Ministry of Machinery.

Activities: Research, development and design of agricultural machines - Testing of new machines and implements - Measuring of stresses in various parts of agricultural machines - Standardization of agricultural machines and their parts - Experimental investigations of the principles governing the design of agricultural machines - Assessment of economic efficiency of new equipment - Automated information systems - Maintenance of comprehensive reference library and information service.

Publications: "Selected Abstracts" and "Ausgewählte Auszüge" (abstracts in English and German from Czechoslovak literature on agricultural engineering), "Informacní zpravodaj" (in Czech, an information bulletin), occasional reports.

Correspondence: Czech, English, French, German, Russian.

Státní zkušebna zemědělských, potravinářských a lesnických strojů (State Test Station of Agricultural, Food-processing and Forestry Machines)
163 04 Praha 6-Řepy (Czechoslovakia)

Vopálenský, V.	Engineer, CSc.	Director
Pastor, B.	Engineer	Technical Manager
Miratský, J.	Engineer	Economical Manager
Horaček, J.	Engineer, CSc.	Head of Branch office in Praha
Schmidt, R.	Engineer	Head of the Branch office in Brno-Lisen
Zacharda, J.	Engineer, CSc.	Head of the Branch office in Rovinka pri Bratislave 900 42

Branch offices:
632 00 Brno - Líšeň
900 42 Rovinka pri Bratislave

Status: Established 25 June 1956. Governed, directed and financed by Federal Ministry of Agriculture and Nourishment of Czechoslovak Socialist Republic.

Activities: - Testing work: STSAFFM tests and approves the agricultural, food-processing and forestry machines of both inland and imported production. The obligatory evaluation of agricultural machines from inland production and the quality examination of above mentioned are also carried out. Furthermore, the STSAFFM carries out the international testing of agricultural and forestry machines for the Council of Economic Mutual Aid (for its Permanent Agricultural Commission). It secures the special services for agriculture, forestry and foodstuff sector.

Publications: The STSAFFM issues the Volume of summarized information on the procedure and results of testing tractors, soil preparation implements, drilling, plant protection, harvesting and forestry machines as well as animal production and food processing machines. The volumes are distributed to 33 agricultural engineering institutions all over the world.

Correspondence: English, German, Russian.

DENMARK

Brede-Lyngby

Kommissionen til Udforskning af Landbrugsredskabernes og Agerstrukturernes Historie, Det kgl. danske Videnskabernes Selskab (The Commission for Research on the History of Agricultural Implements and Field Structures, The Royal Danish Academy of Sciences and Letters)
National Museum, Brede, DK-2800 - Lyngby.

Steensberg, A.	Prof. Dr.Phil.	Chairman
Lerche, G.	M.A.	Secretary

Status: Established in 1967 under the Royal Danish Academy of Sciences and Letters.

Activities: The Commission works in close connection with the International Secretariat for Research on the History of Agricultural Implements (same address), and mainly on preindustrial agricultural implements – Measuring of fossil field-structures, testing by experiments medieval ploughing methods.

Publications: "Tools and Tillage", a yearly journal, first issue 1968, published in English and German – (the International Secretariat for Research on the History of Agricultural Implements)– "Atlas of the Fields of Borup Village, Borup Ris wood in Zealand from about 1000 – 1200 A.D." by A. Steensberg a.o. Copenhagen 1968, 95 pages + Atlas, published by the Commission – "Geräte der Atányer Bauern" by Edit Fél and Tamás Hofer, published by the Commission 1973, c. 900 pages.

Correspondence: English, German, French, Swedish, Norwegian, Danish.

Horsens

Statens jordbrugstekniske Forsøg (National Agricultural Engineering Institute)
Bygholm, 8700 Horsens, Denmark

Klausen, K.G.	B.Sc.agr.	Director

Department of work study:

Keller, P.	B.Sc.agr.	Head of Department
Nielsen, V.	B.Sc.agr.	

Department of barn and stall equipment:

Madsen, N.P.	B.Sc.agr.	Head of Department
Thellesen, H.	B.Sc.agr.	
Thomsen, K.E.	B.Sc.agr.	

Department of field machinery and equipment:

Olsen, V.	B.Sc.agr.	Head of Department
Høy, J.J.	B.Sc.agr.	

Department of buildings and stall fixtures:

Pedersen, S.	B.Sc.agr.lic.agr.	Head of Department
Møller, F.	B.Sc.agr.lic.agr.	
Hansen, K.	B.Sc.agr.	
Tønnesen Aa.	B.Sc.agr.	Implement and machines for horticulture
Simonsen, F.G.	Mech.E.	Testing methods and standardization
Madsen, P.	Mech.E.	Instrumentation

Status: The institute is established on October the 1st 1978 by joining of the former "Statens Redskabsprøver" (Government Machinery Testing Station established 1914) and "De landbrugstekniske Undersøgelser" (Institute for work study and farm buildings established 1956). The Institute is governed by a board responsible to the public and to the Ministry of Agriculture under which the Institute is administered.

Agricultural Engineering activities: Testing, investigations and research in regard to labour requirements, farm machinery, farm buildings and farm equipments.

Publications: Test reports.

Correspondence: Danish, English, German.

København

Jordbrugsteknisk Institut, Den kgl. Veterinaer- og Landohøjskole (Institute of Agricultural Engineering, The Royal Veterinary and Agricultural University)
Rolighedsvej 23, DK 1958 Copenhagen V.

Pedersen, T.T.	B.Sc.agr., M.Sc.agr.eng.	Professor Chairman

Research Institute: Agrovej 10, 2630 Tåstrup

Kofoed, S.S.	B.Sc., M.Sc.mech.eng.	Vice-chairman - power machines
Berthelsen, L.	B.Sc., M.Sc.mech.eng.	Farm machinery
Christiansen, Sv.Aa	Aa. engineer	Technology of materials
Have, H.	B.Sc.agr. lic.agr.	Research
Jørgensen, B.	B.Sc.agr.	Research
Matzen, R.	B.Sc.agr. lic.agro.	Instrumentation
Nielsen, H.	B.Sc., M.Sc.elek.eng.	Research
Thostrup, P.	B.Sc.agr.	Research

Activities: Courses leading to a B.Sc. degree in agriculture and licentiate degree (Ph.D.degree) in agricultural engineering - Research and experiments - Conservation of forage crops. Seedbed preparation. Power take off driven tillage tools. Standardization of agricultural machinery. Alternative energy from biomass and wind.

Publications: Self emptying flat bottom grain drying bin. Seed bed preparation for and seeding of sugar beets, onions and carrots. Regaining of heat from air in the barn. Energy from solid and liquid manure by aerobic fermentation. Harvesting, conservation and storage of paddy rice. Hot water generator for wind mills. Energy from biomass.

Correspondence: Danish, English, German.

Den danske Afdeling af Nordiske Jordbrugsforskeres Forening (Danish Branch of the Scandinavian Society for Agricultural Research)
Rolighedsvej 26, 1958 København V.

Hjortshøj Nielsen, A.	Lektor	Chairman

Activities: The Organization is a union of agricultural research workers carrying out research work in the Scandinavian countries.

Publications: "Nordisk Jordbrugsforskning"(Danish) and "Acta Agricultura Scandinavica" (English), periodicals.

Correspondence: Danish, English.

Hydroteknisk Laboratorium op Klima- og Vandbalancestationen, Den kgl. Veterinaer- of Landbohøjskole (Hydrotechnical Laboratory and Climate Station, The Royal Veterinary and Agricultural University)
Bülowsvej 23, 1870 København V.

Hydrotechnical Laboratory:	Climate Station:
Aslyng H.C., Professor, Ph.D.	Jensen Sv.E., Ph.D.
Jensen H.E., Ph.D., D.Sc., Chairman	Kristensen K.J., M.Sc.
Friis-Nielsen B., Ph.D., D.Sc.	Mogensen V.O., Ph.D.

Climate Station: Højbakkegaard, DK-2630 Taastrup.

Status: The Hydrotechnical Laboratory and Climate Station, established 1950, are organized as a department within the Faculty of Agriculture of the Royal Veterinary and Agricultural University which is under the jurisdiction of the Ministry of Education.

Course	Number of attending students, app.
Agricultural Physics I	80
Agricultural Physics II	50

Course I includes soil physics, agricultural meteorology, soil technology, and land reclamation as well as crop production in relation to soil, water, and climate.
Course II includes soil drainage and irrigation of agricultural land and selected environmental aspects of agriculture.

Activities: Research activities include climate and water balance in Danish agriculture - potential and actual evapotranspiration - soil physics - crop production in relation to soil, water, and climate and environmental aspects of potential crop production.

Publications: Water balance of soils in the Danish agriculture - Shelter belts and their effect on the climate and on the water balance of soils - The influence of tillage on soil temperatures - Soil tillage and water requirements of plants - The influence of the soil-moisture relations on the growth of crops under natural conditions - Crop yields in relation to moisture supply in the soils - Water intake of plants in relation to the moisture condition of the soil - Temperature and heat balance of soils - Climatic aspects of supplemental irrigation - Radiation and energy balance - Soil/water/plant relationships - Photosynthesis - Respiration - Potential crop production.

Correspondence: Danish, English, German.

Statens Byggeforskningsinstitut, afdelingen for landbrugsbygninger (Danish Building Research Institute, Department of Farm Buildings)
P.O. Box 119, DK-2970 Hørsholm.

Huld, T.	M.Sc.(Chem.)	Head of Department
Andersen, K.T.	M.Sc.(Eng.)	
Ellum, J.C.	M.Sc.(Build.)	
Feenstra, A.	Veterinary, Ph.D.	
Holmgaard, P.	B.Sc.(Agr.)	
Møller, S.	M.Sc.(Eng.)	
Pedersen, J.	Dr.Agro.	
Pedersen, J.L.	Pharmaceutical chemist	
Petersen, H.W.	Engineer	
Radum, A.	B.Sc.(Agr.)	
Scheel, B.	Mechanical Engineer	
Skov, O.	Engineer	
Strøm, J.S.	M.Sc.(Eng.)	
Traberg-Borup, S.	M.Sc.(Build.)	

Branches: Climatic laboratory, Roskilde - Experiment stables, Trollesminde, Hillerød.

Status: Established 1947. Directed and financed by the Ministry of Housing and the Ministry of Agriculture.

Activities: Constructional drawings for farm buildings - Radiation and hot-air heating in weaner houses - Livestock houses: Design and equipment - Computerized cost calculation of farm buildings - Tests of experimental cattle housing systems on Danish pilot farms - New and experimental housing systems for pigs - Ambulant testing of climate and ventilation in farm buildings - Performance specifications and recommendations for rooms for development of cuttings – Development of elements and structures for green-houses – Service rooms for green-houses – Cheap and simple buildings for growing pigs – Alternative feedstuffs and feeding technique in pig houses – Table work for grain silos – Dimensioning of steel frames in farm buildings – Outdoor test plant for fullscale experiments with farm building structures – Reduction of condensation nuisances in uninsulated farm buildings.

Publications: SBI Farm Buildings; Landsudvalget for svineavl og-produktion.

Correspondence: English, German.

Afdeling for Landbrugsbyggeri, Kunstakademiets Arktitektskole (Department of Farm Buildings, The Academy of Fine Arts)
Tordenskjoldsgade 1B, København K.

Activities: Teaching about farm buildings at the School of Architecture at the Academy of Fine Arts – Research on Farm buildings.

Correspondence: Danish, German, English.

Afdeling for Landbrugsøkonomi, den kgl. Veterinaer- og Landbohøjskole (Department of Agricultural Economics, Royal Veterinary and Agricultural College)
Bülowsvej 13, København V.

Status: The Royal Veterinary and Agricultural College was established in 1958.

Activities: Teaching undergraduates and graduate students – Degrees obtainable are Agronom (B.Sc.), lic.agro.(M.Sc.) and Dr. agro. (Ph.D.)

Research on: Linear programming – Production functions – Factor analysis – Location of agricultural production.

Publications listed in: "World Agricultural Economics and Rural Sociology Abstracts".

Kolding

Bioteknisk Institut (Biotechnical Institute)
Holbergsvej 10, 6000 Kolding

Sonne-Frederiksen, P.   Lic.agro.   Director

Department of Feed Technology:

Larsen, Chr. Skov.	Lic.agro.	Head of Department
Israelsen, M.	Agronomist	Chief of Research
Jacobsen, E.E.	Chem. engineer	Chief of Research
Blåbjerg, J.	Chem. engineer	Chief of Research
Larsen, N.	B.Sc.	
Therkildsen, N.	Veterinary	
Rexen, B.	Chem. engineer	
Petersen, E.	Chem. engineer	

Department of Biotechnology:

Rexen, F.P.	Chem. engineer	Head of Department
Ravn, T.	Agronomist	
Rasmussen, P. Skovmand	Chem.engineer	
Bentsen, Th.	M.Sc.	

Department of Technology of Vegetables:

Vendelbo, P.	Chem. engineer	Head of Department
Holm, F.	M.Sc.	
Eriksen, Sv.	Chem. engineer, Ph.D.	

Technical Department:

Rasmussen, R.	Construct. engineer	Head of Department
Jensen, J.	Mechanical engineer	Chief engineer
Mogensen, Sv.N.	Mechanical engineer	
Sørensen, P.	Mechanical engineer	
Roed, P.	Mechanical engineer	
Søberg, K.	Mechanical engineer	
Kjeldsen, V.	Agronomist	

Status: Established on 1 January 1959 as an independent institution, affiliated to the Danish Academy of Technical Sciences. 50% self-financing, 50% supported from the State.

Research on: dehydration, pelleting, storage, processing, transport and chemical composition of green crops, pulp, straw, grain, potatoes, fishmeal etc.

Publications: Reports and articles on above research subjects.

Correspondence: English, German.

Lyngby

Internationalt Sekretariat for Udforskning af Landbrugsredskabernes Historie (International Secretariat for Research on the History of Agricultural Implements)
National Museum, Brede, DK-2800 Lyngby

Lerche, G. Dr.	M.A.	Secretary of the International Committee and Secretary of the Secretariat

Status: Established 1954. Branch of the Danish National Museum, supported by the Ministry of Cultural Affairs and National funds and supervised by an International Committee in Scientific Affairs.

Activities: The Secretariat acts as a connecting link between research workers in different countries and collects information on pre-industrial agricultural implements – A library and a photographic archive have been established.

Publications: Newsletters – A journal on the history of the implements of cultivation and other agricultural processes (since 1968).

Correspondence: English, Danish, Swedish, Norwegian, German, French.

Instituttet for Strømingsmekanik og Vandbygning, Danmarks tekniske Højskole (Institute of Hydrodynamics and Hydraulic Engineering, Technical University of Denmark)
Bygning 115, DK-2800 Lyngby, Danmark

Status: Governmental institution

Activities: Courses on: Open channel flow including sediment transport, physical hydrology and water resources, statistical hydrology; degree awarded: M.Sc.; average number of students attending the courses: 50 per year. – Rainfall-run off models, water balance studies, the mechanics of groundwater reservoirs.

Publications: Investigations of flow in alluvial streams – Numerical analysis of unsteady open channel flow – Hydraulic resistance of alluvial streams – The occurrence of suspended loads – Bed load investigation in Skive-Karup river – Instability of erodible beds – Pumping from leaky artesian aquifers – Three-dimensional stability analysis of open channel flow over an erodible bed – Application of crossing theory in hydrology – Mechanics of groundwater reservoirs – Pumping from elastic artesian aquifers – Groundwater development and its influence on the water balance – Risk evaluation of a tidal-influenced flood control reservoir – Numerical simulation of the rainfall runoff process on a daily basis.

Textbooks on: Sedimental Transport in Alluvial Streams – Analysis of Hydrologic Time Series.

Correspondence: Danish, English, German, French.

Viborg

Det danske Hedeselskab (Danish Land Development Service)
Klostermarken 12, DK-8800 Viborg, Danmark

Skov, K. Sandahl	M.Sc.Agric.	Director

Forestry Division (Skovbrugsafdelingen):
Grossen, S.	M.Sc.	Forestry, Head of Division

Soil Improvement, Drainage, and Engineering Division (Grundforbedringsafdelingen):
Vennov, N.	M.Sc., Civil Eng.	Head of Division
Lundager Jensen, J.	M.Sc., Civil Eng.	Head of Hydrological Surveys
Larsen, V.	M.Sc.Agric.	Head of Research Department

Laboratory:
Frederiksen, J.	M.Sc. Chem.Eng.	Head of Laboratory

Outposted Branches: Approx. 40 branch offices distributed all over the country.

Status: Non-profit, independent organization, established in 1866. The executive committee is appointed by representatives of the membership of the organization by the Ministry of Agriculture, and by the main agricultural organizations, respectively.

Activities: Forestry - Shelter belts - Nurseries - Tree improvement - Soil surveying - Drainage - Cultivation - Irrigation - Flood control - Hydrological surveys - Research in soil improvement and drainage - Chemical and physical soil analysis - Sewage and Fresh Water Analysis - Sewage treatment.

Publications: "Hedeselskabets Tidsskrift" (Periodical); Reports.

Correspondence: Danish, English, German.

DOMINICAN REPUBLIC

Instituto Superior de Agricultura (Superior Institute for Agriculture)
Apartado 166, La Herradura, Santiago de los Caballeros, Rep. Dominicana

Mecanización Agrícola:	Director,	
Carrasco, D.	M.S. Profesor	Mecanización Agrícola
Jiménez, A.	Ingeniero Agronomo, Encargado	Especialidad en Mecanización Agricola Maestría en Mecanización Agrícola
Ferrer, P.	Ingeniero Agronomo	Especialidad en Mecanización Agrícola Maestría en Mecanización Agrícola
Rodríguez, L.	Ingeniero Electromecánico	
Sistemas de Riego:	Encargado	
Gonell, Z.	Ingeniero Agrónomo	Especialidad en Mecanización Agrícola Maestría en Riego
Cruz, C.	Ingeniero Agronómo	Especialidad en Mecanización Agrícola Maestría en Riego

Carácter y estructura: Esta institución fue establecida en enero de 1964. Esta institución es dirigida por el sector privado y financiada por los sectores privado y gubernamental.

Actividades: Cursos ofrecidos: (1) Mecanización Agrícola: Taller Agrícola I y II - Topografía - Maquinaria Agrícola I, II, III y IV - Control de Agua - Sistemas de Riego - Potencia en la finca - Electrificación Agrícola - Drenaje - Construcciones Agrícolas I y II - Ingeniería de Alimentos. Título conferido: Ingeniero Agrónomo con concentración en Mecanización Agrícola. (2) Sistemas de Riego: Topografía - Suelo I (Química y fertilidad) - Hidrología - Flujo y Medida del Agua - Relaciones Agua-Planta - Estructuras de Riego - Suelo II (Física) - Sistemas de Riego - Fotogrametría y Fotointerpretación - Conservación de Suelo y Agua - Drenaje - Calidad del Agua y Control de Salinidad - Operación de Proyectos de Riego. Título conferido: Ingeniero Agrónomo con concentración en Sistemas de Riego. - Investigación: Construcción y Evaluación de un Digestor de Biogás - Uso de la Paja de Arroz en el Secado del Arroz - Construcción y Evaluación de una Desfibradora de Cabuya o Sisal para el Pequeño y Mediano Agricultor.

Publicaciones: Trabajos de Investigación publicados: Programa de Tecnología Apropiada. Boletín informativo. - Tecnología Mecánica y Modernizacion del Sector Agricola. - Esfuerzo del Instituto Superior de Agricultura en la Promoción de la Tecnología Apropiada. - Construcción de un Ariete Adaptable a las Condiciones y Necesidades del Pequeño Agricultor de la República Dominicana (Tesis de Ingeniero Agrónomo). - Construcción y Evaluación de dos Prototipos de Medidores de Régimen Crítico en Canales Abiertos (Tesis de Ingeniero Agrónomo). - Construcción y Evaluación de un Motocultor de 7 HP (Tesis de Ingeniero Agrónomo). - Construcción y Estudio de la Factibilidad de Introducción de una Cosechadora de Yuca en República Dominicana (Tesis de Ingeniero Agrónomo.

Correspondencia: Español.

E C U A D O R

L o j a

Facultad de Agronomia y Veterinaria de la Universidad de Loja (Agronomy and Veterinary Faculty, National University of Loja)
Casilla Letra "B" - Loja

Dependencias en otras localidades: Estación Experimental "El Padmi , ubicada en la Provincia de Zamora Chinchipe, a 130 km de la ciudad de Loja, en la República del Ecuador.

Carácter y estructura: La actual Facultad de Agronomía y Veterinaria de la Universidad Nacional de Loja, fue creada el 8 de enero de 1945, con el nombre de Facultad de Ciencias, compuesta por las Escuelas de Ingeniería Agrícola, de Minas y de Arquitectura. Posteriormente, el H. Consejo Universitario dispuso el funcionamiento de las Escuelas de Química Industrial y Agronomía; pero - por razones de índole económico, así como por falta de técnicos especializados, tuvo que suprimirse la Escuela de Química Industrial. Cinco años más tarde, en 1950, se creó la Escuela de Medicina Veterinaria. En la actualidad, esta Facultad cuenta con dos Escuelas, la de Ingeniería Agronómica y la de Medicina Veterinaria. La Universidad Nacional de Loja y sus Facultades son autónomas, asi como lo son todas las Universidades del Ecuador.

Actividades en Ingenieria Rural: (a) Enseñanza: SECAP (Servicio Ecuatoriano de Capacitación Profesional) dictará un Curso de Maquinaria Agrícola y Motores a los Estudiantes del Segundo Curso de la Escuela de Ingeniería Agronómica, como completación a la enseñanza de dicha materia, concediéndoles certificado de promoción al Curso inmediato superior.
(b) Investigación: Actualmente se encuentran en ejecución trabajos de tesis de experimentación e investigación, previos a la obtención del Título de Ingenieros Agrónomos; y, trabajos de investigación en los Laboratorios y Gabinetes de la Facultad de Agronomía y Veterinaria.

Publicaciones: Bibliografía de Investigaciones Agropecuarias - Boletín de la Biblioteca de la Facultad - Publicación de la Tesis de Grado previas a la obtención de los Títulos de Ingeniero Agrónomo y de Doctor en Medicina Veterinaria - Inventario del Jardín Botánico de la Facultad.

Correspondencia: Español, inglés.

Quito

Departamento de Ingeniería Agrícola, Ministerio de Agricultura y Ganadería
Quito, Ecuador

Aldeán Ayala, N.E.	Doctor Ingeniero Agrónomo	Especializado en Mecanización Agraria - Jefe del Dpto. de Ingeniería Agrícola
Idrobo Idrobo, J.G.	Ingeniero Agrónomo	Especializado en Mecanización Agraria - Jefe de la Sección Mecanización Agrícola
Loor Palma, W.	Ing.Agrónomo	Jefe de Centro
Barona, G.	Ing.Agrónomo	Jefe de Centro
Zambrano, F.	Ing.Agrónomo	Jefe de Centro
Guerrero, L.	Ing.Agrónomo	Jefe de Centro
Rivadeneira, W.	Ing.Agrónomo	Jefe de Centro
Suárez, C.	Ing.Agrónomo	Jefe de Centro
Naranjo, L.	Ing.Agrónomo	Jefe de Centro
Valencia, B.	Ing.Agrónomo	Jefe de Centro
Sarango, D.	Ing.Agrónomo	Jefe de Centro
Martínez, P.	Ing.Agrónomo	Jefe de Centro
Añazco, F.	Ing.Agrónomo	Jefe de Centro

Dependencias: Las Jefaturas de los Centros de Mecanización Agrícola están ubicadas en las siguientes provincias: Carchi, Imbabura, Pichincha, Tungurahua, Cañar, Azuay, Loja, El Oro, Guayas, Manabí y Los Ríos.

Carácter y Estructura: El Departamento de Ingeniería Agrícola depende de la Dirección General de Desarrollo Agrícola del Ministerio de Agricultura y Ganadería. Tiene a su cargo la Sección de Mecanización Agrícola que administra los Centros de Mecanización anotados en el numeral anterior.

Actividades: El Departamento con sus Centros de Mecanización, ofrece los servicios de Mecanización Agrícola a precios cómodos, con el objeto de incentivar e introducir la tecnología mecánica en los cultivos, como medio de elevar la producción y mejorar la situación socio-econoómica del agro.

Correspondencia: Castellano y/o Inglés.

Instituto Ecuatoriano de Recursos Hidraulicos (INERHI) (Ecuatorian Institute of Water Resources)
Juan Larrea n. 543 - Quito, Ecuador

Sotomayor, V.J.	Ing.Agrónomo	Riego y Desarrollo Rural	Jefe de Departamento Explotación, Quito
Guzmán, Z.M.	Ing.Agrónomo	Riego y Desarrollo Rural	Jefe de Sec.Desarrollo Integral, Quito
Orquera, C.A.	Ing.Agrónomo M.Sc.	Edafólogo	Jefe de Sec.Unvestigación y Normas, Quito
Reyes, P.C.	Ing.Civil	Riego y Drenaje	Jefe Engdo.Sec.Operación y Manten., Quito
Muñoz, T.C.	Ing.Agrónomo	Edafólogo	Jefe de Laboratorio de Agrometeorología, Quito
Enríquez, P.A.	Ing.Agrónomo	Agrometeorólogo	Jefe de Programa de Agrometeorología, Quito
Abad, S.G.	Ing.Agrónomo	Edafólogo	Ayudante de Edafología, Quito

Bautista, C.H.	Ing.Agrónomo	Desarrollo Rural	Asistencia Técnica, Distrito Cotopaxi
Benalcazar, W.N.	Ing.Agrónomo	Desarrollo Rural	Asistencia Técnica, Distrito Chimborazo
Bustamante, G.R.	Ing.Agrónomo	Desarrollo Rural	Jefe de Asistencia Técnica, Distrito Milagro
Erazo, M.J.	Ing.Agrónomo	Desarrollo Rural	Asistencia Técnica, Distrito Pisque
Escobar, C.R.	Ing.Agrónomo	Desarrollo Rural	Asistencia Técnica, Distrito Pisque
Félix, C.R.	Ing.Agrónomo	Riego y Des.Rural	Jefe Distrito de Riego, Distrito Motúfar
Freire, J.J.	Ing.Agrónomo	Desarrollo Rural	Asistencia Técnica, Distrito Pisque
Gonzabay, P.R.	Ing.Agrónomo	Desarrollo Rural	Asistencia Técnica, Distrito Milagro
Jaramillo, M.O.	Ing.Agrónomo	Edafólogo	Ayudante de Edafología, Quito
Jaramillo, C.B.	Ing.Agrónomo	Desarrollo Rural	Asistencia Técnica, Distrito del Pisque
Mateus, R.M.	Ing.Agrónomo	Desarrollo Rural	Asistencia Técnica, Distrito Tumbaco
Loaiza, L.F.	Agrónomo	Desarrollo Rural	Asistencia Técnica, Distrito El Tablón
Refisch, I.W.	Ing.Agrónomo	Desarrollo Rural	Asistencia Técnica, Distrito M.J. Calle
Rodríguez, V.R.	Ing.Agrónomo	Desarrollo Rural	Asistencia Técnica, Distrito M.J. Calle
Rosero, P.H.	Ing.Agrónomo	Desarrollo Rural	Asistencia Técnica, Distrito Milagro
Sandoval, C.J.	Ing.Agrónomo	Riego y Des.Rural	Asistencia Técnica, Distrito Ambuquí
Sarmiento, C.C.	Ing.Agrónomo	Desarrollo Rural	Asistencia Técnica, Distrito Pisque
Segarra, A.E.	Ing.Agrónomo	Desarrollo Rural	Asistencia Técnica, Distrito Pisque
Soria, S.T.	Ing.Agrónomo	Desarrollo Rural	Asistencia Técnica, Distrito Milagro
Uzca, R.M.	Ing.Agrónomo	Desarrollo Rural	Asistencia Técnica, Distrito Milagro
Valarezo, L.A.	Ing.Agrónomo	Desarrollo Rural	Asistencia Técnica, Distrito Pisque
Vallejo, E.E.	Ing.Agrónomo	Desarrollo Rural	Asistencia Técnica, Distrito Montúfar
Vásquez, L.F.	Ing.Agrónomo	Desarrollo Rural	Asistencia Técnica, Distrito Chimborazo
Velásquez, V.F.	Ing.Agrónomo	Desarrollo Rural	Asistencia Técnica, Distrito Salinas
Vásquez, E.J.	Ing.Agrónomo	Desarrollo Rural	Asistencia Técnica, Distrito Pisque

<u>Dependencias en otras localidades</u>: 10 Distritos de Riego: Montúfar, Provincia Carchi; Salinas, Provincia Imbabura; Ambuquí, Provincia Imbabura; Pisque, Provincia Pichincha; Tumbaco, Provincia Pichincha; Latacunga-Salcedo-Ambato, Provincia Cotopaxi; Chimborazo, Provincia Chimborazo; Milagro-Chilintomo-Banco de Arena, Provincia Guayas; M.J. Calle, Provincia Cañar; Tablón, Provincia El Oro.

Caracter y Estructura: La Institución fué creada por Decreto Ley No. 1551 del 10 de Noviembre de 1966, es un Organismo de Derecho Público, con personería Jurídica, adscrita al Ministerio de Agricultura y Ganadería. Para su financiamiento tiene fondos propios y participa de las asignaciones anuales del Presupuesto General del Estado.

Actividades en Ingeniería Rural: a) Enseñanza: La Institución realiza cursos no periódicos en colaboración, con organizaciones internacionales, son de capacitación en aspectos de Desarrollo Rural, Riego a Nivel Parcelario, Drenaje. Concede Certificaciones de Aprobación y/o Asistencia. El Promedio de alumnos es de 50. - b) Investigación: Programa de Granjas Experimentales en colaboración con el Ministerio de Agricultura y Ganadería, y el Instituto Nacional de Investigaciones Agropecuarias en los Distritos de Riego - Programas y Convenios de Asistencia Técnica en Riego, Drenaje, Maquinária Agrícola, Agronomía a los usuarios de los Distritos de Riego en explotación (Convenio INERHI-IRYDA; Convenio Belga-Ecuatoriano) - Programa de Mejoramiento de la red de conducción en el Proyecto de Riego del Chambo con la Comunidad Económica Europea - Programa de Investigación Agrometeorológica en 10 Distritos de Riego - Programa de Investigación en Riego e Investigación Social en colaboración con FAO - Programa de Investigación con la Universidad Ecuatoriana (Facultades de Ciencias Agrícolas) en problemas de sequía que afecta al Ecuador.

Publicaciones: En el Desarrollo Rural, Instrucciones y publicaciones mimeografiadas para los usuarios del riego sobre diferentes temas: Integración de la agricultura y la ganadería - Publicaciones para los profesionales que laboran en los Distritos de riego - Publicaciones en el campo de los estudios edafológicos con fines de riego.

Correspondencia: Castellano, Inglés, Francés.

Departamento de Ingeniería Agrícola, Facultad de Ciencias Agrícolas, Universidad Central del Ecuador (Department of Agricultural Engineering, Faculty of Agricultural Sciences, Central University, Quito, Ecuador)
Ciudad Universitaria, Casilla Postal A-46-07

Estrella, N.J.	Ing.Agr.	Decano de la Facultad - Topografía
Peñafiel, R.N.	Ing.Agr.	Construcciones Rurales
Cisneros, F.G.	Ing.Agr.	Agrometeorología
Marcial, N.B.	Ing.Agr.	Conservación de Suelos
Rodríguez, P.M.	Ing.Agr.	Avalúos y Peritajes
Paredes, R.N.	Ing.Agr.	Maquinaria Agrícola
Benítez, L.	Ing.Agr.	Riegos y Drenajes
Romero, R.F.	Arquitecto	Matemáticas
Vega, C.	Ing.Civil	Física
Vélez, N.S.	Arquitecto	Dibujo Técnico

Dependencias en otras localidades: Campo Docente Experimental "La Tola", Tumbaco, Provincia de Pichincha - Campo Docente Experimental "Rumipamba", Salcedo, Provincia de Cotopaxi - Pueblo Viejo, Guayllabamba, Provincia de Pichincha - Bellavista, Calderon, Provincia de Pichincha.

Carácter: El Departamento se creó en el año 1978 y depende directamente de la Facultad y su financiamiento a través de la Universidad Central del Ecuador, Institución de Educación Superior Autónoma. Los cursos que se ofrecen están supeditados a los programas de estudios de la Facultad y el título general es el de Ingeniero Agrónomo

Publicaciones: son eminentemente didácticas a cargo de la Facultad de Ciencias Agrícolas. Artículos periodísticos varios sobre experimentaciones.

Correspondencia: Español.

Instituto de Investigaciones Tecnológicas de la Escuela Politécnica Nacional
Apartado 27-59, Quito, Ecuador

Velásquez T., J.   Ingeniero   Director

Divisiones relacionadas con Ingeniería Rural -
Equipos metal-mecánicos para el agro:
Flores, P.   Ingeniero Mecánico   Diseño y construcción
Jaramillo, A.   Ingeniero Mecánico   Diseño y construcción
Jaramillo, J.   Ingeniero Mecánico   Diseño y construcción
Landázuri, O.   Ingeniero Mecánico   Diseño y construcción
Lozada, M.   Ingeniero Químico   Construcción

Alimentos de origen agroindustrial:
Acuña, O.   Ingeniero Químico   Alimentos
Avila, Margoth   Ingeniero Químico   Alimentos
Barrera, R.   Químico   Laboratorio
Carrillo, Cecilia   Químico   Laboratorio
Dávila, J.   Ingeniero Químico   Alimentos
Díaz, J.   Ingeniero Químico   Alimentos
Hinojosa, Oliva   Químico   Laboratorio
León, L.E.   Ingeniero Químico   Alimentos
Paredes, Irma   Química   Laboratorio
Pólit, P.   Ingeniero Químico   Alimentos
Fuertes, Susana   Química   Laboratorio
Vásconez, G.   Ingeniero Químico   Alimentos

Carácter y estructura: Año de iniciación, 1968. Objetivo: El Instituto de Investigaciones Tecnológicas constituye el elemento de vinculación de la Escuela Politécnica Nacional al exterior y realiza funciones de servicio directo a la industria, agro y a organismos estatales de desarrollo. Orienta la investigación de la Escuela Politécnica Nacional en lo referente a investigación aplicada hacia los campos que sean de prioridad nacional y de servicio a la comunidad, en áreas claves de desarrollo y de aplicación immediata. - Financiación: Fondos provenientes de la Escuela Politécnica Nacional (Organismo Estatal); Fondos provenientes de convenios internacionales.

Actividades relacionadas con Ingeniería Rural: Investigaciones que se están desarrollando actualmente - Diseño y construcción de equipos para el agro: Equipos de apicultura, remolques agrícolas, sembradora de maíz, descascadora de chocho - Tecnología de alimentos de origen agroindustrial: Industrialización de frutas, farináceos, proteínas tradicionales.

Publicaciones: en el boletín de la Escuela Politécnica Nacional (E.P.N.)

Correspondencia: Español.

Portoviejo

Facultad de Ingeniería Agrícola, Universidad Tecnica de Manabí (College of Agricultural Engineering, Technical University of Manabí)
Casilla 82, Portoviejo-Manabí

Departamento de Mecanización Agrícola

Departamento de Aguas y Tierras

Departamento de Planeamiento y Obras Rurales

Actividades en Ingeniería Rural: (a) Enseñanza. (c) Investigaciones.

Correspondencia: Español, inglés.

EGYPT
(Arab Republic of)

Alexandria

Agricultural Engineering Department, Faculty of Agriculture, Alexandria University
Chatby, Alexandria, Egypt   Tel. 75405

Korayem, A.Y.	Ph.D. Professor of Farm Power and Machinery	Head of Agricultural Engineering Department
Ibrahim, A.A.	Ph.D. Prof. Emeritus	Farm Building
Abou-Sabe, A.	Ph.D. Part-time Professor	Farm Power and Machinery
Sabbah, M.A.	Ph.D. Associate Professor	Food Engineering
Younis, S.M.	Ph.D. Associate Professor	Farm Machinery
Abou Zeid, S.F.	Ph.D. Assistant Professor	Farm Machinery
Sheibon, M.A.	Ph.D. Assistant Professor	Farm Power
Shoukr, A.Z.	Ph.D. Assistant Professor	Food Engineering
Hassan, M.A.	Ph.D. Assistant Professor	Irrigation and Drainage
Soliman, N.S.	Ph.D. Assistant Professor	Food Engineering
Ghali, A.M.	M.Sc. Assistant Lecturer	
Aly, O.S.	M.Sc. Assistant Lecturer	
Sharaf El-Din Y.I.	M.Sc. Assistant Lecturer	
Abd El-Ghaffar, E.	M.Sc. Assistant Lecturer	
Abou El-Kheir, M.	M.Sc. Assistant Lecturer	
Ismail, K.M.	M.Sc. Assistant Lecturer	
Ismail, S.	M.Sc. Assistant Lecturer	
Kassem, A.S.	B.Sc. Demonstrator	
Telba, M.H.	B.Sc. Demonstrator	
Nawar, F.M.	B.Sc. Demonstrator	

Status: Governmental Institution. The Department was established in 1956, within the Faculty of Agriculture, Alexandria University.

Activities: The Department offers a programme of four years leading to B.Sc. Degree in Agricultural Engineering, and offers graduate studies leading to M.Sc. degree. An average of 300 students will attend the courses. Research on mechanization of Egyptian Agriculture - Use of different kinds of ploughs - Experiments on spraying - Village planning - Tests of Egyptian made tractors to suit local conditions - Utilization of solar energy for drying different crops - Grain drying and dryaeration - An optimum design of peanut sheller - Irrigation by syphons - Drip and sprinkle irrigation - Corn Planters - Minimum tillage - Hydraulic sprayers - Stationery threshers - Design of small traction unit.

Publications: Textbook on "Planning Farm Buildings" in Arabic - Textbook on "Tractors" in Arabic - Textbook on "Farm Machinery" in Arabic - Unpublished Master theses on: Factors affecting draught to ploughs - Haymaking - Factors affecting atomization and spray equipment and machinery for planting.

Correspondence: Arabic, English.

Cairo

Undersecretariat for Engineering Affairs, Ministry of Agriculture
Dokki, Cairo, Arab Republic of Egypt

headed by: Prof.Dr. Aly El Hossary, Undersecretary of State

and includes the following Departments:

- Civil Engineering Department
- Agricultural Engineering Department
- Testing and Research Station for Tractors and Agricultural Machinery (Alexandria)

Undersecretariat for Engineering Affairs, Executive Authority for Land Improvement Projects, Ministry of Agriculture, Dokki, Cairo, Arab Republic of Egypt

Agricultural Engineering Division, Faculty of Agriculture, Cairo University
Giza, Cairo, Arab Republic of Egypt

Agricultural Engineering Division, Faculty of Agriculture, Ain Shams University
Shoubra El Kema, Cairo, Arab Republic of Egypt

Irrigation Research Centre, Ministry of Irrigation
30 Ramsis St., Cairo, Arab Republic of Egypt

- **Research Institute for Water Distribution and Methods of Irrigation**
  22 El-Gallaa St., Cairo, Egypt

  Activities: Water consumption and requirements - Development of the existing methods of irrigation in Egypt - Development of Water distribution devices at the farm level - Study of losses in waterways.

- **Research Institute for Drainage**
  15 Giza St., Irrigation Bldg., Giza, Egypt

  Activities: Development of tile drainage system in Egypt - Development of the general open drainage system in Egypt - Study of the economic effects of drainage systems on crop yield - Study of methods to reuse drainage water for irrigation.

- **Research Institute for Water Resources and its Development**
  15 Ghiza St., Irrigation Bldg., Giza, Egypt

  Activities: Introduction of new methods and research on: compilation and analysis of hydrological data for the Nile Basin - Development of the existing natural surface water resources - Study of long-term storage projects in the Nile - Study of water losses and water budget in the Nile Basin.

- **Research Institute for the Side Effects of the High Aswan Dam** (El-Sadd El Aali)
  1, Safia Zaghlol St., Cairo, Egypt

  Activities: Studying the degradation problem in the Nile channel within the reach between Aswan and Cairo (946 river kilometers) - Studying the necessary methods to protect the existing hydraulic structures and bridges across the Nile from local scour - Studying the sedimentation problems in Lake Nasser - Studying the underground water table around Lake Nasser - Studying long-term water storage with reference to water quality.

- The Hydraulic and Sediment Research Institute
  Delta Barrage, Egypt

  Activities: Model studies for field problems related to irrigation and drainage engineering, navigation and river training — In-situ calibration of pumps in the whole country — Flow measurements along the main canals to produce an actual calibration of the various regulators — Development and modification of various instrumentation used in the experimental work.

- Research Institute for Weed Control and Maintenance of Waterways
  22 El-Gallaa St.,Cairo, Egypt

  Activities: Investigation and study of: Effects of weed on irrigation and drainage systems — Using different kinds of chemicals for weed control in Egypt — Biological methods, using special kinds of fish for weed control in waterways — Various manual and mechanical methods for weed control.

- Research Institute of Groundwater
  15 Giza St., Irrigation Bldg., Giza, Egypt

  Activities: Compilation and analysis of geo-hydrological data in Egypt — Development of existing groundwater supply and investigation of new sources — Study of the effects of the High Aswan Dam water table downstream Aswan — Study of the quality of the groundwater.

- Research Institute of Soil Mechanics and Foundations
  Delta Barrage and Aswan, Egypt

  Activities: Study of soils at the foundations of the hydraulic structures and others — Study of different kinds of filters for the hydraulic structures in view to select the more suitable one for the soils in Egypt — Study of the seepage problems in waterways and around the hydraulic structures.

- The Mechanical Engineering Research Institute
  Mechanical Department, Ministry of Irrigation, Cairo, Egypt

  Activities: Study suitable mechanical equipment for irrigation purposes and its application — Finding methods for protecting the pump components from deterioration by cavitation, corrosion and erosion, and methods to protect electric equipment from deterioration and insulation failure — Carrying out acceptance tests for stationary, floating and portable pumping stations.

- Institute for Survey Researches
  El-Haram St., Aerial Survey Bldg., Giza, Egypt)

  Activities: Develop and practise the new technology of survey in Egypt — Photogrammetry — Cartography — Geodesy — Statistical analysis of errors in geodetic measurements.

- Department of Research Services
  30 Ramsis St., Cairo, Egypt

  Activities: To print and distribute the research works — To organize training courses for beginner trainees — To supervise the somputer section — To take part in international symposium and conferences — To provide the technical libraries of the Centre and the Institutes with recent publications.

Agricultural Engineering Division, Faculty of Agriculture, Mansoura University
Mansoura Dakahlya Governorate, Arab Republic of Egypt.

EL SALVADOR

<u>Centro Nacional de Tecnología Agropecuaria</u>
Apartado Postal 885, San Salvador, El Salvador, C.A.

<u>Personal científico</u>: Cuestiones de Ingeniería Rural (en carácter de encargado o jefe del Departamento).

Garcia Rivera, F.A.   Jefe del Departamento de Ingeniería Agrícola. Especializado en riego y equipo agrícola.

<u>Dependencias en otras localidades</u>: No existen otras dependencias o sucursales en otras localidades. Es una unidad central.

<u>Carácter y estructura</u>: La institución se estableció en 1942 y es una dependencia del Ministerio de Agricultura y Ganaderia del Gobierno de El Salvador.

<u>Actividades en Ingeniería Rural</u>: Enseñanza: No. Investigación: Trabajos de experimentación en ejecución: Evaluación de diferentes tipos de laboreo en maíz (chuzo, bueyes y maquinaria). - Estudio de cultivares de arroz resistentes a sequía. - Conservación de agua y suelo por medio del sistema henequénpasto - Bomba de sustracción de agua (manual)- Sistema de tiro adecuado a un animal - Comparación de dos sistemas de laboreo en maíz con dos densidades de siembra y dos variedades - Estudio sobre requerimientos de agua en principales cultivos del país - Diseño y construcción de un aplicador de insecticida granulado para el control del cogollero en sorgo - Mejoramiento de suelos con materia orgánica - La dobla de maíz y los rendimientos en la producción de maíz para semilla - Diseño y construcción de una sembradora manual.

<u>Publicaciones</u>: Producción de maíz bajo riego superficial en Atiocoyo. Publicado en la Revista Agricultura de El Salvador, El Salvador. 15 (2). 1976 - Producción de frijol bajo riego superficial en Atiocoyo. Publicado en la Revista SIADES. El Salvador. 6 (2-4). 1977 - Importancia de la Ingeniería Agrícola en El Salvador. Publicado Memoria Primer Congreso de Ingenieros Agrónomos. El Salvador. pag. 334-343. Febrero 1978. - El arado zapato, nuevo arado para laboreo en maíz. Publicado memoria XXV Reunión Anual del PCCMCA. Honduras. Vol.1 pag.M7/1-M7/11. 1979. - Construcción de un medidor de pendientes y criterios sobre siembra de cultivos en laderas. Publicación No.6 del ciclo de seminarios sobre tecnología agropecuaria. El Salvador. 13 pag. Febrero 1980 - Bomba hidráulica de Bambú, una alternativa como fuente de agua para maíz y otros cultivos. Publicado en Vol. 1 de resumenes de la XXVI Reunión Anual del PCCMCA. Guatemala. Marzo 1980 y en Tercer Simposio Ingeniería. UCA. El Salvador. Julio 1980 - Los sistemas de riego en terrenos inclinados de El Salvador. Publicación del CENTA, Departamento de Ingeniería Agrícola. El Salvador. 20 pag. Julio 1980.

<u>Correspondencia</u>: Español.

<u>Departamento de Ingeniería Agrícola, Facultad de Ciencias Agronómicas, Universidad del El Salvador</u>
El Salvador, Centro América.

ETHIOPIA

Addis Ababa

Institute of Agricultural Research - Department of Agricultural Engineering
P.O. Box 2003, Addis Ababa

Irdata Mastebaberia Ina Makuakuamia Commission (Relief and Rehabilitation Commission)
P.O. Box 5686, Addis Ababa

Major Mlugeta Kebede	B.Sc. in Ag. Engineering	Head, Engineering Department
Ato Getachew Belaineh	B.Sc. in Ag. Engineering	Head, Irrigation and Rural Water Supply Division

Branches with other locations: 60 (sixty) Settlement Farm units in 10 different administrative regions of the country.

Status: Engineering Department was established in 1975 and is being financed by the Government.

Agricultural Engineering activities: a) Irrigation - Designing and implementing suitable irrigation systems for settlers big scale farms. b) Rural Water Supply: Designing and implementing water supply systems like hand-dug well equipped with lift pump, spring development, etc. for settler farmers. c) Feeder and farm roads - Construction of feeder and in-farm roads. d) Buildings - Farm Buildings like stores (grain, chemical, etc.), Barns, Residential Buildings, etc. are designed and implemented.

Publications: on irrigation water management.

Correspondence: Amharic and English.

FINLAND

Helsinki

Helsingin yliopisto, Maatalousteknologian laitos (Department of Agricultural Engineering, University of Helsinki)

Pehkonen, A.I.	Dr.Agr. Act.Ass.Prof.	Head of Department
Kares, M.	M.Sc.	Instructor in Farm Energy Problems
Sarin, H.	M.Sc.	Instructor in Farm Machinery

Status: Department of Agricultural Engineering established in 1969. Financed mainly by the State.

Activities: Courses of one to four years. M.Sc., Lic.Agr., Dr.Agr. awarded. 10 to 100 students in different courses, about 10 specializing in agricultural engineering yearly. Research in the use of farm machinery, energy, ergonomics, work methods and methodology of work study.

Publications: Articles in the use of farm machinery, energy in farm production, ergonomics in agricultural jobs, functional planning of dairy buildings. Series of publications: Tutkimustiedote. Textbook: Maamiehen koneoppi 2 (Machines and equipment), Kirjayhtymä 1977.

Correspondence: English, Finnish, German, Swedish.

Teknillinen Korkeakoulu (Technical University)
Otaniemi, Helsinki

Agricultural Engineering activities: Training courses of one to two years' duration, with 4 hours lectures and 4 hours practice a week, attended by about 50 students - The courses are part of the curriculum for civil engineers in the Branch for Civil Engineering Research, experiments and tests are also carried out.

Työtehoseura (Work Efficiency Association)
Melkonkatu 16A, SF - 00210 Helsinki 21

Kantola, M.	Prof. h.c., Dr.For.	Managing Director
Oksanen, E.H.	Dr.Agr.	Vice Man.Dir., Head Dept. of Agric. and Rural Structures
Alanko, A.	Architect	Dir. Planning Bureau of Buildings
Backman, S.	M. Home Econ.	Home Economics Research
Haataja, P.	M. For.	Director, Forest Experiment Station
Iisakkila, H.	Tech.Home Econ.	Home Economics Research
Janhonen, H.	M. Home Econ.	Director, Home Economics Department
Laitinen, A.	M.Agr.	Director, Research Bureau of Farm Buildings
Levanto, S.	M.For.	Forest Research and Advising
Liskola, K.	M.Agr.	Agricultural Research
Luoma-Juntunen, P.	M. Home Econ.	Home Economics Research
Mäkelä, J.	Dr.For.	Dir., Forest Department
Nurmisto, U.	Techn.Constr.	Planning Farm Buildings
Orava, R.	M.Agr.	Dir., Field Research Station
Palho, K.	H.Home Econ.	Home Economics Advising
Pokki, J.	M.Agr.	Agricultural Research
Rukko, L.	Home Econ.	Home Economics Research and Advising
Sillanpää, M-L.	M. Home Econ.	Home Economics Research
Turkkila, K.	M.Sc.	Ergonomic Research
Uski, R.	M. Home Econ.	Home Economics Research
Wartiovaara, L.	M.Sc.	Home Economics Research

Outposted Branch: Research and Training Centre, SF-05200 Rajamäki.

Status: Private association, supported by the State and supervised by the Ministry of Agriculture and Forestry. Established in 1924.

Activities: Teaching: In Rajamäki there are about 400 young and adult people on courses of 1-12 months getting basic or advanced technical training. No degrees awarded. - Research: Study of efficient and economical working methods - Establishment of time standards of working operations in agriculture - Development of machines and methods in forestry and agriculture - Plans for farm buildings - Rationalization projects for individual farms - Research in ergonomics, energy and environment.

Publications since 1973: Work studies on field growing of vegetables - Handling and utilization of manure - Handling and ensiling of moist grains - Utilization of wind energy - Silo stores for fertilizers - Manure removal equipment for animal houses - Effect of field size and shape on working time - Technics and economy of making, handling and feeding of grass silage - Norms and standard times for farm work - Handling, storing and utilization of straw on farms - Peat production on farms - Pig and swine houses - Inner wall plates for farm buildings - Animal houses of light construction - Energy in farm buildings - Storages for fuel wood, straw and peat - Plans for farm heating center with own fuel - Handbook for heating with domestic fuel - Way to safe and healthy work in forests - Work clothing in forest work - Multipurpose forestry machines - Job satisfaction of forestry workers - Ergonomic properties of tractors in forest work - Use of energy in homes - Energy need in drying textiles - Domestic fuel in kitchen ranges - Associations monthly journal TEHO.

Correspondence: English, German, Swedish, Finnish.

Salaojakeskus (before 1978 Salaojitusyhdistys ry) (Finnish Field Drainage Centre)
Simonkatu 12A 11, 00100 HKI 10

Perho, H.   M.Sc.Agr.   Chairman
Saavalainen, J. M.Sc.Eng.   Managing Director

Status: Private semi-official association, subsidized by the Government and supervised by the Board of Agriculture.

Main activities: Planning sub-suface drainage covering all of Finland. Activity covers all works concerned - Supervising the execution of subdrainage and overseeing the correct use of Government granted funds for subsurface drainage - Planning farming roads, irrigation of fields and peat production. Training activities: One year long courses for drainage technicians. The programme includes both tehoretical and practical studies. - Lectures at universities and at schools for agricultural education - Advising activity for farmers. - Research activities: Research on the operation of subsurface drains and on the economics and efficiency of subsurface drainage.

Publications: Annual report - Handbook of Drainage - Farming roads - Rural water supply.

Correspondence: Finnish, Swedish, German, English.

Vihti

Valtion Maatalouskoneiden tutkimuslaitos (Finnish Research Institute of Engineering in Agriculture and Forestry)
SF-03450 Olkkala   Tel. (9) 13-46211

Reinikainen, A.	Prof.Dr.	Director
Ahokas, J.	Dipl.Ing.	Head of Department – Technical Dept.
Haber, P.	Eng.	Tractors, tractor diggers, type approval
Piltti, M.	Techn.	Heating, crop drying
Kara, O.	Dr.Agr.	Head of Department – Equipment for working the soil, crop protection, irrigation
Räisänen, L.	B.Sc.Agr.	Equipment for soil
Hänninen, M.	Techn.	Equipment for soil, ploughs
Kiviniemi, J.	B.Sc.Agr.	Head of Department – Equipment for harvesting, horticulture
Laurola, H.	B.Sc.Agr.	Equipment for harvesting, transporting
Turtianen, K.	B.Sc.	Forester Head of Department Equipment for Forestry
Nieminen, L.	B.Sc.Agr.	Head of Department – Equipment for farmstead, home economics
Karhunen, J.	Dipl.Eng.	Equipment for farmstead, instruments
Olkinuora, P.	B.Sc.Agr.	Standardization

Status: Farmers Extension Union: Inspection Committee 1798, Testing Station 1902, Governmental Institution 1912, supervised by the Board of Agriculture.

Activities: Research, development, testing and standardization and safety of machinery and equipment used in agriculture, forestry, horticulture, home economics and home industry.

Publications: Test reports, study reports, information booklets.

Correspondence: Finnish, Swedish, German, English.

FRANCE

Antony

Centre National d'Etudes et d'Expérimentation de Machinisme Agricole (CNEEMA)
(National Centre of Agricultural Machinery Research)   Tél: 666.21.09 Poste 338
Parc de Tourvoie, 92160 Antony   Télex: 204565

Barlet, J.	Ingénieur Agronome,	Ingénieur en Chef du Génie Rural des Eaux et des Forêts (GREF)
Achart, J.	Ingénieur Agronome,	Ingénieur en Chef du Génie Rural des Eaux et des Forêts, Chef du Département Essais

Alix, G.	Ingénieur du Génie Rural des Eaux et des Forêts
Bournas, L.	Ingénieur en Chef du Génie Rural des Eaux et des Forêts, Directeur Technique
Carillon, R.	Ingénieur Agronome, Ingénieur en Chef du GREF, Chef du Service de la Documentation, de l'Information et de la Formation
Davenel, A.	Ingénieur Agronome, Ingénieur Civil du GREF
Jattiot, A.	Ingénieur Agronome, Ingénieur en Chef du GREF, Directeur de l'Echelon d'Auvergne
Lucas, F.	Ingénieur Agronome, Ingénieur du GREF
Lucas, J.	Ingénieur du GREF, Chef du Département Recherches
Mas, D.	Ingénieur Agronome, Ingénieur du GREF, Chef des Services Administratifs
Molle, J.F.	Ingénieur du GREF, Chef de Division
Murat, H.	Ingénieur du GREF, Chef de Division
Pechiné, M.	Ingénieur du GREF
Rantz, J.	Ingénieur en Chef du GREF
Raynaud, P.	Ingénieur Agronome, Ingénieur en Chef du GREF, Directeur de l'Echelon du Midi
Sevilla, F.	Ingénieur du GREF
Spiteri, A.	Ingénieur du GREF
Srour, S.	Ingénieur Civil du GREF
Vellinger, P.	Ingénieur en Chef du GREF
Zwaenepol, P.	Ingénieur Civil du GREF

<u>Annexes situées dans d'autres localités</u>: Echelons – Dans le Midi de la France, le CNEEMA dispose d'un échelon sis à (Gard) et au Tholonet près d'Aix-en-Provence (Bouches du Rhône) et devant être regroupé prochainement à Montpellier. Cet échelon étudie principalement les problèmes de mécanisation des cultures légumières ou maraîchères et d'arboriculture pour aider à la mise en valeur d'importantes régions de la côte méditerranéenne française. Un autre échelon régional a été installé en 1968 à Montoldre dans l'Aller pour traiter principalement des problèmes de mécanisation en matière de récolte des fourrages et d'élevage. Cet échelon est associé à une exploitation agricole de plus de 200 hectares.

<u>Structure</u>: Le CNEEMA a été créé en 1955 sous forme d'un établissement public. Il est doté de la personnalité civile et de l'autonomie financière et rattraché à la Direction de l'Aménagement du Ministère de l'Agriculture. Un Conseil Supérieur de la Mécanisation et de la Motorisation de l'Agriculture, interministériel et interprofessionnel, en approuve les programmes de travaux et les soumet à l'approbation du Ministre de l'Agriculture.

<u>Activités concernant le génie rural</u>: a) <u>Formation</u>: Un Centre de Formation "Machinisme Agricole" existe au CNEEMA. Ce centre dont chaque session dure 9 mois, est réservé à un certain nombre d'élèves de 3ème année de l'Institut National Agronomique et des Ecoles Nationales Supérieures Agronomiques qui désirent recevoir une spécialisation dans la branche Machinisme Agricole. Il reçoit 10 à 15 stagiaires chaque année et le stage est sanctionné par un certificat adjoint au titre normal d'Ingénieur Agronome du stagiaire. Des stages de perfectionnement sont organisés chaque année à l'attention d'ingénieurs, de techniciens ou d'enseignants à raison de 200 à 300 stagiaires par an pour une vingtaine de stages de 2 à 5 jours.

b) <u>Recherches</u>, expérimentation, vulgarisation ou autres activités actuellement en cours, employ nt des mots clefs pour désigner les sujets traités. - Documentation, information, vulgarisation, essais, recherches et expérimentation sur le machinisme agricole. Le CNEEMA est en outre particulièrement engagé dans des expérimentations et des recherches sur les économies d'énergie et les énergies nouvelles en agriculture, la récolte et la conservation des grains et des fourrages, sur le travail du sol, sur les traitements antiparasitaires, sur la récolte mécanique des fruits et des légumes et sur la mécanisation des opérations d'élevage. Le CNEEMA réalise également des enquêtes économiques, détermine les critères de choix des matériels et diffuse des éléments utiles aux vulgarisateurs soit directement au cours des stages qu'il organise, soit par ses publications et son travail de documentation.

<u>Publications</u>: De très nombreuses études et d'importants articles on été publiés dans les périodiques (Etudes, Bulletin d'Information) édités par le CNEEMA sous les rubriques suivantes:Etudes générales sur le machinisme agricole - Etudes statistiques - Etudes économiques - Tableaux de caractéristiques de matériels. Le CNEEMA publie, en outre, un Bulletin Bibliographique mensuel, avec supplément trimestriel international, et des Bulletins d'Essais de matériels agricoles. Le CNEEMA a publié depuis 1960, dans sa collection du Livre du Maître, quatre ouvrages importants illustrant de nombreuses planches hors-texte: Tome I (dernière édition 1978): Moteurs et tracteurs; Tome II (en course de réédition): Principales machines de culture et de récolte; Tome III (1974): Autres matériels et problèmes économiques; Tome IV (1966): Réglage et entretien des matériels agricoles, avec des planches illustrées. En outre le CNEEMA a publié en 1971 un Dictionnaire Technique Pentalingue de la Mécanisation Agricole (français, anglais, allemand, italien et espagnol) qui comprend trois tomes: Tome I, sujets agricoles (cultures, élevages, bâtiments, gestion, économie) mécanique, moteurs, tracteurs; Tome II, machines agricoles et leurs organes; Tome III répertoires alphabétiques des 17.000 termes, par langue qui figurent dans le texte. Enfin différentes brochures de vulgarisation sont diffusées.

<u>Correspondance</u>: Français, allemand, anglais, espagnol, italien, russe.

<u>Centre d'Etudes et d'Experimentation du Machinisme Agricole Tropical (CEEMAT)</u>
(Agricultural Machinery Centre for Tropical Regions)
Parc de Tourvoie - 92160 Antony - France

Tél. 668 61 02
Télex: 204565 F

Uzureau, C.	Ingénieur en Chef d'Agronomie, Directeur
Herblot, G.	Ingénieur en Chef d'Agriculture d'Outre-Mer, Directeur-Adjoint
Chèze, B.	Ingénieur du Génie Rural des Eaux et des Forêts, Chef du programme Etudes et Recherches Techniques
Le Moigne, M.	Ingénieur Agronome, Chef du Programme Economie et Développement
Dutartre, J.	Ingénieur Agricole, Chef du Programme Stages et Formation
Troude, F.	Ingénieur Agronome, Chef du Programme Documentation - Information.

Outre ces personnes, le Centre compte environ 20 ingénieurs ou chercheurs qui sont répartis dans les quatre programmes mentionnés ci-dessus.

<u>Annexes situées dans d'autres Localités</u>: - Antenne Technologie Alimentaire (responsable Griffon, D.) Centre GERDAT B.P. 5035, 34032 Montpellier Cédex, Tél. 63 91 70.
- Antenne CEEMAT à la Réunion (responsable Derevier, A.) Parc de la Providence - 97489 Saint-Denis, Tél. 21-39-73.

<u>Structure</u>: Le CEEMAT a été créé en 1962. Depuis 1972 il constitue un Service extérieur commun au Groupement d'Etudes et de Recherches pour le Développement de l'Agronomie Tropicale (GERDAT). Le GERDAT coordonne l'activité des huit Instituts français de recherche agronomique spécialisés outre-mer, à savoir: - Le Centre Technique Forestier Tropical (CTFT) - L'Institut d'Elevage et de Médecine Vétérinaire des Pays Tropicaux (IEMVT) - L'Institut Français du Café, Cacao et autres plantes stimulantes (IFCC) -

- L'Institut de Recherches Agronomiques Tropicales et de Cultures Vivrières (IRAT)
- L'Institut de Recherches sur le Caoutchouc (IRCA) - L'Institut de Recherches du Coton et des Textiles Exotiques (IRCT) - L'Institut de Recherche sur les Fruits et Agrumes (IRFA) - L'Institut de Recherches pour les Huiles et Oléagineux (IRHO).

Le GERDAT (dont le siège est 42 rue Scheffer 75016 Paris - Tél. 53-56-41) ainsi que le CEEMAT et les Instituts membres sont contrôlés par le Ministère français de la Coopération.

Activités concernant le Génie Rural: (a) Formation: Si le CEEMAT n'est pas un Organisme d'Enseignement, il organise des stages de formation en se basant sur ses activités propres et en prévoyant des cycles de formation auprès des Ecoles, des Coopératives, des Organismes de Recherche et des Constructeurs. Il intervient comme relai du Centre International des Etudiants et Stagiaires pour organiser des stages individuels à des niveaux variables (technicien à chercheur). Il intervient aussi comme organisateur de stages groupés sur un thème bien défini. - Ainsi, en 1979, le Centre à reçu en stages individuels 21 stagiaires provenant de 10 pays d'Afrique ou d'Asie. Il a en outre organisé des sessions de formation groupés, par exemple un stage de 3 mois pour former des responsables de structures de stockage en vrac pour les régions chaudes (16 participants provenant de 11 pays). - Le CEEMAT ne décerne pas de diplôme mais des certificats de stage. - Le CEEMAT contribue aussi au programme d'Ecoles basées en Haute-Volta: Ecole des Ingénieurs de l'Equipement Rural, Ecole de Techniciens Supérieurs de l'Hydraulique et de l'Equipement Rural. (b) Recherches (récemment réalisées ou en cours). - Caractérisation mécanique des sols prenant en masse pour contribuer à une meilleure préparation des sols en zone tropicale sèche (sémi-aride) - Etude de matériels pour traitements phytosanitaires à très bas volume (ULV-CDA) avec mise au point d'un banc d'essai spécifique à cette technologie. - Recherche sur la mécanisation de la récolte du manioc; destruction des parties aériennes et arrachage - chargement des tubercules - Recherche sur le stockage en silos métalliques: incidence des transferts de chaleur à travers des parois métalliques sur la qualité de conservation des grains stockés - Etude des contraintes posées à l'intégration de la mécanisation dans les systèmes de production agricoles sahéliennes - Etude de la conservation sous vide des produits agricoles - Etude sur l'amélioration des caractéristiques des groupes motopompes pour l'irrigation - Mise au point de tracteurs de faible puissance pour les pays en développement - Etude et réalisation d'une moissonneuse-batteuse automotrice à riz spécialement conçue pour les pays en développement - Etude économique sur la motorisation agricole intermédiaire en Afrique de l'Ouest - Mécanisation de la canne à sucre: épierrage, plantation, récolte, transport, cultures intercalaires - Expérimentation de matériels modernes à traction animale au Brésil - Participation à un projet de recherche - développement sur la motorisation paysannale en Côte d'Ivoire - Missions de consultants dans de nombreux pays.

Publications: a) Publications régulières: Revue trimestrielle "Machinisme Agricole Tropical" publiée en français avec résumés en anglais et espagnol - Lettre d'Information trimestrielle diffusée gratuitement - Bulletin bibliographique - b) Autres publications: Aide-mémoire du moniteur de culture attelée - Aide-mémoire pour l'utilisation des motoculteurs dans la zone intertropicale - Titres diffusés par le Centre de Documentation du Ministère de la Coopération: - Manuel de conservation des cultures tropicales - Manuel de culture avec traction animale, réimpression - Maintenance du matériel agricole.

Correspondance: Français, Anglais, Espagnol, Portugais.

Grignon

Institut National Agronomique Paris-Grignon
Centre de Paris: 16 rue Claude Bernard, 75231 Paris Cedex 05   Tél.: Paris 570 15 50
Centre de Grignon: 78850 Thiverval-Grignon   Grignon 0564510

Delage, J.	Ingénieur agronome	Directeur

Chaire de Machinisme Agricole:

Aubineau, M.	Ingénieur agricole	Maître de conférences
De Fournas, M.	Ingénieur agricole	Maître-assistant
Mignotte, F.	Ingénieur agricole	Maître-assistant

Chaire de Mathématiques appliquées et Hydraulique agricole:

Montes, P.   Professeur, Ingénieur Agronome, Ingénieur Civil du Génie Rural
La Chaire est "hors département" dans l'Institut National Agronomique.

Enseignement de Génie Rural:

Carlier   Professeur à la vacation, Ingénieur agronome, Ingénieur du Génie Rural
Enseignement "hors département" au sein de l'Institut National Agronomique

Structure: Date de fondation de l'Institut National Agronomique Paris Grignon: Janvier 1971. Fusion de deux etablissements préexistants: l'Institut National Agronomique de Paris fondé en 1875 et l'Ecole Nationale Supérieure Agronomique de Grignon fondée en 1828.
- Organe de contrôle et de financement direct: Ministère de l'Agriculture (gouvernamental).

Activité concernant le Génie Rural: a) Formation: Contribution au Diplôme d'Agronomie Générale (180 étudiants de 1ère année, 6 Unités de Valeur de 2ème année comportant en moyenne 15 étudiants chacune) - Contribution au Diplôme d'Agronomie Approfondie dans diverses Spécialisations - Thèses de Docteurs Ingénieurs (Sciences Agronomiques: Machinisme agricole) - b) Recherche: Expérimentation en cours sur le "Travail du Sol" en collaboration avec l'Institut National de la Recherche Agronomique (INRA) - Etude de Systèmes Agraires dans l'Oise en collaboration pluridisciplinaire (Département "Systèmes agraires et Développement" de l'INRA).

Publications: Le Machinisme agricole 1974 Encyclopoedia Universalis - Le Tracteur agricole, brochure - Le Semis Direct, Perspectives agricoles.

Correspondance: Français ou Anglais.

Montpellier

Ecole Nationale Supérieure Agronomique de Montpellier (College of Agriculture)
34060 Montpellier Cedex
Tél.: (67) 63 12 65
Télex : INRAMON 490.818 F

Chaire de Génie Rural et Hydraulique - Machinisme Agricole:
Professeur: M. F. de Chabert
Maitre de Conférences: M. Manière
Maitre-Assistant: M. Luc

Structures: Institution gouvernementale dépendant du Ministère de l'Agriculture.

Activités: Cours de Génie rural et de machinisme agricole répartis en 3 ans - Expérimentation en matière de machinisme agricole et en matière de génie rural - Recherches sur l'irrigation et le drainage - Bilans hydriques - Utilisation des sondes - Recherches sur l'irrigation et le drainage - Recherches sur la protection des cultures (moyens terrestres et aériens) - Mécanisation des vendanges et des récoltes fruitières.

Publications: Prix de revient - Motoviticulture et promotion sociale - Motoviticulture et plantations - Evolution des techniques de la Motoviticulture - Vers une mécanisation des vendanges - Motoviticulture - Rapports annuels du Centre pilote de motoviticulture - La pulvérisation pneumatique aérienne et perfectionnements récents, traitements aériens viticoles - Techniques modernes de protection antiparasitaire de vignoble - Machinisme agricole et viticole - Manutention et épandage des engrais organiques dans le vignoble - Facteurs influençant la régularité d'épandage des distributeurs d'engrais chimiques solides et liquides - Facteurs influençant la régularité d'épandage des distributeurs d'engrais chimiques solides- Etude in-situ des écoulements en zone non saturée - Ecoulement dans les drains en matière plastique.

Correspondance: Français.

Nancy

Ecole Nationale Supérieure d'Agronomie et des Industries Alimentaires (College of Agriculture and Food Industry)
38, rue Ste Catherine - 5400 - Nancy
Tél.: (8) 332 95 97

Jacquin, F.   Professeur   Directeur - Science du Sol

Génie rural:
Corda, R.   Ingénieur en chef, Ing. GREF Aménagement agricole et hydraulique

Nogent

Centre Technique Forestier Tropical (Tropical Forest Research Center)
45 bis, avenue de la Belle Gabrielle - 94130 Nogent-sur-Marne
Tél.: 873 32 95

Huguet, L.   Ingénieur Général du Génie Rural des Eaux et des Forêts
Chardin, A.   Ingénieur diplômé de l'Ecole Polytechnique et Ingénieur forestier. Chef de la Division de mécanique et usinage. Specialité: recherches et essais sur l'usinage des bois tropicaux.

Esteve, J.	Ingénieur de l'Ecole supérieure du bois. Chef de la Division des exploitations. Spécialité: recherches et essais sur le matériel d'exploitation forestière, de construction et d'entretien des routes forestières.
Goudet, J.P.	Ingénieur agronome et ingénieur forestier. Division des recherches forestières. Specialité: recherches et essais sur les techniques sylvicoles et l'aménagement forestier.

Stations affiliées:
C.T.F.T. B.P. 764, Pointe Noire, République Populaire du Congo
C.T.F.T. B.P. 116 97310 Kourou, Guyane française
C.T.F.T. B.P. 33, Abidjan 08, République de Côte d'Ivoire
C.T.F.T. B.P. 303, Ouagadougou, République de Haute-Volta
C.T.F.T. B.P. 411, Noumea, Nouvelle Calédonie

Représentants du C.T.F.T.:
B.P. 2102 Yaoundé et B.P. 2022 Maroua, République Unie du Cameroun
B.P. 149 Libreville, République du Gabon
B.P. 745 Antananarivo, République Démocratique de Madagascar
B.P. 2312 Dakar, République du Sénégal
B.P. 225 Niamey, République du Niger

Statut Juridique: Institution gouvernementale, placée sous la double tutelle du Ministère de l'Agriculture et du Secrétariat aux affaires étrangères, chargé de la Coopération, fondée en 1950.

Activité concernant le génie rural: Celle-ci est partagée entre 3 services: Recherches forestières, exploitations forestières, Mécanique et usinage. - Recherches études et essais sur le matériel mécanique utilisé dans les plantations forestières, sur l'équipement des exploitations et industries forestières. Elaboration de projets et d'études techniques d'usinage des bois tropicaux.

Publications: Bois et Forêts des Tropiques, revue bimestrielle publiant des articles et des informations sur la sylviculture, l'exploitation et l'usinage des bois tropicaux.

Correspondance: Français, Anglais.

Paris

Le Centre Technique du Génie Rural des Eaux et des Forêts (Technical Centre of Agricultural Engineering of Water and Forests)
19, avenue du Maine - 75732 Paris CEDEX 15

Foulhouze, R.	Ingénieur général du GREF	Directeur
Martinot, R.	Ingénieur en chef du GREF	Directeur Adjoint

Le Centre Technique a pour mission: notamment dans le domaine de l'Agriculture des Forêts, des Eaux et de l'Equipement Rural et la Protection de la nature, et des loisirs en milieu rural. D'assurer l'information et l'appui technique des Services - De participer à la formation permanente du personnel de Génie Rural des Eaux et des Forêts et d'exécuter les contrôles techniques, soit en appui des services compétents, soit directement en l'absence d'autres organismes spécialisés. Il veille à la cohérence méthodologique des études techniques directement menées par les services.

Structure: Le Centre Technique est organisé en divisions ayant chacune compétence au plan national dans un secteur d'activité des Services - Ces divisions sont regroupées à Antony - Aix-en-Provence - Bordeaux - Clermont-Ferrand - Nogent/Vermission - Grenoble - Nancy et Rennes.

Activité de Formation: Le CTGREF assure la formation de techniciens destinés aux Services de l'Agriculture, les Techniciens de Génie Rural sont formés au Groupement de Nancy - 156 Bd d'Austrasie 54000 Nancy. - Le Centre peut recevoir dans les laboratoires, un nombre limité de jeunes ingénieurs qui préparent des thèmes de l'Université (Docteur Ingénieur - Docteur 3° Cycle) ou qui suivent des cours dans des établissements normaux d'enseignement - Le Centre ne délivre pas de diplôme.

Activité de Recherche: En principe le Centre n'a pas de mission de recherche. Ces activités sont réservées à d'autres organismes, notamment à l'Institut National de Recherches Agronomiques (INRA). Toutefois en l'absence du département Génie Rural à l'INRA, le Centre est appelé à poursuivre et parfois à entreprendre des travaux de recherche qui lui sont nécessaires pour mener à bien son activité essentielle qui est l'Appui Technique aux Services.

Publications: Le Centre publie un bulletin technique du Génie Rural, non périodique qui peut être envoyé à titre d'échange à tous les établissements de recherches ou d'études du Génie Rural. En outre il publie chaque année les comptes-rendu de ses études sous forme de notes ou de brochures.

Correspondance: Français.

Centre Technique du Bois (Wood Technical Centre)
10, Avenue de Saint-Mandé - 75012 Paris

Quiquandon, B.	Ingénieur en Chef du Génie Rural, des Eaux et des Forêts
Hochart, B.	Ingénieur de l'Ecole Supérieure du Bois
Loiseau, P.	Chef de Division "Bois dans la construction"
Crubilé, P.	Section "Charpente"
Corne, M.	Section "Constructions industrialisées"

Annexes situées dans d'autres localités: Station expérimentale à Champs-sur-Marne (77420)

Structure: Organisme d'intérêt public, créé en application de la loi du 22 juillet 1948 sur les Centres Techniques Professionnels, par un arrêté interministériel du 15 février 1952. Il est soumis à la tutelle du Ministère de l'Agriculture et du Ministère de l'Industrie.

Activités concernant le génie rural: Recherche, expérimentation et vulgarisation concernant l'utilisation du bois dans les constructions rurales.

Publications: Cahiers, brochures, périodiques.

Institut de Recherches sur les Fruits et Agrumes (IRFA) (Institute for Research on Fruits and Citrus)
6 Rue du Général Clergerie - 75116 - Paris   Tél.: 553 16 92

Cuille, J.	Directeur Général de l'IRFA	Paris
Mellin, Ph.	Agronome	Martinique
Lassoudiere, A.	Agronome	Côte d'Ivoire
Cassin, J.	Agronome	Corse
Beugnon, M.	Agronome	Côte d'Ivoire
Ganry, J.	Agronome	Guadeloupe

Fouque, A.	Agronome	Guyane
Lenormand, C.	Agronome	Sénégal
Soulez, P.	Agronome	Côte d'Ivoire
Guillemot, J.	Agronome	Sénégal
Moreau, B.	Agronome	Réunion
Meyer, J.P.	Agronome	Haute-Volta
Haury, A.	Agronome	Niger

Stations de Recherche Affiliées:

Cameroun	IRFA - B.P. 13 - Nyombe
Cote d'Ivoire	IRFA - 01 - B.P. 1740 - Abidjan 01
Corse	Station de Recherches Agrumicoles - San Giuliano 20230 San Nicolao
Guadeloupe	IRFA - Station de Neufchateau - Sainte Marie 97130 Capesterre Belle Eau
Guyane	IRFA - B.P. 1125 - 97304 Cayenne CEDEX
Haute-Volta	IRFA - B.P. 136 Ouagadougou
Martinique	IRFA - B.P. 153 - 97202 Fort-de-France CEDEX
Niger	IRFA - B.P. 886 - Niamey
Réunion	IRFA - B.P. 180 - 97455 Saint Pierre CEDEX
Sénégal	Mission IRFA au Sénégal - B.P. 1716 - Dakar

Statut juridique - Institut Privé (loi du 1er Juillet 1901)

Activités en matière de génie rural: Recherches expérimentales et études sur la mécanisation des cultures fruitières sous les tropiques - Etude des techniques d'irrigation par aspersion appliquées à la culture bananière en Côte d'Ivoire, au Cameroun, aux Antilles françaises, à la culture de l'ananas en Côte d'Ivoire, à la culture des agrumes en Corse - Lutte contre l'érosion - Etude des techniques d'irrigation à débit réduit en zones tropicales sèches.

Publications: "Fruits" Revue mensuelle avec résumés des articles en français, allemand, anglais, espagnol, russe - Rapports annuels - Ouvrages en français et en espagnol.

Correspondance: Français, Anglais, Allemand, Espagnol, Italien, Russe.

Institut de Recherches pour les Huiles et Oléagineux (IRHO) (Research Institute for Oils and Oil Seeds)
11, Square Pétrarque, 75016 Paris

Bourges-Maunoury, M.	Président
Fleury, J.C.	Directeur Général
Ollagnier, M.	Directeur des Recherches
Jounanique, R.	Secrétaire-Général
Surre, C.	Directeur du Développement et du Département Palmier à Huile
Fremond, Y.	Directeur du Département Cocotier
Gillier, P.	Directeur du Département des Oléagineux annuels
Gascon, J.P.	Directeur du Département Sélection
Ochs, R.	Directeur du Département Agronomie
Mariau, D.	Directeur du Département Entomologie
Renard, J.L.	Directeur du Département Phytopathologie

Structure: Association sans but lucratif, fondée en 1942, sous tutelle du Ministère de la Coopération et faisant partie du GERDAT.

Activités Générales: Toutes recherches et expérience d'ordre agronomique, chimique, technologique et coopération technique en vue du développement des oléagineux tropicaux, de leur exploitation, de leur transformation et de la valorisation de leur production. L'action de l'IRHO concerne une cinquantaine de pays en Afrique, dans le Sud-Est asiatique, en Océanie et en Amérique latine. Elle s'exerce sur une quinzaine de stations expérimentales et points d'essais, et dans le cadre des programmes de développement par des missions permanentes ou temporaires, des études techniques, un soutien logistique, etc.
Activités en matière de Génie Rural: Recherches et expérimentations sur les équipements agricoles pour la culture des oléagineux tropicaux et le matériel d'extraction des huiles de graines et de fruits; travaux d'aménagement des plantations, drainage, irrigation, réseaux d'exploitation, infrastructure, etc.

Publications: "Oléagineux", revue internationale des corps gras, mensuelle; "Manuel de l'Huilerie de Palme" et autres ouvrages, rapports et études divers.

Correspondance: Français, Anglais, Espagnol.

Bureau Commun du Machinisme et de l'Equipement Agricole
21, rue Chaptal - 75009 Paris -   Tél. 285.46.16

Beauvois, R.	Directeur
Bassez, J.	Ingénieur d'Etude II
Dubalen, J.	Ingénieur d'Etude II
Fievet, G.	Ingénieur d'Etude III
Grillot, M.	Ingénieur d'Etude II
Guillet, J.	Ingénieur d'Etude III
Jacquet, P.	Ingénieur d'Etude II
Siret, P.	Ingénieur d'Etude II
Thoiron, C.	Ingénieur d'Etude II

Structure: Organisme rattaché à l'I.G.E.R.

Bureau Central d'Etudes pour les Equipements d'Outre-Mer (BCEOM)
15, Square Max-Hymans - 75741 Paris CEDEX 15   Tél.: (1) 320-14-10

Segretain, P.   Directeur Général

Rennes

Chaire du Génie Rural - Ecole Nationale Supérieure Agronomique
65, Rue de St. Brieuc 35042 Rennes-CEDEX   Tél.: 59 02 40 Poste 40

Station affiliée: Station expérimentale d'Hydraulique Agricole (Gouvernementale)

Activités: Enseignement de Génie Rural de degré supérieur aboutissant à la délivrance du diplôme d'ingénieur agronome. - Enseignement spécialisé: 10 étudiants hydraulique et climatologie - cartographie et stéréoscopie aboutissant à un diplôme d'agronomie approfondie. - Recherches et expérimentations concernant l'hydraulique agricole, la perméabilité des terres, les drainages, l'écoulement en canaux, les irrigations - Détermination de l'évapo-transpiration.

Correspondance: Français, Anglais, Allemand.

Strasbourg

Ecole Nationale Supérieure des Arts et Industries de Strasbourg (ENSAIS) (The Strasbourg College of Engineers)
24, boulevard de la Victoire - 67084 Strasbourg CEDEX -   Tél.: (88) 35 55 05

Pichoir, J.   Directeur
Frey, P.   Ingénieur civil des Ponts et Chaussées - Agrégé en Génie Civil

Structure: Ecole dépendant de la Direction des Enseignements Supérieurs du Ministère des Universités.

Activités: Enseignement de Génie Civil et Rural conduisant au Diplôme d'Ingénieur.

Ecole Nationale des Ingénieurs des Travaux Ruraux et des Techniques Sanitaires (National College of Agricultural Engineering and Sanitation Engineers)
1, quai Koch - B.P. 1039 F - 67070 Strasbourg CEDEX Tél.: (88) 35 67 72

Hirtz, J-M.   Directeur,
                   Ingénieur en Chef du Génie Rural, des Eaux et des Forêts
Kersauze, R.   Sous-Directeur,
                   Ingénieur des Travaux Ruraux
Desmartin, P.   Directeur des Etudes,
                   Ingénieur en Chef du Génie Rural, des Eaux et des Forêts

Corps enseignant: Cours de première année:

Achart, J.	Professeur Ingénieur en Chef du G.R.E.F.	Technologie des Tracteurs Agricoles et engins de génie civil.
Boux, M.	Professeur Ingénieur du G.R.E.F.	Matériaux et constructions
Caron, J-M.	Professeur Docteur ès sciences	Géologie
Conesa, A.	Professeur Maître de Recherches I.N.R.A.	Agronomie
Delord, P.	Professeur Ingénieur en Chef du G.R.E.F.	Aménagement Foncier
Heil, E.	Professeur Professeur Agrégé	Ecologie
Heinrich, A.	Professeur Docteur Vétérinaire	Zootechnie
Hery, M.	Maître de projet - Ingénieur Divisionnaire des travaux ruraux	Constructions rurales
Hoffert, M.	Professeur Docteur ès-sciences	Géologie
Horny, R.	Professeur Ingénieur	Topographie
Kersauze, R.	Professeur Ingénieur des travaux ruraux	Statistiques
Koechlin, F.	Professeur Professeur agrégé	Mathématiques
Loeffler, J.	Ingénieur des travaux ruraux Chef de travaux pratiques	Topographie
Loudiere, D.	Professeur Ingénieur du G.R.E.F.	Résistance des matériaux et Mécanique des sols
Lucchini, P.	Professeur Ingénieur des travaux ruraux	Constructions rurales

Marilly, D.	Professeur Ingénieur	Dessin
Migault, B.	Professeur et Chef de travaux pratiques Docteur ingénieur	Informatique
Nidenberg, Simone	Professeur Docteur és Sciences	Economie
Pernes, P.	Professeur Ingénieur en Chef du G.R.E.F.	Hydraulique générale
Philippoteaux, Y.	Professeur Conseiller au Tribunal Administratif de Strasbourg	Droit
Rioe, Monique	Chef de travaux pratiques Maîtrise d'informatique	Informatique
Sellier, J-F.	Ingénieur Chef de travaux pratiques	Electrotechnique Générale et Automatismes
Thellier, D.	Professeur Professeur agrégé	Sciences biologiques
Thirion, A-G.	Ingénieur des travaux ruraux Chef de travaux pratiques	Topographie
Thirion, A-G.	Professeur Ingénieur des travaux ruraux	Evaluation des ouvrages de Génie Civil et Terrassements
Thouvenin, G.	Professeur Chef de travaux pratiques d'ENSAM	Electrotechnique Générale et automatismes
Villaneuva, H.	Professeur Maître de projet Professeur d'ENSAM	Béton armé
Weber, P.	Professeur Professeur certifié d'anglais	Langues vivantes (Anglais)
Wieser, A.	Professeur Professeur d'éducation physique	Sports

Cours de deuxième année:

Adam, J.	Professeur Ingénieur des travaux ruraux	Equipements touristiques
Belleville, J.	Professeur Ingénieur du G.R.E.F.	Calcul économique
Bertrand, F.	Professeur Maître de projet Ingénieur	Industries agricoles animales
Boulet, P.	Professeur Responsable du Département Financier au C.A.M.	Comptabilité des Entreprises
Carlier, D.	Maître de projet (ouvrages d'art) Ingénieur	Ouvrages d'art et voirie
Colin, E.	Professeur Maître de projet Ingénieur du G.R.E.F.	Hydraulique fluviale
Deguin, A.	Professeur Ingénieur chimiste	Chimie et Bactériologie des Eaux
Delaunay, J.	Professeur Ingénieur en Chef du G.R.E.F.	Industries agricoles animales
Desmartin, P.	Professeur Ingénieur en Chef du G.R.E.F.	Industries du froid
Dunglas, J.	Professeur Ingénieur en Chef du G.R.E.F.	Ouvrages d'art et voirie
Fischer, E.	Ingénieur en Chef du G.R.E.F.	Expression orale et documentation

Galindo, J.	Professeur Ingénieur	Electrotechnique appliquée et électrification rurale
Garidel Thoron, R.de	Professeur Ingénieur du G.R.E.F.	Ouvrages d'art et voirie
Gilbert, B.	Professeur Maître de projet Ingénieur des travaux ruraux	Assainissement des agglo- mérations (épuration des eaux usées)
Graff, A.	Maître de projet Ingénieur Divisionnaire des travaux ruraux	Hydraulique agricole irrigation
Herve, J-J.	Professeur Maître de projet Ingénieur du G.R.E.F.	Hydraulique agricole Assainissement des terres agricoles
Jacquin, G.	Maître de projet (ouvrages d'art) Ingénieur du G.R.E.F.	Ouvrages d'art et voirie
Javor, E.	Chef de travaux pratiques (voirie) Chef du Laboratoire Régional du C.T.G.R.E.F.	Ouvrages d'art et voirie
Koenig, Hélène	Chef de travaux pratiques Docteur en médecine	Parasitologie
Kremer, M.	Professeur Professeur titulaire	Parasitologie
Laborde, J-P.	Professeur Ingénieur hydraulicien	Hydrologie
Lavillaureix, J.	Professeur Professeur titulaire	Prophylaxie générale et législation sanitaire
Leonce, R.	Professeur Ingénieur du G.R.E.F.	Hydraulique agricole irrigation
Leveau, J-Y.	Professeur Ingénieur	Chimie Organique appliquée et Biochimie
Lieb, J.	Professeur Ingénieur	Electrotechnique appliquée et électrification rurale
Portier, J-C.	Maître de projet Ingénieur	Industries du froid
Protsche, G.	Maître de projet Ingénieur Géologue	Hydrologie
Roche, E.	Professeur Ingénieur en Chef du G.R.E.F.	Adduction d'Eau
Roche, E.	Professeur Ingénieur en Chef du G.R.E.F.	Assainissement des agglo- mérations (réseaux)
Salado, J.	Professeur Ingénieur Hydrogéologue	Hydrogéologie
Schaffer, P.	Professeur Docteur en Médecine	Prophylaxie générale et législation sanitaire
Thirion, A-G.	Maître de projet Ingénieur des travaux ruraux	Assainissement des agglomérations (réseaux)
Vellaud, J-P.	Professeur Ingénieur du G.R.E.F.	Traitement des déchets solides
Weber, P.	Professeur Professeur certifié d'anglais	Langues vivantes (anglais)
Wendling, A.	Maître de projet Ingénieur des travaux ruraux	Adduction d'eau
Wieser, A.	Professeur Professeur d'éducation physique	Sports

Structure: Date de creation: 1960. Ecole Publique d'Enseignement Supérieur, dépendant du Ministère de l'Agriculture, Direction Générale de l'Enseignement et de la Recherche.

Activités concernant l'Etablissement: Cours Scientifique et de formation générale de l'Ingénieur, Cours de techniques générales et appliquées (Aménagements Fonciers, constructions rurales, machinisme agricole, Electrotechnique et Electrification rurale, hydraulique agricole, adduction d'eau, assainissement, ordures ménagères, industries agricoles animales et végétales). Cours de formation juridique, économique, administrative. Enseignement sanitaire. - Sanction des études: diplôme d'Ingénieur des Techniques de l'Equipement Rural, reconnu par la Commission des Titres d'Ingénieur (avec Mention Génie Sanitaire pour les Elèves ayant accompli la troisième année de scolarité à l'Ecole Nationale de la Santé Publique à Rennes). - Durée des études: 3 ans. - Recrutement normal sur concours (deux années de préparation au moins, dans les classes préparatoires aux Grandes Ecoles). Admission exceptionnelle sur titres; l'admission au concours d'entrée est reconnue équivalente au diplôme universitaire d'études scientifiques (D.U.E.S.) - Nombre d'élèves diplômés par promotion: 50 environ: Ingénieurs Travaux Ruraux (fonctionnaires); Ingénieurs des Techniques de l'Equipement Rural (civils) - Activités de formation professionnelle continue dans de nombreux domaines de l'équipement Rural.

Publications: Cours et documents réservés aux besoins internes du service et mémoires de diplôme correspondant à des études effectuées dans des centres de recherche et bureaux d'études.

Correspondance: français.

GAMBIA

Yundum Experimental Station, Western Division, Republic of the Gambia

The Station's programme includes applied research into farm power and machinery, especially simple animal drawn equipment and storage of agricultural products with emphasis on the entomological aspects.

GERMAN DEMOCRATIC REPUBLIC

Berlin

Institut für landwirtschaftliches Maschinen- und Bauwesen der Humboldt-Universität zu Berlin (Farm Machinery and Building Institute, Humboldt University)
Invalidenstrasse 42, 104 Berlin

Outposted Branches: Borkholz and Blumberg.

Status: Governmental Institution supervised by the Ministry of Higher and Professional Education.

Activities: Study courses attended by about 200 students per year - Investigations on the mechanization and technology of the harvesting threshing of cereal crops.

Publications: Textbook on Farm Machinery - Reports issued in the periodicals "Archiv für Landtechnik", "Deutsche Agrartechnik" and "Feldwirtschaft".

Correspondence: German, Russian, English, Bulgarian.

Kammer der Technik – Fachverband Land-, Forst- und Nahrungsgütertechnik (Chamber of Technology, Association for agricultural, forestry and foodstuff engineering)
DDR - 1086 Berlin, Clara-Zetkin-Str. 115-117

Bostelmann, O.	Dr.Agr.	President
Böldicke, H.	Chief Engineer	Vice-President and Secretary

Status: Central Branch organ of the Presidency of the Chamber of Technology founded 1953. The Chamber of Technology (KDT) is the socialiste union of the scientists, engineers, economists and innovators in the German Democratic Republic and was founded in 1946.

Activities: Organization of socialist collective work in the interest of solving scientific and technical tasks. Organization of scientific technical conferences, symposia and courses in the field of agricultural machine building and mechanization of plant and animal production.

Publications: Editor of the periodical "Agrartechnik".

Correspondence: German, Russian, English.

Bauakademie der Deutschen Demokratischen Republic – Institut für Landwirtschaftliche Bauten (Building Academy of the German Democratic Republic - Institute for Agricultural Constructions)

Lammert,, T.	Professor Dr.-Ing.	Director of the Institute
Heinig, W.	Professor Dr.-Ing.	Deputy Director

Status: Research Institution of the German Building Academy since 1952.

Activities: Research and development work in building in rural areas – Functional, constructive technical foundations for agricultural plant production and storage – Investigations on development and re-designing of rural settlements – Housing development in rural areas – Norms and standardization.

Publications: Functional solutions for new buildings and re-construction of cattlesheds and piggeries – Project systems for new buildings and for the reconstruction of agricultural production plants and installations – Functional solutions for constructions and parts thereof, especially for barn-buildings, silos and containers; Project systems for the erection of buildings and installations for production plants – Guidelines for construction plans – Guidelines for safety measures for the installations – Directives for village planning – Projections for the erection of rural housing and institutional buildings – Handbooks on: construction plans for rural buildings and installations – Planning and constructive solutions for milking parlours – Feedlots for cattle and pigs.

Correspondence: German, Russian, English, French.

Dresden

<u>Technische Universität Dresden - Sektion Kraftfahrzeug-, Land-, und Fördertechnik</u>
(Technical University Dresden - Section for Farm Machinery, Land and Development Techniques)
8027 Dresden, Mommsenstrasse 13, DDR

Soucek, R.	Professor Dr.-Ing.	Head, Scientific Section for Farm Machinery
Thurm, R.	Professor Dr.agr.habil.	Scientific Section for Agr.Technology
Götz, I.	Professor Dr.sc.techn.	Section for Maintenance
Hofman, K.	Professor Dr.-Ing.habil.	Section for Tractors and Agricultural Machines.

<u>Status</u>: Founded in 1953 as part of the Faculty for Engineering of the Technical University.

<u>Activities</u>: Courses leading to a degree of Dipl.Ing. in farm machinery construction and in engineering installations, agricultural mechanization and agricultural machines maintenance - Theoretical and experimental research in the field and on the farm.

<u>Publications</u>: Issued yearly about 30 publications, mainly on: "Agrartechnik" (agricultural techniques) and scientific bulletins of the Technical University of Dresden. Textbook on construction of agricultural machinery.

<u>Correspondence</u>: German, Russian, English, Spanish.

<u>Institut für Landtechnische Betriebslehre, Technische Universität</u> (Institute of Agricultural Engineering Management, University of Technology)
Bergstrasse 120, 8020 Dresden

<u>Status</u>: Governmental institution supervised by the Ministry of Higher and Professional Education, established in 1953.

<u>Activities</u>: Study courses concerning the technology and engineering management of agricultural production processes.

<u>Research on</u>: Mechanization of grain harvest - Farm building construction and equipment, including the use of electricity - Cost calculations of farm machinery use.

<u>Correspondence</u>: German, English, Russian.

Halle

Institut für landwirtschaftliche Maschinen- und Gerätekunde der Martin-Luther Universität
(Institute for Agricultural Maschinery and Implements of the Martin-Luther University)
Ludwig-Wucherer-Str. 80/81, 402 Halle (Saale)　　　　　　　　Tel.: 38061

Status: Governmental institution established in 1920 and supervised by the Ministry of Higher and Professional Education.

Training activities: 4 semesters of courses and seminars on all problems of agricultural engineering attended by an annual average of 150 students of agriculture at the Faculty of Agriculture of the Martin-Luther-University, Halle-Wittenberg. Successful examination in "Agricultural Engineering" is one prerequisite for graduation in Dipl.Agr.

Research activities: Mechanization of beet production (tillage seedbed preparation, conditioning of seeds, drilling, crop cultivation, harvesting, processing of harvested crops).

Publications: Seedbed preparation for sugar beets - Spacing of sugar beets with appropriate seeding techniques and thinning and singling methods - Mechanized crop cultivation - New sugar beet topping and lifting mechanism - Sugar beet harvesting in one storage with two machines - Sugar beet transport and transport means - Problems of sugar beet storage - Harvesting and storage of maize - Mechanized removal of manure - Problems of agricultural machinery maintenance - Motor testing - Tractor testing service.

Correspondence: German, English, French, Russian.

Leipzig

Institut für Landtechnik der Karl-Marx-Universität (Farm Machinery Institute of the Karl-Marx University)
Johannisallee 19, 701 Leipzig

Activities: Training courses and seminars leading to the Degree of Dipl.agr. - Farm mechanization - Mechanization in animal husbandry - Mechanization of agriculture in tropical and subtropical regions. - Research work on: Automation of mechanical milking - Mechanization of manure handling in large barns without litter - Tractor and farm machinery control and maintenance service stations.

Correspondence: German, Russian, English.

Institut für Landmaschinentechnik (Agricultural Machinery Institute)
am Lausner Weg, 7031 Leipzig　　　　　　　　　　　　　　　　Tel. 44701

Status: State institution, supervised by the Association of publicly owned Enterprises for the Manufacture of Agricultural Machinery (Vereinigung volkseigener Betriebe des Landmaschinenbaues).

Activities: Fundamental research, experiments and testing - Research on implements and machines for the mechanization of land development work - Mechanization of tillage work - Investigations on and development of machines for sowing, fertilizing, intercultivating and plant protection - Mechanization of work on the farmstead - Equipment for the processing of agricultural products.

Publications: Reports issued in the periodicals "Deutsche Agrartechnik", "Archiv für Landtechnik", "Der Maschinenbau" and "Deutsche Landwirtschaft".

Correspondence: German, English, French, Russian.

Schlieben

Forschungszentrum für Mechanisierung der Landwirtschaft Schlieben-Bornim
7912 Schlieben, Gartenstrasse 30, DDR

Prof.Dr.Agr. Algenstaedt   Director

Status: Governmental institution, established in 1951 and supervised by the Government Academy of Agricultural Sciences.

Activities: Research on mechanization of land development, potato growing, forage crops, animal husbandry and the use of radioactive isotopes in agricultural engineering.

Publications: Reports on the above subjects.

Correspondence: German, Russian, English, French.

Rostock

Institut für Landtechnik, Universität (Institute of Agricultural Engineering, University)
Satower Strasse, 25 Rostock, DDR

GERMANY
(Federal Republic of Germany)

Aachen

Institut für Wasserbau und Wasserwirtschaft, Rheinisch-Westfälische Technische Hochschule Aachen (Institute of Water Resources and Hydraulic Engineering, Technical University Aachen)
5100 Aachen, Mies-van-der-Roche-Strasse 1

Rouvé, G.	Prof.Dr.Ing.	Director
Evers, P.	Dipl.Ing.	
Gerber, K-H.	Dipl.Ing.,	Baudirektor
Indlekofer, H.	Dr.Ing.	
Kerzel, Chr.	Dipl.Ing.	
Krause-Klein, T.	Dipl.Ing.	
Nehlisen, W.	Dipl.Ing.	
Ostrowski, M.	Dipl.Ing.	
Pelka, W.	Dipl.Ing.	
Rhode, F.G.	Prof.Dr.Ing.	
Sacher, H.	Dipl.Ing.	
Stein, U.	Dipl.Ing.	
Stössinger, W.	Dr.Ing.	
Terhoeven, S.	Dipl.Ing.	
Traut, F-J.	Dipl.Ing.	

Branch: Laboratory of Hydraulics Research (Tech.University), Kreuzherrenstrasse, 5100 Aachen.

Status: Established 1955. Governmental, Ministry of Science, Land Nordrhein-Westfalen.

Agric.Engineering activities: Teaching: Study courses (lectures) in agricultural engineering, water resources, irrigation and drainage, hydraulics, system engineering. Research: Experiments with Hele-Show Model, experiments for construction of reservoirs, hydraulics research, mass- and heat transport, groundwater optimization.

Publications: Mitteilungen Institut für Wasserbau und Wasserwirtschaft.

Correspondence: German, English.

Berlin

Normenausschuss Bauwesen in DIN Deutsches Institut für Normung e.V. (Building Division of the German Standards Institute)
D-1000 Berlin 30, Burggrafenstrasse 4, Postfach 1107   Tel.: (030) 2601 501

Pohlmann	Dipl.Ing.	Chairman
Bub	Prof.Dr.Ing.	President of Institut für Bautechnik, Berlin
Boxberger	Dr.-agr.	Stable equipment and floors
Stietenroth	Dr.Ing.	Stable climate
Widderich	Dipl.Ing.	Greenhouse building and horticultural glass
Tietze	Architect	Cribs and troughs
Kirstein	Professor	Type construction with wooden poles
Grimm	Dr.Ing.	Silos

Status: Building Division of the German Standards Institute.

Agricultural Engineering activities: Setting up of standards.

Publications: Horticultural glass; horticultural sheet glass (DIN 11 525) – Horticultural glass; horticultural cathedral glass (DIN 11 526) – Greenhouses; directives for design and construction (DIN 11 535) – Greenhouse of 12 m nominal width in hot-galvanized structural steel work (DIN 11 536) – Silos for ensilage, dimensioning and construction; general directives for tower and underground silos (DIN 11 622) – Stoneware elements for stable construction; mangers, tiles, troughs; sizes, requirements, testing (DIN 18 902) – Floors for stable plants; construction, requirements (DIN 18 907) – Floors for stable plants, slatted floors; dimensions, requirements, execution (DIN 18 908 draft) – Climate in closed stables; water vapour and heat conditions during the winter; ventilation, lighting (DIN 18 910).

Correspondence: German, English, French.

Technische Universität Berlin, Institut für Maschinenkonstruktion, Bereich Landtechnik und Baumaschinen (Technical University of Berlin, Institute of Mechanical Engineering, Department of Agricultural Engineering)
D-1000 Berlin33, Zoppoter Str. 35

Gülich, H.	Dr.Ing.Prof.	Director of Department
Jensen, U.	Dipl.Ing.Prof.	Fluid power Div.
Poppy, W.	Dr.Ing.Prof.	Construction Machinery Div.
Dankwardt, U.	Dipl.Ing.	Safety of Construction Machinery
Dörries, U.	Ing.(grad.)	Instrumentation and Measurement
Höck, J.	Dipl.Ing.	Farm Machinery
Jungerberg, H.	Dipl.Ing.	Computation
Mertins, K-H.	Dipl.Ing.	Tractor Performance
Schmidt, M.	Dipl.Ing.	Farm Machinery
Schütz, F.	Dipl.Ing.	Tractor Performance
Selcan, Z.	Dipl.Ing.	Farm Machinery
Stanev, D.	Dipl.Ing.	Farm Machinery
Ulrich, A.	Dipl.Ing.	Tractor Performance
Weigelt, H.	Dipl.Ing.	Tractor and Machinery

Status: Institution of the Land Berlin (West).

Agric.Engineering activities: Teaching: Agricultural machinery – design and application – Terrain vehicle systems – Tractor and earth moving machinery – Hydraulics. Degree: Dipl.Ing. Dr.Ing. Dr.-agr.
Research: Vehicle dynamics – whole body vibration – Power utilization and fuel consumption – Crop protection, spraying techniques – hydraulic components evaluation.

Publications: Fertilizer distribution and handling – Plant protection engineering – Particles analyzing methods – Bioengineering – Physical properties of liquids – Dynamic behaviour of tractor and driver seats – Tractor performance – Mechanization studies in underdeveloped countries.

Correspondence: German, English.

Institut für Kulturtechnik und Grünlandwirtschaft (Institute of Land Improvement, Land, Water and Grassland Management)
1 Berlin 33, Lentzeallee 76

Status: Established in 1913. Financed by Technische Universität, Berlin.

Activities: Techniques of landscape building (engineering surveys, soil mechanics, water regulation, soil improvement, engineering biology, Waste water and sewage treatment, Water conservation, special buildings). Degree: Diplom-Ingenieur, 20.

Publications: More than 50 publications since 1965 (articles, monographs, dissertations) mainly published or mentioned in "Zeitschrift für Kulturtechnik und Flurbereinigung".

Correspondence: German, English.

Bonn

Referat "Technik und Bauwesen in der Landwirtschaft", Bundesministerium für Ernährung, Landwirtschaft und Forsten (Agricultural Engineering and Farm Building Section, Federal Ministry of Food, Agriculture and Forestry)
P.O.B. 14 02 70   D-5300 Bonn 1

Stutterheim, W.	Dr.agr.	Regierungsdirektor
Dittrich, H.	Dipl.Arch.	Regierungsdirektor (Farm Building)
Reiser, A.	Dipl.Ing.agr.	Regierungsdirektor (Agricultural Engineering)

Other Branches: Bundesforschungsanstalt für Landwirtschaft (FAL) D-3300 Braunschweig-Völkenrode: - Institut für landtechnische Grundlagenforschung - Institut für Betriebstechnik - Institut für Technologie - Institut für landwirtschaftliche Bauforschung. -- Kuratorium für Technik und Bauwesen in der Landwirtschaft (KTBL) D 6100 Darmstadt 12 -- Deutsche Landwirtschaftsgesellschaft (DLG) Fachbereich Landtechnik, D-6000 Frankfurt.

Status: FAL: financed by the Government - KTBL: financed by the Government - DLG: financed by the Government and DLG.

Activities: See FAL.

Publications: - Agricultural Engineering in the Federal Republic of Germany - Farm Buildings in the Federal Republic of Germany.

Correspondence: German, English, French.

Institut für Städtebau, Bodenordnung und Kulturtechnik der Rheinischen Friedrich-Wilhelms-Universität Bonn (Institute for Town Planning, Land Organisation and Land Reclamation, Friedrich-Wilhelm University)
Nussallee 1, 5300 Bonn   Tel.: 0228 / 732619; 737498; 732619

Borchard, K.	Dr.Ing.Prof.	Director, Town Planning and Settlement
Seele, W.	Dr.Ing.Prof.	Director, Real Estate Affairs and Land Economics
Strack, H.	Dr.Ing.Prof.	Traffic, Town and Regional Planning
N.N.	Prof.	Land Reclamation

Fritz, J.	Dr.Ing.	Town Planning and Land Settlement
Radermacher	Dr.Ing.	Land Reclamation
Rieser, A.	Dr.Ing.	Land Reclamation
Bartel, G.	Dipl.Ing.	Land Organisation
Clever, J.	Dipl.Ing.	Land Organisation
Gehrke, W.	Dipl.Ing.	Town Planning and Land Settlement
Müller, R.	Dipl.Ing.	Land Organization
Streich, B.	Dipl.Ing.	Town Planning and Land Settlement

Status: Governmental institution, established in 1903, and supervised by the Ministry of Science of the Land Nordrhein-Westfalen.

Activities: Study courses in town and country planning, road planning, land economics, real estate affairs, land reclamation, development of water resources. Degree awarded: Dipl.Ing. in surveying. Average number of students: 400. Post-graduate seminars. – Research on: Town planning problems in rural and urban areas – Regional planning – Village renewal – Development of natural resources – Traffic planning – Real estate affairs – Land economics – Land Organization – Hydraulics of weed-infested drainage canals – Rationalization of the maintenance of drainage canal systems – Improvement of water-logged soils.

Publications: Numerous publications on above research subjects.

Correspondence: German, English.

Institut für Landtechnik der Rheinischen Friedrich-Wilhelms-Universität Bonn (Institute of Agricultural Engineering, Friedrich-Wilhelms-University of Bonn)
Nussallee 5, D-5300 Bonn 1　　　　　　　　　　　　　　Tel.: 0228/732395
　　　　　　　　　　　　　　　　　　　　　　　　　　　　Telex: 886657 unibo d (707)

Brinkmann, W.	Prof.Dr.Ing.	Head of Institute	Precision drills, mechanization of sugar beet production. Grading and calibration of maize seeds and sugar beet seeds.
Flake, E.	Dipl.Ing.agr. Asst.		Precision drills for sugar beet seeds. Seedbed preparation.
Gehlen, W.	Dr.agr.	Asst.	Mechanization of seedling and harvesting of sugar beets.
Tabesch, F.	Prof.Dr.agr.	Lecturer	Agricultural engineering in tropic and subtropic regions.

Subdivision: Agricultural production systems:

Heege, H.J.	Prof.Dr.agr.	Lecturer	Planting of small grains. Methods of handling fertilizers. Heatpumps in Agriculture.
Simons, D.	Dr.agr.	Asst.	Energy recovery from slurry treatment processes.

Subdivision: Household Equipment:

Schätzke, M.	Prof.Dr.agr.	Lecturer	Energy in household. Home economics. Working methods in household.
Ritterbach, U.	Dr.troph.	Asst.	Methods of cooking.

Branches with other Location: Kiel, Göttingen, Braunschweig, Giessen, Stuttgart-Hohenheim, Freising-Weihenstephan.

Status: Governmental Institute, supervised by the Ministry of Science of the Land Nordrhein-Westfalen.

Agricultural Engineering activities: a) Teaching: Study courses within the curriculum for students of Agricultural Science and Domestic Science leading to the degrees Dipl.Ing.agr., Dipl.Ing.-troph., Dr.agr. and Dr.troph. b) Research: Mechanization of seeding and harvesting sugar beets, Planting of small grains, Seedbed preparation. Methods of handling fertilizers. Energy recovery from slurry treatment processes. Working methods in household. Economizing of energy in household. Methods of cooking.

Publications: Articles concerning the research are published in the following periodicals: Landtechnik - Grundlagen der Landtechnik - Berichte über Landwirtschaft - Zuckerindustrie - Die Zuckerrübe - Hauswirtschaft und Wissenschaft - Haustechnik.

Correspondence: German, English, French.

B r a u n s c h w e i g

Fachgruppe für Anwendungstechnik in der Abteilung für Pflanzenschutzmittel und Anwendungstechnik der Biologischen Bundesanstalt für Land- und Forstwirtschaft (Application Techniques Division of the Department for Plant Protection Products and Application Techniques. Federal Biological Research Centre for Agriculture and Forestry)
Messeweg 11/12, D-3300 Braunschweig   Tel.: 0531/399291

Kohsiek, H.	Dr.Ing.	Director of the Application Techniques Div.
Rietz, S.	Dipl.Ing.	Scientific staff member

Status: Federal institution established in 1950 as testing station for plant protection equipment, directed and financed by the Federal Ministry of Agriculture.

Activities: Investigation on problems concerning the application of plant protection products and the use of plant protection equipment. - Federal testing station of equipment for plant protection.

Publications: List of approved equipment for plant protection: "Pflanzenschutzmittel-Verzeichnis, Teil 6, Anerkannte Pflanzenschutz-und Vorratsschutzgeräte", published annually. Test reports on approved plant protection equipment. - Investigations on the application of plant protection products.

Correspondence: German, English.

Institut für landtechnische Grundlagenforschung der Bundesforschungsanstalt für Landwirtschaft (FAL) (Institute of Basic Research in Agricultural Engineering of the Federal Research Centre of Agriculture (FAL)
Bundesallee 50, 3300 Braunschweig-Völkenrode, Federal Republic of Germany

Batel, W.	Prof. Dr.Ing.	Protection of labour and environment, energy research
Graef, M.	Dipl.Ing.	Protection of labour and environment, energy research
Hardegen, B.	Dipl.Ing.	Protection of labour and environment, energy research
Hinz, T.	Dipl.Ing.	Protection of labour and environment, energy research
Jahns, G.	Dr.Ing., Dipl.-Wirtsch.-Ing.	Protection of labour, automatic control, accident research
Janssen, J.	Dipl.Ing.	Protection of labour and environment
Krause, K.-H.	Dr.Ing.	Protection of labour and environment, simulation methods
Mejer, G.-J.	Dipl.Ing.	Protection of labour and environment, energy research
Paul, W.	Dr.Ing.	Automatic control, simulation methods
Schoedder, F.	Dr.sc.agr. Dipl.Ing.	Energy research, scientific journal: "Grundlagen der Landtechnik"
Speckmann, H.	Dipl.Ing.	Automatic control
Witte, E.	Dr.Ing.	Protection of labour and environment
Vellguth, G.	Dipl.Ing.	Energy and accident research

Status(old): Federal institution, established in 1948, supervised and financed by the Ministry of Food, Agriculture and Forestry.

Agriculture engineering activities: (completed) - (a) Teaching: Prof.Dr.Ing. W. Batel Lectures on "Technology of dust removal" at the Technical High School in Aachen, 10 Students. (b) Research: - on fundamentals of agricultural engineering and engineering design (kinematics, stresses, automation), - on protection of labour and environment (air cleaning, loads by particulate and gaseous emissions, prognoses of pollutant immissions, protection from climatic, noise and vibration load, humanization of work, accident research), - on reduction of energy requirements (process technology, rational energy use).

Publications: More than 230 articles and reports, the bibliographical notes of which are contained in the annual reports on the activities of the Institute for 1965-1980. 15 of these reports are monographs, such as dissertations. Books on: Introduction to grain measurement techniques - Technology of dust removal - Mechanism-atlas for adjustable oscillating motions - Atlas for mechanism design.

Correspondence: German, English.

Institut für landtechnische Grundlagenforschung der Bundesforschungsanstalt für Landwirtschaft (FAL) (Institute of Basic Research in Agricultural Engineering of the Federal Research Center of Agriculture (FAL)
Bundesallee 50, 3300 Braunschweig-Völkenrode, Federal Republic of Germany

Batel, W.	Prof. Dr.Ing. (mech.)	
Ahlgrimm, H.-J.	Dr. rer.nat (Physics)	Properties and behaviour of materials
Dernedde, W.	Dipl.Ing. (mech.)	mech. treatment (plants)
	Dr. sc.agr.	solar energy

Heine, E.	Dipl.Biol.	bioconversion (biogas)
Kloss, R.	Dipl.Ing.	bioconversion (biogas)
Krause, R.	Dr.Ing. (mech.)	mech.treatment (soil)
Orth, H.W.	Dr.Ing. (mech.)	process engineering, heat transfer
Schuchardt, F.	Dr.agr.	bioconversion (biomass)
Sonnenberg, H.	Dipl.Ing. (mech.)	mech. treatment, drying
Würch, H.-H.	Dipl.Ing. (agr.)	bioconversion (biogas)

Status: Federal institution, established in 1948, supervised and financed by the Federal Ministry of Food, Agriculture and Forestry.

Activities: Research: engineering problems in Material treatment in agricultural production: disposal and incorporation of organic wastes into the soil, mechanical and thermal treatment of forage, improvement of processes with regard to energy demand. Processing and conversion of biomass: animal wastes, sewage sludge and plant material for pollution control, and to produce fertilizer, animal feed, fuels (e.g. biogas) and raw material for non-food purposes. Applicability of non-fossil energy: bio-energy, solar energy.

Publications: Brief review of the recent activities in the fields mentioned above and a summary of the results of the actual research work are published in the annual report of the Research Centre; origin publications in scientific periodicals; papers and reports are presented at national and international meetings.

Correspondence: German, English.

Institut für Betriebstechnik der Bundesforschungsanstalt für Landwirtschaft Braunschweig-Völkenrode (FAL) (Institute for Production Engineering, Federal Research Centre for Agriculture Braunschweig-Völkenrode)
Bundesallee 50, D-3300 Braunschweig

Schön, H.	Dr.agr. Prof.	Director	Work and production methods
Artmann, R.	Dipl.-Ing.agr.		Animal production methods
Biller, R.	Dipl.Ing. Dipl.-Wirtsch-Ing.		Plant production methods
Fischer, F.-W.	Dipl.-Ing.agr.		Animal production methods
Hammer, W.	Dr.agr.		Animal production methods
Olfe, G.	Dipl.-Ing.agr.		Plant production methods
Schlünsen, D.	Dr.agr.		Animal production methods
Sourell, H.	Dipl.-Ing.agr.		Plant production methods
Steinkampf, H.	Dr.Ing.		Plant production methods
Thomé, F.-J.	Dipl. Psych.		Ergonomics

Status: Federal institution, established in 1948, supervised and financed by the Ministry of Food, Agriculture and Forestry.

Activities: Physiological assessment of workers in relation to various agricultural production processes - cost of using farm machinery - equipment and economy of irrigation - animal husbandry - methods of dairy farming - traction of farm tractors - power transmission - tractor implement.

Publications on the above mentioned subject matters are issued in the following periodicals: "Grundlagen der Landtechnik", "Landtechnik", "VDI-Zeitschrift", "Mitteilungen der DLG", "Landbauforschung Völkenrode".

Correspondence: German, English, French.

Institut für Pflanzenbau und Pflanzenzüchtung der Bundesforschungsanstalt für Landwirtschaft (Institute of Crop Science and Plant Breeding, Federal Research Centre of Agriculture)
Bundesallee 50, D-3300 Braunschweig, Federal Republic of Germany   Tel.: 0531/596307

Dambroth, M.	Prof. Dr.	Director of the Institute

Divisions of waterhousehold of cultural plants and soil tillage

Bramm, A.	Dr.	Waterhousehold
El Bassam, N.	Dr.	Physiology of yield
Sommer, C.	Dr.	Waterhousehold and physics of soil tillage
Zach, M.	Dipl.Ing.	Soil tillage and plant growth

Status: Federal institution, established in 1948, supervised and financed by the Federal Ministry of Food, Agriculture and Forestry (Bundesminister für Ernährung, Landwirtschaft und Forsten - BML).

Activities: Plant-water relationships - Research on irrigation scheduling - Investigations of soil tillage effects with respect to plant growth - Interactions between soil properties, soil water content and root growth.

Publications: on the above mentioned subject matters are issued in the following periodicals: "Landbauforschung Völkenrode", "Soil and Tillage Research", "Mitteilungen der DLG", "Landtechnik", "Zeitschrift für Acker- und Pflanzenbau".

Correspondence: German, English, French, Spanish, Czech.

Institut für landwirtschaftliche Bauforschung der Bundesforschungsanstalt für Landwirtschaft Braunschweig-Völkenrode (FAL) (Institute for Farm Building Research of the Federal Research Centre of Agriculture Braunschweig-Völkenrode (FAL)
Bundesallee 50, 3300 Braunschweig, Federal Republic of Germany   Tel.: 0531/596-525

Piotrowski, J.	Prof. Dr.	Director
Achilles, S.	Bau-Ing., Architekt	Planning and construction
Borchert, K.-L.	Bau-Ing., Architekt	Buildingphysics and ventilation of livestock buildings
Gartung, J.	Bau-Ing., Architekt	Building costs
Hagemann, D.	Dr.iur.	Building law
Herms, A.	Dr.agr.	Influence of regional development
Hillendahl, W.	Dipl.Ing., Architekt	Planning and construction
Irps, H.	Dr.Ing., Dipl.-Phys.	Animal behaviour and farm buildings equipment
Krentler, J.-G.	Dipl.Ing., Dipl.Wirtsch.-Ing.	Building costs

(Total staff: 35)

The Institute for Farm Building Research is part of the Federal Research Centre with 14 institutes including the Research Centre for Small Animals (poultry) in Celle and the Institute for Animal Husbandry and Animal Behaviour in Mariensee, with research farms in Mecklenhorst, Trenthorst and Wulmenau.

Status: Federal institution, established in 1948, supervised and financed by the Federal Ministry of Food, Agriculture and Forestry.

Activities: The activities of the Institute are directed to research only, there are no teaching activities. Research: 6 scientific working groups with close internal connection and corresponding tasks. They are assisted by a laboratory and a workshop. - function, planning and construction of farm buildings - building costs - building physics and ventilation of livestock buildings - equipment in farm buildings, especially livestock buildings - influence of regional development to farmsteads - building-law (regulations and standards). Planning and design of farm buildings, low-cost constructions and do-it-yourself activities, animal behaviour as the key for designing the outfittings of buildings for animal husbandry, inside climate and ventilation of animal shelters, storage buildings for farm purposes including animal-wastes, investment-costs and expenses for buildings, location of farmsteads in villages and in the countryside, the law and regulations for farming and farm buildings.

Publications: Approximately 60 publications on the above mentioned research subjects are presented every year. The list of publications is part of the annual report of the "Bundesforschungsanstalt für Landwirtschaft Braunschweig-Völkenrode (FAL)" and part G of the BML annual report. Books and reprints may be ordered via KTBL-Schriftenvertrieb im Landwirtschaftsverlag 4400 Hiltrup-Münster.

Correspondence: German, English.

Leichtweiss-Institut für Wasserbau, Technische Universität Braunschweig (Leichtweiss-Institut for Water Research, Technical University, Braunschweig)
P.O. Box 3329, (33) Braunschweig   Tel.: 0531/ 391 3940

Department of Hydraulics, Water Resources and Irrigation
Garbrecht, G.   o.Prof. Dr.Ing.

Department of Hydrodynamics and Coastal Engineering
Führböter, A.   o.Prof. Dr.Ing.

Hydrology Section
Maniak, U.   Prof. Dr.Ing.

Irrigation and Drainage Section
Collins, H.J.   Prof. Dr.Ing.

Soil Science Section
Schaffer, G.   apl.Prof. Dr.agr.

Scientific staff in the five groupings: about 25

Status: Institute of the Technical University, Braunschweig.

Teaching activities: Study courses of 8 semesters duration in Civil Engineering with specialization in Water Resources Development, Graduation to Dipl.Ing., Ph.D. Studies.
Research on: Irrigation methods - irrigation of salts in soils under arid climates - effects regional planning - influence of various treatments on quality and quantity of seepage out of waste deposits - testing devices for drain pipes and filter material - optimization of irrigation projects.

Publications: Numerous publications on the above research subjects.

Institut für Landmaschinen der Technischen Universität Braunschweig (Agricultural Machinery Institute, Technical University)
Langer Kamp 19a, 33 Braunschweig   Tel.: 0531/3912670

Status: Governmental Institution of Land Niedersachsen

Teaching: Agricultural Machinery; Agricultural Tractors; Pneumatic Conveying (Diplom Ingenieur, Doktor Ingenieur courses).

Research: Physical properties of agricultural products; silo and storing problems; tractor hydraulics and gears; obstruction of grain flow in spouts.

Publications: Stress analysis of grain in bins and silos – Physical properties and compression behaviour of roughage – Compression of roughage: fundamentals, high axial pressure, rolled wafers – Overhead throw of bodies from a belt conveyor – Air flow resistance of roughage – Investigation on thrower, blower, conveying – Drying of rolled wafers with heated air – Power losses and efficiency of mechanical and hydrostatic gears for tractors and self propelled agricultural machines.

D a r m s t a d t

Institut für Wasserbau und Wasserwirtschaft, Technische Hochschule Darmstadt (Institute for Hydraulic Engineering and water Resources, Technical University, Darmstadt)
Rundeturmstr. 1, 6100 Darmstadt.   Telex: 419579   Tel.: 06151/162523

Mock, F.J.	Dr.Ing.Prof.	Head of Department Hydraulic Engineering and Water Resources, Irrigation, Drainage Flood Control.
Schröder, W.	Dr.Ing.Prof.	River Training, Coastal Engineering
Börner, R.	Dr.Ing.	Hydro-Power

Status: Governmental Institution, established in 1961 and financed by governmental and semi-governmental organizations.

Activities: a) Teaching: Courses in engineering (water resources, irrigation, drainage, river training, flood control, hydraulic structures) leading to the degree of Dipl.-Ing. (civil and survey), consulting in the mentioned subjects. b) Research: Biological river training measures, water in the non saturated soil zone, monitoring of irrigation and drainage systems operation and maintenance, hydraulic model tests.

Publications: Series "Wasserbau-Mitteilungen", published since 1966 twenty volumes. Content: all activities of the Institute.

Correspondence: German, English.

Deutsche Lehranstalten für Agrartechnik (German Farm Machinery Schools)
6100 Darmstadt-Kranichstein, Bartningstr. 49   Tel.: 06151/75006

Otto, F-K. Dr.   Secretary
Lohde, H-H. Dipl.Ing.agr. Deputy Secretary

Affiliated Schools: School Managers:
2370 Rendsburg, Am Kamp, Te.: 04331/88026   Dir. Dirk Kraemer
2910 Westerstede, Max.Eyth-Str., Te.:04981/3066/7   Dir. Manfred Hüniken

4452 Freren, Bahnhofstr. 67, Tel.: 05902/309 and 1722	Dir. Dirk Sauer
3070 Nienburg, Max-Eyth-Str. 2, Te.: 05021/3015	Schulleiter Wulf Petram
3200 Hildesheim, Lerchemkamp 42, Te.: 05121/52690	Dir. Wilfried Querfurth
4410 Warendorf, An der Tönneburg 2, Te.: 02581/1551	Dir. Friedrich Grothus
4152 Kempen, Krefelder Weg 41, Tel.: 02152/2363	Dir. Gerhardt Schalm
3430 Witzenhausen, Am Sande 20, Te.: 05542/4026	Dir. Gerhard Kunigk
6762 Alsenz, Schulstr., Te.: 06362/465[x]	Dir. Erich-Johann Kühne
8785 Hammelburg, Von-der-Tann-Strasse, Tel. 09732/2205	Dir. Bernhard Olbrich
8462 Neunburg, Amberger Strasse 23/25, Tel.: 09672/1349	Dir. Peter Görke
8050 Freising, Liebigstr. 4, Tel.: 08161/13431[xx]	Dir. Siegfried Rudnick
7312 Kirchheim, Hahnweidstr. 101, Tel.: 07021/54831	Dir. Günther Krapkat

Status: The 13 DEULA Schools in various parts of the Federal Republic of Germany belong to the "Kuratorium für Technik und Bauwesen in der Landwirtschaft", Darmstadt. Established in 1928.

Activities: Practical training on farm machinery to farmers and related occupations, as well as training of teachers in agricultural engineering.

Teaching languages: German, English, French.

Correspondence: German, English, French.

---

[x]) as from 1 April 1982: Bad Kreuznach, Hüffelsheimerstr.
[xx]) as from 1 Jan. 1983: Freising-Weihenstephan, Wippenhauserstr.

Max-Eyth-Gesellschaft für Agrartechnik (Mac-Eyth Society of Agricultural Engineering)
6100 Darmstadt-Kranichstein, Bartningstr. 49,   Tel.: 06151/75002 to 75006

Eichhorn, H.	Prof.Dr.	Chairman
Kämmerling, H-J.	Dr.	Managing Director

Status: Registered association whose organs are: the Board of Directors, the General Assembly of the members.

Activities: Bring together persons and circles concerned with agriculture, farm buildings, improvement of land, labour economics and related subjects to promote science and technologies in the overall field of agricultural engineering, cultivate international contacts to analogous foreign organizations, carry on Max Eyth's, the poet-engineer's tradition. MEG organizes meetings, conferences, technical visits to allow experts and affiliated national and foreign organizations to exchange experiences and foster mutual stimulation in research and development work. MEG comprises five study groups:

Study group:	Name of head:
Juniors Promotion	Dr.-Ing. W. Rau
International Cooperation in Agric.Engineering	Dr. K. Lampe
C.I.G.R.	Dr. F. Meier
Research and Teaching	Prof.Dr.Ing. Kutzbach
History of Agricultural Engineering	Prof.Dr. H. Winkel

MEG represents the Federal Republic of Germany in the International Commission of Agricultural Engineering (C.I.G.R.)

Correspondence: German, English, French, Spanish.

Kuratorium für Technik und Bauwesen in der Landwirtschaft (KTBL) (Board for Agricultural Engineering and Farm Buildings)
Bartningstr. 49   D-6100 Darmstadt-Kranichstein   Tel.: 06151   75 002 - 75006

Reisch, E.	Prof.Dr.Dr.h.c.	Chairman
Gummert, H.	Dr.agr.	Secretary-General
Dohne, E.	Dr.agr.Dipl.-Ing.	Farm Machinery
Pohlmann, H.	Dipl.Ing., Architekt	Farm Buildings
Brundke, M.	Dr.agr.	Agrarian Economics
Otto, F.-K.	Dr.agr.	Secretary of DEULA
Kühner, H.	Dipl.-Ldw.	Vice-Secretary-General Public Relations

Heads of Project Groups:

Dohne, E.	Dr.agr.,Dipl.Ing.	Energy
Schirz, St.	Dr.Ing.	Pollution Control
Achilles, A.	Dr.agr.	Plant Production
Van den Weghe, H.	Dr.agr.	Animal Production
Marten, J.	Dipl.Ing.	Farm Buildings
Kadner, K.	Dipl.Ing.agr.	Multiple Farm Machinery Use
Gerlach, P.	Dipl.Ing.	Planning in the Rural Area
Brundke, M.	Dr.agr.	Calculation Elements

Branches: Deutsche Lehranstalten für Agrartechnik - DEULA (German Training Centres for Agricultural Engineering) - KTBL-Versuchsstation (KTBL Experimental Station), D-3042 Dethlingen/Munster, Te.: (05192)2282. - KTBL Fachgrouppe: Technik und Bauwesen im Gartenbau (KTBL Special Field: Horticultural Technology and Buildings), Godesberger Allee 142-148, D-5300 Bonn 2, Tel.:(0228)376878.

Status: KTBL as successor organization of the Reichskuratorium für Technik in der Landwirtschaft (RKTL), founded in 1928, and of the Kuratorium für Technik in der Landwirtschaft (KTL) - reinstated in 1946 - originated in 1969 from the integration of Bundesarbeitsgemeinschaft landwirtschaftliches Bauen (ALB-Bund). Independent, non-profit making, registered association, comprising the General Assembly of its members, the Steering Committee and the Board of Directors.

Activities: Coordination of work between institutes, organizations and individuals engaged in agriculture, science and industry for engineering and farm building development and its economic application in agriculture, horticulture, fruit-growing and viticulture. KTBL examines proposals and, where appropriate, promotes, coordinates and subsidies development projects. It assists and advises public authorities, and co-operates with international institutions on agricultural engineering and farm building subject matters.

Publications: Publication list annually issued on: "Schriften", "Arbeitspapiere", "Arbeitsblätter", "Kalkulationsunterlagen für Betriebswirtschaft". The "Schriften"(bound) cover the fields of plant and animal production as well as special fields such as horticulture, home management, planning in the rural area, pollution control, viticulture, etc. The "Arbeitspapiere"(bound) are reports on preliminary studies. The "Arbeitsblätter" (loose-leaf publications) deal with agricultural technology, plant and animal production, farm buildings, the do-it-yourself-system in agriculture, planning in the rural area. The "KTBL-Kalkulationsunterlagen für Betriebswirtschaft"(calculation elements for farm management) consist of volumes on: Estimation of work requirement in Agriculture - Costs of farm machinery and farm buildings - Calculation of costs for crop and livestock production – Farm management – Costs of household equipment and buildings – Estimation of work requirement in home economics – Estimation of work requirement in horticulture – Calculation of costs for machinery, equipment and buildings in horticulture.

Correspondence: Dutch, English, French, German, Portuguese, Rumanian, Spanish.

Frankfurt

Deutsche Landwirtschafts-Gesellschaft e.V.(DLG), Fachbereich Landtechnik (German Agricultural Society, Agricultural Engineering Division)
Zimmerweg 16, D-6000 Frankfurt am Main 1   Tel.: 0611 - 7168-0

Sonnen, F.J.	Dr.Ing.	Manager
Diemel, H.	Dr.agr.	Deputy Manager
Böttcher, G.	Dr.agr.	Advisory Services
Vegeley, F.	Dipl.Ing.agr.	Test reports

DLG-Prüfstelle für Landmaschinen Gross-Umstadt (Farm Machinery Testing Station, Gross-Umstadt)
Max-Eyth-Weg 1, D-6114 Gross-Umstadt   Tel.: 06078 - 2021

Freidank, K.A.	Dipl.Ing.	Chief
Rilling, K.E.	Dr.agr.	Deputy Chief: Field Equipment
Voss, U.	Dr.agr.	Animal production, yard and barn machinery
Lober, O.	Dipl.Ing.	OECD Tractor Test
Schäfer, E.	Dipl.Ing.	Measuring Equipment

Status: The Agricultural Engineering Division with its outposted testing station is an autonomous body within the German Agricultural Society. It is supervised by the State and subsidized by the Ministry of Food, Agriculture and Forestry.

Activities: Users test for tractors, farm machinery and implements – OECD Standard tractor tests – Farm machinery and labour management investigations – Advisory services – Lectures and meetings – Dissemination of reference material.

Publications: DLG Test Reports – OECD Test Reports – small test reports – Test experiences – Advisory scripts – Information in press, broadcasting and television.

Correspondence: German, English, French.

Geisenheim

Institut für Technik, Forschungsanstalt für Weinbau, Gartenbau, Getränketechnologie und Landespflege (Technical Department, Institute for Viticulture and Fruit and Vegetable Growing)
Geisenheim

Rühling, W.	Dipl.-Ing. Dr. Prof.	Head of Department
Mackroth, K.	Dipl.-Ing. agr., Dr.	Horticultural Engineering
Bäcker, G.	Dipl.-Ing. agr., Dr.	

Status: The Technical Department was established in 1951. The Institute is supervised by the Hessian Ministry of Culture, Wiesbaden.

Activities: (a) Teaching: training courses and seminars on: Technical design – Engineering materials – Machinery and implements – Greenhouse technique – Horticultural and viticultural engineering fundamentals – Applied engineering of orchards, vegetables and vineyards – Courses attended by an average number of 40 students. Degree awarded: Dipl.Ing.(FH). (b) Research, testing and extension work – Vineyard machinery and power – Mechanization for steep slope vineyards – Horticultural machinery and power – Greenhouse devices – Plastic materials in horticulture – Methods and machinery for harvesting fruit, vegetables and grapes – Field transport devices – Field and greenhouse watering devices.

Publications: Numerous publications on above research subjects.

Correspondence: German, English.

Göttingen

Institut für Landmaschinenkunde (Institute for Agricultural Machinery, University of Göttingen)
Gutenbergstr. 33, 34 Göttingen.

Wieneke, F.	Dr.-Ing.	Prof.	Director
Claus, H.G.	Dr.agr.	Prof.	Head of Division
Eimer, M.	Dr. sc.agr.	Prof.	Senior Research Officer
Gallwitz, K.	Dr.-Ing.	Prof.	emeritus
Prigge, H.	Dr. sc.agr.		Research Associate
Grimm, W.	Dipl.-Ing.		Research Assistant
Asiedu, J.J.	Dr.-Ing.		Temporary Researcher
Heier, W.	Dipl.-Ing.		Temporary Researcher
Lücke, W.	Dipl.-Ing.		Temporary Researcher

Activities: Training: Lectures, seminars and study courses on: Fundamentals of agricultural engineering, field machinery, livestock and farm buildings, applied agricultural machinery, processing and storage of agricultural products, measuring techniques in agriculture, agricultural engineering in the tropics. – Research: Forage harvest and conservation, solar energy, solar drying, postharvest technology, dehydration of forage and fractionation of proteins, optimizing barn drying process with a minimum of energy input and desired nutrition value, processing and storage of tropical agricultural products.

Correspondence: German, English, (French).

Hannover

Institut für Wasserwirtschaft, Hydrologie und landwirtschaftlichen Wasserbau der Universität Hannover (Institute for Water Resources, Hydrologie and Hydraulic Engineering in Agriculture, University of Hanover)
Callinstrasse 32, D-3000 Hannover 1
Tel.: 0511/7622237

Lecher, K.	Dr.sc.techn., Prof.	Director
Hoffmann, B. and 4 assistants	Dr.-Ing., Prof.	Goundwater research, soil physics
Mull, R. and 6 assistants	Dr.-Ing., Prof.	Groundwater research, systems analysis by simulation models
Sieker, F. and 6 assistants	Dr.-Ing., Prof.	Urban hydrology
NN for 14 assistants		Water resources research, Hydrology, Hydraulic engineering in agriculture

Outposted Branch: 3201 Sarstedt, Ruther Str. 40

Activities: Study courses leading to the degree of Dipl.-Ing., viz Dr.-Ing., attended by about 60 students per year. Groundwater research – Simulation of water movement – Water use management – Unsaturated zone – Soil physics – Hydrological statistics – Hydrology of open channel – Urban hydrology – Runoff-models – Flood control-models – Irrigation and drainage schemes – Water resources in semiarid zones – Water management planning – Project economics.

Publications: Reports and studies in the "Instituts-Mitteilungen" (Communications of the Institute) and in the technical press.

Correspondence: German, English, French, Spanish, Portuguese.

Institut für Technik in Gartenbau und Landwirtschaft der Universität Hannover (Institute for Horticultural Engineering, University of Hannover)
Herrenhäuser Str. 2, 3000 Hannover 21
Tel. 0511/7622647

von Zabeltitz, C.	Prof.Dr.Ing.	Director
Tantau, H.J.	Dr. habil	Lecturer
Damrath, J.	Dipl.-Ing.	Assistant
von Elsner, B.	Dipl.-Ing.	Assistant
Klein, Fr. J.	Dipl.-Ing. agr.	Assistant
Meyer, J.	Dipl.-Ing. agr.	Assistant
Strauch, K.H.	Dipl.-Ing. agr.	Assistant

Status: Governmental

Activities: Teaching: 3 study courses leading to the degree of Dipl.Ing. agr. – Research: Greenhouse construction, Energy saving, Climate control, Thermal screens, Double glazing, Solar energy, Reject heat, Desert greenhouse, Solar desalination, Growing rooms, Microcomputer control system, Light transmission, Micro climate, Heating systems, Leaf temperature control, Heat pumps.

Publications: Energy saving, Solar energy, Thermal screens, Reject heat, Double glazing, Climate control, Heating systems, Heat pumps.

Correspondence: German, English.

Institut für Bauforschung e.V. (Building Research Institute e.V.)
An der Markuskirche 1 - 3000 Hannover 1   Tel.: 0511/66 10 96

Menkhoff, H.   Dr.-Ing.   Director

Branch with other location: Institut für Bauforschung e.V., Zweigstelle München, Gräfelfinger Str. 128, 8000 München 70.

Status: Registered private association, established in 1946. Its members are: the Universities, representatives of the Construction and Housing Industry.

Activities: Economic efficiency of different times of farm building

Publications: Expert evidence

Correspondence: German.

Lehrstuhl für das ländliche Bau- und Siedlungswesen der Technischen Hochschule Hannover
(Chair of Rural Buildings and Land Settlement, Technical University)
Schlosswenderstr. 1, 30 Hannover   Tel.: (0511) 7622127

Status: The Chair was established in 1959. Supervised by the Minister of Culture and Education, Niedersachsen.

Activities: Study courses attended by about 15 students per year - Planning of farmsteads - Development of rural houses on slopes.

Correspondence: German, English.

Jork

Abteilung Technik im Obstbau, Obstbauversuchsanstalt der Landwirtschaftskammer Hannover
(Engineering Section of the Fruit Growing Experiment Station)
2155 Jork, Bez. Hamburg   Tel.: 04162/411

Activities: Research, experiment, testing and extension work - Orchard tilling - Mowing machines for mulch production in orchards - Implements for orchard protection against pests and diseases - Experiments with fruit graders - Fruit packing - Storage of fruits - Fruit storage buildings and storage equipment.

Publications: "Mitteilungsblatt des Obstbauversuchsrings Jork, Technische Mitteilungen", a monthly review.

Karlsruhe

Fakultät für Maschinenbau, Universität Karlsruhe (Faculty of Mechanical Engineering)
7500 Karlsruhe, Kaiserstrasse 12   Tel.: 0721/6082430

Jungbluth, G.   Dipl.Ing. o.Prof.   Internal Combustion Engines
Bauer, P.E.   Dipl.Ing. Prof.   Farm power and agricultural machinery

Status: Institution of Baden-Württemberg, supervised by the Ministry of Culture and Education.

Agricultural Engineering activities: Study courses during two semesters to students of engineering.

Correspondence: German, English.

Institut für Wasserbau und Wasserwirtschaft, Versuchsanstalt für Wasserbau und Kulturtechnik, "Theodor-Rehbock-Flussbaulaboratorium", Universität Karlsruhe (Institute for Hydraulic Engineering and Water Resources Planning, Research Institute for Hydraulic Structures and Agricultural Engineering, Theodor-Rehbock Laboratory for River Improvement, University Karlsruhe)
Kaiserstrasse 12, D - 7500 Karlsruhe 1, Federal Republic of Germany

Phone: (0721) 608 - 2194
Telex: 07 - 826 521 Uni

Mosonyi, E.   Prof.Dr.Dr.Dr.h.c.Dr.-Ing. E.h.   Head of the Institute

Scientific staff for hydraulic engineering: 10 persons

Scientific staff for water resources planning and hydrology:
Buck, W.   Dr.Ing.
Hauck, E.   Dipl.Ing.

Scientific staff for agricultural hydrotechnics:
Eggers, H.   Dr.Ing. (on leave to AIT, Bangkok)
Lindner, A.   Dipl.Ing.

Status: Date of establishment: 1901. The Institute is supervised and financed by the Minister of Public Worship and Education of the State of Baden-Württemberg (Stuttgart).

Agricultural Engineering activities: a) Teaching: Study courses provided by the Institute: - Introduction to Hydraulic Engineering - Hydraulic Structures - River Training and Flood Control - Water Power Stations and Dams - Water Resources Development in Arid and Tropical Regions - Agricultural Engineering (incl. Irrigation, Drainage, Erosion Control) - Pedology - Economic Analysis for Water Resources Development - Course of Practical Work in Hydrology - Case Studies in Hydraulic Engineering.
Some study courses provided by other institutes of the university: Agriculture - Construction of Agricultural Roads - Land Organization - Soil Assessment - Meteorology.
Type of degree awarded: Dipl.-Ing.; possible graduation to Dr.-Ing.
Average number of students attending the courses: 20 to 50.
b) Research: Basic research: Hydraulics, Hydrology, Water Resources Management, Erosion Control in Irrigation, Sheet Erosion - Practically oriented research: Investigations in connection with model tests for solving practical problems encountered in the planning of construction or in the operation of hydraulic systems.

Publications: Numerous publications on the above mentioned topics in German and international journals, congress proceedings, etc. - Papers published in the "Mitteilungen" (Reports) of the "Versuchsanstalt für Wasserbau und Kulturtechnik" since 1973.

Correspondence: German, English, French.

Kiel

Institut für Landwirtschaftliche Verfahrungstechnik der Christian-Albrechts-Universität
(Department of Agricultural Engineering, Christian-Albrecht-University)
Olshausenstrasse 40 - 60, 2300 Kiel   Tel. 0431/880-2355

vacant	Prof.Dr.agr.	Director	Livestock mechanization
Isensee, E.	Prof.Dr.agr.	Vice-Director	Field mechanization
Mannebeck, H.	Dipl.-Ing.Dr.agr.	Assistant - electrical control equipment	
Schuster, J.	Dipl.-Ing.agr.	Assistant	
Wagner, M.	Dipl.-Ing.agr.	Assistant	
Wenzlaff, R.	Dipl.-Ing.agr.	Assistant	

Status: Institution of the Land Schleswig-Holstein, supervised by the Ministry of Culture and Education.

Activities: Study courses within the agricultural engineering curriculum, attended by 200 students per year. - Investigations of high capacity harvesters; experiment and testing works on: stone removal machinery, cultivation of quick set hedges, handling and drying of big bale haye, digestation of straw by NH_3, distribution of semi-liquid manure, biological treatment of semi-liquid manure, heat pumps; environmental control for animal production.

Publications since 1973: Harvesting machinery, stone removal machinery, transport of agricultural goods, computerized distribution of feed for dairy cattle, conservation of haye, biological treatment of manure, biological energy.

Correspondence: German, Finnish, French, English.

Institut für Wasserwirtschaft und Meliorationswesen der Christian-Albrechts-Universität
(Institute of Hydraulics and Land Reclamation, Christian-Albrecht University)
Olshausenstrasse 40 - 60, 2300 Kiel

Status: Government institution, established in 1959 and supervised by the Ministry of Culture and Education of Schleswig-Holstein.

Activities: Study courses on water development in agriculture, land reclamation and consolidation and irrigation - Decree awarded after successful examination at the Agricultural Faculty: Dipl.agr., and after further specialization: Dr.agr. - Average number of 25 to 30 students attending the courses - Investigation into the water balance of small catchment areas - Drainage experiments - Studies of the behaviour of groundwater movement - Designing drainage projects for the agricultural practice - Water resources planning and development for irrigation purposes - Determination of areas which require drainage.

Publications: Weather-yield relationship - Plant-water relationship - Behaviour of groundwater movement - Water balanced in small catchment areas - Evapotranspiration - Water permeability - Drainage experiments - Irrigation in Mediterranean countries and in North Germany.

Correspondence: German, English.

Köln

Fachbereich Landmaschinentechnik, Fachhochschule Köln (Department of Agricultural Machinery)
Reitweg 1, 5000 Köln 21
Tel.: 0221/8275 2391

Husemann, R.	Prof. Dipl.-Ing.	Head of Dept., Tractors, hydraulics
Pak, M.	Prof. Dr.-Ing.	Substitute of Head, Mechanics, thermodynamics
Althoetmar, Th.	Prof. Dipl.-Ing.	Agricultural engineering, conveying implements
Flörkemeier, K.H.	Prof.Dr.-Ing.	Machinery parts
Nolte, L.	Prof. Dipl.-Ing.	Materials, testing methods
Schilling, E.E.	Prof.Dr.-Ing.	Machinery for field production
Thiel, M.	Prof.Dr.rer.nat.	Mathematics

Status: Governmental institution, established in 1949.

Activities: 7 Semester, leading to degree of Dipl.Ing.

Correspondence: German, English.

München

Institut für Landmaschinen, Technische Universität München (Institute of Agricultural Machinery, Technical University Munich)
Arcisstrasse 21 - D-8000 München 2
Tel.: 089/2105-2555/2556

Söhne, W.	Prof.Dr.-Ing.	Director
Bacher, R.	Dipl.-Ing.	Asst. - Tractor noise
Bolling, I.	Dipl.-Ing.	Asst. - Soil wheel mechanics
Onderka, G.	Dipl.-Ing.	Scientist - Dust collection
Acholtysik, B.J.	Dipl.-Ing.	Asst. - Feed dosing
Schwanghart, H.	Dr.-Ing.habil.	Akad.Dir. - Tractor safety tests
v. Sybel, H.	Dipl.-Ing.	Scientist
Wessel, J.	Prof.Dr.-Ing.	Processing of solid materials

Status: State institution, supervised by the Bavarian Ministry of Education and Culture, Munich

Agricultural Engineering activities: a) Teaching: Study courses to students of Mechanical Engineering - Agricultural Engineering - Tractors and earthmoving machinery - Mechanics of soil-vehicle systems - Sizing, sieving and sifting machinery - Dust technology - Processing technology. Degrees awarded: Dipl.-Ing; Dr.-Ing. Average number of students: 80.
b) Research: Power transmission between tire and soil - Tractor overturning and continuous rolling - Tractor noise - Soil tool interactions - Processes and feed dosing - Sizing of granular materials by dry and wet methods - Dust collection.

Publications: Tractor concepts - Traction force and slip of tractor tires - Tractor safety against tipping - Tractor overturning and continuous rolling - Strength of tractor frames and cabs - Tractor noise reduction - Drain plow - Mechanics of soil tillage - Vacuum of milking-machines - Feed dosing device - Sifting and sieving of granualar materials - Dust collection.

Correspondence: English.

Institut für Technik im Gartenbau, Fachhochschule Weihenstephan (Institute of Horticultural Engineering, Fachhochschule Weihenstephan)
D 8050 Freising 12 Tel.: 08161/71369

Hege, H.	Dr.Ing.Prof.	Head of Institute
Achatz, A.	Ing.grad.	Operating Engineer
Lecker, F.	Ing.grad.	Glasshouses engineering
Rannertshauser, J.	Ing.grad.	Horticultural equipment
Weber am Bach, K.	Dipl.Ing.Prof.	Technical principles in horticulture.

Status: Governmental Institution, supervised by the Bavarian Ministry of Education and Culture, Munich.

Activities: Teaching: Horticultural engineering curriculum as part of the study courses of 4 years duration for Ing.grad. About 450 students in different curricula for horticultural engineers and landscape gardeners. - Research: Spaced seed of vegetables. Plant protection equipment. Steam sterilisation of soils. Fertilizer application by irrigation and trickle irrigation. Climate control and CO_2-enrichment in glasshouses. Heat isolation of glasshouses and energy saving in horticulture.

Publications since 1973: Spaced seed of vegetables. Young plant production. Sprayers for plant protection. Steaming of soils. Fogging technique for plant protection in glasshouses. Trickle irrigation. Fertilizer solution injectors. Electricity in horticulture.

Correspondence: German, French, English.

Institut für milchwirtschaftliches Maschinenwesen, Süddeutsche Versuchs- und Forschungsanstalt für Milchwirtschaft (Institute of Dairy Equipment, Dairy Research Institution of Southern Germany) Technische Universität München (Technical University of Munich)
D-8050 Freising-Weihenstephan

Kessler, H.G.	o.Prof.Dr.Ing.	Director
Eibel, H.	Dipl.Ing.	Assistant
Fiedler, J.	Dipl.Ing.	"
Hege, W.	Dipl.Ing.	"
Horak, P.	Dipl.Ing.	"
Kammerlehrer, J.	FStR	"
Walenta, W.	Dipl.Ing.	"

Status: State institution, established 1921 and supervised by the Bavarian Ministry of Education and Culture.

Agricultural Engineering activities: a) Teaching: Thermodynamics, Food Engineering, Dairy Technology, Manufacturing of Milk Products, Food Technological Workshop, Dairy Technological Workshop. b) Research: Residence Time Distribution in Tube Systems and Plate Heat Exchangers, Cleaning and Rinsing, Ultrafiltration, Evaporation, Investigation on Control of Recontamination in pipe-line-systems in dairies, Ultra-High Temperature Heating, Colour changes in Milk Products, Cheese Technology, Butter Technology, Heat Transfer and Temperature-Balance and their Technological applications.

Publications: Deutsche Milchwirtschaft; ZUL; DMZ; Milchwissenschaft; Deutsche Molkerei-Zeitung; Die Molkerei-Zeitung Welt der Milch; Ernährungswirtschaft - Lebensmitteltechnik.

Correspondence: German, English.

Stuttgart

Universität Hohenheim

Institut für Agrartechnik (Institute of Agricultural Engineering)
Garbenstrasse 9, 7000 Stuttgart 70   Tel : 0711/45011 - Telex: 7-22959 uniho-d

Lehrstuhl für Grundlagen der Landtechnik am Institut für Agrartechnik (Division of Fundamentals of Agricultural Engineering)

Kutzbach, H.D.	Dr.-Ing., Prof.	Head of Division - Farm tractors, combines
Grobler, W.H.	Dipl.-Ing.agr.	Grain harvesting, combines
Eissen, W.	Dipl.-Ing.	Solar drying
Hofacker, W.	Dipl.-Ing.	Drying
Hutt, W.	Dipl.-Ing.	Pneumatic conveying, drying
Kustermann, M.F.	Dipl.-Ing.	Physical properties of cereal grains
Mühlbauer, W.	Dr.-Ing.	Energy in agriculture, drying and storage
Reisinger, G.	Ing.grad.	Drying, heat recovery systems
Schrogl, H.	Dipl.-Ing.	Tractors, farm machinery
Wacker, P.	Dipl.-Landw.	Ergonomics, combines

Activities: Teaching - Study courses, seminars and institute practica of 2-4 semesters duration within the agricultural engineering curriculum, attended by an average of 70 students. Postgraduate studies in agricultural or mechanical engineering.
Degrees awarded: Dipl.-Ing.; Dipl.-Ing.agr.; Dr.-Ing.; Dr.agr. — Research: Possibilities and limits of energy saving and energy costs reduction in grain drying - thermal and mechanical properties of grain - threshing, separation and cleaning of grain - injectors for pneumatic conveying of solid materials - pneumatic conveying of solid materials - technical problems with use of trailed implements - optimization of drying and cooling of grain - ergonomics of tractors and farm machinery - servicing problems of tractors and combine harvesters.

Lehrstuhl für Verfahrenstechnik in der Tierproduktion (Division of Animal Production)

Bischoff, Th.	Professor, Dr.agr.
Adam, M.	Dr.agr.
Breitenbücher, K.	Dipl.Ing.agr.
Dürr, H-D.	Dipl.Ing.agr.
Lindenberg, G.	Dipl.Ing.agr.
von Oy, K.J.	Dipl.Ing.agr.
Rüprich, W.	Dr.agr.
Vetter, R.	Dipl.Ing.agr.
Wandel, H.	Ing.agr.(grad.)

Activities: Teaching - Fundamentals of Animal Production System Techniques - Technical Description of Buildings - Building Materials and Elements - Buildings and Environment - Livestock Machinery - Animal Production Planning Techniques - Animal Production System Techniques - Degrees awarded: Dipl.Ing.agr, Dr.agr, Dr.oec. — Research: Livestock Building Design - Preservation of Grains and Forage - Storage of Grains and Forage - Environmental Design for Livestock - Feed Preparation, Handling and Storage - Waste Management and Biogas.

Lehrstuhl für Verfahrenstechnik in der Pflanzenproduktion (Division of Plant Production)

Stroppel, A.	Dr.-Ing.,Prof.	Head of Division - Soil tillage
Beek, van der, A.	Dipl.-Ing.	Load in soil tillage
Blümel, K.	Dipl.-Ing.	Rotary tillage
Britzius, U.	Dipl.-Ing.agr.	Density and size distribution of soil particles

Reich, R.	Dipl.-Ing.	Chisel plough
Schäfer, W.	Dipl.-Ing.agr.	Energy and work required in soil tillage

<u>Activities</u>: Lectures and seminars within the agricultural engineering curriculum. Degrees awarded: Dipl.-Ing.agr.; Dr.agr. - Research: Modern soil tillage and harvest techniques and their influence on emergence and yield - technical problems of working in of straw - investigations on cultivators - investigations on rotary cultivators in the soil bin - required work, energy and efficiency and costs for soil tillage in semiarid climates - sorption kinetics of soil water at germinating seeds - connection between density and size distribution of soil particles in seed beds concerning emergence.

<u>Verfahruenstechnik für Intensivkulturen</u> (Engineering in Special Crop Production)

Moser, E.	Prof.Dr.-Ing.
Bernhardt, B.	Dipl.-Ing.agr.
Locher, B.	Dipl.-Ing.agr.
Schmidt, K.	Dipl.-Ing.
Siglinger, M.	Dipl.-Ing.agr.
Sinn, H.	Dipl.-Ing.

<u>Activities</u>: Teaching: Properties of Agricultural Materials - Irrigation and Sprinkler Systems - Agricultural Machinery for Intensive Crops with Application Techniques- Intensive Crops System Techniques - Greenhouse Techniques. Degrees awarded: Dipl.-Ing.agr., Dr.agr. — Research: Harvesting of Special Crops (fruits, vegetables, grapes, hops, tobacco) - Physical properties of Special Crops - Irrigation (trickle, micro-jet, sprinkler) - Plant Protection (application, sprayers).

<u>Landesanstalt für landwirtschaftliches Maschinen- und Bauwesen</u> (Department for Agricultural Engineering and Buildings - Extension service)

Bischoff, Th.	Dr.agr., Prof.	Supervisor
Bewer, E.	Dr.agr.	
Maurer, K.	Dipl.-Ing. Akad.Oberrat	Head of Department

<u>Activities</u>: Research and practical improvement of haydrying systems, enginedriven fans, heat pumps for livestock; Anaerobic fermentation of animal wastes, biogas, storage systems. Extension service in the mentioned lines.

<u>Status</u>: The Institute belongs to the University of Hohenheim which is a governmental institution, supervised by the Ministry of Culture and Education of the land Baden-Württemberg.

<u>Publications</u>: Agrartechnische Berichte (agricultural engineering reports), Tätigkeitsberichte des Instituts für Agrartechnik (annual activity reports of the Institute of Agricultural Engineering).

<u>Correspondence</u>: <u>German</u>, <u>English</u>, French, Czech.

<u>Universität Hohenheim - Lehrstuhl für Wirtschafts-, Sozial- und Agrargeschichte</u> (Department of Economic-, Social- and Agricultural History)
Postfach 106 (P.B. 106) - 7000 Stuttgart 70   Tel.: 0711/4501-2618/2621

Winkel, H.	Prof.Dr.	Director
Hermann, K.	Dr.	Assistant

<u>Status</u>: Governmental institution

Activities: History of economic, social and agricultural development. History of agricultural Engineering.

Publications: Agricultural Engineering History in "Zeitschrift für Agrargeschichte und Agrarsoziologie"; "Geschichte der Landtechnik", "Landtechnik", "Scripta Mercaturae".

Correspondence: German, English, French

Institut für Ländliche Siedlungsplanung, Fakultät Architektur und Stadtplanung, Universität Stuttgart (Institute of Rural Settlement Planning, Faculty of Agriculture and Urban Planning, University of Stuttgart)
Keplerstr. 11, 7000 Stuttgart, Fed. Republic of Germany   Tel. 0711/2073 637

Simons, D.	Dipl.-Ing.	Head of the Institute
Maier, W.F.	Dipl.agr.oec.	

Status: Established 1962. Governmental finance.

Activities: Courses of agricultural building and planning as part of the studies of architecture - Research in the field of community and rural development.

Publications: occasional.

Correspondence: German, English, French.

GHANA

Accra

Mechanization and Transport Department (Ministry of Agriculture)
P.O. Box M.82, Accra　　　　　　　　　　　　　　　　Tel.: 77786 and 77789

Gyarteng, O.K.	B.Sc., M.Sc., M.GH.IE	Director
Batsa, H.P.	B.Sc.Post.grad.Dip. A.M.D.E.	Deputy Director
Quansah, S.S.	N.D.Agric.E.	Mechanization Officer on secondment to Food Production Corporation
Coleman, S.A.	M.Sc.	On course
Abanyie, G.A.	N.D.Agric.E.	Reg. Mechanization Officer
Nketiah, A.K.	N.D.Agric.E.	Reg. Mechanization Officer
Apau, N.A.	N.D.Agric.E.	Reg. Mechanization Officer
Amoako-Atta, E.E.	N.D.Agric.E.	Reg. Mechanization Officer
Agboka, V.K.	N.D.Agric.E.	Reg. Mechanization Officer
Adjei, A.E.	M.Sc.	On course

__Status__: State Department under the Ministry of Agriculture.

__Activities__: Advises government on agricultural machinery policy for the country. Gives mechanized services to the farming community. Maintains repairs and maintenance facilities for all vehicles of the Ministry of Agriculture and all agricultural machinery in the country.

__Publications__: Annual Reports and Hand-outs on Mechanization Extension.

__Correspondence__: English.

__Branch__: __Mechanization Training and Extension Unit - Nyankpala__
P.O. Box 14, Tamale

Kyei, N.　　　　B.Sc.(Agric.); M.Sc.(Agric.Eng.)　　Officer-in-Charge,
　　　　　　　Branch: Mechanization Training and Extension Unit, Nyankpala.

Agricultural Engineering Division, Faculty of Agriculture, University of Ghana
Legon　　　　　　　　　　　　　　　　Tel.: 75381 ext. 8422

Gyasi, S.	B.Sc.(Mech.Eng.), M.Sc.(Agric.Eng.), Ph.D.	Power and Machinery - Ag. Head of Department
Osafo King	B.Sc.(Agric.Eng.)	Soil and Water - Lecturer
Ampratwum, D.B.	B.Sc.(Agric.), M.Sc.(Agric.Eng.), Ph.D.	Drying and Storage - On secondment to Agric.Dev. Bank

__Branches at__: University of Ghana Agricultural Research Stations: a) Kpong - Irrigation; b) Nungua - Power and Machinery.

__Status__: Established 1956. Governing Body: a) Ghana Government, b) USAID (1967-1973).

__Activities__: Teaching: Courses in Power and Machinery - Soil and Water - Farm Buildings; Processing - Engineering Experiment design. - Research: Grain Drying and Storage - Engineering Properties of some Ghanaian Vegetables.

Publications since 1965: numerous publications on agricultural engineering subjects.

Correspondence: English.

Irrigation Development Authority
P.O. Box M.53, Accra   Tel.: 64352

Wiafe	B.Sc.(Eng.)	Deputy Chief Executive Engineering
Aryeetey, A.N.	B.Sc.	Deputy Chief Agronomy
Kemevor, E.D.	B.Sc., M.Sc.	Irrigation Engineer
Awuku, S.M.	B.Sc., M.Sc.	Irrigation Engineer
Shah, K.M.	B.Sc.(Eng.)	Irrigation Engineer
Opoku-Mensah, A.	B.Sc.	Irrigation Engineer
Lawson, M.V.O.	B.Sc.	Irrigation Engineer
Osei, R.K.	B.Sc.	Irrigation Engineer
Mensah, J.W.	B.Sc.	Irrigation Engineer
Obuobie, A.L.S.	B.Sc.	Irrigation Engineer

Status: A semi-autonomous state organization under the Ministry of Agriculture.

Activities: Advises government on irrigated crop production policy. Constructs and maintains dams for agricultural production.

Publications: Annual Reports and Irrigation Hand-outs.

Correspondence: English.

Ghana State Farm Corporation
P.O. Box 299, Accra   Tel.: 66941

Woode, R.   M.Sc.(Soviet)   General Manager - Engineering

Status: State Corporation.

Activities: Engaged in the cultivation of tree crops and livestock production on large scale basis.

Publications: Annual Reports.

Correspondence: English.

K u m a s i

Department of Agricultural Engineering, University of Science and Technology
University Post Office, Kumasi - Ghana

Djokoto, I.K.	B.Sc.,M.Sc.,Ph.D. M.Gh.I.E.	Senior Lecturer and Head, Farm Power and Machinery
Twum, A.	B.Sc.,M.Sc., M.Gh.I.E.	Lecturer, Farm Power and Machinery
Aklaku, E.D.	Dip.Ing.	Lecturer, Farm Structures, Environmental Control, Material Handling and Processing

Status: Established in 1961 as a Department in the Faculty of Agriculture, University of Science and Technology, Kumasi. The institution is financed by the government.

Agricultural Engineering activities: (a) Teaching: - Degree in Agricultural Engineering through Faculty of Engineering, administered by a Board of Studies; with representatives from both Faculties of Engineering and Agriculture. B.Sc. Agricultural Engineering. 2-4 students graduating per year. - Specialization in Mechanization during fourth year of Agricultural Degree Course. B.Sc.(Agriculture); 2-3 students graduating per year.
(b) Research: Processing of plantain and other tropical crops; Design and testing of small farm implements; Storage of some agricultural produce.

Publications: Several publications.

Correspondence: English.

Grains and Legumes Development Board
P.O. Box 4000 - Kumasi				Tel.: 4778

Wilson, T.		B.Sc.(Eng.) Post-grad.Cert.		Agricultural Engineer
			in Agric. Engineering

Status: State Board under the Ministry of Agriculture.

Activities: Development and production high yielding varieties of grains and legumes

Publications: Annual Reports.

GREECE

Athinai

Direction Générale des Améliorations Foncières, Ministère de l'Agriculture
(General Division of Land Reclamation, Ministry of Agriculture)
46, rue Chalcocondyli, Athènes (102) - Grèce

Zacharopoulos, A.	Ing.agronome		Ingénieur civil du Génie Rural - Directeur-Général
Inspecteurs:
Saltapidas, C.		M.Sc. in Irrigation	Inspecteur général
Mountzidis, G.		Agronome		Inspecteur des Organismes d'Irrigation et de Drainage
Avratos, P.		Ing.mécanicien-		Inspecteur technique
			électricien

Service de Planification et d'Exploitation des Projets:
Mavrikios, P.		Agronome		Chef, Directeur Général Adjoint

Direction de Planification et des Etudes Agro-techniques:
Courcoulis, E.		Agronome		Directeur

Section A: De Planification et de Surveillance des Projets
Papazoglou, N.		Agronome		Chef de Section
Pitis, A.		Agronome		Chef de bureau
Fatouros, A.		"			" "
Haratzoglou, M.		"			" "

Section B: Etudes agrotechniques, agroéconomiques et pédologiques des projets
Papaconstantinou, D.				Agronome Chef de Section, specialisé en Hydraulique Agricole

Castanis, D.	Dr., Agronome	Pédologue
Comarima, Mlle J.	Agronome	
Rachoutis, E.	Agronome	
Korelis, L.	Dr., Agronome	Ing. d'Améliorations Foncières

Section C: Développement et Protection des ressources en eau et en sols
Spanopoulou, Mme Marie Ing.Chimiste   Chef de Section

Géologie-Hydrologie:
Stoubos, J.   Géologue   Directeur

Section d'Hydrologie:
Papadopoulos, A.   Agronome, M.Sc. in   Chef de Section
                    Irrigations
Zervoyannis, Mme.Marie Géologue, DEA 3ème cycle, Micropaléontologie

Section d'Hydrogéologie:
Letsios, A.   Géologue   Chef de Section
Kabouridis, Mme.Elisabeth   Géologue

Section d'Etudes Géologiques:
Papadopoulos, N.   Géologue Certificat   Chef de Section
                    d'Interprétation P/A
                    en Géologie

Laboratoire des Eaux:
Malefakis, J.   Chimiste   Chef de Section
Zepos, L.   Agronome   Chimiste
Dionysopoulou-Spyrelli, Mme Stavroula   Chimiste

Mécanisation de l'Agriculture, Exploitation de l'Equipement Mécanique:
Marmaras, B.   Agronome   Directeur

Section de Mécanisation de l'Agriculture:
Karkanis, E.   Agronome   Chef de Section
Koutsovitis, N.   Agronome

Section d'Exploitation de l'Equipement Mécanique:
Konstantinidis, A.   Agronome   Chef de Section
Vatsakis, E.   Agronome

Section d'Electrification de l'Agriculture:
Vekios, P.   Agronome   Chef de Section
Kapogiannis, P.   Agronome

Administration et Exploitation des Projets d'Améliorations Foncières et des Eaux d'Irrigation
Botsoglou, P.   Agronome   Directeur

Section: Questions des Eaux:
Carayannakis, C.   Agronome   Chef de Section
Antonopoulos, P.   Agronome

Section: Organisation et Surveillance des Organismes d'Irrigation:
Malagardis, C.   Juriste   Chef de Section

Section: Exploitation et Gestion des Projets d'Améliorations Foncières:
Salapas, C.   Agronome   Chef de Section
Papavassiliou, C.   Agronome
Tsangarakis, G.   Agronome
Megremis, P.   Agronome
Natsis, T.   Agronome

Service des Projets Techniques et des Constructions:
Stavridis, E.   Agronome, Ing.Civil   Chef de Section

Direction d'Etudes Techniques:
Vlassopoulos, X.   Agronome, Ing.civil   Directeur
                       du Génie Rural

Section A: Etudes:
Bathas, C.   Ing.civil   Chef de Section
Dinos, J.   Agronome, Ing. civil du Génie Rural
Michail, M.   Ing.civil
Chryssafinos, E.   Agronome, Ing. civil du Génie Rural
Tzotzolakis, A.   Ing.civil
Calitsis, V.   Ing.civil

Section B: Etudes de Bâtiments:
Maragoudakis, D.   Ing.civil   Chef de Section
Kefallinos, P.   Ing.Topographe
Toulias, P.   Ing.civil
Moraitou, Mme Danai   Ing.Architecte
Bathas, Mme Eleana   Ing.Architecte
Kouris, S.   Ing.Archit.

Section C: Etudes de Mécanique de sol et géotechniques:
Rigas, N.   Ing.chimiste

Constructions:
Philippou, N.   Ing.civil   Directeur

Section constructions des Projets d'A.F.:
Michail, D.   Ing.civil   Chef de Section
Cantartzis, D.   Ing.civil
Matrakidis, J.   Ing.civil
Kifiotis, R.   Ing.civil
Kyriakopoulos   Ing.Topographe

Section: Construction des Bâtiments et des Installations Agricoles:
Kyriazidis, A.   Ing.civil   Chef de Section

Section: Entretien et Completion des Projets Techniques:
Daskas, A.   Agronome
                       Ing.Hydraulicien

Section: Travaux urgents:
Roumanas, A.   Ing. Topographe

Direction VII:
Vettas, P.   Ing.mécanicien   Directeur

Section A: Equipement mécanique et Ateliers de Réparations:
Sinopoulos, P.   Ing.mécanicien   Chef de Section

Section B: Installations Electromécaniques des Projets d'A.F.
Lykopoulos, G.   Ing.mécanicien

Section C: Installations Electro-mécaniques des Bâtiments:
Tsetis, A.   Ing.mécanicien   Chef de Section

Section C: Sondages et Forages:
Loukopoulos, G.   Ing.mécanicien   Chef de Section
Goussios, D.   Ing.mécanicien

Annexes situées dans d'autres localités: Les annexes sont reparties comme suit sur toute la Grèce. Il y a 7 Directions Régionales, une à chaque Departement Géographique du pays et notamment:

1) La D.R. d'A.F. de Macédoine de l'Est et de Thrace, siègeant à Kavalla.
2) La D.R. d'A.F. de Macédonie Centrale et Occidentale, siègeant à Thessaloniki.
3) La D.R. d'A.F. de la Grèce Centrale, siègeant à Larissa.
4) La D.R. d'A.F. du Péloponnèse et de la Grèce Occidentale siègeant à Patras.
5) La D.R. d'A.F. d'Epire, siègeant à Jannina.
6) La D.R. d'A.F. d'Iraklion, siègeant à Iraklion, Crète.

Il y a 27 Branches d'A.F. situées à Alexandropoulis, Comotini, Serres, Thessaloniki, Alexandria, Edessa, Kozani, Caterini, Volos, Larissa, Carditsa, Trikala, Lamia, Pirée, Thibes, Agrinion, Patrai, Pyrgos, Nafplion, Tripolis, Calamata, Corinthe, Jannina, Preveza, Iraklion, Chania. Il y a 9 Secteurs d'A.F. situés à Arta, Igoumenitsa, Corinthos, Orestias, Rhodes, Cavala, Xanthi, Rethymnon, Aghios, Nicolaos. Il y a aussi 24 Bureaux de Génie Rural aux Directions d'Agriculture de 24 Départements.

Structure: La Direction Générale d'A.F. (ex Service d'A.F.) a été fondée comme S.A.F., en 1958. Il s'agit d'un Service d'Etat dépendant du Ministère de l'Agriculture.

Activités concernant le Génie Rural: a) Formation: L'Institut National Agronomique d'Athènes (durée d'études: 5 années) à partir de la quatrième année d'études offre une option d'Améliorations Foncières (Irrigation, Drainage, Machinisme Agricole, Constructions, Pédologie) menant au Diplôme d'Agronome, option Am.Foncières, Pédologie. En moyenne 15 étudiants suivent les cours chaque année. L'Université de Thessaloniki, Faculté d'Agronomie et des Forêts, à partir de la quatrième année et pendant la cinquième année offre une option de Génie Rural et de Pédologie (Nombre moyen d'étudiants par an: 15). - Un cours de 3ème cycle de Génie Rural, d'une durée de deux ans, est ouvert aux agronomes, ing.civils ou topographes. Après passage des examens et la présentation d'une thèse ce cours est sanctionné d'un diplôme d'Améliorations Foncières, niveau Maitrise (M.Sc.) Nombre moyen annuel d'étudiants: 10-15. - A l'Université Technique d'Athènes, ainsi qu'aux Ecoles d'Ingenieurs civils et Ing.topographes de l'Université de Thessaloniki, de Patras et de Thrace sont donnés des cours d'irrigation et de drainage et aux Ecoles d'Ing. architectes sont enseignés les constructions rurales Recherche-Expérimentation, Vulgarisation. - Des recherches sur les Améliorations Foncières sont effectuées à l'Institut d'Am.Foncières de Sindos qui n'appartient plus à notre Service mais au Service des Recherches du Ministère de l'Agriculture. - Il y a aussi un Institut de Machinisme Agricole, fondé par nous mais appartenant actuellement au Service de Recherche Agronomique, s'occupant de l'essai de toute machine agricole fabriquée ou importée en Grèce ainsi que des recherches sur le machinisme agricole. - Le Service assure la vulgarisation agricole dans les nouveaux réseaux d'irrigation dans le but d'assurer de plus grands rendements par une utilisation judicieuse de l'eau en évitant le gaspillage de l'eau et de l'énergie électrique.

Publications: Rapports et bulletins sur l'irrigation et le drainage.

Correspondance: Grec, français, anglais.

Institution Georgikon Mikhanon ke kataskevon (Institute of Agricultural Machines and Constructions)
61, Dimokratias str., Agii Anargiri Attikis, Athens, Greece

Souvatzis, G.	Mechanical and Electrical Engineer	Director
Georgiou, G.	Chemical Engineer	
Kafetzakis, N.	M.Agr.Sc., Agriculturist	
Kyriakopoulos, J.	Mechanical and Electrical Engineer	
Nikopoulos, T.	Agriculturist	
Pothos, P.	Mechanical and Electrical Engineer	

Rigas, A.　　　　　Mechanical and Electrical Engineer
Vaindirlis, Neoc.　Agriculturist

Status: Governmental, under the Ministry of Agriculture.

Activities: Agricultural Machinery Tests. Research. Popularization.

Publications: Agricultural machinery test reports.

Correspondence: Greek, English.

Ministère des Travaux Publics (Ministry of Public Works)
182, rue Char. Tricoupis, Athènes

Direction Générale de Construction de Travaux

Direction d'Etudes de Travaux Hydrauliques

Direction de Construction de Travaux des Améliorations Foncières

Activités: responsable pour la construction des grands projets d'Améliorations Foncières (barrages, réseaux d'irrigation et routiers)

Institut d'Améliorations Foncières, Sindos, Thessaloniki

Institut de Machinisme Agricole (Lykovrissi, Attique)

Institut de Pédologie et de Climatologie, (Lykovrissi, Attique)

Institut de l'Olivier et des Cultures sous-tropicales

Activités: Cette Station en plus de ces objectifs horticoles, s'occupe aussi de la recherche appliquée sur les methodes d'irrigation (irrigation à gouttes et irrigation fertilisante, utilisation des eaux saumatres).

Thessalonike

Agricultural Engineering Department of the Aristotelian University of Thessaloniki

Status: The Department of Agricultural Engineering was founded a few years before the Second World War, but has been in actual operation since 1950. The Department is under the control of the University of Thessaloniki.

Activities: The Agricultural School provides training in General Agriculture and Forestry. The number of students has fluctuated between 200-250. Besides the educational programme, scientific staff conducts research in various subjects.

Publications: in Journal of Agricultural Engineering Research; Farm Mechanization; textbooks.

Correspondence: Greek, English.

GUATEMALA

Ciudad de Guatemala

<u>Departamento de Ingeniería Agrícola, Facultad de Agronomía de la Universidad de San Carlos</u> (Agricultural Engineering Department, Faculty of Agronomy of the University)
Ciudad Universitaria, Zona 12, Apartado Postal 1545, Guatemala C.A.

<u>Carácter</u>: Institución Estatal, establecida en 1950 y financiada principalmente por el Gobierno de la República, si bien ingresan fondos de otras fuentes y se gobierna autónomamente.

<u>Actividades</u>: Enseñanza de Ingeniería Agrícola con cursos de Climatología, Topografía, Mecánica, Hidrología, Hidráulica, Resistencia de materiales, Aerofotogrametría, Construcciones Rurales, Máquinaria Agrícola.

<u>Correspondencia</u>: Español, Inglés.

Villa Nueva

<u>Instituto Técnico de Agricultura</u>
Bárcena, Villa Nueva, Guatemala

<u>Carácter</u>: La Institución fue establecida como Escuela Nacional de Agricultura en 1921, y en 1967 fue fusionada con la Escuela Forestal Centroamericana para integrar el Instituto Técnico de Agricultura que continúa siendo hasta la fecha. Su dirección y financiamiento tienen carácter gubernamental y dependen del Ministerio de Agricultura del país.

<u>Actividades</u>: Las actividades en Ingeniería Rural están incluidas en un plan de enseñanza que se imparte en un curso agrícola general con seis ciclos semestrales de duración, a cuyo término se otorga el título de Perito Agrónomo. El promedio de alumnos que actualmente estudian en dicho curso es de 350. Eventualmente, cuando la maquinaria del Instituto no está siendo usada en su totalidad para los trabajos del mismo, se prestan servicios de preparación de tierras, principalmente aradura y rastreo, para la siembra en propiedades de agricultores interesados mediante cobro de un efectivo módico que pasa a formar parte del fondo del Estado. Para consulta de los estudiantes, quienes ordinariamente son de precarios recursos económicos, se sacan a mimeógrafo anotaciones de los catedráticos, los cuales son internamente distribuidos, sobre cada una de las asignaturas que contemplan el plan de estudios.

GUINEA

Conakry

Direction Générale de l'Aménagement Rural (Agricultural Engineering, Ministry of Agricultural Development)
Conakry, B.P. 65   Tél.: 419-73

Structure: La Direction Générale de l'Aménagement Rural est une institution gouvernamentale relevant du Ministère du Développement Rural.

Activités: La Direction Générale de l'Aménagement Rural assure l'exécution, le contrôle des travaux d'aménagements ruraux et procède à la recherche et à l'expérimentation des variétés de riz adaptées aux périmètres aménagés.

Publications: Rapport sur la rentabilité de la mise en valeur de terres rizicoles en Guinée - Etude d'un programme d'aménagement hydro-agricole des terres rizicultivables de la Basse Guinée.

Correspondance: Français.

GUYANA

Government Technical Institute - Georgetown
Woolford Avenue, P.O. Box 870, Georgetown, Guyana

New Amsterdam Technical Institute - New Amsterdam
(Ministry of Education)
New Amsterdam, Berbice, Guyana

Port Mourant Training Centre (Guyana Sugar Corporation) - Port Mourant
Port Mourant, Corentyne, Guyana

HUNGARY

Budapesti Müszaki Egyetem - Gépszerkezettani Intézet - Mezőgazdasági Géptan Tanszék
(Technical University of Budapest - Institute of Machine Design - Department of Agricultural Engineering)
H-1111 Budapest, Hungary - Bertalan L.u.1.

Thernesz, V.	M.E.	Professor	Head of Department
Balaton, J.	M.E.	Senior Asst.Professor	
Gerencsér, A.	M.E.	Senior Asst.Professor	
Loboda, K.	M.E.	Asst.Professor	
Mészáros, I.	M.E.	Professor, Ph.D.	
Molnár Gy.	M.E.	Asst.Professor	
Salát, J.	M.E.	Research engineer	

Szabó, L.	M.E.	Research engineer
Varga, I.	M.E.	Professor, Ph.D.
Zalka, A.	M.E.	Professor, Ph.D.

Status: The Department was established in 1888 and it is supported by the Ministry of Education.

Activities: Education: University education leading to a diploma in Mechanical Engineering, with specialization in agricultural machinery design and development. - Research: Experimental and testing work, development and construction in the field of: Soil mechanics, Physical properties of agric. materials, Terrain-vehicle systems; Internal combustion engines; Hydraulics, pneumatics, automatic elements and systems; Tractors and their safety cabs; Tillage machines; Spraying machines; Harvesting and processing machines; Machines for animal production; Horticultural machinery.

Publications: Numerous publications in connection with each of the above mentioned research activities.

Correspondence: Hungarian, English, German, Russian, Spanish.

Mezőgépfejleszto Intézet/MEFI (Research and Development Institute for Agricultural Machinery)
H-1016 Budapest, Krisztina krt 55, Hungary

Bodolai, I.	Dr., candidate in agr. studies	Director
Szalka, A.	M.E.	Technical Adjoint Director
Tischler, M.	M.E.	Scientific Adjoint Director
Hegedüs, F.	Dr., Cert.Econ.	Adjoint Director Economics
Borszéki, L.	M.E.	Chief Engineer, Development Division of horticultural machines
Jerkó, S.	M.E.	Chief Engineer, Dev. Division of machines for arable plant growing and animal husbandry
Zováth, M.	M.E.	Chief Engineer, Dev. Division for agricultural transport and material handling, hydraulics, transmission and standardization
Kostyál, M.	M.E.	Chief Engineer, Dev. Division for machine testing and technical services.

Status: Governmental Institute, established in 1949, belonging to the Trust of Agricultural Engineering/MEZOGEPTROSZT/superintended by the Ministry of Metallurgy and Machine Industry.

Agricultural engineering activities: Research and development of machines and machine systems, principally for the following purposes: - vegetable growing in field conditions, harvesting and processing - Forage growing, harvesting, processing, storing and conservation - Fruit growing, harvesting, processing and viticulture - Animal husbandry: pig breeding, cattle breeding plants, etc. - Tillage, fertilization, plant protection and irrigation - Transmissions, hydraulic equipment for agricultural machines - Experiment and testing of the prototype machines.

Correspondence: Hungarian, Russian, English, German.

Gödöllő

MÉM Müszaki Intézet (National Institute of Agricultural Engineering)
2101 Gödöllö, Tessedik S. u. 4, Hungary

Bánházi, G.	Mech.Eng.Dr.Eng. candidate of agr.sc.	Director
Balázs, F.	Mech.Eng.	Deputy Director
Bölöni, I.	Mech.Eng.Dr.Eng. doctor of techn.sc.	Scientific Deputy Director
Vigh, P.	Mech.Eng.	Technical Deputy Director

Basic research and metrological department:

Szatura, I.	Mech.Eng.	Head, basic researches
Fekete, A.	Mech.Eng. candidate of agr.sc.	Developing of instrumentation
Kovács, L.	Mech.Eng.	Developing of instrumentation
Papp, Z.	Mech.Eng.	Measuring techniques
Fábián, Zs.	Mech.Eng. Dr.Eng.	Automatization of agricultural machines
Földesi, I.	Mech.Eng.	Automatization of agricultural machines
Kulcsár, P.	Electr.Eng.	Instrumentation of measuring processes
Borsa, B.	Mech.Eng.	Reliability and strength investigation of agricultural machines.

Horticultural mechanization department:

Bakos, I.	Mech.Eng.	Head, Mechanization of horticultural production
Dimitrievits, G.	Mech.Eng.	Mechanization of plant protection
Köszegi, G.	Mech.Eng.	Mechanization of viticulture
Szepes, L.	Hort.Eng.	Mechanization of vegetable production
Velich, S.	Mech.Eng.	Mechanization of fruit production
Göblös, G.	Mech.Eng.Dr.Eng.	Mechanization of grape harvesting
Berczi, L.	Mech.Eng.	Mechanization of fruit harvesting
Némethy, L.	Mech.Eng.Dr.Eng.	Mechanization of grape harvesting
Jakovácz, F.	Mech.Eng.	Mechanization of vegetable production
Tatár, S.	Agr.Eng.	Mechanization of vegetable production
Varga Józsefiné (Mrs. Varga)	Hort.Eng.Dr.Eng.	Mechanization of plant protection

Field crop mechanization department:

Fülöp, G.	Agr.Eng.Dr.Eng.	Head, Mechanization of field crop production
Bányai, Zs.	Mech.Eng.	Mechanization of fertilization
Jován, D.	Mech.Eng.	Mechanization of cereal- and roughage-harvesting
Szüle, Zs.	Mech.Eng.Dr.Eng. candidate of agr.sci.	Mechanization of industrial plant production
Zambori, J.	Mech.Eng.	Tractor-testing
Demes, G.	Mech.Eng.	Mechanization of fertilization
Hajdu, J.	Mech.Eng.	Mechanization of hay production
Jóri, J.I.	Mech.Eng.Dr.Eng.	Soil tillage mechanization

Kiss, A.	Mech.Eng.Dr.Eng.	Transport and loading
Majkuth, J.	Mech.Eng.	Mechanization of maize production
Sörös, I.	Mech.Eng.	Mechanization of cereal production
Szabó, D.	Mech.Eng.Dr.Eng.	Mechanization of irrigation
Papp, T.	Agr.Eng.	Mechanization of grain production
Salamon, S.	Agr.Eng.	Organization of field crop producing
Radványi, G.	Mech.Eng.	Development of agricultural tractor park
Kelemen, Zs.	Mech.Eng.Dr.Eng.	Agricultural transport material handling

Livestock mechanization department:

Velez, D.	Mech.Eng.	Head, Slurry treatment, mechanization of animal husbandry
Csermely, J.	Mech.Eng.Dr.Eng.	Mechanization of drying and storing, feed processing technologies
Csoma, M.	Mech.Eng.	Mechanization of pig keeping
Pongrácz, K.	Mech.Eng.	Mechanization of keeping small animals
Tóth, L.	Mech.Eng.Dr.Eng.	Mechanization of cattle- and sheep keeping
Flieg, J.	Mech.Eng.	Mechanization of feed processing, utilization of biomass
Komka, G.	Mech.Eng.	Mechanization of storing of cereals
Vajk, J.	Mech.Eng.	Drying technologies
Herdovics, M.	Mech.Eng.	Mechanization of grain separation
Mészáros, G.	Mech.Eng.	Climate technique
Tikos, S.	Mech.Eng.	Mechanization of pig keeping
Czinkóczky, A.	Mech.Eng.	Mechanization of slurry treatment
Kardos, J.	Mech.Eng.Dr.Eng.	Mechanization of cattle feeding
Zágoni, G.	Mech.Eng.	Mechanization of sheep keeping
Eörssy, M.	Mech. Eng.	Mechanization of technologies for feed mixing

Repair-technological department:

Horváth, J.	Mech.Eng.Dr.Eng. candidate of Techn.sci.	Head, Technology of repairing agricultural machines
Hartmann, V.	Mech.Eng.	Technology of spare-part renovation
Mélykuti, Cs.	Mech.Eng.Dr.Eng.	Repair and maintenance of agricultural machines
Antal, G.	Mech.Eng.Dr.Eng.	Repairing technology of agricultural machines
Boór, F.	Mech.Eng.Dr.Eng.	Diagnostics
Fodor, I.	Mech.Eng.	Technical economics
Horváth, A.	Mech.Eng.	Techniques and coating and anticorrosion protection
Felker, J.	Mech.Eng.	Repairing and production technology
Molnár, G.	Chem.Eng.	Techniques of coating and anticorrosion protection

Department for business management:

Lehoczky, M.	Agr.Eng.	Head, Economic evaluation of agricultural mechanization
Gockler, L.	Agr.Eng.Dr.Eng.	Cost analysis
Tátrai, G.	Mech.Eng.Dr.Eng. candidate of agr.sc.	Development of machinery systems

Agrotechnical control department:

Szabó, L.	Mech.Eng.	Head, Control of labour safety
Buri, I.	Mech.Eng.	Subsequent control of machinery
Molnár, E.	Mech.Eng.	Labour safety evaluation of agricultural machines
Sitkei, J.	Mech.Eng.	Labour safety of food industrial machines
Haba, J.	Mech.Eng.	Labour safety testing of agricultural machines

Engineering Bureau in Gödöllö:

Sárközi, K.	Mech.Eng.	Head, Development of agricultural machines and techniques
Vargha, G.	Mech.Eng.	Development of technology for potato storing
Bárdos, L.	Mech.Eng.	Economics of mechanization of agriculture
Szlavkovszky, I.	Mech.Eng.	Storage and processing of grains.

Bureau of Computing Techniques:

Nyulas, I.	Economist	Head, System organization
Gonda, I.	System Eng.	Programming
Sabau, J.	Mathematician	Programming

Outposted branches: Engineering Bureau in Szolnok; its task is the introduction of research results into practice. Six mechanized pilot farms, i.e. the State farm of Balatonboglár, Hidashát, Komárom, Szekszárd and Siófok, and agricultural co-operative "Szabad Élet" of Székesfehérvár, for the trial and commercial evaluation of new technologies.

Status: The National Institute of Agricultural Engineering is a state institution established in 1949 to succeed the Agricultural Machinery Experimental Station of Magyaróvár founded in 1869; its superior authority is the Ministry of Agriculture and Food.

Activities: The activities of the Institute are concentrated primarily on solving the following tasks: carrying out research for preparing agricultural machinery systems, for developing mechanized technologies of agricultural production branches (like field crop production, horticulture, animal husbandry) for testing and evaluation of new machines - Exploitation of machinery and development of maintenance, and repair works - Technical and economic analysis and prognostics of development of agricultural mechanization - Improving health protection of workers and labour safety - Agrotechnical supervision of new agricultural machines, equipment and production facilities, carrying out tests of labour safety - Promoting introduction of new processes into production practice.

Publications: The most important results of research carried out by the Institute are published in the proceedings "Mezögazdasági Gépesitési Tanulmányok" (Studies on Farm Mechanization) of which 4 to 8 numbers are issued yearly. The results of investigations of new production-technologies are published in the series "Értesitö Termelési Technológiák Vizsgálatáról" (Reports on Production Technologies). The results of machinery tests are reported in the series "Gépvizsgálati Értesitö (Machinery Test Reports).

Correspondence: Hungarian, English, Russian and German.

Debreceni Agrártudományi Egyetem Kutató Intézete (Research Institute of the University for Agrarian Sciences Debrecen)
H-5301, Karcag, Hungary   Pf: 11.   Telex: 72473 datek h

Borsos, J.	Cand. of agr.sci. agr.eng.	Director
Nyiri, L.	Cand. of agr.sci. agr.eng.	Dep. Director
Kapocsi, I.	agr.eng.	Head of Dep.
Kazó, B.	Cand. of agr.sci. agr.eng.	Head of Dep.
Patócs, I.	Cand. of agr.sci. agr.eng.	Head of Dep.
Csontos, I.	agr.eng.	Head of Dep.

Department for Amelioration of Salt-affected Soils:

Karkalik, A.	agr.eng.	Junior member
Karkalik, Zs.(Mrs.)	agr.eng.	Junior member
Kerékgyártó, I.	agr.eng.	Junior member
Mihályi, I.	agr.mech.eng.	Sci.off. in charge
Patócs, B.(Mrs.)	agr.eng.	Research fellow
Prettenhoffer, I.	Acad. doctor of agr.sci. chem.eng.	Consultant

Department for Amelioration of Sour Soils:

Blaskó, L.	agr.eng.	Research fellow
Balogh, I.	agr.eng.	Research fellow
Kocsis, I.	agr.eng.	Junior member

Department for Soil Cultivation:

Andrási, I.	agr.eng.	Sci.off. in charge
Molnár, M.(Mrs.)	agr.eng.	Research fellow
Pál, K.	agr.eng.	Research fellow
Szilágyi, Gy.	production eng.	Sci.off. in charge

Department for Plant Breeding:

Nagy, K.	agr.eng.	Research fellow
Nagy, B.	prod.eng.	Sci.off. in charge
Lazányi, J.	agr.eng.	Research fellow
Balajthy, K.	agr.eng.	Junior member
Sántha, L.	agr.eng.	Sci.off. in charge

Department for Grass-farming:

Csontos, I.	agr.eng.	Research fellow
Nagy, J.	agr.eng.	Research fellow

Department for Economy:

Nagy, P.J.	agr.eng.	Junior member

Central Laboratory:

Karuczka, A.	chem.eng.	Leader of the Lab.

Number of Experiment Stations: 4, at: Ecsegfalva, Hosszuhát, Kisujszállás, Kunhegyes.

Status: Date of the establishment of the Institute: 1947. Directing and financing body: Scientific Research Major Department of the Ministry for Agriculture and Food.

Agricultural Engineering activities: a) Teaching: - b) Research: Elaborating of soil cultivating methods for the main soil types of the country with irrigation and without irrigation. Improving of soil ameliorating methods for salt-affected soils and sour soils.

Publications: Series "Talajtermékenység" Soil Fertility, Bulletin on research results of soil cultivation, soil amelioration and fertilisation, free of charge for Agr. Cooperative farms, State farms and agr. institutions. Our researchers publish their research results in different periodicals.

Correspondence: English, Russian, German, Hungarian.

Mezötur

Debreceni Agrártudományi Egyetem Mez"ogazdasági Gépészeti F"oiskolai Kar (Debrecen University of Agricultural Sciences - Faculty of Agrarian Engineering)
5401 Mez"otur Tolbuchin ut 2. P.O.B. 27 - Tel. 70 - Telex: 23264

ICELAND

Hvanneyri

Bútaeknideild Rannsóknastofnunar landbúnaðarins (Agricultural Engineering Department, Agricultural Research Institute)
Hvanneyri, Borg

Godmundsson, O.	Cand.agr.	Head of Department Testing of machines
Gudmundsson, B.	Lic.agr.	Research Leader, Grass conservation
Einarsson, G.	Lic.agr.	Research Leader, Farm buildings

Status: Governmental institution.

Activities: Research, experiment and testing work. Testing of farm machinery. Methods for harvesting and conservation of grass. Losses and quality in hay and silage making. Forced air drying of hay. Methods of land cultivation. Studies of farm work simplification. Investigations on stable and sheep shed installation. Engineering in livestock production.

Publications: Official test reports. Articles and papers on farm work organization and methods for hay- and silage making.

Correspondence: Icelandic, Scandinavian languages, English, German.

Reykjavík

Byggingastofnun landbúnaðarins (Farm Building Institute)
Laugavegi 120, Reykjavík

INDIA

Andhra Pradesh

The Andhra Pradesh State Agro Industries Corporation Limited
10-2-3, A.C. Guards, Hyderabad - 500004 - A.P.

Status: The Agricultural Engineering activities of the Department of Agriculture, Government of Andhra Pradesh have been transferred to the Corporation from 1.7.69. The Andhra Pradesh State Agro Industries Corporation is a Government undertaking.

Agricultural Engineering activities: The Corporation maintains: bulldozers for land reclamation of the command areas under irrigation projects in the State; crawler tractors for periodical deep ploughing operations; wheeled tractors for seasonal cultural operations and contract spraying operations; rigs for construction of tube wells both for irrigation and drinking purposes; well revitalization units for rejuvenating the dried up wells.

Correspondence: English

Assam

Department of Agricultural Engineering, Assam Agricultural University
Jorhat : 785013

Majumder, M.	B.E.(Mech), M.Tech.(Agr.Eng.)	Head of the Department - Specialization: Farm Power and Machinery
Dutta, P.K.	B.E.(Mech.), M.Tech.(Agr.Eng.)	Specialization: Post Harvest Technology
Sarma, S.K.	B.E.(Mech), M.Sc.(Agr.Eng.)	Specialization: Post Harvest Technology
Neog, P.K.	B.E.(Mech), M.Tech.(Agr.Eng.)	Specialization: Soil and Water Conservation Engg.
Das, P.K.	B. Tech.(Hons) in Agr.Eng. M.Tech.(Agr.Eng.)	Specialization: Irrigation and Soil and Water Conservation Engg.

Status: Established in 1948, semi-governmental. Since 1969 the Department became a part of the Assam Agricultural University. Financed by the State Government.

Agricultural Engineering activities: a) Teaching: agricultural engineering courses are offered to the students of B.Sc.(Ag.) course and soil and water conservation is offered as an elective subject to the undergraduate students, which is taught in the 3rd and 4th year. Average number of students is 120 for B.Sc. course and 10 for elective course. b) Research: Research and testing is mainly concerned with farm machinery and equipment at present. Extension work is done in collaboration with the extension department of the Assam Agricultural University.

Correspondence: English.

Engineering Research and Development Department, Tocklai Experimental Station, Tea Research Association, Jorhat-8, Assam
Jorhat 8, Assam

Boruah, T.C.	B.Sc.(Hons), M.Sc.(Mech.Eng.)	Head, Engineering Research and Development Department
Dutta, P.K.	B.Sc.	Senior Scientific Assistant

Status: Established in 1951 by the Development Panel of the Indian Tea Association (London). At present it is a branch of the Tea Research Association which is a semi-governmental autonomous association. It is funded privately by subsidies given by the Tea Board and the Council of Scientific and Industrial Research.

Activities: Research is carried out on various processes involved in tea manufacture, viz: harvesting, handling and transport of leaf, withering and processing, fermentation and drying, grading and sorting, storage and packing.

Publications: Articles on all facets of tea culture and manufacture are published in the four quarterly and presently biannual newsletter "Two and a Bud" published by the Tea Research Association.

Correspondence: English.

Bihar

Agricultural Engineering (Research)
Patna, Bihar State, India

Ghosh, M.K.	B.Tech.(Hons) Agril.Eng. Agric.Engineer (Research)	Head of Office - Farm Power and Machinery
Sinha, A.K.	B.Sc. Agril.Engg.	Farm Power and Machinery
Jaiswal, U.S.	B.Sc. Mech.Engg.	" " "
Singh, N.P.	B.Sc. Agril.Engg.	" " "
Singh, I.P.S.	B.Sc. Agril.Engg.	" " "
Kumar, N.	B.Sc. Agril Engg.	" " "

Branch: Regional Centres for field trial of implements: Bhalpur, Muzaffarpur and Ranchi.

Status: Governmental Institution established in 1955. Controlled by the Department of Agriculture, Government of Bihar.

Agricultural Engineering activities: Design, development and testing of various farm machineries and implements, Agricultural Engineering research and field experiment.

Publications: Only performance test data and booklets concerning agricultural implements and machinery are published for departmental use.

Correspondence: English and Hindi.

Ranchi Agricultural College, Rajendra Agricultural University
Kanke, Ranchi, Bihar

Singh, J.K.	B.Sc.Ag.Engg. Associate Professor cum-Sr.Scientist M.Tech.(Agril.Engg.) Ph.D. (Agril.Engg.)	Head, Soil and Water Conservation Engineering
Sinha, A.K.	B.Sc.Ag. Asst.Professor cum-Jr.Scientist M.Sc.Ag. Soil Conservation and Agril.Engg. Ph.D.(Tech)	" " "
Prasad, B.N.	B.Sc.Ag. Asst.Professor cum-Jr.Scientist M.Sc.Ag. Soil Conservation and Agril.Engg.	" " "
Sinha, B.K.	B.Sc.Ag., M.Sc.Ag. Asstt.Professor cum-Jr.Scientist Soil Conservation and Agril.Engg.	" " "
Quraishi, S.S.	B.Sc.Ag. Jr.Scientist cum-Asstt.Prof. M.Sc.Ag. Soil Conservation and Agril.Engg.	" " "
Singh, D.K.	B.Sc.Ag. Jr.Scientist cum-Asstt.Prof. M.Sc.Ag. Soil and Water Conservation Technology	" " "
Singh, V.K.	B.Sc.Ag. M.Sc.Ag. Soil Conservation and Agril.Engg.	" " "

Status: A residential campus of Rajendra Agricultural University, Bihar, P.O. Dholi, Musaffarpur.

Agricultural Engineering activities: a) Teaching: Teaching agricultural engineering subjects to a three year B.Sc. course. b) Research: Engaged in research activities covering some of the aspects of soil and water management. Research on some of the low cost farm implements suited to this region is also going on.

Correspondence: English, Hindi.

Soil Conservation Research Centre
P.O. Jogbani, Distr. Purnea, Bihar

Status: Governmental institution, established in 1956 and supervised by the Ministry of Food and Agriculture, Department of Agriculture, New Delhi.

Activities: Research on hydrology of small watersheds - Various aspects of soil and water conservation.

Correspondence: English.

Delhi

Indian Council of Agricultural Research 'Krishi Bhavan' (ICAR)
New Delhi-1

Randhawa, N.S.	Ph.D.	Deputy Director-General - Soils (Agronomy and Engineering)
Pandya, A.C.	M.S.	Director, CIAE-Bhopal
Nirmal, T.H.	M.S.	Assistant Director-General (Agril. Engineering)
Devnani, R.S.	Ph.D.	Project Coordinator (Farm Implements and Machinery)
Alam, A.	Ph.D.	Project Coordinator (Post Harvest Technology)
Singh, C.P.	Ph.D.	Project Coordinator (Agricultural Energy)
Bhattacharya, A.K.	Ph.D.	Project Coordinator (Wells and Pumps)
Biswas, T.D.	Ph.D.	Project Coordinator (Biogas Technology)
Singh, R.P.	Ph.D.	Project Coordinator (Solar Energy)
Maheswari, R.C.	Ph.D.	Project Coordinator (Power Tillers)

Branches with other locations: The Indian Council of Agricultural Research has 33 Central Institutes under it. Other than the Central Institute of Agricultural Engineering of Bhopal, wholly devoted to Agricultural Engineering aspects, the following ICAR Institutes are also carrying out work in Agricultural Engineering dealing with research problems of Crops and Soils and produces of the area.

1. Indian Agricultural Research Institute, New Delhi
2. Indian Sugarcane Research Institute, Lucknow
3. Central Soil Salinity Research Institute, Karnal
4. Central Rice Research Institute, Cuttack
5. Central Arid Zone Research Institute, Jodhpur
6. Jute Technological Research Laboratories, Calcutta
7. Jute Agricultural Research Institute, Barnackpore
8. Indian Grassland and Fodder Research Institute, Jhansi
9. National Dairy Research Institute, Karnal
10. Indian Veterinary Research Institute, Izatnagar
11. Central Institute of Fisheries Technology, Cochin

Status: The Agricultural Engineering Section at the ICAR-New Delhi was established in 1961.

Agricultural Engineering activities: Research: The Agricultural Engineering Research Programme for the whole country is coordinated and administered by the Indian Council of Agricultural Research. The technical programmes of work for the different centres are formulated and progress of work reviewed at the Annual Workshops of the Coordinated Projects. - The following are the All-India Coordinated Research projects in Agricultural Engineering: Farm Implements and Machinery (FIM) - Energy Requirements in Agric. Production (ER) - Wells and Pumps (W&P) - Post-Harvest Technology (PHT) - Operational Research in Post-Harvest Technology (ORPHT) - Power Tillers (PT) - Solar Energy (SE sponsored by the USDA) - Biogas Technology (BT).

Outposted Branches: ICAR provides financial assistance to Agricultural Engineering Education at the different Agricultural Universities established in the different States of the Country.

Publications: The Council publishes an Annual Report as well as Research Highlights on the work carried out at different Institutes and projects and finances writing of new text books in agricultural engineering subjects. The following journals are also published: Indian Farming - Indian Horticulture - The Indian Journal of Agricultural Sciences - 'Kheti' + Monthly Hindi Journal - 'Phal Phool': Quarterly Hindi Journal.

Correspondence: English/Hindi.

Indian Society of Agricultural Engineers, Water Technology Centre, IARI
Pusa, New Delhi-110012, India

Pathak, B.S. President	Dr., Prof. of Eminence	College of Agril.Engg., Punjab Agril.University, Ludhiana-141004
Verma, S.R. Secretary	Dr., Prof.	Farm Power and Machinery, College of Agril.Engg., Punjab Agril.University, Ludhiana-141004
Zachariah, P.J. Chairman, Industry and Trade Committee	Joint Commissioner	(Machinery), Ministry of Agriculture, Krishi Bhawan, New Delhi
Mehta, M.M. Editor, Agricultural Engineering Today	Prof., Deputy General Manager, Escorts Scientific Research Centre, 25 Mathura Road, Faridabad	
Michael, A.M. Editor, Journal of Agril.Engg.	Dr., Project Director, WTC, IARI	

Status: The Society was established in 1960.

Activities: The activities of the Society are aimed at constant improvement of the individual professional excellence, improvement of the agricultural engineering services provided like the maintenance of agro-machinery, improvement in the designs of farm engineering products, raising of the farm productivity through better equipment and process know-how, agro impplement practices, etc., development of product design and frequent inter-communication between the members, professional organizations, manufacturers, extension agencies farming institutions etc., on matters related to development and use of engineering technology in agriculture.

Publications: Journal of Agricultural Engineering, Agricultural Engineering Today, ISAE-Directory, Special reports.

Correspondence: English and Hindi.

Central Soil and Water Conservation Research and Training Institute, Research Centre
Bellary-2, Karnataka

Chittaranjan, S. Scientist S-2 (Engineering)
Patnaik, U.S.   Scientist S-2 (Agricultural Engineering)

Status: This Centre was established in 1954 by the Ministry of Food and Agriculture, Government of India. However, the administrative control of the Centre was taken over by the Indian Council of Agricultural Research, New Delhi, in 1967.

Agricultural Engineering activities: a) Teaching: Study courses are provided of $5\frac{1}{2}$ months duration each, in Soil and Water Conservation to Agricultural Engineering and Forestry graduates deputed by the State Governments. b) Research: on different methods of irrigation - Terraces - water way and their hydrological problems - Agronomic, Physical, Chemical and Silvicultural aspects of black soils of India.

Publications: The research findings of the Centre are being reported periodically in the Annual Reports of the Centre.

Correspondence: English

Gujarat

Department of Agricultural Engineering, Faculty of Agriculture, B.A. College of Agriculture, Gujarat Agricultural University
Anand Campus, Anand, Gujarat, India-388 110

Mistry, P.D.	M.S., Ph.D.	Head of Department
Savani, M.B.	M.Sc.(Agri.)	
Shekh, A.M.	M.Sc.(Agri.)	

Status: Department is financed by Gujarat Agricultural University which has a semi-governmental status since it is financed by State and Central Government.

Agricultural Engineering activities: a) Teaching of courses in: Agricultural Engineering at the B.Sc. (Agri.) degree course - Meteorology and Agricultural Meteorology at the B.Sc.(Agri.) degree course - Agricultural Meteorology at M.Sc.(Agri.) and Ph.D. level. b) Research: Evapotranspiration - Meteorological aspects of Water Management - Drought climatology - Crop-Weather studies - Radiation Balance in Crop Community.

Publications: 12 Research Papers.

Correspondence: English, Gujarat, Hindi.

College of Agriculture, Gujarat Agricultural University
Junagadh, Gujarat, India

Agricultural Engineering Research and Testing Centre - c/o Research Engineer (Agri. Engineer) Gujarat Agricultural University
Junagadh Campus, Junagadh, Gujarat, India

N.M. College of Agriculture, Gujarat Agricultural University
Navsari Campus, Navsari, Gujarat, India

Department of Dairy Engineering, Faculty of Dairy Science, Sheth M.C. College of Dairy Science, Gujarat Agricultural University
Anand Campus, Anand, Gujarat State, 388110

Shah, U.S.	M.E., Ph.D.(S.P. University)	Head of the Department
Bhadania, P.G.	B.Sc. (Dairy Tech.)	
Patel, B.M.	D.M.E., D.E.E.	
Shah, B.P.	B.E.(Mech), M.E. (Mech)	
Shah, C.M.	B.E. (Electrical), M.Sc. (Dairy Engineering)	

Status: Semi-governmental, financed by Gujarat State and Central Government.

Agricultural Engineering activities: Teaching: Undergraduate level for B.Sc. (Dairy Tech) degree; post-graduate level for M.Sc.(Dairy) Dairy Engineering degree.
- Research: Milk and milk product processing equipment, drying and instantization, storage.

Publications: Eight (of individual staff), Nil for the Department.

Correspondence: English, Gujarati, Hindi.

Department of Agricultural Product Process Engineering, Faculty of Agricultural Engineering and Technology, Sheth M.C. College of Dairy Science, Gujarat Agricultural University
Anand Campus, Anand, Gujarat 388110

Siripurapu, S.C.B.	M.Tech (Agri.Engg), Ph.D.	Head of the Department
Jain, R.K.	M.Tech.(C P E)	

Branches with other location: Gujarat Agricultural University has three campuses located at Dantiwada, Junagadh and Navsari. The other departments of Faculty of Agricultural Engineering and Technology, namely, Department of Farm Engineering, Department of Rural Engineering and Department of Soil and Water Engineering are located at Junagadh, Dantiwada and Navsari respectively.

Status: Semi-governmental financed by the Gujarat Agricultural University.

Agricultural Engineering activities: Teaching: Post-graduate courses in Agricultural Process Engineering leading to M.Tech. and Ph.D. — Research: Pulse processing, optimization of operating parameters of domestic flour mill for pulse milling, testing of Janata Chakhi for cereals, pulses and spices, solar drying of onions.

Correspondence: English, Gujarati, Hindi.

Department of Rural Engineering, Faculty of Agricultural Engineering and Technology Gujarat Agril. University S.K. Nagar
Dantiwada, Gujarat 385506

Haryana

Engineering Division, National Dairy Research Institute
Karnal-132001, Haryana

Aneja, V.P.	M.Tech.	Agric. Engineering
Agrawala, S.P.	Ph.D.	Agric. Engineering (Dairy Engineering)
Anap, G.R.	M.Tech.	Agric. Engineering (Process Engineering)
Bikram, K.	B.Tech.	Agric. Engineering
Kohli, R.K.	M.Sc.(Dairying)	Dairy Engineering
Makkar, S.M.	B.Tech.(Hons)	Agric.Engineering
Rawal, S.R.	B.Tech.	Agric.Engineering (Process Engineering)
Samuel, D.V.K.	M.Tech.	Agric.Engineering (Process Engineering)
Sawhney, I.K.	B.Tech.	Agric. Engineering

Branches with other location: National Dairy Research Institute, S.R.S., Bangalore.

Status: Established in 1955 as an autonomous body under Indian Council of Agricultural Research.

Activities: Teaching: B.Sc. Dairy Technology, M.Sc. Engineering and Diploma Dairy Engineering courses are offered to approximately 65 students every year. - Research: Milk Scale formation in evaporators and pasteurizer - Harnessing locally available energy sources for dairy farming - Utilization of solar energy for dairy process - Compressed air/inert gas agitation in dairy equipment - Animal work efficiency studies - Rheology of viscous milk products - Fluidized bed casein dryer - Development of low cost UHT processing equipment.

Publications: Design and development of continuous chhana making machine - Intermediate cooling of milk for long distance transport - Specific heat of concentrated whole milk at higher temperature - Development and testing of soybean steam blancher - Determination of particle size distribution and density of ghee residue for the development of centrifugal ghee clarifier - Study of ghee drainage rate through residue cake - Its application in the design of filtering centrifuge - Development of dung collector - Development of ghee filling machine - Instantaneous determination of moisture content in ghee - Development of khoa making machine - Development of village level milk dehydrator - Dairy effluent and discharge and equipment for their disposal.

Correspondence: English.

Kandarya Mrida Lavanayata Anusandhan Sansanthan - Central Soil Salinity Research Institute
Karnal, India

Pandey, R.N.	M.Tech. Agril.Engg.	Head, Department of Irrigation, Drainage and Agricultural Instrumentation
Batta, H.K.	M.Sc.(Soil and Water Engg)	Drainage and Irrigation
Dhruvanarayana, V.V.	Ph.D.(Civil Engg)	Hydrology and Water Resources
Gupta, R.K.	B.Tech.(Agril.Engg)	Irrigation and Drainage of Salt affected soils
Kamra, S.K.	B.Tech.(Agril.Engg)	
Luthra, S.K.	B.Sc.(Electronics and Telecommunication)	
Singh, O.P.	M.Tech.(Irrigation and Drainage Engg.)	
Tiwari, A.K.	B.Tech.(Ag.Engg)	
Tyagi, N.K.	M.Tech.(Agril.Engg) with specialization in Soil and Water Engineering.	

Status: Established in 1971 as semi-governmental institute.

Agricultural activities: a) Teaching: Ph.D. classes in Agricultural Engineering. Short-term courses of 4-6 weeks duration for inservice training. - b) Research: Surface, sub-surface and vertical drainage methods. Leaching of saline and alkali soils, Hydrological evaluation. Irrigation methods and instrumentation.

Publications: Engineering aspects of land and water management - Dugout ponds for surface and drainage and irrigation - Problems and methodology in watershed research - Surface drainage methods for saline-alkali soils - Design of surface drainage system - Waterlogging of irrigation areas - Effect of water quality on infiltration - Conveyance of water through underground pipelines in tubewell irrigated areas - Performance of tile drains under different spacing - Analysis of water table fluctuations for the study of aquifer properties and water balance of small catchments - Water table responses due to sub-surface drainage - Management of sloping borders for paddy cultivation in alkali soils - Mathematical modelling of salt movement in soils - Equations to predict the leaching of soluble salts in saline soils - Determination of dispersion and adsorption coefficient - Dispersion of surface applied salts - Simplified equations for reactive solute transport in soils - Procedure to avoid trial and error in land levelling calculations - Crop and water yields as affected by rainwater storage in paddy catchments - A field evaluation - Time average concept in design of surface irrigation system - Estimating of water application efficiency in irrigation checks - Salt balance and leaching efficiency - Percolation losses in check irrigation - Irrigation scheduling in canal command areas - Annual report.

Correspondence: English, Hindi.

Himachal Pradesh

Central Potato Research Institute
Simla-171001 (H.P.), India

Nagaich, B.B.	Ph.D. Plant Pathology	Head of Department
Shyam, M.	M.Tech.(Agric.Engg)	
Singh, V.	M.Sc. (Agric.Engg)	Agril. Structure and Processing

Branches: The Institute has 10 regional stations located at Kufri (H.P.), Jullundr, Modipuram (Meerut), Rajgurunagar (Poona), Ootacamund (T.N.), Morena (M.P.), Patna (Bihar), Darjeeling (W.B.), Shillong (Meghalaya) and Kodaikanal (T.N.), but the engineering cell is located at present at Jullundur only.

Status: Date of establishment 1957. The governing body which directs and finances the institution is the Indian Council of Agricultural Research (Governmental).

Agricultural Engineering activities: a) Teaching mechanization of potato cultivation, processing and storage to the participants of different short-term courses offered by the Institute. b) Research: Design, development, field evaluation and extension of machinery and equipment used for the cultivation, processing and storage of potatoes.

Publications: articles, leaflets on machinery and equipment used for cultivation and processing of potatoes.

Correspondence: English, Hindi.

Central Potato Research Institute - Outposted Potato Research Station
Jullundur, Punjab   Tel.: Simla - 3118, 5545
                                                          2088, 5170
                                            Jullundur   72514

Singh, V.	M.Sc.	Outposted research officer, Agril.Engg.
Shyam, M.	M.Sc.	" " " " "

Status: Autonomous institution under Indian Council of Agricultural Research, New Delhi, supervised by the Director, Central Potato Research Institute, Simla (H.P.)

Agricultural Engineering activities: Design and assessment of new implements, such as a potato planter-cum-ridger with fertilizer attachment, potato harvesters, potato graders - Survey, collection and study of indigenous implements used in potato cultivation - Investigation on the suitability of power-driven machinery for potato cultivation.

Correspondence: English, Hindi.

Kerala

Department of Agricultural Engineering, College of Agriculture, Kerala Agricultural University
Vellayani, Trivandrum, Kerala, India

John, P.J.	B.Sc. Engg; M.Sc. Engg., M.I.S.A.E.	Associate Professor and Head of Dept.	Soil Mechanics
Thomas, M.S.	B.Sc. Engg., M.Sc. Engg., M.I.S.A.E.	Assistant Professor	Structural Engineering
Rema Devi, A.N.	B.Sc. Engg., M.Sc. Engg., M.I.S.A.E.	Assistant Professor	Soil Mechanics
Jippu, J.	B.Sc. Ag.Engg., M.Tech., A.M.I.S.A.E.	Assistant Professor	Farm Power and Machinery

Status: Established in 1955 under the Department of Agriculture, Government of Kerala. Constituent College of the Kerala Agricultural University from 1.2.1972.

Agricultural Engineering activities: a) Teaching: Study courses in agricultural engineering for students preparing for B.Sc. degree in agriculture. The duration of the course for the degree is four years consisting of twelve trimesters. The number of students admitted is about 60 per year. b) Research: Wind powered water pump, seed driers and seed drills.

Publications: Some guidelines for achieving agricultural mechanization in Kerala - Popularisation of pedal paddy threshers, prospects and problems - Development of a seedling picking mechanism - Electrophysiology in plant pathology.

Correspondence: English, Malayalam.

Department of Agricultural Engineering, College of Horticulture
Vellanikkara P.O., Trichur, Kerala State, India

Madhya Pradesh

Tractor Training and Testing Station, Tractor Nagar
Bundi 466 445, Madhya Pradesh, India

Misra, S.K.	Director
Dewan, M.R.	Regional Coordinator
Luthra, H.C.	Chief Instructor
Jain, S.C.	Senior Training Officer
Singh, D.N.	Agronomist
Prasad, J.	Test Engineer

Status: Government institution, established in 1955.

Activities: a) Training: To impart 'on-the-job' training to farmers, rural youth artisans, mechanics, undergraduates in agril. engineering and agro-entrepreneurs on the selection, operation, maintenance and repairs of tractors and allied machinery. - b) Testing: To conduct tests on tractors, power tillers, diesel engines, irrigation pumps, threshers, combine harvesters, plant protection equipment and power operated agril. implements in common use, with a view to assess their performance and durability under Indian farming conditions and to ensure that their quality is maintained.

Correspondence: Hindi and English.

Agricultural Engineering Office, Department of Agriculture
Putalighar, Berasia Road, Bhopal, Madhya Pradesh

Machine stations with workshops in: Bhopal, Indore, Itarsi, Khandwa, Vidisha.

Activities: Supply of tractors, bulldozers, pumping sets, improved and efficient agricultural implements, well sinking appliances, harvesting and threshing machinery, farm implements - Investigation of bullock and small power operated machinery - Training of rural youth in operation, maintenance and repair of improved agricultural implements, tractors and equipment.

Publications: Report on areas in Madhya Pradesh which are suitable for irrigation wells.

Correspondence: English, Hindi.

Agricultural Engineering Section, College of Agriculture
Gwalior, Madhya Pradesh

Agricultural Engineering Section, College of Agriculture
Indore, Madhya Pradesh

College of Agricultural Engineering, Jawaharlal Nehru Krishi Vishwa Vidyalaya
Madhya Pradesh

Status: This college is a constituent unit of the Jawaharlal Nehru Agricultural University, Jabalpur.

Activities: Study courses of 5 years duration are provided for the degree of B.Tech. Agricultural Engineering - Agricultural engineering, physics, mathematics and statistics courses are offered in the affiliated colleges to graduates in agriculture - Research and extension activities are carried out in agricultural engineering.

Maharashtra

Agricultural Engineering Section, College of Agriculture
Akola, Maharashtra State

Status: Governmental institution.

Activities: Study courses in Agricultural Engineering - Measurement of power requirement of bullock-drawn indigenous agricultural implements - Improvement of agricultural implements.

Correspondence: English.

Agricultural Engineering Section, College of Agriculture
Kolhapur, Maharashtra State, India

Ghatge, V.S.   Head of Section

Status: Governmental institution.

Activities: Courses in Agril. Engineering, Meteorology, Statistics, Mathematics; B.Sc.(Agril.) - Research on processing, farm power, farm implements.

Correspondence: English.

Agricultural Engineering Section, College of Agriculture
Pune-5, Maharashtra State

Sinnarkar, V.V.	B.Sc.(Ag) Hons. B.Sc.(Ag.Engg) M.Tech.	Head of Section, Soil and Water Conservation
Bhilare, R.M.	M.Sc.(Agril)	
Bote, N.L.	B.Tech.(Ag.Engg)	
Dhotrey, R.S.	B.Sc.(Ag.) Hons. B.Tech.(Ag.Engg)	
Gutal, G.B.	B.Tech.(Ag.Engg)	
Pacharne, D.T.	B.Tech.(Ag.Engg) M.Tech.(Ag.Engg)	

Status: Semi-Government institution established in 1907.

Activities: Teaching: Courses are taught to undergraduates in soil and water engineering - Agricultural structures and rural electrification - Farm machinery and power - Surveying and levelling - Algebra, trigonometry, measuration and co-ordinate geometry - Statistical methods and design of experiments - Agricultural meteorology - work experience in shop practice and tractor driving - Advance courses in upkeep and maintenance of I.C. engines, electrical motors and pumps - Irrigation engineering - Agricultural process engineering - Upkeep and maintenance of tractor - Advanced soil conservation practice - Upkeep and maintenance of plant protection equipment-calculus - Testing of agricultural implements and machinery for local suitability - Extension work on various agricultural engineering problems.

Publications since 1965: Research articles and extension bulletins.

Correspondence: English.

Department of Agri.Engineering, College of Agriculture, Marathwada Agril. University
Parbhani (Maharashtra), India   431402

Rapte, S.L.	M.Sc.(Agril. Engg)	Farm Machinery
Adkine, B.D.	M.Sc. (Agril.Engg)	
Digrase, L.N.	M.Sc. (Agril.Engg)	Soil and Water Engg.
Gore, K.P.	M. Tech.	Soil and Water Engg.
Holsambre, D.G.	M.Sc.(Agril. Engg)	Soil and Water Engg.
Mahajan, R.B.	M.Tech.	Crop Process Engg.
Quadri, S.A.	M.Tech.	Farm Machinery and Power
Vaishnav, V.G.	M.Sc. (Agril.Engg)	

Status: College established in 1956. Semi-Government.

Activities: a) Teaching: Agril.Engineering subjects viz (Farm Machinery and Power, Soil and Water Engg., Agril. Structures etc.) - Teaching to undergraduate and postgraduate students of Agril.Science, for the B.Sc.(Agril.) and M.Sc.(Agril) degree course - Average n. of students for U.G. courses: 100-150 - Average number of students for P.G. courses: 3 to 6. - b) Research: Design and Development of Bullock drawn implements - Testing of improved bullock drawn implements - Drainage layout experiments - Soil conservation experiments. - c) Extension: Demonstration of new implements to farmers - Training Courses in Agril.Engineering for Agril. Supervisors, Agril. Assistant, School teachers, and cultivators of the region. Average of 40 students attend the courses.

Publications: On research subjects in Agril. Engineering viz Farm Machinery, Soil Water Engineering, Crop Processing.

Correspondence: English.

Mysore

Agricultural Engineering Institute
P.O. Box 24, Raichur-584101, Karnataka State

Maurya, N.L.	B.Sc. Agril.Engg., (Farm Power and Machinery) Ph.D.	Principal

Department of Farm Machinery:

Naravani, N.B.	B.E.(Mech), M.Sc. Agril.Engg. (Farm Power and Machinery)	Head of Section
Guruswamy, T.	B.E.(Mech), M.Sc. Agril.Engg. (Farm Power and Machinery)	Asst. Prof. F.M. Section
Chowdegowda, M.	B.E.(Mech), M.Tech.Agril.Engg. (Agricultural Process Engineering	Instructor F.M. Section
Raj, F.I.	B.E.(Mech)	do.
Ramakrishnappa, V.	B.E.(Mech)	do.

Department of Soil and Water Management:

Chaturbhujanathaiah, K.C.	B.Sc., B.E.(Civil), M.S. (Tech)	Head of Department, S.W.M. Section
Nisar Ahmed	M.Sc.(Agril.Chem. and Soils)	Asst.Prof. in S.W.M. Section
Jahagirdar, S.V.	B.Sc., D.A.E., B.Tech. Agril. Engg.	Instructor in S.W.M. Section

Status: Established 1969. Governing Body: University of Agricultural Sciences, G.K.V.K. Bangalore.

Agricultural Engineering activities: a) Teaching: Study course in Agricultural Engineering. Diploma in Agric. Engineering. Average number of students: 70. -
Research: in progress on the following subjects: Farm Power and Machinery: Studies on energy required for selected farm operations - Location-allocation of Agro-Service Centres under selective Mechanization - Draftability of bullocks - Effect of different parameters on field efficiency - Development of improved country seed drill - Comparative study of puddling equipment - Improvement of a power tiller operated blade harrow - Development of an all-purpose machine to power the farm appliances - Improvement of manually operated cotton dibbler - Survey of agricultural implements for Raichur Region - Development and evaluation of hoe groundnut crop - Testing and evaluation of plant protection equipment. Post-harvest Technology and Agricultural Process Engineering: Studies on the crop and machine parameters on the performance characteristics of groundnut pod separator - Design, development and evaluation of groundnut pod separator-cum-cleaner - Development of composite plastic grain storage structures - Evaluation and testing of winnowing equipment for different crops - Modification of grain thresher for safflower. Soil and Water Management Engineering: Study of conveyance performance of field channels using different lining materials for Raichur Region - Study of soil loss using run-off plots for the soils of Raichur Region.

Publications: Articles and bulletins.

Correspondence: English.

College of Agriculture, University of Agricultural Sciences
Dharwar 5, Mysore State

Branches with other location: Department of Agricultural Engineering, College of Agriculture, Hebbal, Bangalore.

Status: The Department of Agricultural Engineering was started in 1949 under the Government. Since October 1965 it is part of the University of Agricultural Sciences.

Agricultural Engineering activities: Teaching: Courses are offered to undergraduate and Post-graduate students of the College of Agriculture. - Research: Work on development of devices to reduce labour in cultivation of crops and in harvesting - Threshing and crop drying experiments.

Publications: Reduction of cultivation costs by use of improved implements - The use of farming machinery in agriculture in Mysore State - Studies on the effect of different tillage practices on the rates of infiltration of water into red soils - Influence of tractor and tractor-drawn implements on the physical properties of soils - Sedimentation in reservoirs and precaution to reduce sedimentation - Influence of cushioning materials on the critical height of fall of eggs - An automatic egg collector-cum-counter for use in poultry houses.

Correspondence: English, Kannade.

Department of Agricultural Engineering, College of Agriculture
Hebbal, Bangalore 560024, Mysore State

Status: The Department was established in 1964, under the University of Agricultural Sciences, Hebbal, Bangalore.

Agricultural Engineering activities: Training of students of B.Sc.(Agr) in three compulsory courses viz. surveying, farm power and machinery, and soil and water conservation. Also offers coaching of Final Year students of B.Sc.(Agr) to specialize in agric. engineering in any four of the subjects listed under Farm mechanics, Servicing and maintenance of farm tractors, Irrigation engineering, Farm electrification, Drainage engineering, Measurement and conveyance of irrigation water, tillage machinery and implements. Farm workshop, tractor operations, crop protection equipment, techniques in farm mechanics, pumps and pumping, soil erosion and its control.
- Research: Design and development of an adjustable bullock-drawn seed-cum-fertilizer drill for all low crops - Study of the problems of storage of farmers in Mysore State to evolve suitable structures for storage - Studies on sprayers - Design and development of cheap and light turnover plough - bullock drawn. Study of the different methods of grain drying: (a) batch type dryer, (b) solar heated air - Effect of peripherical velocity of beater or drum on the performance of power thresher - Variation of ambient temperature in poultry houses due to solar radiation - Design and development for rub type thresher for crops like Ragi, Jowar, etc.

Correspondence: English.

Orissa

College of Agricultural Engineering and Technology, Orissa University of Agriculture and Technology
Bhubaneswar-751003, Orissa

Lal, R.	B.Sc.(Ag.Engg), M.Sc.(Agri.Engg) Ph.D.(Agr.Engg)	Dean, Faculty of Agril.Engineer. and Technology, Irrigation and Drainage Engg.
Sharma, S.D.	B.Sc.Agl.Engg. and Tech., M.Tech.(Agril.Engg), Soil and Water Cons.Engg.	Soil and Water Engineering
Das, G.K.	B.Sc., B.Sc.(Engg) Mech., M.E.Mech(Applied Thermo Science)	Mechanical Engineering
Sharma, K.N.	B.Sc.Agril.Engg. and Tech., M.Tech. (Ag.Engg) Soil and Water Cons.Engg.	Soil and Water Engineering
Ray, A.C.	B.Sc.Engg.(Civil and Mechanical Engg). Ph.D. Agricultural Process Engg.	Civil Engineering and Post-harvest Technology
Senapati, P.C.	B.Sc.(Agril.Engg. and Tech.) M.Tech.(Agril.Engg.) Soil and Water Cons.Engg.	Soil and Water Engineering
Das, D.K.	B.Sc.(Agril.Engg.Tech) M.Sc.(Agril.Engg.)	Farm Machinery and Power

Sahoo, R.K.	B.Sc.(Engg.)Mech.	Mechanical Engineering
Dash, S.K.	B.Sc.(Agril.Engg. and Tech.)	Agril. Engg.
Nayak, S.C.	B.Sc.(Agril.Engg. and Tech.)	Agril. Engg.
Mohanty, S.N.	B.Sc.(Agril.Engg. and Tech.) M.Tech.(Ag.Engg.) Crop Process Engg.	Agril. Process Engg.
Sahu, S.K.	B.Sc.(Agril.Engg. and Tech.)	Agr.Engg.
Pradhan, S.C.	B.Sc.(Agril.Engg. and Tech.) M.Tech.(Agril.Engg.) Farm Machinery and Power	Farm Machinery
Nanda, S.K.	B.Sc.(Ag.Engg. and Tech.) M.Tech.(Ag.Engg.) Farm Machinery and Power	Farm Machinery and Power
Satpathy, G.C.	B.Sc.(Ag.Engg. and Tech.)	Farm Machinery and Power

Status: Established in 1966 as a constituent College at Orissa University of Agriculture and Technology. The University is an autonomous Organization and it is financed by the Government of Orissa and Indian Council of Agricultural Research.

Agricultural Engineering activities: Four year study course leading to B.Sc.(Ag.Engg. and Tech.) degree after I.Sc. and attended by an average number of 40 students. – Design and development of open wells and cavity tube wells – Surface irrigation systems – Paddy drying – Grain storage structures – Development of a puddler – Evaporation and seepage studies – Surface investigation of ground water – Extension of improved agricultural implements and irrigation practices.

Correspondence: English, Oriya.

Agricultural Engineering Division, Central Rice Research Institute
Cuttack 6, Orissa

Status: Semi-governmental institution under the control of Indian Council of Agricultural Research, New Delhi – The Agricultural Engineering Division of the Institute was established towards the end of the year 1959.

Activities: Agricultural engineering is part of a curriculum for post-graduate training of rice research workers, each course lasting six months – Short refresher courses for village level extension workers and subject matter specialists lasting one to two weeks – Design of rice implements – Testing of field performance of implements and machinery used in rice cultivation – Studies of methods of land preparation, sowing transplanting, weeding, harvesting and threshing – Studies on mechanization of rice cultivation.

Publications since 1960: Power tillers in rice cultivation – Improved implements for rice cultivation – A technique for determining the degree and depth of soil puddle – Harvesting rice with the reaper.

Correspondence: English, Hindi.

P u n j a b

College of Agricultural Engineering, Punjab Agricultural University
Ludhiana 141004

Status: Established in 1965. The Governing body is the Board of Management and financing bodies are the Punjab Government and the Indian Council of Agricultural Research.

Agricultural Engineering activities: Teaching: Study courses at undergraduate level leading to a B.Sc.Ag.Eng. degree – Study courses at postgraduate level with specialization in Farm Machinery and Power, Processing and Structures, and Soil and Water Engineering, leading to M.Sc.Ag.Eng. and Ph.D. degrees. – Research: Experimental work in Farm machinery design and development – Storage of potato and onion – Grain storage – Wells and tubewells – Mechanized grain handling – Field drainage – Testing centre engaged in testing and evaluating farm machinery – Extension activities in popularization of farm machinery.

Publications: Articles in ISAE.

Correspondence: English.

Rajasthan

Dayanand College
Ajmer, Rajasthan

Status: Private institution.

Agricultural engineering activities: Study courses as part of B.Sc. degree training in agriculture – Testing and extension work.

Correspondence: English, Hindi.

Rajasthan Agricultural Engineering Board, Government of Rajasthan
Jaipur 6, Rajasthan

Outposted workshops at: Chittorgarh, Hanumangarh, Kotah, Nagaur, Pali, Sojat.

Status: Governmental institution, established in 1966.

Activities: Research, experiment and testing work – Manufacture of improved farm implements and machinery – Introduction of improved farm equipment to the farmers – Contractual and custom work with bulldozers and other agricultural machinery.

Correspondence: English, Hindi.

Agricultural Engineering Department, College of Agriculture
Jobner, Rajasthan

Status: Semi-governmental institution, affiliated with the College of Technology and Agricultural Engineering in Udaipur.

Activities: Courses in Agricultural Engineering are offered, as prescribed under the curriculum leading to B.Sc.(Ag) degree – Research on irrigation methods – Soil and water management – Design and testing of animal as well as tractor-drawn farm implements – Comparative studies on water lifting devices.

Publications: ASAE and ISAE journals.

Correspondence: English, Hindi.

College of Technology and Agricultural Engineering, University of Udaipur
Udaipur, Rajasthan

Department of Farm Machinery
Department of Processing Engineering
Department of Soil and Water Conservation
Department of Civil Engineering
Department of Mechanical Engineering
Department of Physics
Department of Mathematics

Status: University and government institution established in 1964.

Agricultural Engineering activities: Study courses of 5 years duration, leading to B.E.Ag. degree – Research, experiment and extension work on: Agricultural hydrology, Irrigation systems, performance of agricultural machinery and implements, processing of agricultural products.

Publications: Annual reports – Student project – College magazine.

Correspondence: English.

Tamil Nadu

Department of Agricultural Engineering
Madras-5

Radhakanth, P.K.	B.E.	Chief Engineer, (AE)

College of Agricultural Engineering, Tamil Nadu Agricultural University
Coimbatore-641003, Tamil Nadu, India

Sivanappan, R.K.	B.E. M.Tech. FIE, MISAE, AM.ASAE	Dean, College of Agricultural Engineering, and Agricultural Engineering leader

Department of Soil and Water Conservation:
a) Main

Sivanappan, R.K.	B.E. M.Tech. FIE, MASE, AM.ASAE	Professor and Head
Chandrasekaran, D.	B.E., M.Sc.(Engg.), Ph.D.	Associate Professor
Kumaravel, S.	B.E.(Ag.)	Assistant Professor

Padmakumari, O.	B.Sc.(Engg.)	Assistant Professor
Arumugam, K.	B.E.(Ag.)	Research Associate
Senthilvel, S.	B.E.(Ag.)	Research Associate

b) **P.L. 480 Scheme on Soil conservation measures on dryland based on soil loss equation:**

Nachappan, K.M.	B.E.(Civil)	Associate Professor
Gandhi, M.	B.E.(Civil)	Assistant Professor
Santhana Bose, S.	B.E.(Ag.)	Research Associate
Murugeson, S.M.	B.E.(Ag.)	Research Associate
Newman Bright, A.	B.E.(Ag.)	Research Associate

c) **P.L. 480 Scheme on Land and Water use planning of watershed:**

Kotteswaran, M.	B.E.(Civil)	Assistant Professor
Mylsamy, C.	B.E.(Ag.)	Research Associate
Balasubramaniam, F.	M.E.(Ag.)	Research Associate

d) **ICAR Scheme on "Methods of construction of open and tube wells and deepening the existing open wells under quick sand condition to increase yield":**

Kumar, V.	M.E.(Ag.)	Research Associate

Department of Farm Machinery:

a) **Main:**

Swaminathan, K.R.	B.A., B.Tech(Hons) M.Tech.	Professor and Head
Job, T.V.	B.Sc.(Engg)	Assistant Professor
Chinnan Chetty, G.	B.E.(Ag.)	Assistant Professor
Shanmugasundaram, C.R.	L.A.E.	Supervisor
Venkatsubramaniam, M.	B.E.(Elec.)	Instrumentation Engineer
Thangavelu, S.	B.E.(Ag.)	Research Associate
Kamaraj, S.	B.E.(Ag.)	Research Associate

b) **ICAR Scheme: Energy Requirement in Intensive Agric.Production:**

Ramakrishnan, S.	M.Sc.(Ag.)	Assistant Professor
Kombairaj, S.	M.Sc.(Ag.)	Assistant Professor
James Martin, G.	M.Sc.(Ag.)	Assistant Professor

c) **ICAR Scheme on Biogas Technology:**

Rajasekaran, Dr.P.	M.Sc.(Ag.), Ph.D.	Associate Professor
Vijayaragavan, N.C.	M.Sc.(Engg.)	Assistant Professor
John Felix Fernando	B.E.(Ag.)	Assistant Professor

d) **Department of Agro-Industries:**

Kunhi Mohammed, K.T.	M.Sc.(Engg)	Assistant Professor
Shanmuganathan, K.V.	M.Sc.(Engg)	Assistant Professor
Manian, R.	B.E.(Mech)	Assistant Professor

Department of Agricultural Structures:

Karai Gowder, K.R.	B.E.(Civil)	Associate Professor
Natarajan, P.	M.Sc.(Engg)	Assistant Professor
Samraj, R.	B.E.(Ag.)	Assistant Professor
Mohan, M.	B.E.(Ag.)	Research Associate
Vaigundamoorthy, M.	B.E.(Ag.)	Research Associate
Rangasamy, M.V.	B.E.(Ag.)	Research Associate

Department of Agricultural Processing:

Dr. Subramanian, V.	M.Sc.(Engg),Ph.D.	Professor
Balasubramanian, M.	B.Tech,	Assistant Professor
Murugesan, V.	B.E.(Mech.)	Assistant Professor
Muniraj, K.G.	B.E.(Ag.)	Research Associate
Selvaraj, P.K.	B.E.(Ag.)	Research Associate
Arjunan Muthiah, R.	B.E.(Ag.)	Research Associate

b) **ICAR Scheme Studies on Harvest and Post-harvest Technology:**

Latchmanane, S.	M.Sc.(Ag.)	Assistant Professor
Selvararaj, K.V.	M.Sc.(Ag.)	Assistant Professor
Heeralal, B.	B.E.(Ag.)	Assistant Professor
Duraisamy, G.	M.Sc.(Engg)	Assistant Professor
Parvathy, K.	M.Sc.	Research Associate

c) **ICAR Scheme on Operational Research in the field of Post-harvest Technology:**

Ramanathan, M.	B.E.(Mech)	Assistant Professor
Srinivasan, S.	B.E.(Mech)	Assistant Professor
Tamilmani, D.	B.E.(Ag.)	Research Associate

d) **Zonal Research Centre:**

Swaminathan, K.R.	B.A., B.Tech(Hons), M.Tech.	Senior Research Engineer
Shanmugam, A.	M.Sc.(Engg)	Assistant Professor
Sambathrajan, A.	B.E.(Mech)	Assistant Professor
Rangasamy, K.	B.E.(Mech)	Assistant Professor
Kailappan, R.	B.E.(Ag.)	Assistant Professor
Sathyamoorthi, H.M.	B.E.(Ag.)	Assistant Professor
Padmanaban, R.M.	B.E.(Ag.)	Research Associate
Devendran, C.	B.E.(Ag.)	Research Associate

e) **ICAR Scheme - Manufacture of Prototype of selected farm machinery:**

Karunanidhi, R.	B.E.(Mech)	Assistant Professor

f) **P.L. 480 Scheme on Solar Energy:**

Devadoss, C.T.	B.E.(Mech)	Assistant Professor
Sankara Narayanan, M.	B.E.(Ag.)	Research Associate

Status: Date of establishment of Agricultural Engineering College: 1-6-1972; it is a semi-governmental body.

Agricultural Engineering activities: a) Teaching: The courses offered in Agricultural Engineering are: Under-graduate programme: Five-year B.E.(Ag.) degree after completion of pre-university course and four year after (10+2) course under Trimester system.
- Post-graduate programme: M.E.(Ag.) course in two subjects, i.e. Farm Power and Machinery and Soil and Water Conservation Engineering are offered from the year 1977. —
b) Research: Evaluation of different irrigation methods for cash crops and principal crops - Study of different lining materials - Studies on effect and efficiency of soil and water conservation measures - Water optimization studies in drip and sprinkler irrigation - Studies on evaporation reduction - Saline water for irrigation - Soil conservation measures for dry lands based on USLE - Evaluation of an irrigation system and formulation of an operational manual for major irrigation projects - Studies on water use planning of

watershed – Research works on method of construction of open and tube well under quick sand conditions – Pilot project for water management in LBP project – Study of consumptive use for sorghum, pulse, crops and vegetable crops using nutron probe – Investigation of effect of subsurface drainage on crop production under saline water in black cotton soils – Development of water production functions for water resource planning – Current research on standardization of type designs for thrashing floors, drying floor, farm roads, cattle sheds, poultry sheds, storage structures – Design and development of sewage gas plant – grain storage bins – Economical cases for rearing of poultry – Planning and arrangement of layout of stalls in dairy barns – Design and development of cheap outlet – Structures for low pressure pipe lines – Pre-fabricated toilet, on-farm pre-fabricated irrigation distribution system – Poultry waterer, feeder and feeder development – Design and development of special type of sprayers – Power weeder for dryland – Biogas operated engine – Wind mill – Sugarcane harvester – Bullock drawn pebble collector – Tree climber – Pilot plant for furfural extraction from agricultural wastes – Study on microbial retting on jute and mesta – Energy requirement of crop production – Microbial degradation of farm wastes for biogas generation – Utilization of agricultural wastes for paper production – Multipurpose tool bar – Power operated ground digger and stripper – Dry farming implement design for moisture conservation – Design of pump for shallow tube wells – Development of tractor drawn bed former – Solar energy system for drying equipment – Self-propelled paddy harvester – Bullock drawn multipurpose tool carrier – High pressure sprayer – Different types of seed drills – Design and development of different types of threshers and harvesters – Study and evaluation of solar cabinet dryer – Solar still – Solar stream generator – Development of different types of weeders – Driers for tapioca chiles and chillies – Processing by mechanical low cost storage structures and bins – Chemical control of store pests – Biochemical changes in pearl millet, finger millet during storage – Design of farm level parboiling unit – fibre extractor for banana sheeth – Processing techniques in dhall making of pigeon pea – Utilization of agri. waste materials for cattle feed – Making insulation boards from paddy husk – Grading arecanut and potato – Study of bulk density of cereal crops at various soil moisture contents.

Publications: in Madras Agricultural Journal; Irrigation Era, Madras; Tamil Arasu; ISAE; Kisan World; Journal of South Indian Horticulture; Indian Farming; The Agricultural Engineer.

Correspondence: English.

Agricultural College and Research Institute, Madurai
Madurai-625 104 – Tamil Nadu, India

Muthusamy, K.	B.E., M.E.	Head of Department – Associate Professor Specialization: Water Resources Development
Balasubramanian, V.	M.E., M.S. (M.E. (Hydraulics and Hyd. structure)– M.S. (Mech. and Hydraulics))	Divisional Engineer, Water Management Scheme
Gothandapani, L.	B.E., M.Sc.(Eng.)	Assistant Professor – Production Engineering
Kumar, V.J.F.	B.E.(Ag.)	Research Associate – Agricultural Engineering
Sankaralingam, S.	B.E.(Ag.)	Research Associate – Agricultural Engineering
Singaravelu, M.	B.E.(Ag.)	Research Associate – Agricultural Engineering

Status: Established in 1966 as semi-Government Institute.

Agricultural Engineering Activities: Teaching: Courses: Farm Machinery and Implements; Soil and Water Conservation Engineering; Irrigation and Drainage. Degree: B.Sc.(Ag.). Students: 80 per batch. - Research: Evaluation of irrigation methods for different horticultural crops. Moisture conservation methods for dry farming. Studies on different devices of grain storage for small farmers. Effect of puddling implements on percolation loss and water use efficiency in rice crop. Design and development of a manually operated centrifugal sprayer. To study and develop the design criteria of furrow irrigation for efficient water use. To study various irrigation efficiencies in sprinkler irrigation. Design and development of equipment for spot application of fertilizers and pesticides. Effect of drift of Agro chemicals used in different spraying systems.

Publications: Articles on preservation of paddy and reduction of losses in storage. Articles on various Agricultural Engineering subjects were published in local vernacular.

Correspondence: English.

Central Soil and Water Conservation Research and Training Institute - Research Centre 'Patna House' - Ootacamund
Ootacamund - 643 001 (The Nilgiris), Tamil Nadu, India

Jayakumar, M.	B.Sc.(Eng.), Dip in Water Resource and Land Use Planning	Scientist (Eng.)
Padmanabhan, M.V.	A.M.I.E.	Senior Technical Assistant
Pooranachandran, G.	Diploma in Civil Engineering	Senior Technical Assistant
Jeevarathnam, K.	Diploma in Civil Engineering	Technical Assistant
Mohanraj, R.	Diploma in Agricultural Engineering	Mechanic
Chandran, B.	Diploma in Civil Engineering	Overseer-cum-Draftsman

Status: Established in 1955 under Government of India. Transferred to the Indian Council of Agricultural Research with effect from 1-10-1967.

Agricultural Engineering activities: Teaching: Inservice training course of $5\frac{1}{2}$ months for assistants with 25 participants in each course. - Inservice training also given now in soil survey and soil conservation to agricultural and engineering graduates employed under the Scheme for Rural Engineering Surveys. Duration 3 months and 1 month respectively. - Research: Field experiments on conservation measures such as bench terracing, trenching - development of methods and criteria for selection of types and dimensions. Rainfall studies with reference to Soil and Water management. Run-off and soil loss studies estimation - evaluation of land use practices. Infiltration studies. Estimation of soil moisture from climate. Hydrologic studies on catchments.

Publications: Articles: Estimation of area lost under bench terracing and contour bunding - Estimation of rainfall erosion index from mass and maximum intensity - Estimation of soil moisture from rainfall and temperature: two parts. Estimation of soil moisture from antecedent precipitation index. Analysis of rainfall and climatic data for soil and water management. Effectiveness of bench terracing. Conservation evaluation of land use practices - two articles. Inter-relationship of run-off, soil loss, rainfall, etc. Optimum length of bench terrace - Infiltration characteristics - Annual Reports.

Correspondence: English.

Uttar Pradesh

Agricultural Engineering Department, Allahabad Agricultural Institute, Allahabad
Allahabad - Uttar Pradesh   Tel.: 7210

Status: Private institution, established in 1910. Supervised by a Board of Directors, and affiliated to the Allahabad University.

Activities: Teaching: B.Sc. Course in Ag. Engineering, M.Sc. Course in Ag. Engg. - Design and Development activities: One of the major activities of the Department is Design and Development of Agricultural Implements and Machines, Crop processing machinery, dairy machinery and machines used in soil and water conservation engineering etc. The department supplies blue-prints and trial models of successful prototypes to manufacturers of agricultural machinery at a nominal cost. - Research: In addition to the M.Sc.Ag.Engg. students research projects there are several minor projects of the staff of the department such as: bin drying of grains, design parameters and soil parameters on the traction of wheels on deformable surface, liquid fertilizer applicator and water flow measuring device; studies on sprayers and driers for unthreshed crops. - The Agricultural Engineering Department also conducts short term refresher in-service courses on agricultural engineering subjects.

Publications: The Allahabad Farmer.

Correspondence: English, Hindi.

Central Soil and Water Conservation Research and Training Institute
Dehra Dun, Uttar Pradesh - 248195

Sastry, G.	B.Tech(Hons), M.Tech. Dipl. in Hydrology	In charge of Hydrology and Engineering - - Soil and Water Conservation and Watershed Hydrology and Management
Katiyar, V.S.	B.Tech(Hons) Ag.Engg. Certif.course in Soil and Water Conservation	Scientist S-1 - Engineering
Sikka, A.K.	B.Sc.(Agril.Engg), M.Tech. Certif.course in Soil and Water Conservation	Scientist S-1 - Soil and Water Engg.
Juyal, Gopal Prasad	B.Tech.(Agril.Engg) Certif.course in Soil and Water Conservation	Scientist S-1 - Engineering
Joshi, B.P.	B.Sc.(Engg) Water and Soil Engg.	Scientist S-1 - Engineering
Bansal, R.C.	Assoc. M.I.E.(India) (Civil Engg.) Soil and Water Conservation and Watershed Hydrology and Management	Scientist S-2 - Engineering
Sharda, V.N.	M.Tech. Water and Soil Engg.	Scientist S-1 - Engineering
Chandra Prakash	M.Tech. Water and Soil Engg.	Scientist S-1 - Engineering
Rao, D.H.	Dipl. in Civil Engg. Water and Soil Engg.	Jr.Sci. (Engg)
Nema, J.P.	M.Tech. Water and Soil Engg.	Scienstist S-1 - Engineering
Chittaranjan, S.	B.Tech(Hons); M.Tech. Water and Soil Engg.	Scientist S-3 - Engineering

Patnaik, U.S.   B.Sc.(M.P.C.), B.E.(Civ.)   Scientist S-2 - Engineering
                 M.Tech.(Ag.Engg)
                 Soil and Water Conservation
                 and Watershed Hydrol. and Management
Jayakumar, M.   B.Sc.(Engg)   Scientist S-2 - Engineering
                 Water and Soil Engg.
Tiwari, A.K.   M.Tech.   Scientist S-1 - Engineering
                 Soil and Water Engg.

Branches at Bellary, Ootacamund, Kota, Vasad, Agra, Chandigarh.

Status: Semi-governmental institution under the Indian Council of Agricultural Research.

Activities: Research: Hydrological studies on experimental watersheds, hydrometeorology vis-a-vis land uses - effectiveness of mechanical practices - evaluation of gross-erosion vis-a-vis land uses - water balance - lysimeter studies, methods of surface drainage - farm pond vis-a-vis seepage and water use - stabilization of land slides and correction of torrents. - Training: i) at Central Soil and Water Research and Training Institute, Dehra Dun: Advance study courses in Soil and Water Conservation of 5½ months duration for Supervisory level personnel. Average number of trainees 25 per course. Short specialized courses on specific topics, on request. - ii) At the Research Centres of the Institute located at Kora (Rajasthan), Bellary (Karnataka), Ootacamund (Tamil Nadu) and D.V.C. Hazaribagh (Bihar): Advance study courses in Soil and Water Conservation of 5½ months duration for Technician level personnel. Average number of trainees 25 per course.

Publications: Consolidated Annual Half-yearly and Quarterly reports.

Correspondence: English.

Agricultural Engineering Division, Indian Institute of Sugarcane Research
Lucknow-226 002 U.P., India

Sharma, M.P.   B.Sc. Ag.Engg. and Tech.   Leader - Scientist Grade S-2
                                             Farm Machinery and Power
Singh, P.R.   M.Tech. Farm Machinery   Scientist Grade S-1
                 and Power

Status: Established 1952. It is a Central Government Organization under the overall administrative control of the Indian Council of Agricultural Research, New Delhi.

Agricultural Engineering activities: a) Teaching: Conduct short-term training courses for sugarcane research and development workers - Conduct training for undergraduate and postgraduate students of Agricultural Engineering having specialization in Farm Machinery and Power as a requirement of their degree programme. - b) Research: Design and development of animal-drawn/tractor/power tiller drawn implements for sugarcane and sugarbeet cultivation - Testing and field evaluation trials under different agro-climatic soil and regions of equipment designed and developed at this Institute - Manufacturing of prototypes - Studies on irrigation and drainage requirement of sugarcane and sugarbeet crops under Indian conditions.

Publications: Semi-automatic and fully automatic sugarcane planters - Plant protection equipment for spraying tall cane crop - Tractor drawn/Animal drawn sugarbeet drills - Moist hot air treatment plant and relevant basic studies on heat treatment with special reference to control of diseases in sugarcane and deep trench making equipment.

Correspondence: English and Hindi.

Department of Agricultural Engineering - Pant College of Technology - G.B. Pant University of Agriculture and Technology
Pantnagar (Nainital), U.P.

Status: The Department of Agricultural Engineering was established in 1962. It is a part of the College of Technology. Financial assistance is received from the Indian Council of Agricultural Research and the Uttar Pradesh State Government.

Activities: Courses leading to B.Tech. Agril. Engineering - M.Tech.(Irrigation and Drainage) - M.Tech. (Farm Machinery and Power) - M.Tech. (Soil and Water Cons. Engineering) - M.Tech. (Processing and Storage). - Research: projects in progress: Electrical analogy studies on hydraulics of wells - Advance and recession studies in border irrigation - Investigations of surface and sub-surface drainage - Resistance of saturated soil and design of traction devices - Testing, design and development of Soybean machinery - Moisture transfer from soil surface under different conditions - Drying of grains and other food material - Storage Engineering studies - New cropping pattern and water use (Engineering part) - Investigations of length of run on runoff - Runoff from small watersheds - Coordinated research scheme on energy requirement - Handling of granular food materials - Study of gravel filters for well and tile drains. - Farm Equipment Testing on: different types of farm implements, equipment and water pipes. These tests are conducted at the request of firms concerned against payment. The department is developing better facilities for testing of farm equipment and devices to guide manufacturers. - Extension: The department is carrying out an Agril. Engineering Extension programme in 9 districts of U.P., India. The programme consists of: Demonstration with improved agricultural implements - Conducting of training on farm machines and equipment - Publication of technical literature - Organisation of farmers exhibition, Agro-industrial exhibition and Agricultural Machinery parade - Carrying out technical advisory work on the manufacture of agricultural implements - Development of improved implements - Imparting technical guidance in the area of farm development, system layout of irrigation, drainage etc. to the farmers of selected districts.

Publications: Research papers.

Correspondence: English.

West Bengal

Patshan Krishi Gabenshanagar (Jute Agricultural Research Institute)
Barrackpore - 743101, West Bengal, India

Mandal, T.C.	B.Tech(Hons), M.S., Ph.D.	Head of Section, Agricultural Engineering (Farm Machinery and Processing)

Branches with other location:
Ramie Research Station, P.O. Sorbhog, Dist.Kamrup, Assam, India. - Sisal Research Station, P.O. Bamra, Dist.Sambalpur, Orissa, India. - Sunnhemp Research Station, P.O. and Dist. Pratapgarh, U.P., India. - Central Nucleus Jute Seed Multiplication Farm, P.O. Bud Bud, Dist. Burdwan, West Bengal, India.

Status: Established in 1964, financed by Indian Council of Agricultural Research, New Delhi.

Agricultural Engineering activities: Research on the development, modification and testing of seeding, intercultural, harvesting and processing equipment for jute and allied fibres like sisal, ramie, sunnhemp and flax.

Publications in J.Agric. Engg.

Correspondence: English, Bengali and Hindi.

Directorate of Agricultural Engineering, Government of West Bengal
5 Mustaque Ahmed St. Calcutta 16

Bhattacherjee, B.K.	B.Sc. Engg.(Agri.) F.I.E., M I A H	Chief Engineer (Agriculture) West Bengal - Agricultural Engg. Hydrology of Ground Water
Dasgupta, G.S.	B.M.E.	Superintending Engineer (Agri-Mechanical)
Banerjee, P.K.	M I E, M I A H	Superintending Engineer (Agri-Irrigation)
Bandopadhyay, N.K.	B.Sc., B.Tech.(Hons), M.Tech.	Superintending Engineer (Agri-Mechanical)
Bhuti, H.S.	M I E	Superintending Engineer (Agri-Irrigation)
Roy, T.P.	B.Tech(Hons), M.Tech. Diploma in Farm Mech.	Superintending Engineer (Agri-Mechanical)
Moulik, B.	B.E. (M.E.)	Superintending Engineer (Agri-Mech.)
Pal, N.K.	B.Sc.(Cal), B.E.(C.E.)	Superintending Engineer (Agri-Irrigation)
Banerjee, B.K.	D T R P, B.E.(C.E.), M.E.(P H)	Superintending Engineer (Agri-Irrigation)
Bhattacherjee, M.M.	B.M.E.	Superintending Engineer (Agri-Mech.)

Activities: Hydrological study of potential Ground Water in the two southern districts of West Bengal - Aquifer parameters of Lower Damodor belt of West Bengal, as revealed from pumping tests - Long-term measures against draught - Systematic geohydrological survey - Improvement of strength of concrete by using cut mild steel wires as reinforcement - An introduction of submersible pump at R.L.I. centre - Safe exploitation of Ground Water resource - Deep Tubewells - Development of wells by compressed air - Use of Gamma Transmission for soil moisture determination.

Publications: in Annual Souvenir of S A E A, West Bengal.

Post-Harvest Technology Centre, Indian Institute of Technology
Kharagpur - 721302, West Bengal, India Tel. Hijli 221 to 224

Ojha, T.P.	B.Sc., AE, M.Tech., Ph.D., M.I.S.A.E., M.I.S.T.U.S.	Professor and Head
Bal, S.	B.Sc., B.Tech(Hons), M.S., Ph.D.	Assistant Professor - Processing
Bhole, N.G.	B.Sc.Agri., B.Sc. AE, M.Tech., MIE, M.I.S.A.E.	Professor - Processing
Chakravarty, A.	B.E.(Chem.Engg.), Ph.D.	Lecturer - Processing
Chattopadhyay, P.K.	B.Tech., Ph.D.	Lecturer - Processing
Mukherjee, K.K.	M.A.(Eco)	Marketing Economist - Economics and Marketing
Mukherjee, R.K.	M.Sc., D.Phil.	Rice Technologist - Plant Physiology and Food Technology

Status: Established 1970. Governmental Institution.

Agricultural Engineering activities: Teaching: The Centre offers a post-graduate degree course (M.Tech.) in Post Harvest Technology. No. of students per year: 10 approx. Scholarship under Colombo Plan with GOI for International studentship. Undergraduate and postgraduate courses to the students of Agri.Engineering Department. - Training: The Centre organizes three short duration professional training courses for rice mill engineers, rice mill managers and rice mill operators. Summer and winter schools are also organized for University teachers and research scholars on various aspects of post-harvest technology. On request of governments and international bodies of other countries a few specialized short term training courses have also been organized in specific areas of interest. No. of trainees as of 1973: 729. - Research activities: The Centre provides facilities for advanced research on all aspects of post-harvest technology of cereals, pulses and oilseeds leading to the degree of Ph.D. No. of scholars working per year: 10. - Collaborative research projects with the state governments, cooperative bodies, rice mills associations and other private organizations. Need base projects to demonstrate the use of appropriate technology to various agencies are taken up every year. - Research Areas: Harvesting Studies; Drying and Dryers; Parboiling; Milling; Storage; Energy; By-products Utilization; Agricultural Economics and Management.

Publications: The Centre publishes a biannual journal "RPEC Reporter" and in addition small bulletins on various aspects of post-harvest technology.

Water Management Practices, Kalimpong
P.O. Kalimpong, Dist. Darjeeling, West Bengal, India

Adhikary, B.K.	B.Sc.Engg.(Civil) MIE, India	Agricultural Engineering leader

Status: The Institution started in 1973 and the Executive Engineer (Agri-Irrigation) joined in 1977. The Institution is a Govt. body. It is directed and financed by the Indian Council of Agricultural Research, New Delhi and Govt. of West Bengal.

Agricultural Engineering activities: Research - This institution is inter-disciplinary in approach and mainly aims at developing integrated technology for solving specific problems pertinent to the area with the object to utilize both land and water more efficiently for sustained agriculture.

Publications: Results of practical utility emanating from research works are published annually.

Correspondence: English.

Agricultural Engineering Department - Indian Institute of Technology
Kharagpur - 721 302 - West Bengal

Datta, R.K.   B.Tech.(Hons), Ph.D.   Professor and Head of Department

Status: Government Institution, established in 1952.

Teaching activities: The Department offers the following regular courses: Agricultural Engineering (5 years leading to B.Tech.) - Postgraduate degree courses in soil and water conservation engineering, Farm machinery and power, Dairy and food engineering and Crop process engineering (2 years after B.Tech. leading to M.Tech).Post-grad.diploma courses in Farm management technology, Soil science, and Applied botany (1-year after M.Sc. leading to D.I.I.T.). The undergraduate course is attended by about 30 students per year and the post-graduate courses are attended by 45 students per year. -
Research activities: The Department provides facilities for intensive research in different areas of agricultural engineering and agricultural sciences leading to the degree of Ph.D. To date 99 candidates earned Ph.D. degree. 30 candidates are working for their Ph.D. degrees in various disciplines. Research areas: Soil and water conservation - Farm machinery - Dairy and food engineering - Agricultural sciences.

Publications: Annual issues of Research Summaries, and a technical journal: 'The Harvester'.

Correspondence: English.

Rice Process Engineering Centre - Indian Institute of Technology
Kharagpur, West Bengal

Status: Governmental Institution, established in 1970.

Teaching Activities: The Centre offers three short duration courses in rice processing for rice mill engineers, rice mill managers and rice mill operators. It teaches agricultural processing courses to the B.Tech. and M.Tech. students of Agricultural Engineering Department. - Research activities: The Centre provides the facilities for research in the fields of processing of food grains, i.e. parboiling, drying, milling, storage, grain properties and by-products utilization leading to the degree of Ph.D.
- Research in drying, parboiling, milling, storage, properties of paddy grains and by-products, harvesting, marketing economics.

Publications: Research summaries and technical bulletins on rice processing.

Correspondence: English.

INDONESIA

Bandung

Direktorat Penyelidikan Masalah Air (Directorate of Hydraulic Engineering)
Jl. Ir. H. Juanda 193, Bandung   Tel.: 84553, 84554

Rachmat Tirtojondro   Director

Hydrology Division:	Head :	Mr. Muhadi	
Hydraulic Division:	"	Ms. Sulastri Djainuddin	
Water Engg. Division:	"	Mr. Taulu	
Technology Extension Division:	"	Mr. Murwanto Martadimono	
Administrative Div.:	"	Mr. Willy Haryono	

Jakarta

<u>Sub-Directorate Agricultural Mechanization Development – Directorate of Production Development – Directorate-General of Food Crops</u>
Jl. Ragunan, Pasarminggu, Jakarta

Subagyo Wiryosumarto   Chief

<u>Division of Agricultural mechanization for farmland development:</u>
          Head :   Dadang Permana

<u>Division of Agricultural mechanization for processing and storage:</u>
          Head :   R.A. Hamid

<u>Division of Rural technology development:</u>
          Head :   Mr. Suwardjo

<u>Division of Evaluation and standardization of agricultural machinery and mechanization:</u>
          Head :   B. Gultom

Yogyakarta

<u>Bagian Mekanisasi Pertanian, Fakultas Mekanisasi Pertanian, Universitas Gadjah Mada</u>
(Agricultural Engineering Division, Faculty of Agricultural Technology, Gadjah Mada University)
Bulaksumur, Yogyakarta

Dr. Sumangat   M.Sc.   Head

<u>Division of Land and Water Conservation Technique:</u>
Hendro Pawoko Said   Dr.   Head

<u>Division of Agricultural Machinery and Mechanization:</u>
Sunyoto Sumodihardjo   Prof.   Head

Jurusan Mekanisasi Pertanian, Institut Pertanian Bogor (Agricultural Engineering Department, Bogor Agricultural University)
P.O. Box 122, Bogor, Indonesia

Soepardjo, S.	Prof.Ir., MSAE, Ph.D.	Head of the Department
Djojomartono, M.	MSA, Ph.D.	Secretary
Abdullah, K.,	Dr.	Senior Instructor, Product Processing Engineering
Asep, S.	Ir.	Instructor, Soil and Water Engineering
Dastaman, A.	Ir.	Instructor, Farm Power and Machinery
Daywin, F.J.	Ir.	Assistant Professor, Farm Power and Machinery
Dhalhar, M.A.	Ir., MSAE, Ph.D.	Senior Instructor, Soil and Water Engineering
Djojomartono, M.	MSA, Ph.D.	Senior Instructor, Farm Power and Machinery
Eriyatno	Ir., MSAE, Ph.D.	Senior Instructor, Product Processing Engineering
Gardjito	Ir., MSAE	Instructor, Product Processing Engineering
Hardjoamidjojo, S.	Ir.MSAE	Senior Instructor, Soil and Water Engineering
Imam, H.	Ir.	Instructor, Farm Power and Machinery
Irwanto, K.	Ir.	Instructor, Farm Power and Machinery
Kalsim, D.K.	Ir., M.Eng.	Instructor, Soil and Water Engineering
Katu, L.	Ir.	Senior Instructor, Farm Power and Machinery
Kumendong, J.	Ir.	Instructor, Product Processing Engineering
Kusen	Ir.	Instructor, Farm Power and Machinery
Lumintang, T.	Ir.	Instructor, Farm Power and Machinery
Muchlis, A.	Ir.	Senior Instructor, Product Processing Engineering
Pakpahan, D.	Ir.	Associate Professor, Farm Electrification
Partowijoto, A.	Ir.	Associate Professor, Soil and Water Engineering
Pramudya, B.	Ir.	Instructor, Farm Power and Machinery
Pratomo, M.	Ir., M.Sc.	Assistant Professor, Product Processing Engineering
Priyanto, H.A.	Ir.	Assistant Professor, Soil and Water Engineering
Priyanto, S.M.	Ir.	Instructor, Product Processing Engineering
Purwadaria, H.K.	Ir., Ph.D.	Senior Instructor, Product Processing Engineering
Rangkuti, P.	Ir.	Senior Instructor, Farm Electrification
Sarwono, S.	Ir.	Senior Instructor, Farm Electrification
Semat, S.	Ir.	Assistant Professor, Farm Structure
Sembiring, E.M.	Ir.	Instructor, Farm Power and Machinery
Siregar, N.	Ir.	Senior Instructor, Farm Electrification
Sitompul, R.G.	Ir.	Assistant Professor, Farm Power and Machinery
Soepardjo, S.	Ir., MSAE, Ph.D.	Professor, Farm Power and Machinery
Sukartaatmadja, S.	Ir.	Senior Instructor, Soil and Water Engineering
Syarief, ST, R.	Ir.	Senior Instructor, Farm Electrification

Status: Governmental Institution under the Bogor Agricultural University, supervised by the Ministry of Education and Culture. Established in 1964.

Agricultural Engineering activities: Teaching: Four years curriculum leading to "Ir" degree in Agricultural Engineering. Courses also available for graduate study leading to Master and Doctorate Degrees majoring in Agricultural Engineering. The Department of Agricultural Engineering has three divisions: Farm Power and Machinery - Soil and Water Engineering - Product Processing Engineering, Farm Electrification and Farm Structure. - Research, Extensions: Research activities in Department of Agricultural Engineering include experiments, testing, design and extension work related to various fields, such as Farm Power and Machinery, Soil Water Engineering, Product Processing Engineering, Farm Electrification, Farm Structure, Energy in Agriculture, Management and System Analysis.

Correspondence: Indonesian, English, Dutch, German, Japanese.

Balai Penelitian Perkebunan Bogor (Research Institute for Estate Crops)
Jalan Taman Kencana No. 1, P.O. Box 81, Bogor, West Java, Indonesia

Sadikin Sumintawikarta	Ir.-Agronomy	Director

Crop Production in Bogor:
Angkapradipta, P.	Ir.	Soil Science, Head of Division
Hardjono, A.	M.Sc.	Soil Science
Abdul, M.	M.Sc.	Breeding
Sukarja, D.	Ir.	Breeding
Dedy, S.	Dr.	Breeding
Amin, T.	Ir.	Agronomy
Rachmat, W.	M.SC.	Agronomy
Bambang Odang, M.	Ir.	Agronomy
Tirtoboma	Ir.	Agronomy
Wardojo, S.	Dr.Ir.	Entomology
Didik, S.	Ir.	Entomology
Supadmo, B.	Ir.	Plant Pathology
Sukirman, P.	Ir.	Plant Pathology
Soedarsan, A.	Dr.Ir.	Weed Science
Basuki	Dr.	Weed Science
Sumarjono	Ir.	Weed Science
Muharminto	Drs.	Agricultural Economics
Arief, P.	Ir.	Agricultural Economics
Dundawa	Ir.	Social Economics
Jimmy Teh	Ph.D.	Physiology
Etty, W.D.	Dra.	Physiology
Nurita Toruan	Dra.	Physiology

Crop Production Division in Jember:
Sunarjo	M.Sc.	Agronomy, Head of Division
Saleh, M.	Ir.	Soil Science
Hartana, I.	Ir.	Breeding
Sidharta, H.	Ir.	Agronomy
Sulistiyo	Ir.	Agronomy
Situmorang, S.	Ir.	Agronomy
Soedarsono	Ir.	Agronomy
Nano, P.	Ir.	Entomology
Zulkifli, M.	Ir.	Technology

Technology Division in Bogor:

Walujono, K.	Drs.	Chemistry, Head of Division
Abednego, J.G.	Drs.	Technology
Barjono, H.	Ir.	Technology
Soemarno, K.	Ir., M.Sc.	Technology
Budiman, S.	M.Sc.	Technology
Sugianto, S.	Ir.	Technology
Utami, T.	Ir.	Technology
Soewarti, S.	Dr.Ir.	Technology
Eddy Junadi Amir	Ir.	Technology
Oskari, A.	Drs., M.Sc.	Technology
Suharto	Drs.	Technology
Ridha Arizal	Drs.	Technology

Branch Office - Crop Production Division in Jember, Jalan Moh.Seruji 2, Jember, East Java, Indonesia

Status: Semi Governmental institution, established in 1926, directed by the Agency for Agricultural Research and Development, and funded by the Government Estates.

Agricultural Engineering activities: All research work is directed towards the highest possible production and quality of the plantation crops, such as rubber, coffee, cocoa and tobacco at the lowest production cost, including the technical aspects of planting, plant breeding, propagation, soil mapping, optimum fertilization, crop protection, eradication of weeds, yield stimulation, processing of the production and quality control. The Institute conducts also extension work and extends services for analyzing agricultural chemicals.

Publications: Communications of the RIEC (irregular); Menara Perkebunan, a continuation of "De Bergcultures" (bi-monthly); Rubbber Bibliography, Coffee Bibliography, Cocoa Bibliography, Rubber Statistical Report, Coffee Statistical Report, Cocoa Statistical Report and Annual Report (in Indonesian)

Correspondence: Indonesian, English.

Lembaga Penelitian Tanaman Industri (Research Institute for Industrial Crops)
Jalan Cimanggu 1, Bogor, West Java, Indonesia

Lembaga Penelitian Hutan (Forestry Research Institute)
Jalan Gunung Batu, Bogor, West Java, Indonesia

Lembaga Pusat Penelitian Pertanian (Central Research Institute for Agriculture)
Jalan Merdeka 99, Bogor, West Java, Indonesia

Pusat Penelitian dan Pengembangan Ternak (Central Research and Development for Animal Husbandry) Cibedug - Ciawi, Bogor, West Java, Indonesia.

IRAN

Karadj

Department of Agricultural Engineering, Division of Farm Power and Machinery, College of Agriculture, Tehran University
Karadj, Iran

Borghei, A.M.	Dr.Ag.Eng.	Associate Professor and Head
Behrooszi-lar, M.	Ph.D.Ag.Eng.	Assistant Professor
Ehsani, A.	Dr.Electricity	Assistant Professor
Fatehi, D.	Dr.Ag.Machinery	Assistant Professor
Rahbar, M.	Dr.Physics	Associate Professor
Shafii, A.	B.Sc.Ag.Machinery	Assistant
Taher Khorramabadi, A.	B.Sc.Ag.Machinery	Assistant
Tabesh, F.	Dr.Agric.Machinery	Associate Professor
Zahedi Fard, A.	B.Sc.in Physics	Assistant

Status: Governmental institution, established in 1958. Financed and supervised by the University of Tehran.

Agricultural Engineering activities: Teaching: The College of Agriculture has a four years curriculum for B.Sc. and a fifth year for M.Sc. The two first years are general for all students - from the 3rd year students are selected for each of 12 majorings - After finishing the four years curriculum, students are awarded a B.Sc. degree in farm power and machinery engineering - The graduated who have obtained their degree with a good result can study the fifth year after which they are awarded a M.Sc. degree in farm power and machinery engineering. Number of students who are accepted each year in this department: 15.
Research activities: Study of Wear on Agricultural Implements - Study on the Technical and Economical Aspect of Alfalfa Harvesting - Automatisation of Irrigation Systems.

Publications: Bulletins (CEEMA).

Correspondence: Iranian, French, English.

Department of Agricultural Engineering, Division of Farm Power and Machinery, College of Agriculture, Tehran University
Karadj

Correspondence: Iranian, English, French.

Tehran

Agricultural Engineering Department, Ministry of Agriculture
Tehran

Correspondence: Iranian, English.

IRAQ

Abu Ghraib

Agricultural Engineering Section, College of Agriculture
Abu Ghraib

Status: Governmental institution

Activities: Study courses as part of general agricultural course

Publications: "Problems of Farm Mechanization in Iraq and their Proposed Solutions".

Correspondence: Arabic, English.

Baghdad

Department of Irrigation and Agricultural Machinery - College of Engineering - University of Baghdad
Baghdad

Hamad, Safa N.	Ph.D.	Instructor - Head of the Department
Kharrufa, Najib S.	Ph.D.	Professor
Mohammed, Abdul-Ilah Y.	Ph.D.	Instructor
Eloubaidy, Aziz F.	Ph.D.	Instructor
Hindee, Walid N.	Ph.D.	Instructor
Al-Jalil, Hamid F.	Ph.D.	Instructor
Mohammed Ali, Ahmed A.	Ph.D.	Instructor
Al-Masri, Naufal A.	Ph.D.	Instructor
Shukri, Tarik A.	Ph.D.	Instructor
Izzat, W.R.	M.Sc.	Assistant Instructor
Abdullah, Farouk	M.Sc.	Assistant Instructor
Kubba, F.A.	M.Sc.	Assistant Instructor

Status: Governmental Institution, established in 1970.

Directorate-General of Agricultural Machinery, Ministry of Agrarian Reform
Sa'doon Street, Baghdad

Technical Division

Operational Division

Pumps Division

Civil Engineering Division

Machine hire stations in: Abu Ghraib - Mosul - Kirkuk - Kut

Mechanical Centres in: Arbil – Sulaimaniya – Dayali – Ramadi – Karbala – Hilla – Diwaniya – Nassiriya – Amara – Aziziya – Basra – Al-Suwaira – Latifiya.

Mechanical units in: Salman Pak – Balad – Tarmiya – Tikrit (Baghdad) – Hawaija (Kirkuk) – Al-Kassim – Mussayib Project – Kafl (Hilla).

Centre for Agricultural Mechanization – Sweira (established in 1978 with the assistance of UNDP and FAO).

Status: Governmental institution, established in 1943.

Activities: The Directorate is responsible for the planning, coordination, provision and supervision of mechanical work and equipment for larger Government development work – Occasional training courses for tractor drivers and mechanics working in the Directorate and its branches.

Correspondence: Arabic, English.

IRELAND

Carlow

Agricultural Engineering Department, The Agricultural Institute (An Foras Taluntais), Oak Park Research Centre, Carlow
Tel. (0503) 31425   Telex: 33038

Cunney, M.B.	B.E., C.Eng.	Head of Department. Grain storage and drying. Machinery Management. Energy in agriculture.
Comerford, P.J.	B.Agr.Sc., M.S.	Forage Mechanization. Materials handling. Electric fencing.
Fortune, R.A.	B.Agr.Sc.	Soil tillage; crop establishment; fertilizer handling and distribution.
O'Callaghan, C.	B.Sc.(Chem)	Grain drying.
Rice, B.	B.E., M. Eng.Sc., M.I.E., C.Eng.	Root and vegetable mechanization, crop spraying equipment; root crop storage. Automatic data analysis.

Status: Established in 1960; State-sponsored organization: financed mainly from Government funds.

Machinery Testing activities: Originally the main work of the department but now discontinued except for special tests for manufacturers.

Research activities: Tillage for root and cereal crops; direct drilling, costs of grain and sugar beet harvesting systems. Sugar beet storage; fertilizer handling and distribution. Liquid manure handling; comparison of forage harvesting methods; Tests on grain dryers; design of ventilation systems for grain stores. Energy use in Agriculture. Alternative fuels.

Publications: Sixty machinery test reports on individual machines and comparisons. Annual Research Reports. Occasional reports and leaflets on research topics.

Correspondence: English, German, French.

Cork

Dairy and Food Engineering Department, Faculty of Dairy and Food Science, University College
Cork

Synnott, E.C.	BE, MEngSc, PhD, CEng, FIEI, FIFST	Professor
MacCarthy, D.A.	BE, MD(ChemEng), MS(Food Sc), MIEI	Lecturer

Status: Established in 1927. Financed by Department of Agriculture and Fisheries, and operated as a Department of University College, Cork, a constituent college of the National University of Ireland.

Activities: (a) **Teaching**. Dairy Engineering taken as part of a two-year Diploma Course in Dairy Science attended by 35 students per year. – Meat Engineering taken as part of a two year Diploma Course in Meat Science attended by 10 students per year. – Dairy and Food Engineering taken as part of a four year degree course for B.Sc. in Dairy and Food Science attended by 40 students per year. (b) **Research** on: Electrical methods of testing characteristics of foods – Fire and explosion hazards in milk powder dryers – On-farm cooling of milk – Thermal properties of food materials.

Publications: Power demand variation with time during butter churning – Chruning time and energy variation with churn parameters – Power variation during making of butter – Effect of refrigerant charge level on performance of immersion coolers – Developments in dairy effluent treatment methods in Ireland – Effect of some physical properties of milk powder on minimum ignition temperature – Effect of temperature on thermal characteristics of milk powders – Variation in dielectric properties of butter with frequency, temperature and season – Effect of moisture content, working time, pressure on sample and hard and soft fat fractions on dielectric properties of butter at 300 KHZ.

Correspondence: English.

Dublin

Soil Physics and Pedology Department, The Agricultural Institute (An Foras Taluntais),
Kinsealy, Malahide Road, Dublin 5.

Telephone: 460644
Telex: 31479

Burke, W.	B.Agr.Sc., M.S.	Head of Department
Bulfin, M.	B.Agr.Sc. (Forestry), M.F.S.	Soil survey, land use classification, forestry
Galvin, L.F.	M.E., A.M.I.Mun.E., A.M.I.C.E.I.	Land drainage and reclamation
Gleeson, T.	B.Agr.Sc.	Wet land management: land drainage and reclamation
Jelley, R.M.	M.A., Ph.D.	Soil structure, soil moisture.

Status: Established in 1959. State-sponsored organization: financed mainly from Government funds.

Activities: Field studies in land drainage, reclamation and hydrology. Investigation of causes and effects of treading damage to pasture by livestock on heavy soils. Investigations of structure in selected soils. Experimentation in soil structure improvement. Agricultural climatology, soil survey, land use, forestry.

Publications: Climate, wet land management, drainage, peat reclamation, land grading and planing, soil survey, land use.

Correspondence: English, German, French.

Farm Structures and Environment Department, The Agricultural Institute (An Foras Taluntais),
19 Sandymount Avenue, Dublin 4.   Tel.: 684711

Tuite, P.J.	B.Arch., Dip: T.P., M.R.I.A.I., M.I.P.I.	Head, Farm Structures & Environment Department, Economics & Rural Welfare Research Centre.
Daly, O.G.	B.E., D.P.A., C.Eng., M.I.E.I.	
O'Farrell, F.	B.Arch., M.R.I.A.I.	
Kavanagh, A.J.	B.Arch.	

Status: Established in 1960. State-sponsored organization: financed mainly from Government funds.

Activities: Research: <u>Pigs</u>: Housing and environmental requirements of dry sows, farrowing sows and fattening pigs – <u>Cattle</u>: Housing and environmental requirements of calves, breeding stock and fattening stock – <u>Dairy Herds</u>: Housing and environmental requirements of calves, milking cows and replacement stock. Design of milking parlours and installations. Design of handling and treatment facilities. – <u>Sheep</u>: Housing and environmental requirements for fattening stock. Design of handling and treatment facilities – <u>Storage</u>: Storage and handling of animal feedstuffs and implements – <u>Manure</u>: The collection handling, storage and disposal of animal wastes and silage effluent – <u>Farm Dwellings</u>: The design of standard plans for dwellings and ancillaries – <u>Rural Environment</u>: The impact of agricultural practices on rural environment and the mitigation of undesirable results.

Publications: Standard design drawings for various farmyard enterprises and dwelling houses, including the adaptation of existing buildings to new uses, Handbooks on cattle enterprises and dairying. Publications in scientific journals and journals for the industry.

Correspondence: English, French.

Faculty of Engineering and Architecture, Engineering School, University College
Upper Merrion Street, Dublin 2.   Tel: Dublin (01) 761584 Telex: 4114 UCD E1

Professorship of Agricultural Engineering: McNulty, P.B., B.E., PhD, C.Eng.
Power and Machinery, Physical properties.

Dodd, V.A.	B.E., Ph.D., C.Eng.	Soil and Water Engineering
McKenna, B.M.	B.E., M.Eng. Sc., C.Eng.	Food Process Engineering
Ward, S.B.	Agr.Sc., M.Agr.Sc.	Farm Machinery

Status: University Department in the School of Engineering, University College, Dublin, instituted in 1961.

Agricultural Engineering Activities: Teaching: Four-Year course leading to the Degree of Bachelor of Engineering (Agricultural). All students of engineering at University College, Dublin attend a common course in their first year and compete for places in the second year branches on the results of the First University Examination. Fifteen places are offered in Agricultural Engineering in the second year. Service teaching is also provided to the Faculty of Agriculture to students of General Agriculture, Horticulture and Forestry.

Research: In recent years, postgraduate students have worked for the Degree of Master of Engineering Science and the Ph.D. Degree, in the following areas: Vibratory potato digging, Mechanics of grass cutting, Milking machines, Environment control in farm buildings, Reverse osmosis of dairy products, Physical properties of grasses, cereals, and fruit and vegetables, Mechanical oil expression, Disposal of farm wastes, Energy in Agriculture, Surface soil damage, Mechanized feeding of animals.

Publications: Publications in all above research areas.

Correspondence: English, French, German.

Agricultural Colleges, c/o Department of Agriculture
Kildare Street, Dublin 2

Activities: Courses in operation of farm machinery.

Protected Crops Department, Agricultural Institute (An Foreas Taluntais)
Kinsealy Research Centre
Malahide Road, Dublin 5

Tel.: 460644
Telex: 31479

Maher, M.J.	M.Agr.Sc.(Hort.)	Head of Department. Plant-environment simulation models
O'Flaherty, T.	B.E., Ph.D.	Control of environment for protected crops

Status: Established in 1959: State-sponsored organization: financed mainly from Government funds.

Activities: Research and extension work on heating and ventilation of glasshouses, artificial lighting and plant growth, plant-environment interaction in a glasshouse, plastic structures for protected cropping and energy utilization in greenhouses.

Publications: Temperature control in heated glasshouses - Heat requirements of glasshouses - Analysis of glasshouse climate - Performance and design of growing rooms - Efficiency of glasshouse boilers - Plastic structures for horticultural crops. Crop production in hydroponic systems - energy.

Correspondence: English.

Farm Development Service, Department of Agriculture,
Agriculture House, Kildare Street, Dublin 2 — Tel.: 789011 — Telex: 24280 AGRI EI.

Hickey, P.K.	B.E., M.I.C.E.I., A.M.I.A.E.	Land drainage
McGauran, V.	B.E.	Farm buildings

Status: Government Agency.

Activities: Specialist advice and educational courses.

Correspondence: Irish and English.

Comhlucht Siucre Eireann Teo (Irish Sugar Company Ltd.) - St. Stephen's Green House, Dublin-2.

Rowlette, M.	B.Sc.AYR
Comeford, C.	M.Sc.AYR
Walsh, M.J.	B.E. D.PA

Status: Semi-governmental, supervised by the Managing Director of the Company.

Agricultural Engineering activities: Research development and testing work - Design of machinery for mechanizing sugar beet and vegetable crops and peatland reclamation.

Correspondence: English.

Agricultural Engineering Department, Regional Technical College, Tralee, Co. Kerry

Farm Buildings Division, Regional Technical College, Athlone, Co. Westmeath

ISRAEL

Bet Dagan

The Institute of Agricultural Engineering – Agricultural Research Organization
P.O. Box 6, Bet Dagan, Israel   Tel. (03) 940303

Alper, Y.	D.Sc.	Director

Field Crops Harvesting Division:
Alper, Y.	D.Sc.	Head of Division
Margolin, A.	B.Sc.	
Elkin, I.	M.Sc.	
Hetzroni, A.	B.Sc.	
Wolf, I.		
Sagi, I.		
Michai, G.		
Antler, A.		
Eshed, A.		

Fruit Crops Harvesting Division:
Sarig, Y.	Ph.D.	Head of Division – Deputy Director
Grosz, F.	M.Sc.	
Malkin, O.	B.Sc.	
Spitzer, M.	M.Sc.	
Regev, I.	B.Sc.	
Rasis, A.		
Beres, H.		

Sizing and Separation Division:
Feller, R.	D.Sc.	Head of Division
Zaltzman, A.	Ph.D.	
Mizrach, A.	M.Sc.	
Shmilovitch, Z.		
Egozi, H.		
Meir, D.		

Handling and Packaging Division:
Nahir, D.	D.Sc.	Head of Division
Yekutieli, O.	M.Sc.	
Abramovitch, B.	M.Sc.	
Ganmor, S.	M.Sc.	
Ronen, B.		
Grosser, Y.		

Environmental Engineering and Energy Division:
Felsenstein, G.	M.Sc.	Head of Division
Segal, Y.	Ph.D.	
Haas, A.	M.Sc.	
Merbaum, A.	B.Sc.	
Weisblum, A.		
Bar-Lev, E.		
Madan, Y.		
Regev, R.		

Protected Crops Mechanization Division:
Zamir, N.	M.Sc.	Head of Division
Levav, N.	B.Sc.	
Arbel, A.		
Perlstein, G.	M.Sc.	

Testing and Instrumentation Division:

Peiper, U.M.	M.Sc.	Head of Division
Shmilovitch, S.	M.Sc.	
Zaslaver, G.	M.Sc.	
Dranker, A.	M.Sc.	
Weinberg, P.	M.Sc.	
Gazar, M.		
Katz, Y.		

Pesticides Application Division:

Frankel, H.		Head of Division
Austerweil, M.	M.Sc.	
Abadik, F.	M.Sc.	
Riben, Y.		
Shtainer, B.		
Babazada, M.		

Industrial Engineering and Systems Division:

Silberstein, B.A.	M.Sc.	Head of Division
Pasternak, H.	Ph.D.	
Lidror, A.	M.Sc.	
Yosef, S.	B.A.	
Engel, H.	B.A.	
Shamir, N.		

Status: Governmental institution within the framework of the Israel Agricultural Research Organization (ARO). Established in 1965, directed and financed by the Ministry of Agriculture.

Activities: Applied and basic research; Development, adaptation and improvement of farm machinery methods; Testing of locally developed and imported agricultural machinery and equipment; Extension work; Teaching of undergraduate and graduate courses in the Technion, Israel Institute of Technology, Haifa, and the Hebrew University, Rehovot.

Present activities under progress: Development of a tomato harvester for the fresh market, mechanization of plastic tunnels erector for weather protection of early raw crops, development of a harvester for paprica, development of a transplanter, mechanical harvesting of peaches grown in a meadow orchard; Mechanical harvesting of jojoba beans, mechanical harvesting of pecans and almonds, mechanical harvesting of olives, study of mechanical damage incurred in mechanical harvesting, cracking of macadamia nut; Utilization of the fluidized bed principle for separation of agricultural products from clods and stones, separation of peanuts from foreign materials, separation of flower bulbs from clods; Automatic wrapping of citrus fruits, improving waxing techniques, development of various handling aids in packing houses, picking and handling of bananas. Refrigerated storage of agricultural produce, precooling of fruits and vegetables, environmental conditions in refrigerated vessels and containers for export of fruit and vegetables, air flow patterns and distribution of temperatures in cold stores, drying and conditioning of produce such as onions, garlic, tobacco, medicinal and spice plants, etc., solar energy collectors and heat storage for heating of agricultural structures, development and evaluation of fresh air ventilating systems for transport containers, research on thermal properties of agricultural products; Application of pesticides and other chemicals in the form of aerosols, sprays, granulates and dust, development and testing of new methods and equipment for the application of chemicals. New formulations of materials for waxing fruits, use of aerosol chambers for disinfection and other treatments of produce. Ultra low volume periferial spraying of orchards and row crops. Greenhouse climate control, use of solar energy for greenhouse heating and cooling via a water medium, plant growing in artificial beds, solar heating greenhouses adapted to hilly terrain; Application of productivity engineering techniques in agriculture, planning and evaluation of overall systems of production and handling of agricultural produce, studies of handling, packing and transport of fruits and vegetables for export, studies of human engineering and improvement of manual operations with respect to sorting and packing, reduction of hand labour requirements for agricultural operations; Studies of total production systems combining aspects of labor, mechanization economics and management, research at a high level of generalization, to pinpoint bottlenecks in existing or planned systems, feasibility studies to determine priorities in planning research work, marketing research with respect to agricultural export.

Publications: Numerous publications on above research subjects.

Correspondence: English, Hebrew.

Haifa

Faculty of Agricultural Engineering, Technion, Israel Institute of Technology
Technion City, Haifa

Ravina, I.	D.Sc.	Assoc.Professor, Dean. Soil Science, Soil Physical Chemistry, Soil and water quality
Soil and Water:		
Amir, I.	D.Sc.	Senior Lecturer. Agricultural systems management
Avnimelech, Y.	Ph.D.	Assoc.Professor. Soil and Water Chemistry
Benami, A.	Ph.D.	Senior Lecturer. Irrigation Engineering
Hagin, J.	Ph.D.	Professor. Soil Science, Soil fertility and fertilizers
Karmeli, D.	Ph.D.	Assoc.Professor. Irrigation, Soil Survey
Kimor, B.	Ph.D.	Assoc.Professor. Agrobiology, Marine Biology
Naveh, Z.	Ph.D.	Assoc.Professor. Ecology
Nir, D.	D.Sc.	Assoc.Professor. Water Resources Eng., Irrigation and Drainage Eng.
Nir, Z.	D.Sc.	Teacher grade A. Fluid Mechanics and Hydraulics
Newman, P.	Ph.D.	Senior Lecturer. Agrobiology, Plant Physiology
Seginer, I.	D.Sc.	Professor. Agrometeorology
Sinai, G.	D.Sc.	Lecturer. Hydrology, Soil Conservation, Drainage Eng.
Zaslavsky, D.	Ph.D.	Professor. Soil Physics, Hydrology, Drainage Eng.
Zur, B.	Ph.D.	Senior Lecturer. Soil Physics, Irrigation Science
Farm Mechanization:		
Galili, N.	D.Sc.	Senior Lecturer. Machine Dynamics
Kornecki, Z.	Ph.D.	Professor. Strength of Materials
Manor, G.	D.Sc.	Senior Lecturer. Agric. Power and Machinery
Sagi, R.	D.Sc.	Professor. Animal Husbandry
Spector, M.	D.Sc.	Senior Lecturer. Soil-Machine Interaction
Peleg, K.	D.Sc.	Senior Lecturer. Packing of Fruits and Vegetables
Wolf, D.	D.Sc.	Senior Lecturer. Agric.Power and Machinery, Machine-Soil Interaction.

Activities: Teaching: Study courses leading to B.Sc., M.Sc. and D.Sc. degrees in agricultural engineering, with two options: Soil and Water Engineering and Agricultural Power and Machinery. Courses are given in soil and water engineering such as: irrigation; drainage; soil conservation; hydrology; agrometeorology; soil physics; soil reclamation; water resources and agricultural systems; etc., in power and machinery such as: tractors; tillage and harvesting machinery; agricultural crop processing and packaging; controlled environment; husbandry machinery; automation, etc., in general agriculture such as: agrobiology, fundamentals of crop and animal production; plant physiology; soil science; etc., and in general engineering such as: fluid mechanics; geomechanics; thermodynamics; etc.

Research: Agricultural Machinery Research Center: farm machinery engineering, tractors and power units, soil cultivation, climate control in agricultural buildings, handling, transportation and packaging of agricultural products, controlled agriculture, earth moving equipment. Fertilizers and Soils Center: evaluation of the effectiveness of fertilizers, availability of nutrients, organic soils, soil structure, soil stabilization and stabilizers, soil and water salinity, basic research in soil chemistry and physical chemistry, saturated and unsaturated flow, erosion and soil conservation, small water reservoirs, effluents for irrigation, use of marginal soils and water, drainage engineering. Irrigation and Drainage Center: irrigation planning and design, irrigation methods and equipment, soil-water-plant relationship, agrometeorology, automation and control in irrigation, water resources and agricultural systems analysis.

Publications: Agricultural Engineering research. Laboratory assay. Field trials. Wind tunnel layer simulation. Double layer theory. Poisson-Boltzman equation with hydration. Solubility. Fluid mechanics. Vibration of liquid. Pump theory. Agricultural systems. Arid zone agriculture. Agricultural systems. Computerized models. Linear programming. Operation research. Planning agricultural production. Models in dairy cows. Moshav planning. Environmental control. Computational study. Animal shelters and housing. Poultry houses. Environmental conditions in animal housing. Heating. Ventilating. Greenhouses. Environmental control. Thermal conditions. Organic plant growth substrate. Greenhouse drainage. Greenhouse irrigation. Aerial photography. Hydrology. Flow through porous media. Water resources. Optimization. Simulation. Water allocation. Water quality. Aquifer management. Lake Kinneret. Surface runoff. Rivers. Open channels models. Sediment. Fish ponds. Aquaculture. Plant soil water relationship. Soil dynamic in tillage and traction. Tires. Off the road. Tractor trailer relations. Clod pulverizers. Pipes soil interaction. Asbestos cement pipes. Soil physics. Soil profile. Soil stability. Soil conditioners. Impermeable-semi-permeable membrane. Erosion. Desert soil weathering. Mountainous soils. Soil reclamation. Drainage. Sodicity. Salinity. Irrigation. Irrigation hydraulics. Irrigation system operation. Scheduling irrigation water. Trickle-drip irrigation. Center pivot irrigation. Sprinkler irrigation. Irrigation networks. Fertilizers. Calcium phosphate. Calcium carbonate. Foliar fertilizers. Plant nutrition simulation. Root uptake. NPKS fertilizers. Macronutrients. Fertilizer spreaders. Urea-form fertilizers. Windbreaks. Orchard windbreaks. Wheat-fertilizers. Nitrogen fertilizers. Irrigation. Avocado. Banana handling. Fruits. Vegetables. Packaging. Processing produce. Ecology. Air pollution. Effect of air pollutants on vegetation. Plant climatization. Landscape planning. Fire ecology. Israel. Mediterranean uplands.

Correspondence: Hebrew, English.

Rehovot

Settlement Study Centre
Rehovot, Israel - P.O.B. 555

Activities: Research and training Institute engaged in activities related to rural regional development.

Jerusalem

Rural Settlement Department, the Jewish Agency
P.O. Box 92, Jerusalem

Weitz, R.   Professor   Head of Department
             (D.Agr.)

Activities: Regional rural planning. Architectural planning of rural settlements, co-operatives (Moshvim) and communal organizations (Kibbutzim). Planning of regional service centres. Replanning of existing settlements.

Correspondence: Hebrew, English.

ITALY

Bari

<u>Istituto di Meccanica Agraria, Facoltà di Agraria, Università di Bari</u> (Institute for Agricultural Mechanical Engineering, Faculty of Agriculture, University of Bari)
Via Amendola 165/A, 70126 Bari   Tel.(080) 339560

Amirante, P.	Ing.Prof.	Director of the Institute, Chairman for Agric. Mechanization and Agroindustries
Dipaola, G.	Ing.Prof.	Chairman for Agric.Mech.Engineering
Zanna, L.	Dr.Ing.Prof.	Chairman for Agric.Mechanization
Arrivo, A.	Ing.Prof.	Asst.Prof., Agric.Mech.Engineering
Bellomo, F.	Dr.	Research fellow for Ministry of Agric. and Forests, c/o Agric.Eng.Institute
Di Candia, E.	Ing.	Asst., Agric.Mech.Engineering
Giametta, G.	Dr.Prof.	Asst.Prof., Agric.Mech.Engineering
Guarella, P.	Dr.Prof.	Asst.Prof., Agric.Mech.Engineering
Grittani, P.	Dr.	Asst., Agric.Mech.Engineering
Mongelli, C.	Ing.	Research fellow for C.N.R., c/o I.M.A.
Panaro, V.	Ing.Prof.	Asst.Prof.,Agric.Mech.Engineering
Pasqualone, S.	Ing.Prof.	Asst.Prof.
Pellerano, A.	Ing.Prof.	Asst.Prof.
Quartulli, S.	Ing.Prof.	Asst.Prof.

<u>Branches</u>: The Institute has other Branches where teaching and research is carried out, i.e.: 1) Institute for Farm Buildings and Rural Constructions, Bari University; 2) at Villa Sbisà (University Campus) and School farm "Martucci" situated in the agro Valenzano (BA) and other private farms all over Puglia region.

<u>Structure</u>: Established in 1938, when the Faculty of Agriculture was established by the Ministry of Public Education, which supervises and partially finances its activities.

<u>Activities</u>: Teaching in agricultural engineering leading to degrees (Dottore in Scienze agrarie and Dottore in Scienze forestali). Research: 4 phases: basic and theoretical, technological, laboratory and field. Research subjects: vines, olives, beetroot, cauliflower, artichokes, tobacco; forest mechanization and mechanization on experimental farms; wind and solar energy and waste management; testing and evaluation of machinery.

<u>Publications</u>: The Institute has produced, over the years, some textbooks on mechanization, and various publications in "Ingegnería Agraria", "Macchine e Motori agricoli", "Mondo agricolo","Annali della Facoltà di Agraria, Università di Bari", etc.

<u>Correspondence</u>: Italian, French, English.

Bologna

Istituto di Meccanica Agraria, Università di Bologna (Institut de Machinisme Agricole, Université de Bologna)
via Filippo Re, 4 - Bologna 40126
Tel.: 051 223769

Manfredi, E.	Dr.Ing.Prof.	Directeur
Ade, G.	Dr.Agr.	Expérimentateur
Artuso, M.L.	Dr.Agr.	Expérimentateur
Baraldi, G.	Dr.Agr.	Prof.Mécanisation Agricole
Bentini, M.	Dr.Agr.	Expérimentateur
Bonafede, M.	Dr.	Assistant Prof. - Mathématiques
Bosi, P.	Dr.Agr.	Assistant Prof. - Physique
Capelli, G.	Dr.Agr.	Assistant Prof. - Mécanisation des élevages
Casini-Ropa, G.	Dr.Agr.	Prof. - Mécanisation Agricole - Station d'Essais
De Zanche, C.	Dr.Ing.	Assistant Prof. - Station d'Essais
Guarnieri, A.	Dr.Ing.	Expérimentateur
Marocchi, A.	Dr.Agr.	Expérim.Technique
Nicoletti, G.	Dr.	Assistant Prof. Mathématiques

Station affiliée: Centre Expérimental de Cadriano, via Gandolfi 19 - 400557 Cadriano (BO) Tel. 766632.

Structure: Institution gouvernementale, fondée en 1901 et dépendant du Ministère de l'Education.

Activités: Cours de mathématiques, physique, machines agricoles, méthodes de mécanisation en agriculture, machinisme pour l'élevage, fréquentés par 150/200 étudiants. L'achèvement des études porte au diplôme de "Docteur-Agronome" en SciencesAgricoles, en Science de la Production Animale. Homologation de machines agricoles, effectuée sous la direction du Ministère des Transports. Essais OCDE. Recherches sur les semoirs, la récolte des fourrages, du maïs, des cultures industrielles, des fruits et des légumes. Séchage artificiel des produits agricoles. Mécanisation des appareils pour la lutte antiparasitaire. Organisation scientifique du travail en agriculture.

Publications: Les 650 mémoires, publiées depuis 1980, concernent: la motorisation et la mécanisation en colline et en montagne. Etudes sur les caractéristiques dynamiques, cinématiques, fonctionnelles et sur la stabilité des tracteurs. Etudes sur le travail du sol. Recherches expérimentales sur les pulvérisateurs, les semoirs, la récolte des fourrages, du riz, du maïs, de la betterave à sucre, des fruits et légumes. Appareils pour la viticulture et l'oénologie. Bulletin d'essais OCDE.

Correspondance: Italien, Français, Anglais, Allemand, Espagnol.

Istituto di Genio Rurale, Università di Bologna (Institut de Génie Rural, Université de Bologna)
4, via Filippo Re - 40126 Bologna
Tél. 051/233375 237798

Rossini, R.	Dr.Ing.Prof.	Directeur Technique des Installations pour l'Irrigation
Hydraulique Agricole:		
Giari, M.	Dr.Ing.Prof.Ass.	Hydraulique Agricole
Cacchi, D.	Dr.Agr.Prof.Ass.	Technique de l'Assainissement des Sols (Hydraulique et constructions)
Taglioli, G.	Dr.Agr.Chercheur	
Zampighi, C.	Dr.Agr.	
Conti, C.	Dipl.Agr.	
Stabellini, M.	Dipl.Agr.	

Conti, C.	Dipl.Agr.	
Stabellini, M.	Dipl.Agr.	

Constructions Rurales:

Venturi, A.	Dr.Ing.Prof.Ass.	Topographie et Constructions Rurales
Checchi, A.	Dr.Agr.Chercheur	
Simoni, A.	Dr.Agr.	

Technologies Physiques:

Veronesi, G.	Dr.Phy.Prof.Ass.	Physique
Barnabei, M.	Dr.Mat.Prof.	Mathématiques
Caprara, C.	Dr.Phy.	
Pavanelli, D.	Dr.Geol.	

Structure: Institution gouvernementale constituée en 1969 (en rassemblant les Instituts d'Hydraulique Agricole et l'Institut de Topographie et Constructions Rurales), controlée par le Ministère de l'Education. L'Institut comprend un nouveau secteur pour les recherches dans le domaine des Technologies physiques appliquées à l'agriculture, pas encore reconnu officiellement.

Activités: Cours annuels de formation conduisant à l'attribution du diplôme de docteur en sciences agronomiques: cours indiqués à côté des enseignants. Les cours sont fréquentés par environ 250 étudiants par an. Recherche et expérimentation sur le drainage; la technologie de l'irrigation par aspersion, localisée et par installations automotrices (Hydraulique Agricole). Etudes sur le stockage et l'utilisation agronomique du lisier des porcheries. Etudes sur l'amélioration des structures des serres, structures à cables (Constructions Rurales). Appareils de mesure en agriculture: projet, construction et essai. Applications énergétiques traditionnelles et renouvelables en agriculture; analyses énergétiques; recherches sur la géothermie, biogas, énergie solaire, production de biomasses avec buts énergétiques. Banque et élaboration des données météoclimatiques; formulation des data bases, développement des systèmes pour l'acquisition des données et leur connexion en réseau; étude de méthodes pour l'élaboration des informations collectées.

Publications: Problèmes de drainage dans les terrains italiens. Problèmes de la technologie de l'aspersion et de la microirrigation. Etudes sur l'aménagement et l'assainissement hydraulique dans l'environnement du Po. - Etudes sur la séparation des lisiers liquides et successif lagunage avant l'épandage: solutions operatives. Comparaison de normatives européennes pour le calcul des structures des serres. Premieres études de projet de tensostructures dans les serres. - Observations sur les applications des énergies renouvelables Résultats d'analyses énergétiques. Applications dans l'élaboration des données météoclimatiques et de certaines méthodes pour l'élaboration des informations collectées.

Correspondance: italien, français, anglais, allemand.

Catania

Istituti di Idraulica Agraria, di Meccanica Agraria, di Topografia e Costruzioni Rurali, Università di Catania (Instituts d'Hydraulique Agricole, de Machinisme Agricole, de Topographie et Constructions Rurales, Université de Catane)
via Valdisavoia 5 - 95123 Catania

Institut d'Hydraulique Agricole

Institut de Machinisme Agricole

Institut de Topographie et de Constructions Rurales

Structure: Instituts fondés en 1954/55, sous le contrôle du Ministère de l'Education.

Activités: Cours de formation conduisant à l'attribution du diplôme de docteur agronome.

Correspondance: Italien, Français, Anglais.

Firenze

Istituto di Meccanica agraria e meccanizzazione (Institut de Machinisme agricole et de mécanisation, Université de Florence – Agricultural Machinery and Farm Building Institute)
Piazzale delle Cascine, 15 – 50144 Firenze   Tél.: 055/366595 – 367549

Dallari, F.A.	Dr.Ing.Prof.	Directeur, Chaire de Mécanique agricole
Galigani, P.F.	Dr.Prof.	Assist. chargé du cours de Technique de la mécan.agric.
Giuntoli, V.	Dr.Ing.	Assist. chargé du cours de Physique
Mazzanti, R.	Dr.Ing.	Assist. chargé du cours de Mécanique des cultures tropicales
Zoli, M.	Dr.Ing.Prof.	Assist. chargé du cours de Mathématiques
Cioni, A.	Dr.Ing.	Chargé du cours de Mécanique du sol
Spugnoli, P.	Dr.Ing.	Bourse d'étude C.N.R. en Mécanique agricole et chargé du cours de Dessin
Vannucci, D.	Dr.Agr.	Bourse d'étude C.N.R. en Mécanique agricole

<u>Fermes expérimentales affiliées</u>: Monna Giovannella, Antella près de Florence.

<u>Structure</u>: Institution gouvernamentale fondée en 1924, controlée par l'Université de Florence, dépendant à son tour du Ministère de l'Education Nationale (Ministero della Pubblica Istruzione).

<u>Activités</u>: a) Formation: Cours universitaires annuels, faisant part de la Faculté d'Agriculture de l'Université de Florence, et conduisant au Doctorat (Laurea) en sciences agricoles: les cours tenus dans l'Institut sont: Mécanique agricole; Physique pour les agronomes; Mathématique pour les agronomes; Mécanique du sol et actions mutuelles entre les outils et le terrain; Technique de la mécanisation agricole; Mécanisation des cultures tropicales; Dessin. Aux différents cours sont inscrits jusqu'à 400 étudiants. – b) Recherches: sur machines pour la récolte des pommes de la plante; sur machines pour le triage de la tomate; sur l'évolution des tracteurs pour sols en collines; sur simplification des moissonneuses-batteuses pour sols en pente; sur machines pour le travail vibrant du sol; sur la récolte mécanique du raisin; sur l'organisation du travail en agriculture.

<u>Publications depuis 1973</u>: Récolte et triage mécanique de la tomate pour l'industrie. Travail vibrant du sol. Récolte mécanique et interceptation des fruits. Mécanisation des pépinières. Evolution des tracteurs agricoles. Récolte mécanique de la canne à sucre. Triage photooptique de la tomate. Taille des vignobles. Proposition pour l'unification de la méthodologie d'essaie des machines agricoles. Récolte mécanique des fraises. Récolte mécanique des poivrons. Utilisation et prix horaire des machines agricoles.

<u>Correspondance</u>: Italien, Français, Anglais.

Istituto di Costruzioni Rurali e Forestali, Università di Firenze (Institut des Constructions Rurales et Forestières, Université de Florence – Farm and Forestry Building Institute, University of Florence)
Piazzale delle Cascine, 18 – I-50144 Firenze, Italia

Gasperi-Campani, I.	Dr.Ing.Prof.	Directeur de l'Institut – Chargé du cours de Topographie et constructions rurales; chargé du cours de topographie (forestière)
Panero, V.	Dr.Arch.Prof.	Chargé du cours des Constructions Forestières

Gori, P.	Dr.Arch.	Agrégé pour les exercitations de constructions forestières
Pancani, M.	Dr.Ing.	Agrégé pour les exercitations de topographie
Sorbetti-Guerri, F.	Dr.For.	Collaborateur pour les Constructions Rurales

Annexes situées dans d'autres localités: L'Institut va se rendre dans un nouvou siège à Quaracchi, chez Les Cascine, où aura à disposition plusieurs locaux et laboratoires. Maintenant dispose de locaux à les Cascine, d'un laboratoire à Montecatini, chez Pistoia, et de la possibilité d'exercitations à Vallombrosa chez Florence.

Structure: Institution gouvernementale fondée en 1869 en qualité de part de l'Ecole forestière de Vallombrosa; nombreuses transformations ont porté à l'actuelle Facoltá di Agraria de l'Université de Florence qui contrôle l'Institut et qui dépend à son tour du Ministero della Pubblica Istruzione (Ministère de l'Education Nationale).

Activités concernant le génie rural: a) Formation- Cours universitaires annuels, faisant part de la Faculté d'Agriculture de l'Université de Florence, et conduisant au Doctorat (Laurea) en Sciences agricoles, ou bien au Doctorat (Laurea) en Sciences forestières, ou enfin au Doctorat en Agriculture tropicale et subtropicale. Les cours tenus dans l'Institut sont: Topographie et Constructions Rurales, Topographie (pour forestiers), Constructions Forestières. Aux différent cours sont inscrits jusqu'à 220 Etudiants. - b) Recherches: Recherches sur les Structures en bois lamellé collé, recherches sur problèmes d'énergies alternatives, recherches sur les ponts forestiers en bois, recherches sur les agglomérations urbaines dans la Maremma tuscaine chez Grosseto, études sur la conservation des pommes de terre, études sur problèmes de l'écologie rurale.

Publications: Articles: sur la législation écologique italienne sur l'enseignement universitaire italien de matières du génie rural, sur les constructions industrielles pour la culture des champignons, pour la production du vinaigre, sur les épreuves de bois lamellé collé, sur la projection des serres, sur l'énergie qui est nécessaire pour le chauffage des serres de la floriculture.

Correspondance: Italien, français, anglais, allemand, espagnol, portugais, latin.

Istituto di Idronomia e Idraulica Agraria, Università degli Studi di Firenze (Institut d'Hydronomie et d'Hydraulique Agricole, Université de Florence)
Piazzale delle Cascine 18 - 50144 Firenze

Annexes: Champs expérimentals et offices de campagne à Branzolino (Forlì)

Structure: Institut fondé en 1912 et placé sous le contrôle du Ministère de l'Education Nationale.

Activités: Cours de formation dans le cadre de l'enseignement de la Faculté. - Recherches sur: drainage des terrains - hydrologie des petits bassins montagnards, des rivières de montagne, application de la photogrammétrie.

Publications: articles sur le drainage, les constructions hydrauliques, sur les crues des cours d'eau.

Correspondance: Italien, Français, Anglais, Allemand.

Istituto Sperimentale per lo Studio e la Difesa del Suolo (Institut Expérimental pour l'Etude et la Conservation du Sol)
Piazza M. D'Azeglio 30, Firenze   Tél.: 263700/294072

Arcara, P.G.	Directeur de la Section de Biologie du Sol - Agronome
Bazzoffi, P.	Agronome, Section de la phisique du sol
Bidini, D.	Chimique. Section de la chimie du sol
Chisci, G.C.	Professeur d'Agronomie, Directeur de la Section de la phisique du sol
Gregori, E.	Forestier, Section de la Biologie du Sol
Lorenzoni, P.	Géologue, Section de Génèse, Cartographie et Classification du Sol
Lulli, L.	Agronome, Directeur chargé de la Section de Génèse Cartographie et Classification du Sol
Magaldi, D.	Géologue, Section de Génèse, Cartographie et Classification du Sol; Professeur chargé de Minéralogie et Géologie
Miclaus, N.	Agronome, Section de Biologie du Sol
Panicucci, M.	Agronome, Directeur de la Section de Technologie du Sol
Piovanelli, C.	Agronome, Section de Biologie du Sol
Raglione, M.	Géologue, Section détachée Minéralogie du Sol
Rodolfi, G.	Géologue, Section de Génèse, Cartographie et Classification du Sol
Sfalanga, M.	Géologue, Section de la Phisique du Sol: La Mécanique du Sol
Zanchi, C.	Agronome, Section de la Phisique du Sol: Conservation du Sol et Drainage
Ronchetti, G.	Directeur de l'Institut Expérimental pour l'Etude et la Conservation du Sol. Agronome. Professeur de Géopédologie.

Annexes: Section détachée "Technologie du Sol", via Cagliari, 88100, Catanzaro - Section détachée "Mineralogie du Sol", via Casette, 1 - 02100 Rieti.

Structure: L'Institut a été restructuré le 23 Nov. 1967 d'après D.P.R. n.1318 et successivement par le D.P.R. n.245, 1° Avril 1978 en gardant la même dénomination qu'il avait depuis sa fondation en 1952. L'Institut est sous le contrôle du Ministère de l'Agriculture et des Forêts.

Activités: Recherches sur les petits bassins versants pour étudier les relations entre les aspects physiographiques et hydrologiques. Etudes sur l'hydrologie du sol par rapport à sa nature, aux couvertures naturelles, pour l'exploitation agricole, etc. Etude du drainage du sol et de la meilleure utilisation des pluies naturelles et de l'irrigation. Etude sur l'érosion du sol: méthodologie expérimentale, action érosive des pluies, sols sujets à l'érosion en fonction de ses propriétés physiques, chimiques et biologique, des couvertures végétales naturelles et artificielles et des traitements biologiques et mécaniques pour l'exploitation agricole et forestière. Classification, cartographie et génèse du sol. Activité chimique et biologique en rapport à l'utilisation du sol.

Publications: Les recherches sont publiée dans les Annales de l'Institut.

Correspondance: Italien, Français, Anglais.

Milano

Istituto di Ingegneria Agraria, Università di Milano (Institute of Agricultural Engineering, University of Milan)
Via Celoria 2 - 20133 Milano   tel.: 02-292181/02-296851/02-2367625

Gasparetto, E.	Prof.Dr.Ing.	Director

Energy and Mechanization Division:

Pellizzi, G.	Prof.Dr.Ing.	Head of Division
Balsari, P.	Dr.Agr.	
Bersi, P.		
Bodria, L.	Dr.Ing.	
Castelli, G.	Prof.Dr.Ing.	
Pieretti, G.	Dr.Ing.	
Riva, G.	Dr.Ing.	
Rubino, A.		

Testing Division:

Cavalchini, A.	Dr.Ing.	Head of Division
Febo, P.	Dr.Ing.	
Mancastroppa, S.		
Natalicchio, E.	Dr.Agr.	
Pessina, D.		
Trolli, C.		
Viola, L.		

Farm Building and Rural Planning Division:

Sangiorgi, F.	Dr.Agr.	Head of Division
Bonfanti, P.	Dr.Agr.	
Busi, R.	Dr.Ing.	
Crivelli, C.	Dr.Mat.	
Faletti, L.	Dr.Ing.	
Toccolini, A.	Dr.Ing.	

Status: Governmental university institution, established in 1871.

Activities: University courses in: basic agricultural machinery, agricultural machinery; agricultural mechanization; ergotechnics and farm organization; animal husbandry mechanization; farm buildings; rural planning organization; rural electrification; mathematics. Theoretical and experimental (field and laboratory) research on: tractors and implements; mechanization, ergotechnics and work organization; farm buildings; renewable energies; energy saving.

Publications since 1973: 200 theoretical, experimental and testing reports on: tractors, primary and secondary tillage implements; forage, wheat, corn and paddy harvesters; milking machines; drying plants; animal feeding machinery; farm mechanization; rural areas mechanization and organization; farm electrification; farm buildings standardization and design; farm structures reorganization; renewable energies (sun, biogas, biomasses, wind, geothermy); cogeneration and heat pumps; developing countries mechanization and machinery manufacturing.

Correspondence: Italian, English, French, Spanish.

Progetto Finalizzato Meccanizzazione Agricola, Consiglio Nazionale delle Ricerche
(Research Project on Agricultural Mechanization, National Research Council)
via Celoria 2 - 20133 - Milano   Tel.: 02-292181/02-296851/02-2367625

Pellizzi, G.	Prof.Dr.Ing.	Scientific Director
Semenza, C.	Dr.Agr.	

Subproject 1 - Mechanization of harvesting, storage and distribution of forage
Lisa, L.   Dr.Agr.   Head of Subproject

Subproject 2 - Mechanization of industrial and vegetable crop harvesting
Manfredi, E.   Prof.Dr.Ing.   Head of subproject

Subproject 3 - Mechanization of fruit and strawberry harvesting
Baldini, E.   Prof.Dr.Agr.   Head of subproject

Activities: Optimum relationships between machines, structures and crops; uniforming of work calendars; optimization of mechanization levels in consistency with economic and social conditions; minimizing of production costs; increase of the quality of the production of the main crops. The research is carried out by different research units, coordinated amongst each other, and distributed among the various Italian regions.

Publications since 1976: 180 theoretical, experimental and testing reports on: mechanization of forage crops for cows and calves; mechanization of forage crops for dairy cattle; mechanization of forage crops for sheep and goats; mechanical harvesting of sugar beets; mechanical tobacco harvesting and curing; mechanical harvesting of industrial tomatoes; mechanical harvesting of artichokes and cauliflowers; mechanical harvesting of peppers; mechanization of grape harvesting; mechanization of olive harvesting; mechanization of fruit and strawberry picking; mechanical harvesting of citrus fruits.

Istituto di Idraulica Agraria (Institute of Agricultural Hydraulics, University of Milan)
via G. Celoria 2 - 20133 Milano   Tel.: 23 05 12 - 23 61 023

Romita, P.L.	Dr.Ing.Prof.	Directeur, Hydraulique agricole
Giura, R.	Dr.Ing.Prof.	Assainissement hydraulique
De Wrachien, D.	Dr.	Chargé de cours, Géopédologie
Galbiati, G.L.	Dr.	Chargé de cours, Hydrotechnique agricole
Greppi, M.	Dr.Ing.	
Gruppo, M.		
Previtali, F.	Dr.	

Structure: 1964. Institution de l'Université de Milan.

Activités concernant le génie rural: a) Formation: cours universitaire pour le diplôme de Docteur en sciences agricoles suivis par une moyenne de 100 étudiants. b) Irrigation, drainage, assainissement agricole, aménagements hydrauliques-agricoles, gestion de ressources en eau, hydrologie.

Publications: en "Irrigazione", "L'Idrotecnica", "Geologia Tecnica", "Genio Rurale".

Correspondance: Italien, français, anglais, espagnol, allemand.

Padova

Istituto di Meccanica Agraria, Facoltà di Agraria, Università di Padova (Institute of Agricultural Mechanical Engineering, Faculty of Agriculture, Padova University)
Via Gradenigo 6, 35100 Padova

The following subjects belong to the Institute:

<u>Forestal Watershed Management</u>
<u>Forest Hydrology</u>
<u>Forest Constructions</u>
<u>Topography</u>
<u>Agricultural Mechanical Engineering</u>
<u>Topography and Farm Constructions</u>
<u>Farm Mechanization Technique</u>
<u>Land Reclamation Technique</u>
<u>Agrarian Hydraulics</u>
<u>Mathematics</u>

Benini, G.	Engineer, Prof.	Chairman of Forest Watershed Management. Director of the Institute of Agricultural Mechanical Engineering
Cera, M.	Prof.	Chairman of Agricultural Mechanical Engineering
Chiumenti, R.	Prof.	On annual contract of Topography and Farm Constructions
Della Lucia, D.	Eng., Prof.	On annual contract of Agrarian Hydraulics
De Zanche, C.	Eng., Prof.	Chairman of Agricultural Mechanical Engineering
Fattorelli, S.	Eng., Prof.	Chairman of Forest Hydrology and Professor on annual contract of Land Reclamation Technique
Zoppello, G.	Prof.	On annual contract of Farm Mechanization Technique

<u>Branches</u>: Laboratory of the Institute of Agricultural Mechanical Engineering, of Legnaro, via Roma, 35020 Legnaro (Padova)

<u>Status</u>: The Institute of Agricultural Mechanical Engineering was established in 1951 and since then it has belonged to the Faculty of Agricultural Sciences of the University of Padova.

<u>Activities</u>: Teaching: The subjects of Agricultural Engineering form part of the subjects for the university degree in Agricultural Sciences. Research: The Institute periodically conducts several research activities (commitments from the CNR, Regions, etc.)

<u>Publications</u>: in Macchine e Motori Agricoli - Agricoltura delle Venezie - Rivista di Ingegneria Agraria - L'Informatore Agrario - Rivista di Viticoltura e di Enologia di Conegliano - L'Industria Saccarifera Italiana - Terra e Sole - Monti e Boschi - Genio Rurale - L'Italia Agricola . Collana Verde.

Associazione Italiana di Genio Rurale (Italian Association of Agricultural Engineering)
A.I.G.R.
via Gradenigo 6 - 35100 Padova   Tél.: (049) 651285

Bernini, G.	Président, Professeur à l'Université de Padoue de "Sistemazioni idraulico-forestali"
Ramadoro, A.	Président honoraire
Stefanelli, G.	" "
Priorelli, G.	" "
Pratelli, G.	" "
Pellizzi, G.	Vice-Président, Directeur de la Revue
Fattorelli, S.	Secrétaire
Casini-Ropa, G.	Conseiller
Indelicato, S.	"
Mennella, V.	"
Rossini, R.	"

1ère Section   Science du sol et des eaux - Président Giura, R.
2ème Section   Constructions et équipements ruraux - Président Bianchi, A.
3ème Section   Mécanisation et machinisme agricoles - Président Manfredi, E.
4ème Section   Electrification rurale et installations des fermes - Président Paniale, G.
5ème Section   Organisation scientifique du travail en agriculture - Président Cera, M.

Structure: L'Association est la section italienne du C.I.G.R. (Commission Internationale du Génie Rural) qui a son siège à Paris 75015 - 17, Rue de Javel. - La A.I.G.R. fut fondée le 16 mars 1959. Elle est une Association privée de techniciens spécialisés dans les problèmes du Génie rural.

Activités concernant le Génie Rural: A présent, aucun cours de formation est tenu, pourtant ils sont prévus dans le Statut A.I.G.R. A présent, l'Association n'organise pas de recherches (mais elles aussi sont prévues dans le Statut). L'Association s'occupe de promouvoir, organiser et coordonner, avec des congrés et Réunions, les études et recherches réalisées par d'autres Associations (sourtout Universités).

Publications: Rédaction des Actes de Réunions, compte-rendus, etc. Propriétaire et organisme officiel de la "Rivista di Ingegneria agraria" (Revue du génie agricole) publiée par Edagricole, Bologna. Les membres de l'Association Italienne du Génie Agricole collaborent avec la Revue mentionnée.

Correspondance: Italien, Anglais, Français.

P e r u g i a

Istituto di Costruzioni Rurali e Topografia, Università degli Studi di Perugia (Institute for Farm Buildings, University of Perugia)
Borgo XX Giugno 74, 06100 Perugia   Tél.:(075) 30530

Mennella, V.G.G.	Prof.Eng.	Director
Lorusso, A.	Prof.Eng.	Associate Professor
Biscarini, M.	Dr.Eng.	Assistant
Borghi, P.	Dr.Agr.	Technical expert
Tesei, P.	Geom.	Technical expert
Gasperini, F.	Geom.	Technical admin.
Bacchi, P.		Executive

Status: Governmental University institution, established 1887.

Agricultural Engineering activities: a) Teaching: University courses in technical design and materials, farm building and land survey. b) Research: Theoretical and experimental research on farm building design, climatization, rural planning, energy problems, solar energy, utilization of wastewater.

Publications since 1973: 70 theoretical, experimental and testing reports on: farm buildings design, farm structures reorganization, utilization of farm waste, environment and work organization, building cost, materials and building process, landscape planning, solar energy.

Correspondence: Italian, English, Spanish, German.

Istituto di Meccanica Agraria, Università di Perugia (Institut de Machinisme Agricole, Université de Pérouse)
Borgo XX Giugno 74, 06100 Perugia

Structure: Institution gouvernementale, fondée en 1887, placée sous le contrôle du Ministère de l'Education.

Activités: Cours de physique et mécanique agricole faisant partie des cours de formation agricole générale, fréquentés annuellement par 200 étudiants et conduisant au grade universitaire de "docteur" en sciences agronomiques - Recherches sur les machines pour le travail du sol et pour la récolte - Etudes de mécanisation intégrale des entreprises agricoles - Etudes sur les caractéristiques du tracteur - Etudes sur l'organisation du travail en agriculture.

Correspondance: Italien, français, espagnol, anglais.

P i s a

Istituto di Topografia e Costruzioni Rurali dell'Università di Pisa (Institut de Topographie et de Constructions Rurales de l'Université de Pisa)
Via del Borghetto 80, 56100 Pisa

Geri, G.   Dr.   Chargé des cours - Topographie et géodesie

Statut juridique: Institution gouvernementale, fondée en 1917 et placée sous le contrôle du Ministère de l'Education.

Activités: Cours de topographie et constructions rurales, frequentés par environ 70 étudiants par an.

Publications: en   Leçons de topographie et de constructions rurales".

Correspondance: Italien, français, anglais.

Istituto di Meccanica Agraria, Università di Pisa (Institut de Machinisme Agricole, Université de Pisa)
Via del Borghetto, 80 - 56100 Pisa   Tél.: 571560

Di Ciolo, S.	Prof.Ing.	Directeur
Giordani, G.	Dr.Ing.	Assistant
Lucignani, M.	Dr.Ing.	Assistant

Statut juridique: Institution gouvernamentale, fondée en 1840, placée sous le contrôle du Ministère de l'Education.

Activités: Cours de mathématiques, physique, machines agricoles, méthodes de mécanisation en agriculture. L'achèvement des études porte au diplôme de docteur en sciences agronomiques. Homologation officielle de machines agricoles, effectuée sous la direction du Ministère des transports. Recherches théoriques et expérimentales sur champs et au laboratoire de machines agricoles motrices actionnées.

Publications: Mécanisation en agriculture, transports agricoles. Récolte mécanique des olives, tomates, fruits, fraises, légumes. Mécanisation de la viticulture. Labour du terrain avec des machines à outils vibrants et oscillants.

Correspondance: Italien, anglais, français.

Istituto di Idraulica Agraria, Università di Pisa (Institut d'Hydraulique Agricole, Université de Pisa)
Via del Borghetto 80, 56100 Pisa, Italia   Tel.: (050) 57 15 52

Grossi, P.	Prof.Dr.	Directeur, Professeur Assistant, Hydraulique générale et irrigation, Assainissement et conservation du sol.
Ficini, F.	Dr.Agr.	Professeur Assistant, Approvisionnement hydrique pour l'agriculture.
Megale, P.G.	Dr.Ing.	Hydraulique générale et hydrologie.

Stations affiliées: Ferme expérimentale de Tombolo (Pisa). Champ expérimental hydrologique-forestier de Monte Pruno (altitude 900 m), Pisa.

Statut juridique: Institution gouvernamentale, contrôlée par le Ministère de l'Instruction Publique.

Activités: Cours de formation professionnelle, 2 annuels et 1 semestriel, fréquentés par environ 50 étudiants. - Etude du bilan hydrologique. Arrangements hydro-forestiers en montagne. Nouvelles techniques d'irrigation automatique et semi-permanente. Travaux d'assainissement. Diffusion de l'eau dans le sol. Protection des litoraux sableux. Projet optimal.

Publications: Danger d'accumulation des pluies sur les terrains de bonification à écoulement intermittent. Problème du revêtement des canaux d'irrigation. Eléments de base pour la conservation du sol en Italie. Influence des aménagements hydrauliques agricoles et forestiers sur le régime des cours d'eau. Normes pour la recherche du coéfficient d'infiltration des sables avec la méthode de l'alternance de charge. Irrigation automatique dite "goutte à goutte". Projet optimal de grands aménagements hydrauliques.

Correspondance: Italien, anglais, français, allemand.

Roma

Servizio Macchine Agricole e Fertilizzanti Chimici, Direzione Generale della Produzione Agricola, Ministero dell'Agricoltura e delle Foreste (Service de Machines Agricoles et d'Engrais Chimiques, Direction Générale de la Production Agricole, Ministère de l'Agriculture et des Forêts)
via XX Settembre, 20, Roma

Radicioni, A.   Dr.   Chef du service

Istituto Sperimentale per la Meccanizzazione Agricola (Experimental Institute of Agricultural Mechanization)
Via XX Settembre 98 E - 00187 Roma   Tel.: 484767-4755739

Colzani, G.	Dr.Ag.	Professor and Director of the Institute
Bastianoni, M.	Dr.	Research
Guarna, S.	Ing.	"
Marsili, A.	Ing.	"
Nuccitelli, G.	Ing.	"
Santoro, G.	Ing.	"
Cherubini, S.	Per.Agr.	Expert
Manetta, M.	Per.Agr.	"

Branches: Operative Section for Agricultural Mechanization in Treviglio (BG), via Milano 43/A.

Status: The Institute was established in 1967 - Semi-governmental - Supervised and financed by the Ministry of Agriculture and Forestry.

Activities: No teaching. Research on mechanization of agriculture.

Publications: on trials and tests carried out at the Institute and the Reviglio Branch concerning agricultural machines and solar energy.

Correspondence: English, French, Italian.

Scuola Nazionale di Stato per la Meccanica Agraria (National School for Agricultural Mechanical Engineering)
Capannelle - 00178 Roma - tel. 7990743

Romanello, G.	Prof.Ing.	Director
Baldovin, G.	Ing.	Agr. Engineering machinery and agr.hydraulics
Raparelli, F.	Dr.	Economic evaluation of agr. machinery
Simoni, M.	Ing.	Agric.Engineering machinery - General and applied mech.

Status: Established 1959. Supervised and financed by the Ministry of Public Education.

Activities: Teaching: Normally, there are two types of courses: a) Course leading to the qualification of operator mechanic (ag. machines), duration 2 years; b) Course for technicians of 1 year. - Research: Experimentation and promotion of agricultural machines. Publications: on agricultural equipment, tractors and their operation.

Correspondence: French and English, Italian.

Utenti Motori Agricoli (U.M.A.), Ente Assistenziale (Organisation d'Assistance aux Usagers de Moteurs Agricoles)
via Sicilia 194, 00187 Roma   Tel.: 4754251

Bureaux: dans tous les chefs-lieux de province.

Status juridique: Institution sémi-gouvernementale dépendant du Ministère de l'Agriculture.

Correspondance: Italien, français, anglais.

Sassari

Istituto di Costruzioni Rurali e Topografia, Università degli Studi di Sassari
(Institute of Farm Building, Topography and Design)
Via E. De Nicola 1 - 07100 Sassari (Sardegna) - Tél.(079) 217427

Satta, R.	Prof.Dr.Agr.	Director of the Institute, charged with Farm building, Topography and design
De Montis, S.	Prof.Dr.Ing.	Asst.Prof., charged with Zootechnical building
Pisanu, M.	Prof.Dr.Agr.	Asst.Prof.
Porcu, C.	Prof.Dr.Ing.	

Structure: State institution established in 1947 and supervised by Ministry of Education.

Activities: Teaching: Average number of 40 - 50 students getting a degree in Agricultural Sciences - Study course on: fundamentals and generals in farm structure - Physiological principles, comfort and health in livestock and plant housing - Planning and coordinating task and agritourist settlements concerning resources and structures for the development and the improvement of sheep, goat and cattle breeding in Sardinian districts - surveying - rural road planning - generals in rural mapping - Audio-visual support.

Publications: on Shelters for meat cattle - Italian trends in industrialized building for cattle. Observations and suggestions about alternative solutions - Activities and services: dimensional factors in modular planning of breeding stock buildings - Factors and functional parameters for the project of zootechnic buildings.

Correspondence: English, French, Italian.

Torino

Istituto di Meccanica agraria, Università di Torino (Institut de Mécanique et Machinisme Agricole - Institute of Agricultural Machinery and Farm Mechanization)
Via Michelangelo 32, 10126 Torino

Montuoro, P.	Dr.Ing.	Prof. Directeur
Ghiotti, G.	Dr.Agr.	Expérimentateur

Structure: Institution gouvernementale, fondée en 1935, dépendant du Ministère de l'Education.

Activités de formation: Cours de formation de la durée d'un an (environ 250 étudiants) faisant partie du curriculum agricole de la Faculté d'Agronomie (4 ans) et conduisant au diplôme de docteur en sciences agronomiques. Recherches: Recherches sur le labourage du sol avec des instruments traditionnels et avec des instruments particuliers. Recherches sur les tracteurs agricoles et sur les organes de transmission et de propulsion. Recherches sur les machines pour la distribution d'engrais, pour les traitements antiparasitaires et pour le semis. Recherches sur les machines pour la récolte, le fânage et la conservation des fourrages. Recherches sur les machines d'intérieur de ferme. Recherches sur la mécanisation en colline et sur la mécanisation de la riziculture.

Publications: Etudes sur les tracteurs agricoles et investigations sur leurs organes de transmission et de propulsion. Transmissions hydrauliques dans les machines agricoles. Problèmes agronomiques et de sécurité dans les tracteurs. Etude du procédé dynamique des sièges des tracteurs. Sous-soulage et labourage profond. Etude dynamique des charrues portées. Labourage en rizière. Investigations sur l'emploi des outils pour la construction de diguettes en rizière. Essais de distributeurs d'engrais. Semaille mécanique de fourragères mixtes. Interventions mécaniques pour l'engorgement du sol très perméable de rizière. Emploi des machines pour les traitements antiparasitaires. Fauchage mécanique des fourrages en plaine et en colline. Etude dynamique de motofaucheuses. Ramassage-pressage des fourrages et de la paille. Essais de moissonneuses-batteuses pour la récolte du riz. Efficacité des différents groupes batteur et contre-batteur des moissonneuses-batteuses à riz. Récolte mécanique du maïs en plaine et en colline. Problèmes généraux de mécanisation agricole.

Istituto per la Meccanizzazione Agricola (Institute of Agricultural Mechanization)
Via O. Vigliani 104 - 10135 Torino   Tél.:(011) 341015-341016

Lisa, L.	Dr.Agr.	Directeur

Section Agrotechnique:
Elia, P.	Dr.Agr.	Chef de la Section
Ciotti, A.	Prof.Agr.	Expérimentateur
Ferrero, A.	Dr.Agr.	Expérimentateur
Finassi, A.	Dr.Agr.	Expérimentateur

Section Machinisme agricole:
Giaia, R.	Dr.Ing.	Chef de la Section
Gioco, M.	Dr.Ing.	Expérimentateur
Potecchi, S.	Dr.Ing.	Expérimentateur
Vallania, R.	Dr.Phys.	Expérimentateur

Service Documentation et Calcul:
Potecchi, S.	Dr.Ing.	Chef du Service

Service Administratif:
Lisa, L.   Dr.Agr.   Chef du Service

<u>Annexes</u>: Station affiliée pour la mécanisation de la riziculture à Vercelli. Ferme expérimentale pour la mécanisation des terrains en pente à Vezzolano (Asti).

<u>Structure</u>: Institution sémi-gouvernementale dépendant du Conseil National des Recherches, (CNR), fondée en 1951 sous le nom de "Centro Nazionale Meccanico Agricolo" (Centre National de Machinisme Agricole). Depuis mai 1968, cette dénomination a été changée en "Laboratorio per la Meccanizzazione Agricola" (Laboratoire pour la Mécanisation Agricole) et dès décembre 1979 en "Istituto per la Meccanizzazione Agricola" (Institut pour la Mécanisation Agricole).

<u>Recherches en cours</u>: Mécanisation de la récolte, conservation et utilisation des fourrages (ensilage, agglomération, pertes de fenaison, alimentation des bovins) - Energétique dans la mécanisation (emploi de l'énergie solaire: séchage de produits agricoles, pompage de l'eau) - Mécanisation de la viticulture (mode de conduite, taille, opérations en vert, récolte) - Mécanisation de la riziculture (travail du sol, désherbage, séchage) - Mécanisation des terrains en pente - Confort du conducteur des machines agricoles (conditionnement de la cabine et protection contre les vibrations) - Nouvelles installations pour les essais des machines agricoles.

<u>Activités diverses</u>: Essais d'homologation des machines agricoles - Essais de performance et d'aptitude des machines agricoles.

<u>Publications</u> (depuis 1973): Téchnique de cultivation des prairies - Composition de la prairie et développement des mauvaises herbes - Epoque de coupe des fourrages - Récolte, conservation et utilisation des tiges de maïs - Pertes de récolte et de conservation des fourrages - Consommation d'énergie dans la déshydratation des fourrages - Emploi de l'énergie solaire pour intégrer le séchage de produits agricoles - Ensilage des fourrages - Aménagement des terrains en pente en vue de la mécanisation du verger - Mécanisation du travail du sol dans les vignobles en pente - Sur les modes différentes de conduite de la vigne - Taille mécanique de la vigne - Mécanisation des opérations en vert de la vigne - Récolte mécanique du raisin par coupe et sécourage sur vignobles en pente - Nettoyage du raisin après la récolte mécanique - Analyse du coût de vendange par simulation - Travail du sol en rizière avec différents outils - Résistance à l'usure des éléments des houes rotatives - Semis du riz en lignes sur terrain sec - Emploi des engrais liquides en rizière - Désherbage chimique du riz - Automatisation d'installations statistiques pour le séchage du riz - Enfuissement de la paille du riz - Consommation d'eau dans l'irrigation par ruissellement - Mécanisation de la récolte des poivrons - Mécanisation de la récolte des noisettes - Emploi des moteurs à vent en agriculture - Véhicule-laboratoire de mesurage - Etude sur le comportement dynamique d'un siège pour tracteur - Installation pour les essais des cabines et cadres de sécurité - Bulletins d'essais de tracteurs agricoles et des structures de sécurité suivant les codes italiens et OCDE.

<u>Correspondance</u>: Italien, français, anglais.

<u>Istituto di Idraulica Agraria dell'Università di Torino</u> (Institute of Agricultural Hydraulics, University of Turin)
Corso Raffaello n. 8 - 10126 Torino   Tél.: 682789

Tournon, G.   Prof.Ing.   Directeur
Allavena, L.   Dr.Ing.   Asst.
Merlo, C.   Dr.Agron.   Asst.

Structure: Institut gouvernemental, fondé en 1964, dépendant du Ministère de l'Education Nationale.

Activités: Cours d'Hydraulique Agricole et de Technique de l'irrigation et de l'assainissement agricole ayant la durée d'un an (avec environ 60 étudiants) et faisant partie du "Curriculum studii" de la Faculté de Sciences Agricoles. - Recherches: Irrigation souterraine, Irrigations des rizières; Mesures des débits d'eau pour l'irrigation; Evapotranspiration; Etude des caractéristiques hydrodynamiques des sols.

Publications: Irrigation souterraine capillaire; Effets thermiques de l'eau dans les rizières; Evapotranspiration et besoins en eau en Italie; Filtration de l'eau en sols stratifiés irrigués par submersion; Méthode de calcul des besoins en eau d'irrigation; Diffusion de l'eau débitée par l'irrigation souterraine; Besoins en eau des rizières.

Correspondance: Italien, français, anglais.

CEMOTER - Centro di Studio per le Macchine Movimento Terra e Veicoli Fuoristrada
(Research Centre for Earth-moving Machinery and Off-road Vehicles)
C.so Duca degli Abruzzi 24 - 10129 Torino   Tél.011/515891

Rigamonti, G.	Directeur	
Bonini, R.	Ingénieur -	Emissions de bruit et sécurité des engins de travaux publics
Giva Magnetti, L.	Technicien	Transmissions hydrostatiques des engins de travaux publics
Rigamonti, G.	Ingénieur -	Transmissions hydrodinamiques des engins de travaux publics
Zarotti, G.L.	Ingénieur -	Transmissions hydrostatiques et hydrodinamiques des engins de travaux publics

Structure: 1er Septembre 1970 (originé de la restructuration du "Centro Studi di Motorizzazione Agricola", institué en 1964. Consiglio Nazionale delle Ricerche - Roma (Conseil National des Recherches).

Activités: Formation: Thèses de diplôme d'études supérieures sur des sujets concernant les engins de travaux publics, systèmes de locomotion, contact sol/véhicule, transmissions de la puissance. - Domaine des recherches: transmissions hydrauliques, éléments de contact sol/véhicule (pneus, chenilles), ergonomique des engins de travaux publics.

Correspondance: Italien, français, anglais, allemand.

IVORY COAST

Direction des Aménagements Ruraux (Agricultural Engineering Service)
B.P.V. 9, Abidjan

Service de la Promotion du Machinisme Agricole

Service de l'Hydraulique Agricole

Service de l'Equipement Rural

Services extérieurs

Région Centres des Aménagements Ruraux, B.P. No. 9, Bouaké

Région Centre Ouest, Daloa

Région Nord

Région Sud

Statut juridique: Service technique gouvernamental dépendant du Ministère de l'Agriculture.

Activités: Conception et contrôle des programmes d'aménagements ruraux - Préparation et exécution de projets de génie rural - Aménagements hydro-agricoles, aménagements de bassins versants, pistes, barrages de retenue, construction de bâtiments agricoles, expérimentation de machinisme agricole.

Correspondance: Français.

Université Nationale de Côte d'Ivoire (National University of Ivory Coast)
B.P. V 34, Abidjan   Tél.: 439000

Correspondance: Français.

Institut Africain pour le Développement Economique et Social
B.P. 8 - Abidjan 08

Correspondance: Français.

JAMAICA

Agricultural Engineering Division, Ministry of Agriculture
Hope Gardens, Kingston 6, Jamaica W.I.

Pusey, J.E.   Director of Engineering

Branches with other location: Microdam Project and Irrigation Project, Construction Branch, Hague/Meylersfield Drainage Project.

Status: Central Government.

<u>Agricultural Engineering activities</u>: Minor Irrigation Projects - Construction of Farm Building - Repairs to buildings (minor and major) etc. - Central Repair Workshop: Repairs to Agricultural equipment and motor vehicle.

<u>Correspondence</u>:   English.

JAPAN

Abiko-City

<u>Denryoku-Chuo-Kenkyu-sho, Seibutsu-Kankyo-Gijitsu-Kenkyu-sho</u> (Bio-Environment Laboratory, Central Research Institute of Electric Power Industry)
1646 Abiko, Abiko City, Chiba, Japan

Nakamura, H.	Dr.Eng.	Director
Minohara, Y.	Dr.Agr.	Deputy Director
Yamamoto, Y.	Dr.Agr.	Chief, Applied Biology Section

<u>Status</u>: Private (non-profit) institution, established in December 1957.

<u>Agricultural Engineering activities</u>: Research, experiment, testing and extension work. - Major research efforts include: Utilization of thermal effluent for greenhouse heating - Systematization of the soil-air heat exchange greenhouse - Improvement of efficiency of electricity utilization in protected horticulture by: Development and application of a multi-element control system which enables the atmospheric environment of the greenhouse to be adjusted in accordance to solar radiation; Development of optimized environmental control regimes for increasing fruit and vegetable production - Detection of bio-functions (photo-electric reaction, stem enlargement and contraction, electric character of leaves, etc.) as a response to environmental conditions.

<u>Publications</u>: CRIEPI Report.

<u>Correspondence</u>: Japanese, English.

Fokuoka - shi

<u>Kyushu Daigaku, Nogaku Bu, Nogyokogakka</u> (Department of Agricultural Engineering, Faculty of Agriculture, Kyushu University)
Hakozaki-cho, Higashi-ku, Fukuoka-shi, 812 Japan

<u>Irrigation and water utilization</u>

<u>Drainage and Reclamation</u>

<u>Land Improvement</u>

<u>Agricultural Meteorology</u>

<u>Farm Machinery</u>

<u>Agricultural Process Engineering</u>

<u>Status</u>: Established in April, 1942. Responsible to the Ministry of Education.

Activities: Undergraduate Course: Agricultural Civil Engineering Course - Agricultural Machinery Engineering Course - M.Sc. Course - Ph.D. Course. - Research: Irrigation and water utilization - seepage water control, unsaturate percolation, permeability, sprinkler irrigation, groundwater - drainage and reclamation - water utilization, drainage, water replacement of retarding basin - hydraulics, reclamation, channel flow of gut - land improvement - land improvement, conservation, earth structure, soil properties - ground settlement, off-the road locomotion, bearing capacity - soil physical analysis - agricultural meteorology - meteorological calamities, artificial weather control, biometeorology, radar meteorology, flood forecasting, aeral rainfall measurement, air pollution - farm machinery - tractor vibration, hay wafering, grain binder, soil mechanics, plasticity, soil cutting, dynamics wheel, rotary cultivation - agricultural process engineering - drying, cooling, evaporation, thermal properties, mechanical properties, optical properties, storage, packing, transportation, farm structure.

Publications: Ground water, coastal sand, hydraulic model study - Dam, circular channel, waterhammer, intake pipe.

Correspondence: Japanese, English.

Hekikai-shi

Aichi-ken Nogyo-Shikenjyo, Nokigu-bu (Farm Machinery Department, Aichi Agricultural Experiment Station)
Anjyo-machi, Hekikai-shi, Aichi-ken

Activities: Research, experiment, testing and extension work - Construction and testing of animal drawn implements.

Publications: Methods of cultivating uplands for rice crops - Tillage of paddy fields with animal drawn ploughs.

Correspondence: Japanese, English.

Kitamoto-chi

Norin-sho, Joji-shikenjo, Hatasaku-bu, Kikaika-kenkyushitu (Mechanization Laboratory, Upland Farming Division, Central Agricultural Experiment Station, Ministry of Agriculture and Forestry)
Kitamoto-shi, Saitama-ken, Japan

Status: Governmental institution established in 1959.

Agricultural Engineering activities: Studies on mechanization of upland weeds control - Studies on mechanization methods to increase production and reduce labour requirements - Studies on mechanization of multi cultures.

Publications: Annual reports.

Correspondence: Japanese, English.

Ibaraki

<u>Chikusan Shikenjo</u> (National Institute of Animal Industry)
Tsukuba Norindanchi, P.O. Box 5, Ibaraki 305, Japan

Yukio Azuma	B.Sc.	Dept. of Feeding and Management, Chief of Lab. of Animal Husbandry Engineering
Akihiro Gondo	B.Sc.	Dept. of Feeding and Management, Researcher of Lab. of Animal Husbandry Engineering.

<u>Status</u>: Governmental institution, established in 1916, and supervised by the Ministry of Agriculture, Forestry and Fisheries. (The Laboratory is newly established in October, 1978).

<u>Agricultural Engineering activities</u>: Research on livestock raising machinery and facilities.

<u>Correspondence</u>: Japanese, English.

Konosu-shi

<u>Norinsuisan-sho, Noji-shikenjo</u> (Central Agricultural Experiment Station)
Konosu-shi, Saitama-ken

<u>Farm Operation Division</u>:
Tani, K.	Agr.Eng.	Head
Miyazawa, F.	Agr.Eng.	Chief of First Lab. - Mechanization
Irie, M.	Agr.Eng.	Chief of Second Lab. - Mechanization

<u>Upland Farming Division</u>: Kannondai, Yatabe-cho, Tsukuba-gun, Ibaraki-ken
Naka, S.	Agr.Eng.	Chief of Mechanization, Lab. Mechanization

<u>Status</u>: Governmental institution, established in 1959 and 1962, supervised by the Ministry of Agriculture, Forestry and Fisheries.

<u>Activities</u>: Research on utilization and improvement of machinery in rice, wheat, barley, vegetables, forage crops and soybean cultivation.

<u>Publications</u>: J.Cent.Agric.Exp.Stn.: Mechanization of upland crop cultivation under intercropping system. Investigations on plastic seeding-mulching machine. Simulation of harvesting system of grass. Improvement of atmosperic environment for combine and dryer operator. Labour-saving device for daikon (Japanese radish) harvesting work. Mechanization of welsh onion cultivation.

<u>Correspondence</u>: Japanese, English.

Kyoto-shi

Kyoto Daigaku, Nogakubu, Nogyo-Kogakuka (Department of Agricultural Engineering, Kyoto University), Kitashirakawa, Sakyo-ku, Kyoto-shi

Maruyama, T.		Head of Department
Hasegawa, T.	Dr.Prof.	Hydraulic Structure Engineering
Aoyama, S.	Dr.	Instructor, Hydraulic Structure Eng.
Uchida, K.	Dr.	" " " "

Maruyama, T.	Dr.Prof.	Irrigation and Drainage
Mitsuno, T.	Dr.	Associate Prof., Irrigation and Drainage
Maekawa, T.		Instructor " " "
Kobayashi, S.		Instructor " " "
Nishiguchi, T.	Dr.Prof.	Rural Planning
Kitamura, T.	Associate Prof.	" "
Imai, T.		Instructor, Rural Planning
Ushino, T.		Instructor, " "
Minami, I.	Dr.Prof.	Water Use Engineering
Fukuma, J.		Associate Prof., Water Use Engineering
Torii, K.		Instructor " " "
Kawachi, T.	Dr.	Instructor " " "
Tanaka, T.	Dr.Prof.	Farm Power
Yamazaki, M.	Dr.	Associate Professor, Farm Power
Oida, A.	Dr.	Instructor " "
Tano, N.		Instructor " "
Kawamura, N.	Dr.Prof.	Farm Machinery
Namikawa, K.	Dr.	Associate Prof., Farm Machinery
Fujiura, T.		Instructor " "
Ura, M.		Instructor " "
Yamashita, R.	Dr.Prof.	Farm Processing Machinery
Ikeda, Y.	Dr.	Associate Prof., Farm Processing Machin.
Ikeda, Y.	Dr.	Associate Prof. " " "
Kato, K.	Dr.	Instructor " " "
Goto, K.		Instructor " " "

Status: University established in 1897. Department of Agricultural Engineering established in 1923. Government institution.

Activities: Teaching: Four years undergraduate course leading to Bachelor degree attended by 45 students, and graduate course leading to Master and Doctor degrees. – Research: Hydraulic Structure Engineering – Concrete engineering, soil mechanics, dam engineering, foundation engineering, structural mechanics, design engineering – Irrigation – Upland irrigation, paddy irrigation – Drainage – underdrainage, surface drainage – Rural planning – Land consolidation, village planning, land use planning – Farm Power – Dynamics of tractors and soil-machine system, automatic control of hydrostatic transmission tractor, automatic guidance of farm tractors – Farm Machinery – Automatic control of rotary tiller, automatic control of combine, rice transplanter, forage harvester, micro-computer application to farm machinery, deeper tillage with rotary tiller, pest control drainage mole – Farm Processing Machinery – Drying, corting, milling and refrigeration of agricultural products, engineering of system for processing agricultural products.

Publications: Earth Dam (Analysis of-, under vehicular traffic) – Vibration (Analysis, modes, damping, model test) – Earthquake or seismic (Response, accelerogram), Micro tremor polder reclamation embankment, Safety analysis, slope stability analysis optimal design process, Finite element method, Structure – System, Normally consolidated clays CU and CD Test (Triaxial test), Reinforced concrete beam, Dam engineering, Numerical analysis, Geomechanics, Random vibration, Soil mechanics, Stress-strain relation of soils, Soil under cyclic loading, Dynamic properties of soils, Initial stress. Probabilistic study, Ultimate bending movement, Aseismic design.

Correspondence: Japanese, English, German.

Morioka-shi

<u>Tohoku Nogyo Shikenjo</u> (Tohoku National Agricultural Experiment Station)
Morioka, Japan

Honda, T.	Director
Ishihara, S.	Rice cultivation Section
Koizumi, T.	Field crops Section
Murao, S.	Field engineering Section

<u>Status</u>: Governmental institution, established in 1950.

<u>Agricultural Engineering activities</u>: Research and development of agricultural machinery and methods of mechanized work - Irrigation and drainage improvement.

<u>Publications since 1960</u>: Study on cultivation methods using drills - Study on the relations between the characteristics of powdered and granularized fertilizers and the improvement of fertilizer distributors - Study on the economics properties of combine harvesters - Studies on the preparation of grass silage.

<u>Correspondence</u>: Japanese, English, Spanish.

Nagakute-mura

<u>Aichi-ken Nogyo Sogo Shikenjyo, Kiso-Kenkyu-bu, Dai ichi Kenkyushitsu</u> (First Laboratory, Basic Research Division, Aichi Agricultural Research Centre)
Nagakute-cho, Aichi-gun, Aichi-ken

<u>Status</u>: Governmental Research Centre with the First Laboratory (Agricultural Engineering) established in 1967, and financed by the Aichi prefectural office and the Government of Japan.

<u>Agricultural Engineering activities</u>: Testing and improvement of farm machinery - Improvement of rice and vegetable transplanters - Improvement of harvesters and grain dryers.

<u>Publications</u>: Studies on the mechanization of rice transplanting - Studies on harvesting and drying methods with windrowers.

Naganuma

<u>Hokkaido-ritu Chuou Nogyo-Shikenjo Nogyo kikai-bu</u> (Division of Agricultural Machinery, Hokkaido Central Agricultural Experiment Station)
Naganuma, Hokkaido, 069-13, Japan

Saito, W.	Dr. Agr.	Chief of Division

<u>Branches</u>: Hokkaido-ritu Tokachi Nogyo-Shikenjo kikai-ka (Division of Agricultural Machinery, Hokkaido Tokachi Agric.Experiment Station), Memuro, Kasaigun, Hokkaido, 082, Japan.

<u>Status</u>: Established 1950, Governmental.

Agricultural Engineering activities: Training, research, experiment, testing and extension work - Tractor test on the field - Safety flame and cabin - Rice planter - Combine and related machinery on the paddy field - Drier of rice, wheat and beans - Pipeline milker - Feed produced machinery for cattle - Mechanization of vegetables.

Publications: Effect by covering picking tines with rubber and adaptability of wax bean - On the performance test of 100ps class tractors - Performance of the riding-type of 6 row rice transplanter - Studies on the Vegetable Planter-Stanhay spacing drill - Performance of the Onion Packinghouse - Studies on the Compact machine - Research on damage of potatoes and onions.

Correspondence: Japanese, English.

Obihiro

Obihiro Chikusan-Daigaku, Nogyo-kogaku-ka (Agricultural Engineering Department, Obihiro University of Agriculture and Veterinary Medicine)
Inada-cho, Obihiro-shi, Hokkaido 080

Ono, T.	Prof. Ph.D.	Agricultural Power
Taniguchi, T.	Assoc.Prof. Ph.D.	Agricultural Power
Miyamoto, K.	Assoc.Prof. Ph.D.	Field Machinery
Takahata, H.	Prof. Ph.D.	Machinery for Animal Husbandry

Status: Governmental institution, established in 1963.

Activities: Teaching: Undergraduate course with 40 students - Graduate master course with 6 students. - Research activities: Studies on farm wheel by soilbin - Performance of depth control system - Performance of vacuum regulator for milking machine - Physical studies on the internal bruising of potatoes - Fundamental studies on crop cutting - Studies on silage making equipment - Performance of haycuber.

Publications: Traction characteristics of two-wheel drive tractors - Control characteristic on draft controlled plowing - Performance of the stone picker - Operating efficiency of potato harvester - Studies on the wafer formation of dried plant - Studies on the piston-type wafering machine.

Correspondence: Japanese, English.

Okayama-shi

Okayama-ken Nogyo-Shikeniyo, Nokigu-bu (Farm Machinery Department, Okayama Agricultural Experiment Station)
Kitakata, Okayama-shi

Status: Local governmental institution.

Activities: Studies on the most appropriate farming systems, based on the utilization of small tractors - Labour saving with rotary power tillers in the cultivation of rice and wheat - Studies on seed drills and fertilizer distributors - Tests of the crop driers.

Publications: The construction and function of the rolls in the huller - The tillage effect of various rotary tillers - Farm mechanization in Okayama.

Correspondence: Japanese, English.

Omiya-shi

Nogyokikaika Kenkuyusho (Institute of Agricultural Machinery)
1-chome, Nisshincho, Omiya-shi, Saitama-ken

Kogure, M.	Director General
Kawada, N.	Director
Maeda, K.	Director

1st Division for Machinery in Common:

Takenaga, T.	Head	
Fujii, K.		Safety for farm machinery
Miura, K.		Engine, tractor and tillage machinery
Miura, K.		Seeding and fertilizing machinery
Takenaga, T.		Pest control machinery

2nd Division for Machinery for Grain and Beans:

Ban, T.	Head	Drying, storage, processing and handling equipment
Takao, H.		Drying, storage, processing and handling equipment
Hoshino, S.		Transplanting machinery
Ichikawa, T.		Threshing and harvesting machinery

3rd Division for Animal Husbandry:

Okui, K.	Head	Fodder crop producing machinery
Kitamura, M.		" " " "
Kuwana, T.		Livestock raising machinery
Susawa, K.		Feed processing machinery

4th Division for Horticulture:

Hirata, K.	Head	Orchard machinery
Furuya, T.		Vegetable processing and storage equipment
Nagaki, T.		Orchard machinery

Kurata, I.   Horticultural machinery and installation in greenhouse
Yamamoto, K.   Vegetable and commercial crop machinery

Testing Division:
Ariyoshi, M.   Head
Kanatsu, T.   Test of riding tractor (engine displacement > 1500cc)
Yagi, S.   Test of riding tractor (engine displacement ≤ 1500cc)
Kanai, K.   Test of rice transplanter, beet harvester and potato harvester
Tosaki, K.   Test of sprayer (power, airblast and multipurpose type)
Suzuki, M.   Test of combine harvester
Mori, Y.   Test of safety cab and frame for tractor

Planning and Survey Division:
Aiko, I.   Head
Enokido, K.   Survey and information
Nihei, S.   Product information

Experimental Farm at Kawazato-mura, Kitasaitama-gun, Saitama-ken:
Takemura, G.   Chief

Status: Governmental institution, established in 1962, jointly financed by the Japanese Government and private concerns, supervised by the Ministry of Agriculture, Forestry and Fisheries.

Activities: Development, research, experiment and tests of agricultural machinery – Survey and information service.

Publications: Technical reports, Test reports, Annual reports on the activities of the Institute, Yearbook.

Correspondence: Japanese, English, German, French.

S a k a i - s h i

Osaka Furitsu Daigaku, Nogaku-bu, Nogyo-kikai-gaku Kenkyi-shitsu (Agricultural Machinery Laboratory, College of Agriculture, University of Osaka Pref.)
Mozu-ume-cho, Sakai-shi, Osaka-fu

Status: Governmental and prefectural institution, established in 1949, supervised by the Osaka Prefecture.

Activities: Study course in: agric. machinery with B.Sc., M.Sc. and Dr.A.E. degrees. Research on: Dynamics between tractor and tillage implements – Relationship between wheel and soil – Dynamics and auto-design of power sprayer – Pattern recognition and grading of biological products.

Publications: Journal of the Society of Agricultural Machinery.

Correspondence: Japanese, English.

Sapporo-shi

Norinsuisan-sho Hokkaido Nogyo-shikenjo (Hokkaido National Agricultural Experiment Station)
Hitsujigaoka, Toyohira-ward, Sapporo-city, Hokkaido, Japan 061-01

Masuo, Y.	B.Sc.	Director
Ozaki, K.	Dr.	Deputy Director

Research Planning and Coordination Office:
Todani, A.		Head of Office
Kon, T.	B.Sc.	Chief of Planning Section
Sakurai, T.		Chief of Research Coordination Section
Kukutsu, K.		Chief of Information Section

First Crop Division:
Miyasaka, A.	Dr.	Head of Division

Second Crop Division:
Kawamura, E.	B.Sc.	Head of Division
Kasano, H.		Chief of Industrial Crop Laboratory (Engaru, Monbetsu-gun, Hokkaido 099-04)

Sugar Beet Division:
Takase, N.	Dr.	Head of Division

Upland Farming Division (Memuro-cho, Kasai-gun, Hokkaido 082):
Okubo, T.	Dr.	Head of Division

Agricultural Chemistry Division:
Kishita, A.	Dr.	Head of Division
Otowa, M.	B.Sc.	Chief of Soil Survey and Soil Classification Laboratory

Agricultural Physics Division:
Inoue, K.	B.Sc.	Head of Division
Sato, H.		Chief of Agricultural Engineering Lab.
Fukunaka, H.		Agricultural Engineering
Miyama, K.	M.Sc.	" "
Fujioka, S.	B.Sc.	Chief of 1st Farm Mechanization Lab. (Grassland)
Maeoka, K.	B.Sc.	Farm Mechanization
Tamaki, K.	B.Sc.	" "
Honjyo, H.	B.Sc.	Chief of 2nd Farm Mechanization Lab. (Paddy field)
Tomita, M.		Farm Mechanization
Sawamura, N.	B.SC.	" "
Takezono, T.	Dr.	Chief of 3rd Farm Mechanization Lab. (Agricultural Installation)
Amano, K.	B.Sc.	Farm mechanization
Tomari, I.	Dr.	Chief of Agricultural Meteorology Laboratory

Fujiwara, T.	B.Sc.	Agricultural Meteorology
Ishiguro, T.		" "

PLant Pathology and Entomology Division:
Ichinohe, M. — Dr. — Head of Division

Livestock Division:
Omori, S. — Dr. — Head of Division

First Grassland Development Division:
Nitta, K. — B.Sc. — Head of Division

Second Grassland Development Division:
Shimada, Y. — B.Sc. — Head of Division

Farm Management Division:
Morinio, T. — Head of Division

Branches with other location:
Upland Farming Division: Memuro-cho, Kasai-gun, Hokkaido 082
Potatoes Breeding Laboratory: Shinamatsu, Eniwa-shi, Hokkaido 061-14
Industrial Crop Laboratory: Engaru-cho, Monbetsu-gun, Hokkaido 099-61
Peat Soil Laboratory: Kaihatsu, Bibai-shi, Hokkaido 072
Heavy Clay Soil Laboratory: Komukai, Monbetsu-shi, Hokkaido 099-61

Status: This station was established with National funds in 1901. It is financed by the Ministry of Agriculture, Forestry and Fisheries.

Activities: Research: Grassland Agriculture, Livestock Farming, Upland Farming, Horticulture, Rice Culture, Soil and Fertilizers, Agricultural Meteorology in Cold Regions, Plant Protection, Plant Pathology and Entomology, Farm Management, Agricultural Engineering.

Publications: Research Bulletin of the Hokkaido National Agricultural Experiment Station. Miscellaneous Publication of the Hokkaido National Agricultural Experiment Station. Soil Survey Report of the Hokkaido National Agricultural Experiment Station. Annual Report of the Hokkaido National Agricultural Experiment Station.

Correspondence: Japanese, English.

Hokkaido Daigaku, Nogakubu, Nogyotogakka (Agricultural Engineering Department, Faculty of Agriculture, Hokkaido University)
Kita-ku Kita-9 Nishi-9, Sapporo

Chair of Land Improvement:
Kataoka, T. — Dr.Agr.Prof.
Umeda, Y. — Dr.agr.Assoc.Prof.
Sakurada, J. — M.Sc.agr. Asst.
Nagasawa, T. — B.Sc.agr. Asst.

Chair of Agricultural Physics:
Dohkoshi, J. — Dr.agr. Prof.
Horiguchi, I. — Dr.agr. Assoc.Prof.
Hoshiba, S. — B.Sc.agr. Asst.

Chair of Soil Amelioration:
Sasaki, S. — Dr.agr. Prof.
Maeda, T. — Dr.agr. Assoc.Prof.
Yazawa, M. — B.Sc.agr. Asst.
Soma, K. — Dr.agr. Asst.

Chair of Agricultural Machinery:

Nambu, S.	Dr.agr.Prof.
Takai, M.	Dr.agr.Assoc.Prof.
Itoh, M.	Dr.agr. Asst.
Hata, S.	Dr.agr. Asst.

Chair of Farm Power:

Matsui, K.	Dr.agr. Prof.
Terao, H.	Dr.agr. Assoc.Prof.
Matsumi	Asst.
Ohmiya, K.	B.Sc.agr. Asst.

Chair of Agricultural Process Engineering:

Ikeuchi, Y.	Dr.agr. Prof.
Itoh, K.	Dr.agr. Assoc.Prof.
Matsuda, J.	Dr.agr. Asst.

Status: The Department of Agricultural Engineering was established in 1957 within the University of Hokkaido which is the National University of Japan.

Activities: The four year curriculum in agricultural engineering of two courses (Land Improvement Course, Agricultural Machinery Course) leading to B.Sc. degree – Number of students of each course 20 a year – Graduate program leading to M.Sc. and Dr.agr. degree. – Research: Water balance – Irrigation water temperature – Peat land use – Subsurface drainage – Soil erosion – Farm building – Storage of potatoes and vegetables – Green house – Solar energy – Cold weather damage – Cold air – Mass intrusion – Volcanogeneous soils – Weathering of soils – Rheology of soils – Amelioration of volcanogeneous soils – Soil Machine system – Accurate planting – Spraying system – Sugar beet harvester – Hay making – Man-machine system – Farm mechanization planning – Alternative fuels – Vibratory tillage tool – Chain saw – Traction theory – Agricultural terrain – Drying and storage of farm products – Physical properties of farm products – Energy analysis.

Publications: Water balance – Run-off – Peat land – Water temperature – Ground freezing – Environmental design – Potato storage – Solar energy – Green house – Allophen soil – Drainage – Seepage water – Soil structure – Flexible cage wheel – Sprayer nozzle – Hay making – Potato grader – Farmers' fatigue and farm injuries – Vacuum metering device – ROPS – Tractor noise – Vibratory tillage tool – Tractor wheel slippage – Tractor dynamics – Saw chain – Drying of rough rice – Storage of rice – Storage of onion – Drying of baled hay.

Correspondence: Japanese, English.

Miyagi-ken Nogyo-Center, Einoh-kikai-bu (Farm Management and Machinery Department, Miyagi Prefectural Agricultural Research Center)
Takadate, Natori, Miyagi, Japan – Zip code 981-12

Toyama, K.	Chief of Farm Machinery Department
Hiroshima, K.	Staff Scientist

Status: Date of establishment: April 1, 1980.

Activities: Research, experiment, testing and extension work. Research and experiments on the effect of the mechanical tillage and the precision seeders on the growth of rice plants in the paddy fields. Improvement of drying and arranging operations of rice plants. A survey of dust and noise pollutions in the rice processing plants.

Correspondence: Japanese, English.

Tokyo

Tottori Daigaku, Nougaku-bu, Nougyo-Kougaku-ka (Department of Agricultural Engineering, Faculty of Agriculture, Tottori University)
Minami 4-101, Koyama-cho, Tottori-shi, 680

Laboratory of Irrigation and Drainage:
Nomura, Y.	Professor, Dr.Agr.	Head of Laboratory
Hasegawa, K.	Assoc.Professor, B.S.	
Inoue, M.	Research Assoc., M.S.	

Laboratory of Agricultural Hydraulic Structures Engineering:
Tsuge, M.	Professor, Dr.Agr.	Head of Laboratory
Watanabe, S.	Assoc.Professor, M.S.	
Hattori, K.	Research Assoc., M.S.	

Laboratory of Land Reclamation:
Kouno, H.	Professor, Dr.Agr.	Head of Laboratory
Yoshida, I.	Assoc.Professor, M.S.	
Chikushi, J.	Research Assoc., Dr.Agr.	

Laboratory of Farm Equipment and Machinery:
Ishihara, A.	Professor, Dr.Agr.	Head of Laboratory
Higuchi, H.	Assoc. Professor, Dr.Agr.	
Iwasaki, M.	Research Assoc., Dr.Agr.	

Laboratory of Farm Power and Machinery:
Komatsu, M.	Professor, Dr.Agr.	Head of Laboratory
Koike, M.	Assoc. Professor, Dr.Agr.	
Misao, Y.	Research Assoc., M.S.	

Status: Established in 1949. Governmental (State) body.

Activities: Training: Undergraduate and graduate courses for about forty students for B.Sc., fourteen for M.S. - Research: Fundamental studies on the elements of water consumption in a field. On the statistical analysis of hydrological data. Research on the use of neutron moisture meter in studies of the soil water movement. Hydraulic research on the freshening reservoir. Experimental studies on durability of concrete. Fundamental studies on farm roads. Fundamental studies on the corrosion of concrete in sulfuric acid solution. Studies on the physical and engineering properties of organic soils. Studies of the distribution techniques of agricultural granules. Studies on power train and durability of small powered tractors. Studies on the harvesters for the root crops. Development of a rotary-panbreaker. An automatic traction system for farm tractor. Computer aided design of a tractor.

Publications: Bull. Sand Dune Res.Inst (Japan) - Bull. Faculty of Agriculture, Tottori Univ. (Japan) - Trans. JSIDRE (Japan) - Journal of the S.A.M. (Japan).

Correspondence: English, Japanese.

Tottori Daigaku, Nougaku-bu, Fuzoku Sakyu Riyou Kenkyu Shisetsu (Sand Dune Research Institute, Faculty of Agriculture, Tottori University)
1390, Hamasaka, Tottori-shi 680

This institute consists of five divisions. Among these, the following two divisions are related to education and research for the undergraduate and graduate students of department of agricultural engineering.

Meteorological Environment Division:

Matsuda, A.	Professor, Dr.Agr.	Head of Division
Kamichika, M.	Assoc. Professor, M.S.	

Hydrology and Irrigation Division:

Kodani, Y.	Professor, Dr.Agr.	Head of Division
Yano, T.	Assoc. Professor, Dr.Agr.	
Yamamoto, T.	Research Assoc., Dr.Agr.	

Status: Established in 1958. Governmental (State) body.

Activities: Staff belongs to the above division and also participates as a teaching staff in the Faculty of Agriculture. In the Sand Dune Research Institute, undergraduate and graduate students can take credits towards graduation thesis. – Research: On the heat budget in sand dunes and arid land. On the development of preventive method against meteorological hazard. On the water balance in sand dunes and arid land. On agricultural water use in sand dunes and arid land.

Publications: Bull Sand Dunes Res.Inst (Japan) – Bulletin Interdisciplinary Res.Inst. Environmental Sci.(Japan) – Trans. JSIDRE (Japan).

Correspondence: English, Japanese.

Mie Daigaku, Nogakubu, Nogyokikai-gakka (Department of Agricultural Machinery, Mie University)
Kamihama-cho, Tsu 514, Japan

Isa, T.	Ph.D.	Professor and Chairman – Field Machinery
Ichikawa, M.	B.S.	Assoc.Prof. – Machinery Design – Rice transplanter
Ito, N.	Ph.D.	Assoc.Prof. – Power and Machinery – Automatic control
Osita, S.	M.S.	Assist.Prof. – Thermal processing
Sato, K.	M.S.	Assist.Prof. – Field machinery
Takeda, S.	Ph.D.	Professor – Power and Machinery, Biomass Energy
Taijri, I.	M.S.	Assoc.Prof. – Field Machinery – Hillside tractor
Nakagawa, K.	Ph.D.	Professor – Harvesting and processing
Hoki, M.	Ph.D.	Assoc.Prof. – Power and Machinery – Physical properties
Horibe, K.	B.S.	Assoc.Prof. – Processing, Solar drying
Mori, K.	Ph.D.	Professor – Machinery design, Electrical power and processing

Status: Established in 1922. Governmental institution financed by the Ministry of Education.

Activities: Teaching: Study courses in professional agricultural machinery: four year undergraduate leading to B.S. degree, graduate program for M.S. Approximately 160 undergraduate students plus 10 graduate students. – Research: Automatic control of rice combine – Slippage control of tractor – Developmental research of multiple purpose farm vehicle – Direct seeding for wet paddy – Biomass energy production and utilization – Burning speed of internal combustion engine – Pneumatic seeding system – Micro computer control of seeder – Automatic control and simulation of hillside tractor – Measurements of tractor travelling – Mechanical properties of rice and soybeans – Tropical rice mechanization – Ultrasonic measurements – Grain separator – Grain drying by solar energy – Disc cutter blades – Thermal properties of rough rice measured by spherical heat source – Sprinkler nozzle – Simulation of rough rice drying by cool air – Thermal properties of agricultural products – Absorption and dehydration characteristics of agricultural materials – Automatic detection system of grain moisture – Energy efficiency for heat treating system – Dynamic characteristics of rice transplanter.

Publications: Numerous research papers on above subjects (in Japanese or English).

Correspondence: Japanese, English, German.

Norinsuisan-sho, Shikoku Nogyo-shikenjyo, Kikaika-kenkyushitsu (Farm Mechanization Laboratory, Division of Land Utilization, Shikoku Agricultural Experiment Station, Ministry of Agriculture, Forestry and Fisheries)
Zentsuji-cho, Zentsuji-shi, Kagawa-ken, 765, Japan

Furukawa, T.	B.Ag.	
Hasegawa, S.	M.Ag.	
Itokawa, N.	B.Ag.	
Kawasaki, K.	B.Ag.	Chief of Laboratory

Status: Governmental Institution, established in 1947.

Activities: Studies on farm mechanization on slopes. Studies on farm mechanization on paddy fields in south-west warm area.

Activities: Research: Studies on the mechanical harvesting, transportation in slope citrus orchards and in slope grass lands. Studies on the harvesting and drying systems of soybean and barley on paddy fields. Studies on the running performance and adaptability of farm transport vehicles on slopes.

Publications: List available on request.

Correspondence: Japanese, English, German.

Norin-sho, Shinko-kyoku, Kenkyu-bu (Research Council, Development Bureau, Ministry of Agriculture and Forestry)
Kasumigazeki 2-1, Chiyoda-ku, Tokyo

Activities: Research and extension work. Planning and management of research and experimental work carried out in the Governmental and prefectural experiment stations.

Publications: Annual reports on research and experiments in the prefectural experiment stations.

Correspondence: Japanese, English.

Tokyo Daigaku, Nogaku-bu, Nogyo-kogaku-ka (The University of Tokyo, Faculty of Agriculture - Agricultural Engineering Department)
Yayoi-cho, Bunkyo-ku

Status: Government institution supervised by the President of the University of Tokyo.

Activities: Undergraduate course of 4 years. Master course of 2 years. Doctor course of 3 years. Research and experimental work is carried out.

Correspondence: Japanese, English.

Tokyo Noko Daigaku, Npgaku-bu, Nogyo-kikai Kenkyu-shitsu
(Agricultural Engineering Department, Faculty of Agriculture, Tokyo University of Agriculture and Technology)
Saiwai-machi, Fuchu-shi

Status: Governmental institution.

Activities: Study courses of four years duration, leading to B.Sc.Agr. degree (Nogaku-shi). Study on the construction of rice transplanting machines – Research on the methods of transplanting. Study of the influence of mechanical tillage on soil structure. Study of the adherence of different soils to agricultural implements. Study on leaf picking machine for mulberry. Study on the physical properties of cut-flowers. Study on the transportation and storage of cut-flowers.

Publications: Soil crumbling effect of cultivator tines. Construction of rice transplanting machines. The influence of rotary tillage on soil structure.

Correspondence: Japanese. English.

Tokyo Kyoiku Daigaku, Nogaku-bu, Nogyo-Nogaku-ka (Agricultural Engineering Department, Faculty of Agriculture, Tokyo Educational University)
Ikerizi-cho Setagaya-ky

Activities: Study courses leading to a degree in agricultural engineering. Research, experiment, testing work. Studies of motocultivators. Study of the effects of harrowing and tilling on the tilth. The displacement of soil with rotary tillers.

Correspondence: Japanese, English.

Nogyo-kikai Gakkai (Society of Agricultural Machinery, Japan)
c/o Tokyo Daigaku, Nogaku-bu, Yayoi-cho, Hingo, Bunkyo-ku

Status: Private association of agricultural engineering specialists in the Government at Universities, Prefectures and factories.

Activities: Seasonal meetings on research work carried out by the members.

Publications: "Journal of the Society of Agricultural Machinery Japan", issued quarterly.

Correspondence: Japanese, English, German, Russian.

Yatabe-machi

<u>Norinsuisan-sho Nogyodoboku-shikenjo</u> (National Research Institute of Agricultural Engineering)
1-2 Kannondai 2-chome, Yatabe-machi, Tsukuba-gun, Ibaraki-ken

Nakajima, Y.	Dr.Agr.Eng.	Director
Kishimoto, R.	Agr.Eng.	Research Coordinator

<u>Department of Water Management</u>:

Igarashi, M.	Agr.Eng.	Head
Shibuya, K.	Agr.Eng.	Chief, Laboratory
Kimura, S.	Agr.Eng.	" "
Shiraishi, H.	Dr.Agr.Eng.	" "
Mitsuda, M.	Agr.Eng.	" "

<u>Department of Land Improvement</u>:

Chiba, T.	Dr.Agr.Eng.	Head
Sasano, S.	Agr.Eng.	Chief, Laboratory
Iwata, S.	Agr.Eng.	" "
Kawano, H.	Dr.Agr.Eng.	" "
Negishi, H.	Agr.Eng.	" "
Yamashita, S.	Agr.Eng.	" "
Sakurai, K.	Agr.Eng.	" "

<u>Department of Hydraulic Engineering</u>:

Nakahara, M.	Dr.Agr.Eng.	Head
Ishino, S.	Dr.Agr.Eng.	Chief, Laboratory
Iawasaki, K.	Dr.Agr.Eng.	" "
Ueda, M.	Agr.Eng.	" "
Sekiya, T.	Agr.Eng.	" "

<u>Department of Construction Engineering</u>:

Mishina, N.	Agr.Eng.	Head
Takaoka, K.	Agr.Eng.	Chief, Laboratory
Ebina, Y.	Agr.Eng.	" "
Kawaguchi, T.	Agr.Eng.	" "
Nakayama, Y.	Geology	" "

<u>Branches with other location - Saga Branch</u>:
1-1, Hinode 2-chome, Saga-shi, Saga-ken

Miyahara, A.	Agr.Eng.	Head
Hara, T.	Agr.Eng.	Chief, Laboratory
Nakamura, R.	Agr.Eng.	" "
Kamimura, H.	Agr.Eng.	" "

<u>Status</u>: Governmental institution, established in 1961, supervised by the Ministry of Agriculture, Forestry and Fisheries.

<u>Activities</u>: a) Training: The training has continued for 75 years since 1905. The Institute succeeded the training function in December 1961. The following 4 courses are given here for retraining governmental and prefectural engineers. Primary course, Middle course and Specialized course of Agricultural engineering techniques and Construction control techniques. b) Research: Water Management: Four laboratories perform research activities which are concerned with water management in broad area, including basic studies on hydrology, on application of radioisotopes, on systems analysis of water utilization facilities and on environmental engineering for rural area. - Land Improvement: Seven laboratories carry out the research which is concerned with development and improvement of farm land and farm facilities, such basic studies as on rural planning, on irrigation, drainage,

reclamation and consolidation of paddy and upland fields, on the structure and materials of farm facilities, and on proper utilization of water and other fluid in the facilities. - Hydraulic Engineering: Four laboratories engage in the researches concerned with hydraulic structures such as dams, headworks, canals, lakes, lagoons, coasts, and estuaries; and hydraulics, hydraulic design and optimal hydraulic control of these facilities have been studied. - Construction Engineering: Four laboratories perform the researches on structures, on construction materials, on soil engineering, and on engineering geology which are concerned with dams, headworks, canals, roads and other facilities as well as farm land conservation. - Saga Branch: Three laboratories perform the researches on efficient utilization and readjustment of arable land, on foundation and construction engineering and on special soil in the western region of Japan. - Joint Fundamental Research: For increasing agricultural productivity, the Institute is carrying out researches for utilization of natural energies, radioisotope, remote-sensing, etc.

Publications: Bulletin of the National Research Institute of Agricultural Engineering (in Japanese with English summaries), issued once a year. Technical reports of the National Research Institute of Agricultural Engineering (in Japanese), Series (A) Land Improvement, (B) Hydraulics, (C) Construction Engineering, (E) Saga Branch, (F) General matters; issued irregularly.

Correspondence: Japanese, English.

KENYA

Nairobi

Department of Agricultural Engineering, University of Nairobi
P.O. Box 30197, Nairobi, Kenya

Muchiri, G.	B.Sc., M.Sc.	Chairman
Ellen, H.	B.Sc., M.Sc.	
Katyal, A.K.	B.Sc., M.Sc.	
Kijne, J.W.	B.Sc., M.Sc., Ph.D.	
Mburu, S.G.	B.Sc., M.Sc.	
Shukla, B.D.	B.Sc., M.Sc., Ph.D.	
Srivastava, R.K.	B.Sc., M.Sc., Ph.D.	
Thomas, D.B.	BA, M.Sc.	

Status: Department of the University of Nairobi, established in 1974.

Activities: Teaching: Ag.Eng. to B.Sc. students in Agriculture and Agricultural Engineering and Diploma students in Irrigation and Soil conservation. Research: into energy requirements and mechanisation problems of small farmers and problems of irrigation and soil and water conservation.

Publications on soil conservation, erosion, agricultural machinery, irrigation hydrology, agricultural engineering.

Correspondence: English.

Land Development Division, Ministry of Agriculture
P.O. Box 30028, Nairobi, Kenya

Kenya Agricultural Research Institute (KARI)
P.O. Box 30148, Nairobi, Kenya

Njoro

Agricultural Engineering Department, Egerton College, Njoro
P.O. Njoro, Kenya   Tel.: Njoro 27 44 47 - Telegr.: College, Njoro

Misiko, P.A.M.	Ph.D.	Senior Lecturer/Chairman - Soil and Water Resources
Chauhan, A.M.	M.Sc.	Lecturer - Farm Power and Machinery
China Soita	B.Sc.	Lecturer - Soil and Water Conservation
Edwards, E.	M.SC.	Lecturer - Electrical/Processing
Grover, S.K.	M.Sc.	Lecturer - Mathematics and Physics
Kamau, F.		Demonstrator - Farm Power
Kaizer, P.	Ph.D.	Lecturer - Soil and Water/Irrigation
Kihurani, E.G.		Lecturer - Welding/Fabrication
Kimani, P.K.	B.Sc.	Lecturer - Soil and Water/Irrigation
King, A.	M.Sc.	Lecturer - Soil and Water
Lakonen, W.	B.Sc.	Lecturer - Soil and Water Resources
McRota, J.H.	B.Sc.	Lecturer - Farm Machinery
Misiko, E.P.	Dipl.E.A.	Demonstrator - Soil and Water Management
Mugeto, J.	Dipl.E.A.	Demonstrator - Farm Power and Machinery
Ngunjiri, G.M.D.	Dipl.E.A.	Demonstrator - Farm Power and Machinery
Odeph, D.M.		Demonstrator - Drawing/Structures
Olychina, P.A.		Demonstrator - Welding/Electric/Fabrication
Waddale, D.	M.Sc.	Lecturer - Processing
Welikhe, W.	Dipl.E.A.	Demonstrator - Farm Power and Machinery
Weru, S.S.	M.Sc.	Lecturer - Farm Power and Machinery
Wright, C.R.	Ph.D.	Civil Engineering

Status: Private Institution, financed by Kenya and other Governments, overseas agencies and private firms.

Activities: a) Teaching: Three year diploma course in Agricultural Engineering Technology (Options in Soil and Water Engineering, Farm Power and Machinery) with a current enrollment (total of three years in attendance) of 96 engineering students. Service courses on agriculture, animal husbandry, dairy technology, food science, range management, horticulture, home economics, agricultural education and animal health, attended by 750 students over 3 years. Different agricultural engineering in-service training courses attended by 46 students, from other Countries like Zambia, Ghana, Sudan ,etc. - b) Research: Three major areas are currently underway fully financed by the Kenya Government through the Ministry of Energy: Solar Energy, Biogas and Agricultural Mechanization in marginal and semi arid areas.

Publications: Soil and Water Conservation in Kenya. Special issue No.27 University of Nairobi and Egerton College (1978) - Alternative Rural Energy Source: Egerton College Agricultural Bulletin - Biogas: A Future Source of Energy for Rural Kenya: Kenya Farmer.

Correspondence: English.

KAMPUCHEA

<u>Direction du Génie Rural, Ministère de l'Agriculture</u>
Phnom Penh, People's Republic of Kampuchea

<u>Correspondence</u> : French.

KOREA

<u>Nongup Jinheung Gongsa, Shihumso</u> (Agricultural Engineering Laboratory, Agricultural Development Corporation - AEL, ADC)
487, Poil-Ri, Euiwang-Eub, Shiheung-Gun, Gyeonggi-Do, 171-11 Korea - Mail: Anyang
P.O. Box 12, Gyeonggi-Do Cable: AGRIDEVELCORP ANYANG
  Telex: K 24890 Koadc   Tel.: Anyang 2-6131
  "   2-8131

Yeo, Un-Choll	BS	Head of the AEL Hydraulic structure (siphon spillway of irrigation storage dam and turnout of irrigation canal
<u>Research Division</u>:		
Kim, Ju-Chang	BS	
Hoang, Chong-Ser	MS	Upland irrigation
Kim, Kyoo-Jun	BS	
Lee, Yong-Jig	BS	
Nho, Jong-Gu	BS	
Ryu, Ki-Hee	MS	Irrigation
<u>Hydraulics Division</u>:		
Kim, Yeong-Bae	MS	
Park, Han-Gyu	BS	Coastal engineering
Jon, Moo-Kab	BS	
Park, Chang-Gyu	BS	
Lee, Doo-Jae	BS	
<u>Materials Division</u>:		
Hwang, Kyu-Tea	BS	
An, Teak-Yong	BS	
Shin, Eui-Kyun	BS	Chemical materials
Hyun, Mi-Suk	BS	
<u>Soil Mechanics Division</u>:		
Ryu, Ki-Song	MS	
Kim, Ho-Il	BS	
Koo, Yo-Han	BS	
Kim, Byeong-Hee	BS	
Kim, Hyun-Tai	BS	
<u>Soil Division</u>:		
Synn, Sang-Hyuk	MS	
Baeg, Cheong-Oh	MS	
Han, Jae-Woo	BS	
Lee, Hee-Duk	BS	

Hu, Jung	BS
Kim, Sam-Chul	BS
Han, Gyeong-Tae	BS
Ahn, Yeal	BS
Jo, Gug Hyen	BS
Kim, Hyang-Suk	BS

Water Quality Division:

Shim, Jae-Hwan	BS
Park, Kuen-Joe	BS
Kim, Moung-Hwa	BS
Synn, In-Hoe	BS
Lee, Kang-Shik	BS
Lee, Mynn-Tak	BS

Branches: 8 Branch Laboratories in 8 provinces - 7 Branch Laboratories in 7 Project field offices.

Status: Established in 1962 as Agricultural Engineering Research Center. Semi-governmental Organization.

Activities: Research, experiments and testing.

Publications: Planning and design standards in series - Standard drawing in series - Annual report - Technical manual - Technical data book.

Correspondence: English, Japanese.

Suweon

Nongchon-Chinheungchung, Nongup-Kigehwa-Yon guso (Agricultural Mechanization Institute, Office of Rural Development)
249 Sudoong-Dong, Suweon, Korea

Han, S.K.	Ph.D. Ag.Sci.	Director of Agricultural Mechanization Institute - Specialization: Agricultural Mechanization
General Research Division:		
Kim, K.S.	B.S.Agr.Engg.	Adaptation and utilization of agric. machinery
First Research Division:		
Lee, Y.Y.	M.S.Agr.Engg.	Development of agricultural machinery
Second Research Division:		
Lee, Y.K.	M.S.Agr.Engg.	Post-harvest technology

Status: 1962 founded as Institute of Agricultural Engineering and Utilization, Office of Rural Development - 1979 Reorganized as Agricultural Mechanization Institute, Office of Rural Development. Governing bodies: Governmental institute, directed and financed by Central Government.

Agricultural activities: Research: Development of agricultural machinery - Adaptation and evaluation of agricultural machinery - Research on maximization of agricultural machinery utilization - Improvement of performance and quality of agricultural machinery.

Publications: Research reports of the Office of Rural Development - Rural Development (Agr.Engineering).

Correspondence: Korean, English.

The Korean Society of Agricultural Machinery - Department of Agricultural Engineering, College of Agriculture, Seoul National University
Suweon, Korea 170   Tel.: Suweon 6-2721-7

Chung, Chang Joo	Ph.D.	President
Kim, Yong Hwan	Ph.D.	Vice-President
Choi, Kyu Hong	Ph.D.	Vice.President
Koh, Hak Kyun	Ph.D.	Managing Director

Status: Private organization, established in 1976.

Activities: Research on the design and improvement of agricultural machinery - Research on the promotion of agricultural mechanization.

Publications since 1976: The Journal of the Korean Society of Agricultural Machinery - Agricultural Machinery Yearbook.

Correspondence: Korean, English, Japanese.

L E B A N O N

Bureau de Génie Rural, Ministère de l'Agriculture (Agricultural Engineering Office, Ministry of Agriculture)

Office National du Litani (Litani River Authority)
Bayrouth, B.P. 3732

Ziade, A.	Ing.civil	Directeur Général
Chebli, S.	Ing.agr.	
Geadah, A.	Ing.hydr.	Chef du département réseaux d'irrigation
Hachem, B.	Ing.agr.	
Karaa, K.	Ing.agr.	
Rammal, H.	Ing.agr.	
Takch, Z.	Ing.agr.	
Terzibachian, J.	Ing.agr.	Chef de service de l'aménagement rural

Annexes situées dans d'autres localités: 5 annexes: Chtaura - Saida - Nabatiyeh - Sour - Labaa

Structure: Office gouvernemental sémi-autonome sous la tutelle du Ministère des Ressources Hydrauliques et Electriques, institué en 1954 et géré par un Conseil d'Aministration.

Activité de génie rural: Planification, études et exécution des projets d'irrigation du bassin du Litani. Determination des besoins en eau des cultures irriguées à l'échelle de l'exploitation.

Publications: Rapports techniques annuels.

Correspondance: Arabe, Français, Anglais.

Department of Soils, Irrigation and Mechanization, American University of Beirut
Beirut, Lebanon

Henderson, Harry D.	Ph.D.	Head of Department – Mechanization
Ryan, J.	Ph.D.	Soil Conservationist – Soils and Con-Conservation
Nimah, M.	Ph.D.	Agric.Engineer – Irrigation
Shammas, A.	Ph.D.	Soil Scientist – Soils/Environment
Macksoud, Sa.	Ph.D.	C.E. – Irrigation
Bashour, I.	Ph.D.	Soil nutrition – Soils
Hariq, N.	M.Sc.	Soil Scientist – Soils
Bartosik, A.	Ph.D.	Agric.Engineer – Mechanization
Zimmerman, T.	B.Sc.	Agric.Engineer – Mechanization
Jubran, L.	B.Sc.	Agric. – Irrigation/Mechanization

Branches: Includes research and educational center located at Houch Sneid, Bekaa.

Status: Faculty of Agricultural and Food Sciences was formed in 1952. The Agricultural Engineering Department was formed in 1960, but was reorgnized into the Department of Soils, Irrigation and Mechanization in 1978.

Activities: Education: research and demonstration work – Courses in irrigation and soil sciences and agricultural mechanization leading to the B.Sc. and M.Sc. degrees in agricultural sciences – Research in water requirements, irrigation methods, soil conservation, implements suitable for local manufacture, tillage, economics of mechanization.

Publications: Agricultural Engineering Journal.

Correspondence: English.

Ministère des Ressources Hydrauliques et Electriques (Ministry of Hydraulic and Electric Resources)
Chiah

Section de Génie Rural, Institut de Recherches Agronomiques (Agricultural Engineering Section, Agronomic Research Institute)
Tel-Amara, Rayak

LESOTHO

Maseru

Lesotho Agricultural College (LAC)
P.O. Box Ms 829, Maseru 100, Lesotho   Sekolo Sa Temo

Owen, M.T.   N.D.A. Cert.Ed.   Head of Engineering Department

Status: LAC was established thirty years ago. Its main function is to provide courses leading to Certificate in Agriculture and Diploma in Agriculture, both are two-year courses. In September 1978 a two-year course leading to Certificate in Agricultural Engineering was started. - The College is part of the Ministry of Agriculture and Marketing, but has its own Board of Governors composed mainly of Ministry people, which is advised by the Academic Board composed of senior staff. Current expenses are provided by the Ministry, most of the recent capital input was British government aid.

Activities: At present, 16 students are taking the Certificate in Agricultural Engineering course. The aim is to produce agricultural engineers/mechanics. No research is carried out.

Correspondence: English.

Ministry of Agriculture and Marketing, Agricultural Mechanization Section, Crops Division
P.O. Box Ms 92, Maseru 100, Lesotho

Mohloai, N.O.   B.S. Agric.Mech.   Head
                      Dipl.Gen.Agr.

Dependent branches: North (Butha-Buthe and Leribe districts): Leuta, C., Certif. Agric.Mech.; Certif.Gen.Agri. - Central (Berea and Maseru districts): Ramangoaela, N.E. Certif.Gen.Agri., ten years mechanization experience - South (Mafeteng, Mohales'Hoek and Quthing districts): Phakoe, L., 20 years mechanization experience.

Status : This Agricultural Mechanization Section is a governmental organization under the Ministry of Agriculture and Marketing. The section started in 1964 as part of the Division of Soil Conservation. The function was the training of farmers in the proper use and maintenance of ox-drawn equipment. This function shifted to tractor power and equipment with the increasing use of such machinery. The Farmers Training Centres (FTCs) are used for this purpose. This activity expanded further in the 70s when more private people and Credit Union groups bought tractors and needed trained operators. - The arrival of the Thaba Bosiu Project in 1973 helped the Section to intensively train ox-equipment owners in the use and repair of such implements. Though this work was limited to the project area, the attention to this equipment indicated its importance and the need to have it repaired and put into a better operating condition. - The work achieved by this project in that area led to the continuation of the same activity in the Basic Agricultural Services Programme (BASP). This programme has about six teams covering all the northern districts, part of the central ones and all the southern districts - The expansion of these services is indicating the acceptance (by the authorities) of the fact that animal power, ploughing about 80% of the cultivated land country-wide, is still going to stay long in Lesotho. It also indicates that, should tractorization be introduced, appropriate low to medium powered machines should be considered. The Section is at present contacting other immediately involved Divisions of the Ministry to assist in intensifying testing of agricultural machines and equipment and training to all levels of farmers and machine operators.

Activities: No teaching is carried out. – Testing: quick testing of new tractors – field testing (by placing with a farmer and following performance record) ox-carts, maize shellers, etc. – Extension work: Advisory services to machine users on the farms – operating three branches of beginners in tractor driving.

Correspondence: English.

LIBERIA

Monrovia

Central Agriculture Experimental Station – Ministry of Agriculture
Monrovia, Liberia

Jarrett, H.O.   Dr.   Head

Correspondence: English.

University of Liberia – College of Agriculture and Forestry
P.O.B. 9020 Monrovia   Tel.: 221537 – 222515

Correspondence: English.

LIBYA

Faculty of Agriculture, Alfateh University
P.O.B. 13538, Tripoli   Tel.: 36010/8

Faculty of Engineering, Alfateh University
P.O.B. 13589, Tripoli

Correspondence: Arabic, English

Ministry of Agriculture – Agricultural Research Centers at:

Sebha, Fezzan Region
El Marj, Jebel el Akhdar

Garabulli Farm Mechanization Training Centre
Garabulli, Western Region

Status: Governmental institution, established in 1964, and supervised by the Ministry of Agricultural and Animal Wealth and by the Department of Farm Mechanization, Sidi Mesri, Tripoli.

Activities: Agricultural tractor and farm machinery operators courses, including maintenance of tractors, implements and combine harvesters – Special courses on water pumps and irrigation equipment – Special course on land clearing equipment and soil conservation.

The courses are for two and three months duration. Certificates issued upon successful completion of courses.

Correspondence: Arabic, English.

LUXEMBURG

Luxembourg

Service du Génie Rural auprès des Services Techniques de l'Agriculture (Agricultural Engineering Service of the Technical Services of Agriculture)
16 route d'Esch, Luxembourg

Structure: Le service du Génie Rural fut créé en 1883; c'est un service public qui dépend du Ministère de l'Agriculture.

Activités: Adaptation et amélioration des conditions de production des exploitations agricoles - Elaboration de projets de bâtiments ruraux - Etablissement de projets de drainage, d'irrigation et d'aspersion - Entretien et amélioration des cours d'eau non navigables ni flottables - Projets d'ouvrages hydrauliques - Voirie rurale: entretien et nouvelles constructions, élaboration des projets et contrôle des travaux - Mesures de vulgarisation en machinisme agricole.

Correspondance: Français, allemand.

Institut Supérieur de Technologie
Kirchberg, Luxembourg

Lycée Technique Agricole
Ave. Salentiny 72, Ettelbruck

MADAGASCAR

Antananarivo

Centre National de Recherche Appliquée au Développement Rural Cendraderu
B.P. 1690 Antananarivo

Centre Technique Forestier Tropical
B.P. 904, Antananarivo

Département Agric., Université de Madagascar
Campus Universitaire D'Ankatso, B.P. 566, Antananarivo

Direction du Génie Rural, Ministère du Développement Rural (Directorate of Agricultural Engineering, Ministry of Rural Development)
B.P. 1061, Nanisana, Antananarivo

Centre National d'Etudes et d'Essais du Machinisme Agricole

Service de l'Hydraulique Agricole

Service de Gestion des Réseaux Hydrauliques

Service de l'Aménagement Rural et de l'Industrialization Agricole

Service Provincial

Service Provincial- Fianarantsoa

Service Provincial - Tamatave

Service Provincial - Ambatondrazaka

Service Provincial - Majunga

C.OM.E.N.A.

Service Provincial - Tuléar

S.O.D.E.M.O. - Morondava

Service Provincial - Diégo-Suarez

Statut juridique: Institution gouvernementale, dont la structure actuelle a été décrétée le 6 juillet 1972.

Correspondance: Malagasy, français, anglais.

MALAWI

Lilongwe

Bunda College of Agriculture, University of Malawi
P.O. Box 219, Lilongwe, Malawi

Boshoff, W.H.	Ph.D.	Farm Power, Machinery, Structures
Kamanga, H.M.	B.Sc.	Farm Power/Machinery
Kasomekera, Z.M.	M.Sc.	Soil/Water engineering
Mwinjilo, M.L.	M.Sc.	Farm Power/Machinery
Ngwira, S.C.	B.Sc.	Farm Power/Maths/Physics
Sichinga, C.A.	M.Sc.	Soil/Water engineering
Simango, D.G.	M.Sc.	Structures
Sulaimana, R.A.	M.Sc.	Farm Machinery/Maths/Physics
Temple, S.J.	M.Sc.	Farm Power/Machinery.

Status: Established 1965 by the Government of Malawi as an Agricultural Engineering Department in the University of Malawi.

Agricultural Engineering activities: a) Teaching: 3 year Diploma in Agriculture, 120 per year - 5 year Degree in Agriculture, 35 per year - M.Sc.(Agric.Eng) newly established.
b) Research: Runoff, erosion, hydrology studies - Methane gas production, crop storage and structures, simplified tillage, ergonomics of hand cultivations, comparative studies, hand, ox, tractor. Machinery and power unit evaluation.

Publications: Soil compaction, ox cultivation, tractor evaluation.

Correspondence: English.

Farm Machinery Investigation and Testing Unit, Chitedze Agricultural Research Station
P.O. Box 158, Lilongwe

Correspondence: English.

Thondwe

Makoka Agricultural Research Station
Private Bag 3, Thondwe

Nyirenda, G.K.C.   B.Sc.(Malawi)
                      M.Sc.(London) D.I.C.   Officer-in-Charge

Status: Established in 1967. The Station is part of the Agricultural Research Services of the Ministry of Agriculture and Natural Resources since 1975.

Activities: Testing and modification of spraying machinery for small farms.

Publications: Various articles on tiedridging, tailboom Knapsack sprayers, ULV sprayers.

Correspondence: English.

MALAYSIA

Kuala Lumpur

Demeterian Pertanian dan Sharikat Kerjasama (Ministry of Agriculture and Co-operatives)
Kuala Lumpur

Drainage and Irrigation Division

Agriculture Division

Research Station at Ampang

Activities at the Research Station: Hydraulic research – Hydrological work – Soils and materials testing – Training.

Correspondence: Malay, English.

Bahagian Kejuruteraan Pertanian & Teknoloji, Fakulty Pertanian
(Agricultural Engineering and Technology Division, Faculty of Agriculture, University of Malaya)
Pantai Valley, Kuala Lumpur

Status: The Faculty was established in 1960 and is a semi-governmental body and finance comes mainly from local sources.

Activities: a) Teaching, b) Research.

Publications: English, Bahasa, Kebangsaan.

Serdang

Fakulti Kejuruteraan Pertanian, Universiti Pertanian Malaysia (Faculty of Agricultural Engineering, University of Agriculture, Malaysia)
Serdang, Selangor, Malaysia

Choa, S.L.	M.S. (UC Davis)	Dean – Power and Machinery

Department of Farm Power and Field Engineering:

Bardaie, M.Z.	Ph.D. (Cornell)	Head – Power and Machinery, Systems Engineering
Abu Hassan, M.N.	B.Sc.	Structures, Civil Engineering
Bakar, M.R.	M.S.	Soil and Water, Surveying
Kwok, C.Y.	M.Sc.	Soil and Water, Hydrology
Lee, T.S.	B.E.	Soil and Water
Mohammad Soom, M.A.	M.S.	Soil and Water
Wan Ismail, W.I.	M.S.	Power and Machinery
Hj. Jaafar, M.S.	M.S.	Power and Machinery
Zakaria, A.A.	M.S.	Soil and Water

Department of Engineering Science, Processing and Environmental Control:

Husain Salleh	M.S.	Head – Processing
Abang Abdullah A. Ali	M.Sc.	Structural Engineering
Ahmad Jusoh	B.E.	Processing
Abdul Rahman Bidin	B.Sc.	Chemical Engineering
Bahanurdin Hitam	M.Sc.	Mechanical Engineering
Fuad Abbas	M.Sc.	Mechanical Engineering
Hamzah Salleh	M.Sc.	Mechanical Engineering
Hishamuddin Jamaluddin	M.S.	Food Engineering
Johari Endan	M.S.	Processing
Ungku Kamal U. Mohsin	M.Sc.	Electrical/Electronics
Saxena, N.C.	Ph.D.	Farm Structures, Materials Sc.
Syed Mansur S. Junid	M.Sc.	Farm Buildings
Razali Saaban	B.Sc.	Chemical Engineering

Status: Established 1975 as Faculty of Agricultural Engineering – The Ministry of Education, Malaysia, through the University of Agriculture, direct and finance this faculty.

Agricultural Engineering activities: a) Teaching, leading to B.E.(Agricultural) – This is a 4 year course with an average of 35 students per year; Diploma of Agricultural Engineering – This is a 3 year course, at the sub-professional level with an average of 40 students per year. – b) Research: on Energy, Machinery, Water Resources, Farm Structures, Water Treatment, Mill Yield of Padi.

Publications: in Trans. ASAE, Pertanika etc.

Correspondence: English.

MALI

Bamako

Ministère de l'Agriculture - Direction du Génie Rural (Ministry of Agriculture - Agricultural Engineering Dep.)
B.P. 155, Bamako

Bathily, C.B.	Ingénieur hydrotechnicien Master of Sciences	Directeur National
Samake, D.	Ingénieur topographe Master of Sciences	Directeur Adj.

Bureau d'Etudes:
Keita Nancoman	Ingénieur hydraulicien	Chef
Gadelle, F.	Ing. G.R.E.F.	Conseiller Principal
Zalla, O.	Ingénieur des travaux agricoles	
Berthe, D.	Ingénieur topographe	

Machinisme Agricole:
Zerbo, D.	Ingénieur d'Agriculture	Chef
Keita, M.	Ingénieur d'Agriculture	

Mission d'Etudes du Génie Rural a Ségou:
Samake, M.	Technicien supérieur	Chef

Bureau de Surveillance à Mopti:
Konate, S.F.	Ingénieur du Génie Civil	Chef

Activités: Aménagements hydro-agricoles - Hydraulique pastorale - Machinisme agricole - Bâtiments Ruraux.

Correspondance: Français, Anglais, Russe.

MAURITANIA

Nouakchott

Direction du Génie Rural, Ministère du Développement Rural (Agricultural Engineering Department - Ministry of Rural Development)
B.P. 173, Nouakchott, Mauritanie

Fall, O.	Ingénieur Agro-GR	Directeur

Status: 1953 - Ministère du Développement Rural

Activités: Expérimentation - casier rizicole (entretien infrastructure rurale)

Correspondance: Français.

MAURITIUS

Ministry of Agriculture, Engineering Division
Richelieu, Petite Rivière, Mauritius

Satar, A.R.   B.Sc.   Head of Division

Status: Reorganized in 1958. Government Body.

Activities: Advisory and educational work on agricultural engineering questions. Service of tractors, vehicles and agricultural machinery. Field tests of agricultural machinery. Field tests of agricultural machinery to ascertain optimum working conditions and to effect modification, if necessary. Design and construction of buildings required by the Ministry. Design and installation of irrigation systems.

Correspondence: English and French.

School of Agriculture - University of Mauritius
Réduit, Mauritius

Unmole, H.   Dip.Agric. and S. Tech.   Lecturer in Agric. Engineering
                    B.Sc., M.Sc.

Status: This institution was established in 1971 and absorbed the Mauritius Agricultural College which was created in 1921.

Agricultural Engineering activities: Teaching: Agric. Engineering is taught as part of the diploma course in Agriculture and Sugar Technology and of the Degree in Agriculture. About 25 students attend the courses. - Research: Energy analysis, solar dryer, biogas.

Correspondence: English, French.

MEXICO

Chapingo

Universidad Autónoma de Chapingo
Escuela Nacional de Agricultura (National College of Agriculture)
Chapingo, Mexico

Culiacan

Escuela de Agricultura, Universidad de Culiacán (College of Agriculture, University of Culiacán)
Culiacán, Sinaloa

Actividades de ingeniería: Cursos de hidráulica agrícola, maquinaria agrícola y construcciones rurales.

Guadalajara

Facultad de Agronomía -
Escuela de Agricultura, Universidad de Guadalajara (Faculty of Agronomy, School of Agriculture, University of Guadalajara)
Las Aguas, Mpio de Zapopan, Jal. - Apartado Postal No.129

Hidráulica:
Gomez Villarruel, E.   Ing.
Maciel Gutiérrez, A.   Ing.

Maquinaria:
Rodríguez García, A.   Ing.
Hermosillo de la Cerda, H. Ing.
Rivas Martínez, A.   Ing.

Construcción:
Ruíz Alcantar, A.   Ing.

Carácter y estructura: Esta institución fue establecida el 17 de septiembre de 1964, con carácter gubernamental.

Actividades en Ingeniería Rural: a) Enseñanza: Actualmente se cuenta con 5 orientaciones: Desarrollo rural, Ganadería, Bosques, Fitotecnia y Suelos - El promedio de alumnos es de 3,400.

Publicaciones: "Introducción a la Maquinaria Agrícola".

Correspondencia: Español.

Hermosillo

Escuela de Agricultura y Ganadería, Universidad de Sonora (College of Agriculture and Animal Husbandry, University of Sonora)
Hermosillo, Sonora

Carácter: Le Escuela fue establecida en octubre 1953.

Actividades de Ingeniería Rural: Enseñanza.

Correspondencia: Español, inglés.

Monterey

Departamento de Suelos e Irrigación, Escuela de Agricultura y Ganadería - Instituto Tecnológico y de Estudios Superiores de Monterey (Soils and Irrigation Department, College of Agriculture and Animal Husbandry, Institute of Technology of Monterey)
Monterey, N.L.

Sección de Suelos

Sección de Irrigación

Sección de Ingeniería

Sección de Maquinaria Agrícola

Sección de Meteorología

Campo Agrícola Experimental

Carácter: Institución privada fundada en 1948, controlada bajo el Instituto Tecnológico y de Estudios Superiores de Monterey.

Actividades: Enseñanza - Investigaciones.

Correspondencia: Español, inglés.

Saltillo

Departamento de Ingeniería Agrícola, Escuela Superior de Agricultura "Antonio Narro", Universidad de Coahuila (Agricultural Engineering Department, College of Agriculture "Antonio Narro", University of Coahuila)
Buenavista, Saltillo, Coahuila

Carácter: Institución gubernamental fundada en 1923 - Financiamiento privado.

Actividades: Enseñanza y Investigaciones.

Correspondencia: Español, inglés.

MOROCCO

Meknès

Ecole du Génie Rural et de Topographie de Meknès (School of Agricultural Engineering and Topography)
B.P. 4003 - Meknès, Maroc

Tazi, M.A.   Directeur

Structure: placée sous la tutelle du Ministère de l'Agriculture et de la Réforme Agraire.

Activités concernant le Génie Rural: Diplôme d'Adjoint Technique du Génie Rural et de Topographie - 140-150 étudiants.

Correspondance: Français.

Rabat

Direction de l'Equipement Rural, Ministère de l'Agriculture et de la Réforme Agraire (Directorate of Rural Engineering, Ministry of Agriculture and Agrarian Reform)
B.P. 2069, Rabat

Oulad Chrif, B.   Ingénieur du Génie Rural   Directeur

Division de l'Hydraulique Agricole et des Améliorations Foncières:
(Division of Agricultural Hydraulics and Land Improvement) - B.P. 1069 - Rabat

Bastos, A.   Ingénieur de Génie Rural   Chef de Division

Service des Expérimentations d'Hydrauliques Agricoles:
(Agricultural Hydraulic Experiment Service) 461, Avenue Hassan II - Rabat

Yacoubi Soussane, M. Ingénieur de Génie Rural   Chef de Service
Aballagh, M.   Ingénieur des Travaux Ruraux
Corlier, L.   Ingénieur Agronome
Dagnelies, R.   Ingénieur de Génie Rural
Faraj, H.   Ingénieur Pédologue
Handouf, A.   Ingénieur Agricole
Hascoet, C.   Ingénieur Hydraulicien
Larhrafi, M.   Ingénieur des Travaux Ruraux
Mekrane, M.   Ingénieur Biomètre
Touil, M.   Ingénieur des Travaux Ruraux

Stations Expérimentales relevant des Offices Régionaux de Mise en Valeur Agricole

O.R.M.V.A. des Doukkalas	:	Stations: Ouled Frej, El Mechrek, Zemamra
O.R.M.V.A. Basse Moulouya	:	Stations: Bourghiba, Garet, Shouyaya
O.R.M.V.A. du Gharb	:	Station de Sidi Slimane
O.R.M.V.A. du Tadla	:	Station de Ouled Gnaou
O.R.M.V.A. du Haouz	:	Station du Haouz Central
O.R.M.V.A. du Souss-Massa	:	Stations: Taroudant, Ait Amira, Bloc expérimental du Massa
O.R.M.V.A. du Tafilalet	:	Station de Tinjdad
O.R.M.V.A. de Ouarzazate	:	Parcelles expérimentales
O.R.M.V.A. du Loukkos	:	Sakh-Sokh et Guédira (Stations)

Structure: Le Centre de Recherches et d'Expérimentations du Génie Rural fut fondé en 1952. En 1961, il est devenu le Centre des Expérimentations de l'Office National des Irrigations. En 1967 il a été rattaché à la Direction de la Mise en Valeur Agricole puis à la Direction de l'Equipement Rural à partir de 1973. En 1978 a été en Service des Expérimentations d'Hydrauliques Agricoles.

Activités: Etudes des besoins en eau des cultures (quantité d'eau nécessaires, rythme des irrigations, pratiques culturales pour améliorer la productivité de l'eau). Etudes des paramètres de l'irrigation gravitaire(longueur des dispositifs, débits, doses en fonction des types de sol et des pentes). Etude des paramètres de l'irrigation par aspersion. Etude du matériel d'irrigation par aspersion et suivi de ce matériel au champs(Laboratoire d'essai de matériel, Banc d'essai mobile) - Etude des paramètres de drainage et suivi des réseaux de drainage. Etude de l'évolution des sols irrigués en fonction des propriétés physiques et chimiques des sols, des modes d'irrigation, du travail du sol, étude de technique nécessaire pour lutter contre une mauvaise évolution des sols irrigués. Amélioration de diverses techniques culturales.

Publications depuis 1965: Rapports d'Expérimentation.

Correspondance: Français.

Sidi Bouknadel

<u>Ecole Mécanique Agricole</u> (Agriculture Mechanic School)
Sidi Bouknadel, Maroc

Nacaf, M.　　　　　　Institut Agron.méditerr.　　　Directeur
　　　　　　　　　　　Montpellier

<u>Structure</u>: Date de Création: Octobre 1971 - Organe directeur: Ministère de l'Agriculture et de la Réforme Agraire.

<u>Activités concernant le Génie rural</u>: - Formation: Atelier, motorisme, machinisme, électromécanique, sciences humaines, stages - Diplôme d'Adjoint technique dans les options: machinisme agricole, électromécanique. Moyenne d'étudiants: 35 option machinisme, 10 option électromécanique.

<u>Correspondance</u>: Arabe, français.

NEPAL

Katmandu

<u>Agricultural Engineering Section, Jagadamba Krishi Bhawan</u>
Pulchowk, Lalitpur
<u>Agricultural Engineering Development and Research Section</u>

<u>Status</u>: Established in September 1952.

<u>Agricultural Engineering activities</u>: Design and Construction of Farm Structures. Research on Irrigation. Research, Design and Construction of Farm Machineries and Implements.

<u>Publications</u>: Guides to farm machineries and irrigation workers.

<u>Correspondence</u>: English.

<u>Agri Engineering Development and Research Section</u>
Ranighat, Birganj

<u>Status</u>: Governmental - established in 1961.

<u>Agricultural Engineering activities</u>: Research: Research, experiments and testing of improved agri implements, mechanization and modification of indigenous implements and cultural practices.

<u>Correspondence</u>: English.

Department of Irrigation, Hydrology and Meteorology
Panipokhari, Kantipath, Kathmandu

Branches with other location: Four regional directorate offices are established in four different sectors: Dhankuta Development Center, Dhankuta (Eastern Sector) - Katmandu Development Centre, Katmandu (Central Sector) - Pokara Development Centre, Pokhara (Western Sector) - Surkhet Development Centre, Surkhet (Far Western Sector).

Status: Governmental body, directed and financed by HMG, Nepal.

Agricultural Engineering activities: Planning, design, execution and supervision of irrigation projects.

Correspondence: English.

NETHERLANDS

Arnhem

N.V. Heidemaatschappij BeheerLovinklaan
Lovinklaan 1, P.O. Box 33, Arnhem   Telex: 45623 Heidemaatij Tel.: 085-778911
                                      Cables: Heidemaatij Arnhem

Hellema, H.T.   President
Florian, G.J.jr.   Head of Section Soil and Water

Staff: 173 university graduates and 91 college graduates.

Local Branches: Integrated services - Heidemij Nederland B.V., Lovinklaan 1, Arnhem
                                      Tel.: 085-778911 - 085-435156
Contracting: Lareco B.V. Koningsweg 35c
International Branch: ILACO (International Land Development Consultants B.V.)

Status: A holding company with 30 subsidiaries and joint ventures having 5000 people in their employ, is fully owned by the P.U. Society Koninklijke Nederlandsche Heidemaatschappij. Their main objective is to improve the environment.

Activities: Physical planning - Town planning - Project development - House building - Architecture - Landscape architecture - Road construction - Traffic engineering - Hydraulic engineering - Environmental engineering - Utility building - Structural and services engineering - Industrial building - Agricultural engineering: studies on agricultural development and farm management planning - advice on the use of agricultural areas - agricultural classification - advice on and landscape design of agricultural areas - agricultural engineering in the tropics and sub-tropics - Recreation and tourism - Water management: control of irrigation, drainage and infiltration - water winning - water supply - geohydrology - agrohydrology - Forestry and verdant vegetation: study and advice on growth factors - exploitation and management of forests - forest plans - Management and real estate - Soil science: agricultural and horticultural projects - studies on soil salinity and seepage - soil mapping - Soil mechanics - Geodesy - Research and laboratory work: study of physical transport characteristics for: water management, road engineering and environmental engineering -

study of soil samples, chemical and physical analyses - Computer techniques: development of systems and programmes - Development of machinery: development and/or improvement of machinery used in land development and water management projects - Technical registration: inventories of landed property and land use - preparation of cadastral, classification and summarized maps.

Publications: on above activities.

Correspondence : Dutch, English, French, German, Spanish.

Delft

Waterloopkundig Laboratorium (Delft Hydraulics Laboratory - DHL)
Rotterdamseweg 185, P.O. Box 177 - 2600 MH Delft, The Netherlands
Tel.: (015)-569353
Telex: 38176 hydel nl
cables: hydrolab delft

Prins, J.E.   M.Sc.   Managing director
Diephuis, J.G.H.R.   M.Sc.   Deputy managing director

The academic staff consists of about 160 university graduated engineers and scientists (about 10% of them Ph.D.) of various disciplines - engineering (civil, hydraulic, coastal, sanitary, mechanical, agricultural, naval, physical, electronic) - mathematics -chemistry - biology - hydrology - hydrogeology - economy - ecology - geochemistry.

Branches with other location: De Voorst Laboratory (Repelweg 10, Marknesse, tel.: (05274)-2922 - Wageningen Branch (Nieuwe Kanaal 11, Wageningen, tel.: (08370)-82776, cooperating with the Wageningen Agricultural University) - Haren Branch (Oesterweg 92, Haren (Gr.), tel.: (050)-346541, cooperating with the Institute of Soil Fertility).

Status: Foundation (non-profit, semi-governmental) established in 1927, governed by a Board of Trustees appointed by the Ministers of Public Works and Waterways, of Education and Science and of Finance. The Delft Soil Mechanics Laboratory is DHL's sister organization in the foundation. - Capital for investments is borrowed from the Dutch Government, to which interest has to be paid. Research work for clients is charged at cost price, which comprises also a proportional share of general costs, basic research, etc.

Agricultural Engineering activities: (a) Teaching: DHL specialists lecture at the International Institute for Hydraulic and Environmental Engineering at Delft. DHL often accomodates trainees as part of a research contract - (b) Research: DHL's investigations are carried out by the following advisory branches in the form of desk studies and physical or mathematical models: Density currents and transport phenomena - dredging technology - pumps and industrial circulations - maritime structures - Locks, weirs and sluices - environmental hydraulics - rivers and navigation - harbours and coasts - closure works - hydrodynamics and morphology - systems approach. These branches are supported by the Site investigations service, and the mathematical, data processing, and instrumentation departments. - Activities important for agriculture: Salt intrusion, scour and sedimentation, irrigation, bed protection, bank erosion, groundwater flow, hydrometry, flood control, land reclamation, silt quality (heavy metals content), aquatic ecology, hydrology, water intakes, pumps, calibrations, network computations, river training, thermal pollution, environmental impact, optimization of (waste) water, purification systems, statistical analysis of data, river basin development and management, planning of drinking water supply systems, quantitative and qualitative management of groundwater and surface water, advice on multidisciplinary problems. - For hydraulic research modern facilities are available such as: wave and discharge flumes, current basins, test rigs for pumps, calibration rigs for flowmeters and control valves, sediment transportation flumes, computer programmes, sophisticated laboratory and hydrographic equipment.

Publications: Research data are presented to clients in confidential reports or published in DHL publications, conference papers, scientific journals and DHL's quarterly publication "Hydro Delft". The DHL publications comprise about 200 titles and cover most research activities; a complete index can be obtained from the DHL Library Information and Documentation Branch. The publications may be obtained preferably on an exchange basis; single copies and subscriptions to the "Hydromechanic and Hydraulic Engineering Abstracts" on request.

Correspondence: Dutch, English, French, German, (Spanish).

E d e

Centrale School voor Tuinbouwtechniek (Central School of Horticultural Engineering)
Zandlaan 29, Ede

Status: Established in 1961 as a Foundation (private).

Activities: Training of teachers and practical courses in the subject of horticultural engineering.

Publications: Textbooks and manuals covering the entire field of horticultural engineering for use at middle and higher horticultural schools.

Correspondence: Dutch, English, German, French.

L e l y s t a d

Rijkswaterstaat directie Zuiderzeewerken (Rijkswaterstaat Zuyderzee Project Directorate)
1 Maerlant, 8224 AC Lelystad, The Netherlands   Tel.: 03200-41911

Hooning, W.F.	Civil Engineer	Director
Dijkstra, J.	" "	Head of the main division for polder affairs, technical execution of drainage works, set up of parcelling schemes.
Van Ovost, P.	" "	Head of the main division for water management and shipping routes.
Snijdelaar, M.	" "	Head of the division for water management, water/management in polders.

Status: Date of establishment May 31 1919, modified on November 25, 1976. The Ministry of Transport and Public Works is financing and directing our directorate.

Agricultural Engineering activties: Drainage works in newly reclaimed areas - Parcelling of these areas - Water management in the reclaimed polders - Investigation of water and salt balances of surrounding lakes - Experiment and research for the improvement of the water quality.

Publications covering above activities.

Correspondence: Dutch, English, French and German.

Proefstation voor de Akkerbouw en de Groenteteelt in de Vollegrond (Research Station for Arable Farming and Field Production of Vegetables)
Main office: Edelhertweg 1, Lelystad - P.O. Box 430, 8200 AK Lelystad
Tel.: 03200-22714
Office in Alkmaar: Olympiaweg 16, 1816 MJ Alkmaar   Tel.: 072-111944

Termohlen, G.P.	Dr.ir	General Director
van Kampen, J.	Dr.ir.	Technical Director

Number of Staff: appr. 120

Soil physics, water supply, manuring and mechanization:

Hellings, A.J.	Ir.	(detached from the Institute for Land and Water Management Research)
Lumkes, L.M.	Ing.	
Nicolaï, P.	Ing.	
Nierop, A.		
v.d. Schans, D.A.	Ing.	
Titulaer, H.H.H.	Ir.	

Branches with other locations: The main office of the Research Station is situated in Lelystad; there is also an office at Alkmaar. The Research Station guides research at seventeen experimental centers (both for arable farming and outdoor vegetables), located in all parts of the country. By communicating research results to regional extension services, the link with practical farming is being made.

Status: The Research Station was established in 1976, as a result of combining two separate research stations. It is a government institution, which is under the jurisdiction of the Directorate of Arable Farming and Horticulture of the Ministry of Agriculture and Fisheries. This ministry almost entirely finances the institution.

Agricultural Engineering activities: Research: Crop production research on potatoes, sugar beet, cereals and maize, grass seeds, oil- and fibre crops, pulse and vegetables for canning, vegetables fresh market, variety testing. Improving cropping cultivation, increase of yield and quality of the crops.- Farming systems research on crop rotation, soil fertility and mechanization, plant protection. Examining the relation between crops and how to obtain an optimum farming plan. - Farm management research on farm planning, labour use, equipment. Optimizing farming results by tuning farming plan, machinery and labour into each other as well as possible. Examining optimum area per labour and intensification of soil use. - Regional research on tracing and solving regional problems in arable farming and outdoor vegetables; planning research in regional centers. - The experimental farm in Lelystad covers an area of 195 hectares; the experimental garden in Alkmaar (research on vegetable crops only) covers 5 hectares.

Publications: Every year appr. 200 articles are being published in various magazines and papers.

Correspondence: Dutch, English, French, German.

Proefstation voor de Rundveehouderij (Research and Advisory Institute for Cattle Husbandry)
Runderweg 6, 8219 PK Lelystad, the Netherlands   Tel.: 03200-22514

Hiusman, L.H.	Ir.	Director	
Jong, M.P. de	Ir.	Deputy Director	
Boonman, D.C.M.	Dr.Ir.		Farm management
-			Coaching advisory offrs.
Snijders, P.J.M.	Ir.		Farm equipment
Thiemann, P.B.R.	Drs.		Research farm org. and econ.
Kuipers, A.	Dr.Ir.		Farm service systems
Thomas, H.	Ir.		Grassland management, fodder harvesting and conservation
Luten, W.	Ir.		Methods of cultivation and utilization
Schukking, S.	Ir.		Fodder harvesting and conservation
Wieling, H.	Ir.		Grassland management systems
Jong, M.P. de	Ir.		Dairy and beef cattle and sheep
Meijer, A.B.	Ir.		Dairy cattle
Oostendorp, D.	Ir.		Beef cattle and sheep
Kommerij, R.	Drs.		Veterinary aspects
Verboon, M.C.	Ir.		Coordination regional experimental farms
"C.R. Waiboerhoeve"			Experimental farm
Bruggen, C. van	Ing.		Farm engineer
Eldik, J. van			Information and special services.

Farm data: 200 ha grassland on marine clay, 500 dairy cows, 250 heifers and calves, 400 beef cattle, 200 sheep.

Status: 1970. Governmental, Ministry of Agriculture and Fisheries.

Agricultural engineering activities: Grassland management - Fodder harvesting and conservation - Livestock improvement - Animal nutrition - Calf rearing - Milk production - Beef production - Animal health - Labour - Mechanization - Farm buildings - Dung processing - Farm systems - Economy - Training Advisory officers - Environmental Energy - Excursions at experimental farm (6 units: Tuesday, Wednesday, Thursday).

Publications: Reports, publications (Dutch language with English summary) and articles (Dutch) on activities mentioned - Annual report - Annual report with results of experimental farm (Dutch with English summary).

Correspondence: Dutch, English, German, French.

Utrecht

Landinrichtingsdienst, Ministerie van Landbouw en Visserij (Service for Land and Water Use, Ministry of Agriculture and Fisheries)
Rijkskantorengebouw "Westraven" - Griffioenlaan 2 - P.O. Box 20021 - 3502 LA Utrecht
Tel.: 030-859111

Molenaar, N.	Ir.	Director
Segeren, W.A.	Prof.-ir.	Deputy Director
Greve, N.H.A.	Ir.	General affairs
Segers, A.J.A.M.	Ir.	Land development projects
Blom, J.E.	Drs.	Admin. and financial affairs
Slot, P.	Drs.	Research
Jonkers, H.J.	Ir.	Execution of works

<u>Outposted branches</u>: eleven provincial bureaus for geography of the land in Groningen, Leeuwarden, Assen, Zwolle, Arnhem, Utrecht, Haarlem, The Hague, Goes, Tilburg, Roermond.

<u>Affiliated Research Institute</u>: Institute for Land and Water Management Research (ICW); address: "Staring building" - Marijkeweg 11 - P.O. Box 35 - 6700 AA Wageningen.

<u>Status</u>: established in 1935 by the Ministry of Agriculture and Fisheries.

<u>Activities</u>: the Government Service for Land and Water Use promotes, prepares and finances land development projects, such as land consolidation schemes and water management and road construction works - The execution of integrated land consolidation schemes include consolidation of fragmented agricultural holdings, improvement of drainage and water management, road construction, layout and construction of farmsteads, enlargement of undersized holdings, utility works and landscaping. - The land consolidation projects generally cover a fairly large area of several thousand hectares each, totalling about 40.000 hectares yearly. - It usually takes a period of approx. ten years between the approval of a project and its completion. - The execution of the projects is in the hands of private contractors specialised in land development works. Due to the high degree of mechanization of the operations, the majority of the projects is executed on a tender basis, the contractors being supervised by land development companies. - A part of the work in land development projects is carried out on behalf of municipalities and catchment boards.

and

Instituut voor Cultuurtechniek en Waterhuishouding (ICW) - Ministerie van Landbouw en Visserij (Institute for Land and Water Management Research, Ministry of Agriculture and Fisheries)
Marijkeweg 11, Wageningen - P.O. Box 35, 6700 AA Wageningen   Tel. 08370-19100

Oosterbaan, G.A.	Ir.	Director
Bijkerk, C.	Prof.ir.	Deputy Director
Rijtema, P.E.	Dr.ir.	Water quality
Wesseling, J.	Dr.ir.	Hydrology
Wind, G.P.	Ir.	Soil Technology

Bijkerk, C.	Prof.ir.	Land Use Planning
Locht, L.J.	Drs.	) Economics
Rijholt, J.W.	Ir.	
Stol, Th.	Dr.ir.	Mathematics

Responsibility: The institute is a foundation established by the Netherlands''Government to carry out research on the improvement of water management, soils, land layout, land use and environment in non-urban areas.

Activities: Studies are made in the fields of: hydrology, hydrogeology, evapotranspiration, drainage, salinity problems, sprinkling and sub-irrigation, aeration, land treatment of industrial waste water, pollution of surface and ground water, regional water management schemes, land levelling, changing of soil profiles, subsidence and compaction of soils, requirements of size and shape of fields and holdings, accessability of fields and farm buildings, site, size and shape of outdoor recreation projects, urbanization aspects of rural regions, physical planning of a region, execution techniques of land consolidation plans, pilot plans of land development schemes, short- and long-term effects of amelioration measures on the economy of farm and region, benefit-cost analyses. Supervisor is the Netherlands Land Division Survey, a computerized data bank and mapping system devised by the institute.

Publications: ICW announcements, Regional studies, Technical bulletins, Miscellaneous reprints.

Correspondence: Dutch, English, French, German.

Wageningen

Consulentschap voor Landbouwwerktuigen en Arbeid (Advisory Service for Agricultural Machinery and Labour Management)
Mansholtlaan 12, Wageningen

Crucq, J.	Ir.	Chief Mechanization Adviser
Andringa, J.T.	Ing.	
Delbrugge, H.J.	Ir.	
Hoenderken, J.A.	Ir.	
Loo, L. van	Ing.	
Nieuwenhuijse, L.	Ing.	

Status: Governmental service, supervised by the Ministry of Agriculture and Fisheries.

Activities: Advisory work on farm machinery and labour management - The work is carried out through regional advisers.

Correspondence: Dutch, English, French, German.

Wageningen

Vakgroep Cultuurtechniek Landbouwhogeschool (Department of Land and Water Use)
Nieuwe Kanaal 11, 6709 PA Wageningen   Tel.: 08370-89111

Hellinga, F.	Prof.Dr.Ir.	Head of Department

Land Use:

Hellinga, F.	Prof.Dr.Ir.	Head of Sub-division
Lier, H.N. van	Prof.Dr.Ir.	Project planning
Jaarsma, C.F.	Ir.	Traffic research in rural areas
Sparenburg, G.A.	Ir.	Rural allotment and re-allocation

Hydrology:

Molen, W.H. van der	Prof.Dr.Ir.	Head of Sub-division
Hoorn, J.W. van	Dr.Ir.	Land drainage and salinity
Koopmans, R.W.R.	Dr.Ir.	Numerical solutions of ground water flow
Ledeboer, H.F.	Ir.	Land drainage
Schaaf, S. van der	Ir.	Electric analog models of ground water flow
Zeeuw, J.W. de	Dr.Ir.	Relation between precipitation and discharge. Development of precipitation measurement procedures.

Soil Improvement:

Duin, R.H.A.	Prof.Dr.Ir.	Head of Sub-division
Eppink, L.A.A.J.	Ir.	Soil management and improvement, soil erosion control
Schouwenaars, J.M.	Ir.	Technical planning of nature reserves
Voet, J.L.M.	Ir.	Water oriented outdoor recreation.

Status: Governmental institution.

Agricultural Engineering activities: Teaching – Undergraduate and graduate study courses, of two years each, 40 to 50 students per year, final degree "Landbouwkundig ingenieur (Ir)" (equivalent to M.Sc.agr.) with specialization in land and water use. Graduate course "soil and water management" of 2 years for foreign students, in the English language, final degree M.Sc. Research – See under "Teaching".

Publications: Design criteria for drainage, relationship between rainfall and discharge, development in land use, layout of farm fields, consolidation of fragmented holdings, erosion control, traffic research in rural areas.

Correspondence: Dutch, English, French, German, Spanish.

Vakgroep Weg- en Waterbouwkunde en Irrigatie, Landbouwhogeschool (Department of Irrigation and Civil Engineering, Agricultural University)
11 Nieuwe Kanaal, 6709 PA Wageningen, The Netherlands

Horst, L.	Prof.Ir.	Head, Irrigation
Stamhuis, E.	Prof.Ir.	Civil engineering
Genet, W.B.M.	Ir.	Surface irrigation
Hendrickx, J.	Ir.	Rice Irrigation (stationed in Mali)
Knaap, G.J.J. van der	Ir.	Civil engineering, construction
Kijne, J.W.	Dr.	Water use of plants
Laan, J.C. van der	Ir.	Civil engineering, reclamation
Meijer, Th.K.E.	Ir.	Sprinkler, drip irrigation

Roscher, K.	Ir.	Irrigation management
Smaalen, H. van	Ir.	Rural roads, soil mechanics
Vink, N.H.	Ir.	Irrigation agronomy

Status: Governmental Institution.

Activities: a) Undergraduate and graduate courses (Dutch), leading to the University degrees "Kandidaat" (B.Sc.)(± 120 stud.) and "Ingenieur" (M.Sc.)(± 40 stud.); 2-years graduate course (English) leading to M.Sc. degree (± 15 students).
b) Irrigation efficiencies tertiary block layout, irrigation water management, water movement in soils, crop water use, water-yield relations of irrigated crops, supplementary irrigation, health aspects of irrigation, drip irrigation, discharge coefficients of measurement structures, appropriate technology, pump, soil strength and soil stabilization of low volume roads, frost susceptibility of soils, traffic engineering, erosion and sedimentation.

Publications: See b) above.

Correspondence: Dutch, English, French, German, Spanish.

Vakgroep Hydraulica en Afvoerhydrologie, Rijkslandbouwhogeschool (Department of Hydraulics and Catchment Hydrology, Agricultural State University)
De Nieuwlanden, Nieuwe Kanaal 11, Wageningen, The Netherlands   Tel.: (08370) 82897

Kraijenhoff van de Leur, D.A.	Prof. Ir.	Hydraulics and Catchment Hydrology
Gaasbeek, G.H.	Ing.	Electr. Equipment and analogues
Keuning, D.H.	Dr. Ir.	Mathematical models
Leenen, J.D.	Ir.	Mathematical models
Pitlo, R.H.	Ir.	Hydraulic models
Verhagen, J.H.G.	Ir.	Environmental Hydrodynamics
Warmerdam, P.M.M.	Ir.	Hydrological field studies

Status: Governmental organization.

Agricultural Engineering activities:
Research:
representative catchment studies in rural and urbanized areas. Hydraulic model studies on structures for flow measurement. Weed growth regulation in ditches.

Publications: Hydrologic research in the catchment of the Hupelse Beek, Netherlands. IHP project Study Group Hupelse Beek - Brink Depth Method in Rectangular Challen - Modelling urban run-off: a quasilinear approach - Evapotranspiration during a drought in a sandy region of the Netherlands (Journal of Hydrology) - Discharge Measurement Structures (Report of the Working Group on Small Hydraulic Structures) - Regulation of aquatic vegetation by interception of daylight - Wage run-up Influence on Overtopping of Leeves (Journal of Hydraulics Division) - An Advection ridity Approach to Estimate Actual Regional Evapotranspiration (Journal of Water Resources Research).

Correspondence: Dutch, English, French, German.

International Institute for Land Reclamation and Improvement/ILRI
P.O. Box 45, 6700 AA Wageningen, Netherlands

Tel.: 08370-19100
Telex: 75230

Schulze, F.E.	Ir.	Director Land and Water Management
Boonstra, J.	Ir.	Hydrology
Bos, M.G.	Ir.	Irrigation and Drainage Engineering
Jong, C. de	Ir.	Agro-Economy
Jurriëns, M.	Ir.	Irrigation and Drainage Engineering
Kortenhorst, L.F.	Ir.	Tropical Agriculture
Naber, G.	Ir.	Librian. Land and Water Management
Oosterbaan, R.J.	Ir.	Land and Water Management
Ridder, N.A.	Prof.Dr.	Geohydrology
Slabbers, P.J.	Ir.	Land and Water Management
Someren, C.L. van	Ir.	Land and Water Management
Sprey, L.H.	Drs.	Agro-Economy
Steekelenburg, P.N.G. van	Ir.	Rural Sociology
Wolf, J. de	Ir.	Land and Water Management
Zijlstra, G.	Ir.	Land and Water Management

Status: Date of establishment: 1956. The Institute is a foundation under responsibility of the Netherlands Government, Ministry of Agriculture. It is subsidized by the Netherlands Government and supervised by a Management Board.

Activities: The Institute has as its major objectives: the collection of information in the fields of land and water use and the dissemination of knowledge on these subjects, with emphasis on irrigation, drainage, and salinity control. The Institute employs a permanent staff of 16 specialists in various disciplines. It has built up an excellent library, and has established a network of relations for an exchange of information with institutes in many parts of the world.

The Institute's publication program is governed by the practical benefit to be derived by people in the field, particularly those working in developing countries. Annually since 1962 the Institute organizes an International Course on Land Drainage, in cooperation with the International Agricultural Centre. It also organizes courses and workshops on land drainage in developing countries. The Institute provides specialist services, frequently in collaboration with national research institutes, development agencies, or government bodies, in programs of bilateral or multilateral technical cooperation.

These services include project identification, project appraisal, support to project execution, and project evaluation. The director is F.E. Schulze.

Correspondence: Dutch, English, German and French.

Instituut voor Cultuurtechniek en Water huishouding (Institute for Land and Water Management Research)
Marijkeweg 11, 6709 PE Wageningen-(P.O. Box 35, 67 AA Wageningen) Tel.: 08370-19100
Telex: 75230 - start message with: ICW

Oosterbaan, G.A.	Ir.	Director
Bijkerk, C.	Prof.Ir.	Dep. Director

Water Quality:

Rijtema, P.E.	Dr.	Head of Dept. Water Quality
		General Water Quality
Steenvoorden, J.H.A.M.	Ir.	Water Quality in Agriculture
Hamaker, Ph.	Dr.	Water Quality in Horticulture
Kemmers, R.H.	Drs.	Natural Environment
Drent, J.	Ir.	Quality Surface Waters
Hoeks, J.	Dr.	Waste Materials and Soil Protection
Ploegman, C.	Ing.	Experimental Field
Harmsen, J.	Mr.	Laboratory

Water Management:

Wesseling, J.	Dr.	Head of Dept. Water Management
		General Hydrology
Stuyt, L.C.P.M.	Ir.	General Hydrology
Kouwe, J.J.	Ir.	Regional Water Management
Ernst, L.F.	Dr.	Theoretical Hydrology
Pomper, A.B.	Drs.	Geohydrology
Feddes, R.A.	Dr.	Agrohydrology
Wesseling, J.G.	Ir.	Agrohhydrology
Bakel, P.J.T. van	Ir.	Numerical Modeling
Nieuwenhuis, G.J.A.	Ir.	Remote Sensing
Boheemen, P.J.M. van	Ir.	Water Supply
Hellings, A.J.	Ir.	Horticultural Water Management
Stakman, W.P.	Ir.	Laboratory

Soil Technology:

Wind, G.P.	Dr.	Head of Dept. Soil Technology
		General Soil Technology
Wijk, A.L.M. van	Dr.	Nonagrarian Land Use
Bakker, J.W.	Ir.	Soil Aeration
Schothorst, C.J.	Ing.	Bearing Capacity and Subsidence
Boels, D.	Ir.	Execution Techniques
Valk, G.G.M. van der	Ir.	Bulb-growing Soils

Land Use Planning:

Bijkerk, C.	Prof.Ir.	Head of Dept. Land Use Planning
Linthorst, Th.J.	Ing.	General Land Use Planning
Visser, A.C.	Ir.	Parcellation
Kleef, H.A. van	Ing.	Land Division Survey
Kester, J.A.	Ir.	Layout Agricultural Areas
Oostrom, C.G.J. van	Ir.	Layout Horticultural Areas
Oostrom, C.G.J. van	Ir.	Regional Studies
Michels, Th.	Ir.	Physical Planning and Traffic
Alderwegen, H.A. van	Ir.	Outdoor Recreation
Verweij, E.J.	Ir.	Settlement
Rheenen, J. van	Ir.	Layout Models

Specialists Sections:

Locht, L.J.	Drs.	Regional and Project Economics
Vreke, J.	Drs.	Regional and Project Economics
Righolt, J.W.	Ir.	Farm Management and Farm Economics
Stol, Ph.Th.	Dr.	Mathematics
Schierbeek MSF, E.W.	Ir.	Editor

General Affairs:

Green, G.M.	Mr.	Head of Dept. General Affairs

Branches with other location: The experiment field 'Sinderhoeve' is located at Renkum.
Address: Telefoonweg 79, 6871 NJ Renkum. Tel.: 08373-2885.

Status: Government Research Foundation, established in 1955, financed by the Government and supervised by a Board of Government officials and representatives of private engineering consultant firms.

Activities: Department of Water Management Research: hydrology, hydro-geology, evapotranspiration, drainage, sprinkling and subirrigation, water balance, saturated and unsaturated water flow, influence of deep well pumping on water supply to vegetation, regional water management schemes. - Department of Water Quality Research: salinity problems and crop reaction, transport and accumulation of pollutants through soils, agricultural land use and groundwater and surface water quality, land treatment of industrial waste water, sanitary landfills and groundwater pollution, pollution balances, leaching problems, natural chemical composition of groundwater and surface water, dumping of manure, fertilizers balance and leaching in greenhouses, nature conservancy in relation to water management and water quality. - Department of Soil Technology Research: aeration, land levelling, changing of soil profiles, subsidence, compaction of soils and nonagrarian soil improvement for recreation areas and sports fields. - Department of Land Use Planning Research: requirements of size and shape of fields and holdings, accessibility of fields and farmbuildings, rural road requirements, re-siting of farmbuildings, site, size and shape of outdoor recreation projects, urbanization aspects of rural regions, physical planning of a region, execution techniques and land consolidation plans, advancement of the Land Division Survey Netherlands (supervision of the operational Survey included). - Department of Project Regional and Farm Economics Research: benefits - cost analysis and short- and long-term effects of amelioration measures on the economy of farm and region, pilot plans of land development schemes.

Publications: The results of research are published as articles by the Institute in one of the following series, which are equivalent with regard to their scientific value: Technical Bulletins, reprints or original publications in the English language; Mededelingen (Technical Papers), reprints and original publications in Dutch, but (except for the Annual Reports) an extensive English summary is present and all legends, captions and tables are given in English; Miscellaneous Reprints, reprints of scientific articles originally published in English or Dutch in a size other than the size of the first two series, or in another language than English or Dutch. When in the publication lists an English title is mentioned after a Dutch one, an extensive summary in English in the sense as mentioned above is present. - These three series are sent free of charge on an exchange basis to interested parties. A fourth series, Regionale Studies ICW, is made up of books in the Dutch language mostly dealing with regional problems in the Netherlands.

Correspondence: Dutch, English, French, German.

Stichting Technische en Fysische Dienst voor de Landbouw (Technical and Physical Engineering Research Service)
12 Mansholtlaan, Wageningen, tel.: 08370 - 19143 - Corresp. address: P.O.B. 356, 6700 AJ Wageningen, The Netherlands. Telex: 45330 CT WAG.

van Beek, A.M.K.	Drs.	Director
Borel, G.	Ir.	Dep. Director
Department of Electronics:		
Borel, G.	Ir.	Head
van Lopik, R.A.M.	Drs.	
Mazee, A.N.	Ir.	
Department of Technical Physics:		
Schurer, K.	Dr.	Head

Department of Engineering:
Bosch, H.   Ing.   Head

Consulting Engineers and Maintenance Department:

Koppe, R.	Ir.	Head
Meerman, H.J.	Ir.	(consulting)
Talma, C.E.	Ing.	(maintenance)
Leusink, F.J.	Ing.	(building)

Department of Electron Microscopy:

Henstra, S.	Ing.	Head
Boekestein, A.	Ir.	

Department of Economics and General Affairs:
Rozdeiczer, M.K.   Ir.   Head

Status: Semi-governmental institution, established in 1964, financed by the Government and supervised by a Board of Government officials, professors of the Agricultural University and Directors of agricultural institutions.

Activities: Advice is given on the application of mechanical, electric, electronic and physical equipment in the agricultural research and also on the selection of instruments and apparatus. An extensive technical documentation service is at the disposal concerned with these matters. The development and construction of physical, mechanical and electronic measuring and controlling instruments, machines, tools and technical apparatus, which are not on the market. Modification of manufactured equipment. Calibration of instruments such as thermometers, hygrometers, lightmeters, manometers, chemical-analytic weights, etc. The design, realization and maintenance of all types of technical installations and equipment normally encountered in the agricultural research laboratories. The design and realization of buildings for agricultural research such as glasshouses, buildings for animal housing and phytotrons for the horticultural and agricultural institutes. Electron microscopical work is carried out with the transmission electron microscope Philips EM 300 and the scanning electron microscope JEOL-JSM-U3 and JEOL-JSM-35C.

Publications: Reports on the above-mentioned subjects are published in: - A yearbook (in Dutch), including a description of the most important projects and results in that year. - Bulletins (in Dutch, sometimes English). - Articles in Dutch and foreign journals.

Correspondence: Dutch, English, German, French.

Landbouwtechniek, Landbouwhogeschool (Department of Agricultural Engineering, Agricultural University)
Mansholtlaan 12, 6708 PA Wageningen, the Netherlands   Tel.: 08370/19119 or 08370/82980

Moens, A.	Prof.Ir.	Head, Agricultural Machinery and Work Organization
—	Prof.	Head, Mechanical Engineering
—	Prof.	Head, Agricultural and Horticultural Buildings
Bouman-Sweers, M.J.M.	Ir.(Mrs.)	Farm Buildings
Goense, D.	Ir.	Tropical Engineering (located in Surinam, South America)
Heijning, J.J.	Ir.	Transport Engineering and Stress Analysis
Huisman, W.	Ir.	Harvest Engineering and Farm Technology
Loon, J.H. van	Med.	Ergonomics (human aspects)

Meuleman, J.	Ir.	Crop Cultivation Engineering
Sar, T. van der	Ir.	Tropical Engineering
Vos, H.W.	Ir.	Livestock Engineering and Farm Work Organization
Vries, H.C.P. de	Ir.	Mechanical Technology
Zander, J.	Dr.Ir.	Ergonomics (engineering aspects)

Branch with other location: Section of Agricultural Engineering for Tropical Crops, Centre of Agricultural Research in Surinam (Aouth America).

Status: Governmental institution, supervised by the Board of Governors of the Agricultural University.

Agricultural Engineering activities: Teaching - "Kandidaat A-B" courses: mechanical engineering, crop and livestock production engineering and work study, construction of agricultural and horticultural buildings. - "Doctoraal" course: a specialization in one or two of the above-mentioned subjects. 45-60 students per year. - Research: Automatization aspects of farm machinery, appropriate rice thresher, ergonomic research on farm machinery, energetic properties of cycle mowers, cooperative use of farm machinery, spare part distribution.

Publications: Spraying distribution - Rice thresher - Mechanization of Cassave harvest - Ergonomics - Spare part distribution.

Correspondence: Dutch, English, French, German.

Instituut voor Mechanisatie, Arbeid en Gebouwen (IMAG) (Institute of Agricultural Engineering)
Mansholtlaan 10-12, Wageningen. (P.O. Box 43, 6700 AA Wageningen) Tel.: 08370-19119
Telegr.: Imag Research Wageningen
Telex: 45330 CTWAG

Coolman, F.	Ir.	Director
Duinker, A.	Ir.	Senior Deputy Director
Glerum, J.C.	Ir.	Deputy Director
Laurs, J.J.	Ir.	Head, Engineering division, Head, Design dept.
Maring, J.	Ing.	Head, Engineering principles department
Poesse, G.J.	Ir.	Head, Machinery division & Soil tillage/traction dept.
Bouman, A.	Ir.	Head, Field machinery and materials handling dept.
Benders, G.A.	Ir.	Head, Livestock machinery department
Kraai, A.R.	Ir.	Head, Testing department
Krolis, K.E.	Drs.	Head, Labour and work management division and Head, Systems department
Elderen, A. van	Dr.Ir.	Head, Operations research department
Brabander, W.H. de	Ir.	Head, Buildings division and Head, Structures dept.
Jongebreur, A.A.	Ir.	Deputy head, Buildings division and Head, Agricultural department
Zilverberg, H.		Head, Construction department
Mulder, W.P.	Ir.	Head, Climate control division and Head, control department
Heijna, B.J.	Ir.	Deputy head, Climate control division and Head, climate department
Postma, G.	Ing.	Head, Livestock and environment liaison department
Voermans, J.A.M.	Ir.	Deputy head, Livestock and environment liaison dept.
Post, C.J.	Ir.	Head, Horticulture and recreation liaison dept.

Status: The Foundation Institute of Agricultural Engineering is supervised by the Division for Agricultural Research of the Ministry of Agriculture and Fisheries.

Activities: Research is carried out in order to promote: an efficient mechanization and automatization in agriculture and horticulture; - an optimal application of labour and an efficient work organization in agriculture and horticulture. - The realization of: efficient and relative cheap agricultural buildings and green houses, including the climate control in these spaces; - an effective technical contribution of agriculture and rural community to aspects as prevention of pollution, land scape maintenance, open air recreation grounds and liveability.

Correspondence: Dutch, English.

Consulentschap voor Bedrijfsuitrusting en Arbeid in de Tuinbouw (Advisory Service for Horticultural Engineering and Labour Management)
Mansholtlaan 10, 6708 PA Wageningen.

Stender, J.A.	Ir.	Advisory Officer

Activities: Advice on the purchase, improvement and construction of machines, implements, glasshouses, installations, climate control, and also advice on workmethods, organization and labour-management in horticulture.

Sprenger Instituut (Institute for Research on Storage and Processing of Horticultural Produce)
Haagsteeg 6, Wageningen (P.O. Box 17, 6700 AA Wageningen) Tel.: 08370-19013
Telegr.: Sprenger Institut, Wageningen

Rijkenbarg, G.J.H.	Drs.	Director
Meffert, H.F.Th.	Dipl.Ing.	Deputy Director
van Beek, G.	Ir.	Physical engineering
Berkholst, Miss Chr.A.M.		Biology
Boer, W.C.		Production Management
Duvekot, W.S.	Ir.	Development
Gorin, N.	Dr.	Chemistry
Greidanus, P.	Drs.	Economics and Statistics
Klop, W.	Ir.	Biochemistry
Koek, P.C.	Ir.	Microbiology
Meer, M.A. van der	Drs.	Food Chemistry
Nieuwenhuizen, G.H. van	Ir.	Transport
Rudolphij, J.W.	Ir.	Physics
Staden, O.L.	Drs.	Physiology
Schouten, S.P.	Drs.	Storage
Stenvers, N.	Ir.	Storage
Schijvens, E.P.H.M.	Ir.	Food Technology
Steinbuch, E.	Ir.	Food Technology
Tijskens, L.M.M.	Drs.	Process Technology

Status: Governmental institution, financed by the Government and supervised by a Board of Government officials and representatives, growers, merchants, industrialists and consumers.

Activities: Research on all problems connected with the marketing of horticultural produce - grading, packing, transport and storage of vegetables, fruits, flowers and flower bulbs - Preservation value of new varieties, technological research and quality research on processed products - Physical engineering and process engineering - Biochemical, chemical, physiological and biological laboratory research.

Publications: Annual reports in Dutch and English languages - Publications in Dutch and foreign journals and in the series of the Institute entitled "Bulletins" (Dutch) and "Mededelingen" (Dutch or English).

Instituut voor Bewaring en Verwerking van Landbouwproducten (IBVL) (Institute for Research on Storage and Processing of Agricultural Produce)   Tel.(Central Bldg.) 08370-
Bornsteeg 59, Wageningen. (P.O. Box 18, 6700 AA Wageningen)   Telex: 45371   19043

Kappetein, G.C.	Ir.	Deputy Director
Rastovski, A.	Ir.	Potato storage and handling
Haan, P.H. de	Ing.	
Es, A. van	Ing.	Biochemical research on potatoes
Hartmans, Miss K.J.	Drs.	
Hesen, J.C.	Ir.	Quality aspects of potatoes for the processing industry
Meijers, C.P.	Ing.	
Hak, P.S.	Ing.	
Hesen, J.C.	Ir.	Liaison officer for the potato processing industry
Ludwig, J.W.	Ing.	Hygienic and quality aspects at processing potato products
Keijbets, M.J.H.	Dr.Ir.	
Sijbring, P.H.	Ir.	Potato processing technology
Remmen, H.H.J. van	Ing.	
Keijbets, M.J.H.	Dr.Ir.	
Haan P.H. de	Ing.	Drying and storing of grains, seeds and pulses
Sparenberg, H.	Ir.	
Jansen, J.		
Hofenk, G.	Ir.	Processing of straw and agricultural wastes
Rijkens, B.A.	Drs.	
Bois, W.F. du	Prof.Dr.Ir.	
Sijbring, P.H.	Ir.	Production and processing of vegetable protein
Es, A. van	Ing.	
Laarhoven, G.J.M.	Ir.	
Leutscher, H.J.	Ir.	Grass and forage crops
Ogink, J.J.M.	Ing.	
Sparenberg, H.	Ir.	Fibre crops
Leutscher, H.J.	Ir.	
Bois, W.F. du	Prof.Dr.Ir.	
Schild, J.H.W. van der		Information Officer

Status: The Institute was established in 1956. It is financed by the Ministry of Agriculture and Fisheries, various agricultural boards and industrial firms, and from fees on patent rights. It is directed by a Board of Management and supervised by the Ministry of Agriculture.

Activities: Research in connection with the establishment of standards for the conditioning of agricultural products and the development of drying, cooling and storage methods and transport systems; Semi-technical and practical research on processing methods; Research on the chemical, physical and biochemical properties of agricultural products in connection with quality, storage and processing; Application of conservation storage methods in practice; Research on the influence of different storage conditions on the quality and characteristics of processed potatoes and other agricultural products; Research is carried out on potatoes, onions, grain, pulses, grass and forage crops, straw, fibre crops.

Publications since 1973: potatoes: biochemical research – storage – sprout inhibition – disease control – variety research – processing suitability – processing technology – energy saving and environment protection at processing – products quality (crisps, chips, flakes, etc.) – nutrition; grains/seeds/pulses: drying and storage – vegetable protein production; forage crops: artificial dehydration; flax: automated processing; straw processing.

Correspondence: Dutch, English, German, French.

Consulentschap in Algemene Dienst voor Boerderijbouw en Inrichting (C.B.I.) (Agricultural Advisory Service for Farm Buildings and Equipment)
Mansholtlaan 12, Postbus 43, NL 6700 AA Wageningen, The Netherlands

Leidekker, J.L.	Ir.	Advisory Officer
Prinsen, L.	Ing.	Cattle Housing Department
Toren, G.A.	Ing.	(arable land)
Hop, J.		
Freriks, J.H.	Ing.	Pig Housing Department
Weerdhof, A.M. van de	Ing.	Poultry Housing Department (Energy, Environment)
Stuut, H.	Ing.	Construction Department
Maanen, N.B.		
Visser, D.R.	Ir.	Horse Housing (Publications)

Status: Governmental institution, supervised by the Ministry of Agriculture and Fisheries.

Activities: Farm magazines.

Correspondence: Dutch, German, French, English.

Vakgroep Waterzuivering L.H. (Department of Water Pollution Control A.U.)
De Dreijen 12, Postbus 8129, P.O. Box 8129, 6700 EV Wageningen

Section Water Purification:

Bovendeur, J.	Ir.	Chelation of heavy metals, humic acids
Buuren, J.C.L. van	Ir.	Chelation of heavy metals, humic acids
Klapwijk, A.	Dr.Ir.	Denitrification, nitrification
Lettinga, G.	Dr.Ir.	Anaerobic waste water treatment
Rensink, J.H.	Ir.	Bulking sludge, phosphate removal
Velsen, A.F.M. van	Ir	Ammonia removal and regaining

Section Hydrobiology:

Cuppen, J.G.M.	Drs.	Macrofauna, macrofyten
Den Hartog, C.	Prof.Dr.	Head of section, macrofyten
Roijackers, R.M.M.		Algae

Status: The Department of Water Pollution Control has been established in 1965, and is a department of the Agricultural University. The section Hydrobiology has started in 1975.

Activities: a) Teaching: 'kandidaats'-courses: water pollution, water chemistry. Waste water treatment. Hydrobiology. - 'doktoraal'-courses: special subjects on water chemistry, waste treatment and hydrobiology. - practical courses: chemical examination of water and waste water treatment methods. Hydrobiological examination of surface water. b) Research: Biological waste water treatment methods, i.e., aerobic methods (nitrification, prevention bulking sludge, phosphate elimination); anaerobic methods (anaerobic treatment processes by methane fermentation, denitrification, sulphate reduction). - Physical/chemical treatment methods: removal and regaining of ammonia, sludge conditioning, sorption processes. - Study of the interaction between humic acids and heavy metals in surface water. - Hydrobiology: structure and function of aquatic ecosystems dominated by aquatic plants; biological assessment of surface water supply; typology of algae communities.

Publications: activated sludge, bulking sludge, phosphate removal, denitrification, nitrification, anaerobic waste water treatment, anaerobic stabilization, ammonia removal and regaining, chelation of heavy metals, humic acids, upflow (anaerobic) sludge blanket (UASB), (USB) treatment processes, algae communities, Chrysophyta, biological assessment of surface water quality, structure and function of aquatic ecosystems, Ranunculus.

Correspondence: Dutch, English, French, German.

Stichting voor Bodemkartering (STIBOKA) (Soil Survey Institute)
Staringgebouw, Marijkeweg 11, P.O. Box 98, 6700 AB Wageningen.   Tel.: 08370-19100
Telegrams: Stiboka, Wageningen

Schans, R.P.H.P. van der	Ir.	Director
Schelling, J.	Dr.Ir.	Deputy director
Westerveld, G.J.W.	Ir.	Head, Soil mapping
Steur, G.G.L.	Ir.	Coordination national soil survey
Zegers, H.J.M.	Ing.	Commissioned surveys
Vleeshouwer, J.J.	Ing.	Systematic surveys

Holst, A.F. van	Ir.	Scientific coordinators
Lynden, K.R. Baron van	Ir.	
Stolp, J.	Ir.	
Stuurman, F.J.	Ir.	
Wallenburg, C. van	Ir.	
Oosten, M.F. van	Dr.Ir.	General research
Haans, J.C.F.M.	Dr.Ir.	Head, Soil survey applications
Sluijs, P. van der	Ir.	Hydrology
Bouma, J.	Dr.Ir.	Applied soil physics
Poelman, J.N.B.	Ir.	Physical parameters
Smet, L.A.H. van	Dr.Ir.	Arable- and grassland farming
–		Forestry
Dam, J.G.C. van	Dr.Ir.	Horticulture, parks and gardens
Geenen, H.G.M.	Ir.	Town and country planning
Zuilen, E.J. van	Ir.	General interpretations

Regional Offices

Clingeborg, A.E.	Ing.	North (Groningen, Friesland, Drenthe), Engelse
Rutten, R.		Kamp 6, Groningen, 050-169111
Beekman, A.G.		East (Overijssel, Gelderland, Utrecht), Staring-
Hurk, J.A. v.d.	Ing.	gebouw, Wageningen
Wallenburg, C. van	Ir.	West (Noord-Holland, Zuid-Holland, Zeeland), Overslag 4, Boskoop, 01727-3540
Kanters, H.L.		South (Noord-Brabant, Limburg), Nieuwlandstraat 28, Tiburg, 013-425961

Status: The Institute works under a board of representatives from several sections of the Ministry of Agriculture and Fisheries, other Ministries and private institutions, and is supervised by the Ministry of Agriculture and Fisheries.

Activities: – Study of the soil the the Netherlands for classification and mapping; – Systematic survey of the soils of the Netherlands; – Interpretations of soil survey data for various types of land use (e.g. crop production, grassland farming, forestry, urban and recreational land use); – Systematic geomorphological survey of the Netherlands (in cooperation with the Geological Survey of the Netherlands); – Ecological and physiognomic landscape survey of the Netherlands; – Commissioned soil surveys and soil survey interpretation (e.g. for rural development and reconstruction, development of urban areas, rural and urban zoning).

Publication series: De bodemkartering van Nederland (Verslagen Landbouwkundige Oonderzoekingen), with a summary. – De bodem van Nederland (m. kaarten 1:200.000); Bodemkaart van Nederland (1:50.000). – Bodemkundige Studies (Verslagen Landbouwkundige Onderzoekingen), with a summary. – Boor and Spade, with summaries. – Soil Survey Papers. – Geomorfologische kaart van Nederland (1:50.000).

Correspondence: English, Dutch.

NEW ZEALAND

Canterbury

Agricultural Engineering Department, Lincoln College, University of Canterbury
Lincoln College, Canterbury   Tel.: Christchurch 228-029

Ward, G.T.	B.Sc.(Eng), Ph.D., C.Eng., F.N.Z.I.E., F.I.Mech.E., Mem. ASAE	Head of Department. Environmental Engineering, Heat and Energy Transfer Process Engineering.
Ballisat, D.J.	R.E.A., F.N.Z.I.W.	Senior Instructor, Welding Technology.
Calvert, I.	B.E.(Civil), M.N.Z.I.E., A.N.Z.I.M.	Reader, Structures and Services, Engineering Management.
Cherry, N.J.	B.Sc.(Hons), Ph.D., F.Roy.Met.Soc., F.Amer.Met.Soc.	Senior Lecturer, Agrometeorology, Wind Energy.
Chilcott, R.E.	B.Sc.(Eng), D.I.C, M.Sc., C.Eng., A.F.R.Ae.S.	Senior Lecturer, Aerodynamics, Wind Energy Design.
Dakers, A.J.	B.E.(Agr), M.E., M.N.Z.I.E.	Lecturer, Soil and Water, Waste Disposal.
Davies, T.R.H.	B.Sc.(Hons), M.Sc., Ph.D.	Senior Lecturer, Soil and Water, River Morphology.
Douglas, B.	B.Sc., M.A.	Senior Lecturer, Watershed Management, Land Use.
Darch, E.	R.E.A.	Senior Instructor, Welding Technology.
Huber, D.G.	B.A.Sc., M.A.Sc., Ph.D., P.E., F.N.Z.I.E.	Reader, Soil and Water, Hydrology.
Lindsay, G.G.	B.Sc.Agr., M.I.Agr.E., Assoc.Mem.A.S.A.E.	Senior Lecturer, Mechanization.
Mackenzie, D.W.	B.E.(Civil), M.N.Z.I.E.	Senior Lecturer, Structures.
McLellan, A.	M.Agr.Sc.	Lecturer, Mechanization, Energy, Structures.
Robertson, G.L.	B.Arch., A.N.Z.I.A.	Lecturer, Structures.
Short, R.P.	A.N.Z.I.W.	Senior Instructor, Machine Technology.
Seaton, P.M.		Senior Instructor, Power.
Smythe, V.G.	B.Sc.(Hons), M.Sc.	Energy Research Fellow, Wind Energy.

Status: Established in 1946 - Department of College of Agriculture of University of Canterbury - Largely Government financed through University Grants Committee but not subject to direct Government control.

Teaching Activities: Service courses in agricultural engineering topics as part of courses for diplomas in Agriculture (1 year), Farm Management (1 year), Horticultural Management (1 year), Field Technology (1 year), Landscape Technology (1 year), Parks and Recreation (3 years), Wood Technology (1 year) and Natural Resources (1 year). Degree courses are given for degrees in Agricultural Science, Horticultural Science and Commerce. Professional engineering degree courses are offered for the degree of B.E. (Agricultural)(4 years). This programme commenced in 1967 and graduates approximately 15 students annually. Post-graduate degrees are offered as follows: Dip.Ag.Eng., M.E., M.Ag.Sc., M.Hort.Sc., M.Appl.Sc., M.Sc.(Resource Management), Ph.D. Subjects include Agricultural Engineering, Energy, Waste Management, Irrigation, Drainage, Water Resources, Mechanization, Hydrology. Post-graduate students approximate 10 students annually.

Research activities: Largely in association with New Zealand Agricultural Engineering Institute, q.v. agricultural hydrology, soil conservation, irrigation, drainage, farm dams and water supply, farm structures, farm machinery, water, solar, wind energy, water resources.

Publications: Hydrology of small watersheds – Catchment models – Soil erosion – Design and operation of irrigation and water supply systems. River morphology, sediment transport, waste management, groundwater. Extension publications on farm tractors, farm safety, water supply, irrigation.

Correspondence: English.

New Zealand Agricultural Engineering Institute, Lincoln College
Canterbury, New Zealand   Tel.: Christchurch 228-029

Watson, E.M.	B.Sc.(Hons)(Mech.Eng),M.N.Z.I.E.	Director
Dunn, J.S.	B.Sc.(Agric.), T.Eng.(C.E.I.), M.I.Agr.E	Senior Principal Research Officer
Heiler, T.D.	A.S.T.C.(Hons), B.E.(Hons), M.I.E., M.N.Z.I.E.	Principal Research Officer
Bidwell, V.J.	B.E.(Hons)(Civil), Ph.D.,M.N.Z.I.E.	Senior Research Officer
Garden, G.M.	B.E.(Mech), Dip.Ag.(C.A.C.) M.N.Z.I.E.	Senior Research Officer
Painter, D.J.	B.E.(Hons)(Mech), Ph.D.,M.N.Z.I.E.	Senior Research Officer
Carran, P.S.	B.E.(Hons)(Civil), M.E.,N.Z.C.E.	Research Officer
Henderson, C.F.	B.E.(Chem), B.Sc.,Ph.D.(Biochem. Eng)(Hons)	Research Officer
Hirsch, S.J.J.	B.Sc.(Elect.Eng.Sc.), M.N.Z.I.E.	Research Officer
Maber, J.F.	B.Agr.Sc.,Dip.Ag.Eng.(Hons), Assoc.Mem.ASAE	Research Officer
Steele, P.E.	B.Sc.(Ag.Eng.), Dip.Eng.,N.D.A., M.Sc.(Ag.Eng), M.I.Agr.E.	Research Officer
Young, R.W.J.	M.A.(Mech.Sci.)	Research Officer
Antis, L.J.	B.Eng.(Ag.)(D.D.I.A.E.)	Research Officer
Kerr, L.E.	B.E.(Mech.)(Hons)	Research Officer
Harrington, G.J.	B.E.(Hons)(Agr.),M.E.,M.N.Z.I.E.	Research Officer
Martin, G.A.	B.E.(Agr.)	
Robinson, M.D.	B.E.(Agr.)	

Status: Institute established in 1964. Controlled by a Management Committee and funded primarily by Ministry of Agriculture and Fisheries.

Activities: Institute is run in close cooperation with Department of Agricultural Engineering of University of Canterbury, the Director being also University Professor of Agricultural Engineering – The Institute is engaged in research development testing and extension work in all branches of agricultural engineering – Major items at present include water resources and water conservation – irrigation – drainage – soil erosion – field crop mechanization – horticultural crop mechanization – application of fertilizers and pesticides – fencing – aerial spraying – farm management.

Publications: Work is reported in Newsletters, Extension Bulletins, Test Reports, Project Reports and Lincoln Papers in Water Resources.

Correspondence: English.

Palmerston North

Massey University
Palmerston North, New Zealand

A) **Department of Agricultural Engineering:**

Clarke, R.M.	B.E., C.Eng.	Reader	Agricultural Engineering
Dickscon, A.J.		Senior Lecturer	Agricultural Engineering
Studman, C.J.	B.Sc.(Hons.), Ph.D.	Senior Lecturer	Agricultural Engineering
Stinson, G.E.	B.Age.Sci.	Junior Lecturer	Agricultural Engineering
Warburton, D.J.	B.Agr.Sci., Ph.D.	Lecturer	Agricultural Engineering

Status: Massey University was established as Massey Agricultural College in 1928 as a Government financed but independently governed institution.

Activities: a) Teaching — Fourteen Agricultural Engineering Courses are taught for post-graduate and under-graduate degrees and diplomas in Agriculture and Horticulture. Average class size is 40. b) Research Energy — Development of a 'once through' solar water heater — comparative studies of solar water heating systems — Alternative fuels — Energy saving in dairy sheds. Automation — Automation in dairy sheds. Waste — Development and testing of systems for treatment of dairy and piggery waste — Odour control. Farm Structures and Services — Development of water supply systems and equipment — Effects of wind on farm structures.

Publications: Machine Milking (Massey University, 5th ed. 1978). Papers in engineering and scientific journals on: Alternative systems of dairy waste treatment (4 papers) — Pollution legislation — Odour control — A once through solar water heater system for farm dairies — Wind power potential for energy saving — Automation in farm dairies — Indentation and wear of metals — Numerous articles in farming journals on agricultural and horticultural engineering related topics.

Correspondence: English.

B) **Department of Soil Science:**

Bowler, D.G.	Dip.Agr.	Senior Lecturer	Soil Water Management
Climo, W.J.	B.Agr.Sc.(Hons)	Junior Lecturer	Soil Water Management
McAuliffe, K.W.	M.Agr.Sc.	Junior Lecturer	Soil Water Management
Woodgyer, W.R.		Drainage Supervisor	Drainage Extension Service

Status: as above

Activities: (a) Teaching — Introductory Soil Water Management courses in: Soil Science ((a) (110) for B.Agr.Sc. and B.Hort.Sc: Soil Productivity (35) for B.Agr., and Soils and Fertilizers (250) for Dip.Agr. and Dip.Hort. — Advanced Soil Water Management course in Soil Science II(d)(45) for B.Ag.Sc., B.Hort.Sc. and B.Agr.Sc.(Hons). (b) Research — Water harvesting — Spray disposal of dairy factory/dairy shed effluent — Performance of subsurface drainage systems — Iron sludge — Direct drilling.

Publications: Drainage of Wet Soils (D.G. Bowler (ed.) Hodder and Stoughton (publ.) 1980. 300 pages cost $10. — Papers in scientific journals on: Spray irrigation onto pasture of dairy factory and dairy shed effluents — water harvesting — drainage design — sportsturf irrigation.

C) **Department of Agronomy - Agricultural Mechanization**

Baker, C.J.	M.Agr.Sc., Ph.D. M.I.Ag.Eng.	Reader, Agricultural Mechanization
Sims, R.E.H.	B.Sc.(Hons), M.Sc.(Ag.Eng.), M.I.Ag.Eng.	Senior Lecturer, Agricultural/Horticultural Mechanization
Ritchie, W.R.	B.Agr.Sc.	
Choudhary, M.A.	B.Sc.(Agr.Eng.), M.Sc., Ph.D.	Post-Doctoral Fellow, Agricultural Mechanization
Wooding, M.G.	M.Agr.Sc.	Research Fellow, Agricultural Mechanization.

Status: As above.

Activities: (a) Teaching - Agricultural Mechanization component Agronomy 1(b)(100) - Horticultural Mechanization component Horticultural Production IIb(35). - Agricultural/Horticultural Mechanization II(13), Agricultural Mechanization II Special Topic (1) for B.Agr.Sc. and B.Hort.Sc., B.Agr.Sc.(Hons), and B.Hort.Sc.(Hons). - Agricultural Mechanization I (27), Agricultural Mechanization II for B.Agr. - Advanced Agricultural/Horticultural Mechanization for M.Agr.Sc., M.Phil., M.Hort.Sc. (b) Research - Development of techniques and equipment for zero tillage including precision seeding - studies of zero-tillage seed groove formation/seeding emergence interactions - studies of effects of different seedbed preparation techniques on soil and plant responses - development of techniques for rapid assessment and categorisation of soil/environment/machine interactions in zero tillage and development of prediction models - development of condensed tillage techniques and equipment.
Over-drilling of sportsfields - energy conservation in seedbed preparation - wear characteristics of zero tillage tools - insect problems in zero tillage. Studies of machinery requirements for oilseed rape production and harvesting - studies of tallow and vegetable oils as fuel substitutes or extenders in compression ignition engines - investigation of lateral traction characteristics of tractor rear tyres - development of power-take-off shaft guard universal testing apparatus - development of simple closed metering system for agricultural chemicals - Rationalization of machinery ownership and use.

Publications: Proceedings Massey Farm Machinery and Engineering Conference (1st ed. 1969, 2nd ed. 1971, 3rd ed. 1973). Proceedings International Conference on Energy Conservation in Crop Production (1977). Papers in scientific journals on: A tillage bin and tool testing apparatus for turf samples - Experiments relating to the techniques of direct drilling of seeds into untilled dead turf - Some effects of cover, seed size and soil moisture status on seedling establishment by direct drilling - An investigation into the techniques of direct drilling seeds into undisturbed, sprayed pasture - The effects of tillage and zero-tillage systems on soil aggregates in a silt loam - Tractor fuel requirements of two tillage systems and zero-tillage 1977 - Establishment of lucerne on sand country - Developments with seed drill coulters for direct drilling: 1. Trash handling properties of coulters (1979) - Developments with seed drill coulters for direct drilling: 2. Wear characteristics of an experimental chisel coulter (1979) - Developments with seed drill coulters for direct drilling: 3. An improved chisel coulter with trash handling and fertilizer placement capabilities (1979). Developments with seed drill coulters for direct drilling: 4. Band spraying for suppression of competition during overdrilling (1979) - Comparisons of four reduced time and energy seedbed preparation systems: Year one of a perennial experiment - Physical effects of direct drilling equipment on undisturbed soils: 1. Wheat seedling emergence from a dry soil under controlled climates (1980) - Physical effects of direct drilling equipment on undisturbed soils: 2. Wheat seedling emergence from a moist soil under controlled climates. Physical effects of direct drilling equipment on undisturbed soils: 3. Seed groove formation by a "triple disc" coulter and seedling performance - Physical effects of direct drilling equipment on undisturbed soils: 4. Seedling performance and in-groove-micro-environment in a dry soil (1980) - Physical effect of direct drilling equipment on undisturbed soils: 4. Measurement of soil compaction in the vicinity of drilled grooves (1980) - Physical effects of direct drilling equipment un undisturbed soils: 5. Groove compaction and seedling root development (1980) - Physical effect of direct drilling equipment on undisturbed soils: 6. Resistance to seedling emergence and groove cover (1980) – Equipment impact and recent developments (1979). Field studies of direct drilling equipment under a range of climatic and soil conditions – Soil and Tillage Research (in press) – Comparative methods of harvesting oilseed rape – Low energy and minimum time cultivation – Sharing the cost and use of farm machinery – Effects of planting pattern and sowing method on the seed yield of safflower, oilseed rape and lupin.

Wellington

Agricultural Engineering Section, Ministry of Agriculture and Fisheries
P.O. Box 2298 – Wellington, New Zealand

Name	Qualification	Position
Scott, D.F.	Dip.Agr.	Chief Advisory Officer – Section Leader
Alley, P.	B.Agr.Sc.	Farm Advisory Officer
Sllison, W.D.	B.Agr.Sc.	Farm Advisory Officer
Baker, A.	B.Agr.Sc.	Farm Advisory Officer
Bealing, J.D.	B.Agr.Sc.	Farm Advisory Officer
Cartwright, M.J.	B.Agr.Sc.	Farm Advisory Officer
Crosbie, C.J.	B.Agr.Sc.	Farm Advisory Officer
Cross, D.	B.Agr.Sc.	Farm Advisory Officer
Duignan, B.	B.Agr.Sc.	Farm Advisory Officer
Farrent, P.	B.Agr.Sc.	Farm Advisory Officer
Coxhead, G.	B.Agr.Sc.	Farm Advisory Officer
Hutchinson, G.	B.Agr.Sc.	Farm Advisory Officer
Kyle, R.	B.Agr.Sc.	Farm Advisory Officer
Lord, P.D.	B.Agr.Sc.	Farm Advisory Officer
Lrd, P.I.	B.Agr.Sc.	Farm Advisory Officer
McDonnell, R.J.	B.Agr.Sc.	Farm Advisory Officer
Northcolt, R.	–	Farm Advisory Officer
Parata, N.	B.Agr.Sc.	Farm Advisory Officer
Phillips, F.W.	B.Agr.Sc.	Farm Advisory Officer
Piper, G.	B.Agr.Sc.	Farm Advisory Officer
Pottinger, B. Miss	B.Agr.Sc.	Farm Advisory Officer
Warren, A.F.	M.Agr.Sc.	Farm Advisory Officer
Weston, L.	B.Agr.Sc.	Farm Advisory Officer
Wilson, E.W.S.	–	Farm Advisory Officer
Harrington, G.J.	BE.Agr.ME	Senior Engineer
Martin, G.A.	BE.Agr.	Engineer
Giffney, A.	BE.Agr.	Assistant Engineer
Robinson, M.	BE.Agr.	Assistant Engineer
Potts, R.	NZCE	Engineering Technician

Activities: Advisory Services in Agricultural Machinery, Structures Irrigation Water Supply and Drainage.

Publications: Articles in NZ Journal of Agric. Advisory Pamphlets. Technical handbooks for advisers.

Correspondence: English.

Water and Soil Division, Ministry of Works
Vogel Building, Aitken Street, P.O. Box 12-041, Wellington North   Tel.: 729.929

Gibson, A.W.	B.E. AMICE M.N.Z.I.E.	Director
Dixie, R.C.	B.Agr.Sc.	Assistant Divisional Director
Howard, R.K.		Assistant Director, Planning and Technical Services
Cowie, C.A.		Assistant Director, Operations
Taylor, Dr. M.E.U.		Research Director

Activities: Professional, administrative and research services to the National Water and Soil Conservation Organization - Research, experimental and extension work - River control - Drainage - Water and soil conservation projects - Overall supervision of local catchment authorities.

Publications: "Soil and Water", six times a year - Research, management and miscellaneous publications on water and soil subjects - Information leaflets.

Correspondence: English.

NIGER

Niamey

Université de Niamey - Ecole Supérieure d'Agronomie
B.P. 237, Niamey   Tél.: 73-27-13/14/15

Institut National de Recherches Agronomiques au Niger (INRAN)
B.P. 150, Niamey

- Département des Recherches Forestières - B.P. 225, Niamey

Service du Génie Rural
Niamey

NIGERIA

Anambra State

Department of Agricultural Engineering - University of Nigeria
Nsukka, Anambra State, Nigeria

Odigboh, E.U.	Ph.D.	Head of Dep., Agricultural Products Processing
Ahmed, S.F.	Ph.D.	Farm Machinery and Power
Anazodo, U.G.	Ph.D.	Farm Machinery
Udeh, N.C.	Ph.D.	Soil and Water Engineering
Ezeike	Ph.D.	Rural Elec. and Agric. Products Processing
Chukwuma, G.O.	M.Sc.	Soil and Water Engineering
Anigbankpu, S.	M.Sc.	Farm Power and Machinery
-		Farm Planning and Structural Design
-		Soil and Water Engineering

Status: Established in 1962, financed by the Federal Government.

Activities: Teaching: The Department offers a five year and a four year programme for the B.Eng. degree in Agricultural Engineering. In the last two years of the programme, students have the option of majoring in either Power and Machinery and Agricultural Product Processing or Soil and Water Engineering and Farm Structures. The given agricultural engineering courses are supplemented by courses in basic engineering sciences taught in Civil, Mechanical and Electrical Engineering Departments as well as courses given by the Faculty of Agricultural Sciences. - Post graduate studies in both Farm Machinery and Soil and Water Engineering. - Research: Design, development and construction of machinery needed for Tropical Agriculture. Study of the problem of Soil Erosion and Soil Conservation. Introduction of new irrigation systems in Tropical areas during the dry season.

Correspondence: English.

Imo State

Engineering Research Department, National Root Crops Research Institute
P.M.B. 1006, Umudike - Umuahia, Imo State, Nigeria

Nwokedi, P.M.	B.Sc.Agric.Eng (Designing) M.Sc.Agric.	Head, Engineering Research Dept. Teaching Mechanization.(Part-time at School of Agric. Umudike)
Obi, W.U.	B.Sc.Agric.Eng.	
Anyabolu, G.K.	Higher Dip.Agric.Mech.	Field trials
Egbulonu, C.S.	(C&G)	General machine fabrication
Okakpu, M.A.C.	Higher Dip.(Agric)	Teaching Farm structures, Machines efficiency determination (Mechanization)

Branches with other location: Joint Programme with F.A.O. on Cassava Planter and Harvester Fabrication.

Status: The Engineering Research Department was established in 1976 under the National Root Crops Research Institute Umudike with the following objectives: (a) Finding ways of solving root and tuber crops mechanization. (b) Designing, development and fabrication of machines needed for root crops mechanization including field operations and processing. (c) Giving lectures in Agric.Engineering subjects to ordinary and Higher Diploma students at the School of Agric. N.R.C.R.I. Umudike. This Department is financed by the Federal Government as a department under the N.R.C.R.I. Umudike and having joint international projects with other bodies like F.A.O. etc.

Agricultural Engineering activities: Teaching: The Department undertakes Agric. Mechanization and teaching of farm structures to students doing the 2 year Ordinary Diploma course and to students doing another 2 years leading to the Higher Diploma Certificates. - Research: The experiments and testing machines are: Design and construction of seed yam planter/harvester - Design and construction of machine for transplanting propagules - Design, fabrication and evaluation cassava harvester for large and small scale farmers - Determination of optimum spacing for mechanized planting and harvesting - Design and construction of sweet potato planter and harvester - Design and construction of cocoyam planter/harvester - Tillage experiment on yam planting.

Fabrications and Publications: Prototype machine for planting and harvesting cassava have been completed (both large and small scale farming)(Joint Project with F.A.O.). -

Prototype machine for peeling cassava has been completed – Prototype machine for harvesting seed yam, cocoyam and sweet potato has been completed – Machine for chipping cassava has been completed (for human and animal consumption) – Machine for dewatering mash cassava has been completed – A paper has been delivered in Agric. Engineering and Irrigation Seminar in Zaria in 1975 and subsequently various other papers on Agricultural Engineering in Nigeria.

Correspondence: English.

Oya State

Engineering Division, Ministry of Agriculture and Natural Resources
Oyo State Headquarters at the Secretariat, Agodi, Ibadan

'Lere Adeyemo, T.	M.Sc.Agric.Eng.	Head of Division
Ogundipe, A.O.	B.Sc.Agric.Eng.	In-charge of Field Operations and Monitoring
Adelekan, O.	B.Sc.Agric.Eng.	In-charge of Irrigation
Akindele, A.A.	B.Sc.	In-charge of Tillage Investigation and Soil Conservation
Kola Ogedengbe	B.Sc.Agric.Eng.	In-charge of Oyo Zone, on Agricultural Engineering activities
Olufayo, A.	B.Sc.Agric.Eng.	In-charge of Agric. Machinery Training Centre at Fashola
Adeyekun, A.A.	B.Sc.Agric.Eng.	Assisting Chief Agric. Engineer in Tillage and Storage Investigations
Alagbe, S.A.	B.Sc.Agric.Eng.	In-charge of Ilesa Zone for all Agric. Engineering activities
Okediran, M.O.	B.Sc.Agric.Eng.	In-charge of Osogbo Zone for all Agric. Engineering activities
Opakunle, O.O.	B.Sc.Agric.Eng.	In-charge of Fashola – Mechanical.

Status: Established as a Branch of Agricultural Extension Services in 1949. Became a separate Division in April, 1973. The Governing Body is the Oyo State Government through the Ministry of Agriculture and Natural Resources, Oyo State, Nigeria. It directs and finances the Institution through the Permanent Secretary of the Ministry of Agriculture and Natural Resources.

Activities: Teaching: Basic Training of: the Ministry's personnel engaged on Farm Mechanization. Farmers: users of agricultural machinery. Continuing and on-the-job training. Comparative study of three methods of land clearing, preparatory to mechanical cultivations. Investigations of the adaptability of the small walking-type tractors in local agriculture. Studies to determine the comparative economics of the various methods coping with tall grass and fallow bush before mechanical cultivation. Study on the effect of dry-season cultivations on week control and crop establishment. Comparative study of Aluminium and concrete-stove silos in regard to caking of stored grain. Developing of natural air-convection dryer. Study of Minimum and Zero Tillage. Irrigation and soil conservation activities or schemes.

Publications: in Tropical Stored Products Bulletin, 1972.

Correspondence: English.

University of Ibadan - Faculty of Technology - Agricultural and Forestry Engineering
Ibadan, Oyo State   Tel.: 400550

Fashola Mechanization Training School
Oyo, Oyo State

University of Ife (Ile Ife)
Ife   Tel.: 2291

National Centre for Agricultural Mechanization
Ilorin

                                Z a r i a

Ahmadu Bello University - Faculty of Engineering/Institute for Agric. Research
P.M.B. 1044, Zaria, Nigeria

Kaul, R.N.	Ph.D.	Head of Department of Agric.Engineering
Adebija, P.A.	AAS(Mech)	Senior Tech. Officer
Ahmed, A.	B.Eng(Agric)	Graduate Asst.
Arinze, E.A.	M.Sc.	Lecturer
Braide, F.G.	Ph.D.	Senior Lecturer
Choudhury, M.S.	Ph.D.	Prin. Res. Fellow
Duru, J.O.	Ph.D.	Senior Res. Fellow
Kalkat, H.S.	M.Sc.	Research Fellow
Lysak, J.W.	Ph.D.	Senior Lecturer
Maiwada, S.	B.Eng(Agric)	Asst. Res. Fellow
Mittal, J.P.	Ph.D.	Lecturer
Musa, H.L.	M.Sc.	Senior Lecturer
Nwa, E.H.	Ph.D.	Senior Lecturer
Okeoma, M.	B.Sc.(Agric.Eng)	Graduate Asst.
Oladimeji S.O.	Full Tech.Cert.	Senior Technologist
Oni, K.C.	M.Sc.	Lecturer
Yadav, R.C.	M.Sc.	Lecturer

Branches: There are outstation facilities of research at Bakura, Kadawa and Mokwa.

Status: Established as an Agric. Engineering Section in the Institute for Agric. Research and converted to a full Department of Agricultural Engineering in 1975.

Agricultural Engineering activities: Teaching: Bachelor of Engineering (Agric), 10 students - Master of Engineering (Agric), 10 students - Service courses to Faculty of Agriculture students preparing for B.Sc. and M.Sc. in Agriculture. 75 students. - Research: The Department conducts research primarily through the Institute of Agricultural Research which is a National (Federal) Institution but administered through A.B. University. Research covers testing, development and management of equipment especially for small scale farmer situations. Several prototypes fabricated. Also research on soil and water engineering aspects of local crops.

Publications: Reports on Departmental Machinery Development project and Test Report on equipment tested at the Department - Report on experiences on management of equipment at different level - Publications on research based at Masters level. Annual

Reports and Departmental contribution to conference papers and other journal articles.

Correspondence: English.

NORWAY

Ås

Landbruksteknisk institutt (Norwegian Institute of Agricultural Engineering)
N-1432 Ås-NLH Postboks 65   Tel.: 47 2 94 00 60

Aas, K.	Mech.Eng.	Prof. Director
Husum, T.	Cand.Agr.	Office Manager
Nordby, A.	Cand.Agr.	Deputy Director

Farm Machinery Research and Development, Farm Machinery Systems, Ergonomics:

Sjøflot, L.	Cand.Agr.	Head - Dr.Scient.
Elkjaer, K.	Cand.Agr.	
Fladstad, O.	Cand.Agr.	
Kjus, O.	Cand.Agr.	
Reiling, J.	Cand.Agr.	
Romstad, T.E.I.	Cand.Agr.	
Time, K.	Cand.Agr.	

Horticultural Engineering:

Nordby, A.	Cand.Agr.	Head - Assoc.Prof.
Berntsen, R.	Cand.Agr.	Dr.Scient.
Holmøy, R.	Cand.Agr.	Dr.Scient.

Land Reclamation, Subsoiling, Drainage Machinery:

Aamodt, H.	Cand.Agr.	Head

Design, Development and Testing of Specialized Equipment for Mechanization of Soil and Plant Research:

Øyjord, E.	Cand.Agr.	Head
	M.Sc.	

Agricultural Machinery Testing:

Weseth, G.	Cand.Agr.	Prof. - Head
Bardalen, A.	Cand.Agr.	
Bøn, K.	Cand.Agr.	
Heir, J.A.	Cand.Agr.	
Mølnå, B.	Cand.Agr.	Dr.Scient.

Extension Service:

Rød, P.O.	Cand.Agr.	Consultant

Western Norway Station, Voss.:

Mehl, I.	Cand.Agr.	Head
Gjuvsland, M.	Cand.Agr.	

Northern Norway Station, Sortland:

Alhaug, K.	Cand.Agr.	Head
Bjørnstad, E.	Cand.Agr.	

IAMFE: The International Association on Mechanization of Field Experiments (IAMFE) has its temporary Secretariat and Information Centre at the Norwegian Institute of Agricultural Engineering. Mr. Egil Øyjord is President of IAMFE and daily leader of the IAMFE Secretariat and Information Centre.

Status: Governmental institution, established in 1947 and supervised by the Royal Ministry of Agriculture. The Institute is associated with the Agricultural University of Norway through its Director who is also Professor of the Department of Farm Power and Machinery at the University.

Activities: Research on and development of working methods, farm machinery and equipment for harvesting and conservation of forage, harvesting of grain, straw and herbage seed and manure handling and treatment. Research on ergonomics and on automation in agriculture - Investigations on plant protection equipment in horticulture, agriculture and forestry - Research on seed drills, planting equipment and plant lifters - Investigations on vegetable, raspberry and blackcurrant harvesting. Research on land reclamation, subsoiling and drainage - Design, development, testing soil tillage and information on specialized equipment for mechanization of soil and plant research - Investigations on milking machinery - Investigations on tractor accidents - Investigations on three-point linkage implement coupling - Testing of tractors and agricultural machinery, in field and laboratory - Testing of, and research on agricultural machinery and equipment for hilly land - Extension work and short information courses for extension service personnel in governmental and private services - Lecturing by the scientific staff at the Agricultural University of Norway.

Publications: Reports and bulletins.

Correspondence: Norwegian, English, German.

Institutt for maskinlaere, Norges landbrukshøgskole (Department of Farm Power and Machinery, Agricultural University of Norway) - Postboks 65, 1432 Ås-NLH Tel.:(02)940060

Aas, K.	Mech.Eng.	Prof.
Nordby, A.	Cand.Agr.	Prof. (Reader) Head
Bjugstad, N.	Cand.Agr.	
Berentsen, O.	Cand.Agr.	Dr.Scient.
Christensen, S.	Cand.Agr.	
Qvam H.	Mech.Eng.	

Status: Governmental. Date of establishment: 1898/1919.

Activities: Courses of farm machinery and related agricultural engineering. The scientific staff participates in research work at the Norwegian Institute of Agricultural Engineering. - The degree-awarding authority is the Agricultural University of Norway. The degrees are: cand.agr. - dr.scient. - dr.agr. - This department offers altogether fourteen courses in the subject matters of Farm Power, Farm Machinery, Farm Mechanization, some Mechanical and Electrical Engineering. Participation: up to 80 students. Research: Listed under the Norwegian Institute of Agricultural Engineering.

Publications: Listed under the Norwegian Institute of Agricultural Engineering.

Correspondence: Norwegian, English, German.

Institutt for Kulturteknikk, Norges Landbrukshøgskole (Department of Hydrotechnics, Agricultural University of Norway)
P.O. Box 32, 1432 Åas-NLH Norway

Bergedalen, J.	Ass. Professor	Head of Department
Bjerve, L.	Cand.agr.	Pollution
Gröterud, O.	Assoc. Professor	Limnology
Hove, P.	Ass. Professor	Drainage

Lundekvam, H.	Cand.agr.		Pollution
Myhr, E.	Assoc. Professor		Irrigation
Tufte, V.	Cand.agr.		Irrigation

Status: Established in 1950. Governmental institution.

Activities: (a) Teaching: Hydrology, irrigation, drainage, water supply, water pollution and rural sewer. Degrees corresponding: B.Sc. and M.Sc. Students attending the courses: 10 to 60. (b) Research: Runoff from small catchment areas, water quality, drainage, drainage materials, irrigation, water pollution.

Publications: Results are published in "Meldinger fra Norges Landbrukshøgskole", "Reports from Norges Landbruksvitenskapelige Forskningsråd" and in "Stensiltrykk fra Institutt for Hydroteknikk NLH".

Correspondence: Norwegian, English, German.

Institutt for bygningsteknikk, Norges landbrukshøgskole (Department of Farm Buildings, Agricultural University of Norway)
Post Box 15, N-1432 Ås-NLH

Berge, E.	Cand.agr.	Dr.techn.	Building production systems
Gjerde, I.	Cand.agr.	Dr.scient.	Building costs
Gjestang, K.E.	Cand.agr.	Dr.scient.	Cattle housing
Gravås, L.	Cand.agr.	Dr.scient.	Pig housing
Grae, T.	Cand.agr.	M.Sc.	Build. physics, materials and construc.
Hjulstad, O.	Prof.Emerit.		
Høibø, H.	Cand.agr.	Civ.Eng.	Structural design
Løken, K.A.	Cand.agr.	M.Sc.	Build. physics, materials and construc.
Moe, B.A.	Arch.		Architecture
Nygaard, A.	Cand.agr.	Dr.agric.	Animal housing
Roer, P.	Cand.hort.		Horticultural buildings
Tjernshaugen, O.	Cand.agr.	Dr.scient.	Manure handling

Status: Department of farm buildings at the Agricultural University of Norway.

Activities: Responsible of teaching of all aspects of farm building at the university. Central responsibility of extension service on farm buildings in Norway. Responsible for research and investigation concerning farm buildings.

Publications: Reports, reprints and mimeographed papers.

Correspondence: Norwegian, English, German.

Driftsteknisk avdeling, Norsk Institutt for skogforskning (NISK) (Division of Forest Operations and Techniques, The Norwegian Forest Research Institute), P.O.B. 61, N-1432, Ås

Samset, I.	Dr.h.c., professor	Head of division
Fjone, H.	Forest eng.	Planning of transport network, aerial photography
Frønsdal, J.	Forest eng.	Truck and tractor transport, road construction
Gimse, A.	Forest eng.	Cost analysis of logging equipment

Juel, E.	Forest eng.	Teaching
Krogstad, I.	Forest eng.	Thinning operations
Kyllo, O.	Forest eng.	Forest operations in steep terrain, cable operations
Lisland, T.	Mech. eng.	Experimental workshop
Skaar, R.	Forest eng.	Teaching, long distance transport, planning of road network, road maintenance
Strømmes, R.	Forest eng.	Mechanized reforestation, work studies
Sørhagen, O.	Civil eng.	Machine testing
Vik, T.	Forest eng.	Ergonomics

Branches: Silvifuturum, Hurdal: Experiments on forest operations under ordinary terrain conditions. – Silvimontana, Kviteseid: Experiments on forest operations under rough terrain conditions.

Status: Established on 12 May 1947. Financed by governmental bodies and semigovernmental bodies.

Activities: (a) Teaching: Forest operations and techniques (general course) – 30 students, Forest operations field training course – 15 students, Forest operations and techniques (main course) – 7 students. – (b) Research: Tractor transports – Thinning operations – Truck transport – Mechanized reforestation – Ergonomics – Operations in farm forests – Development of logging equipment to be used in steep terrain – Testing of mechanical equipment and machines for forest operations – Work studies – Snow and climate studies.

Publications: in Tidsskr.Skogbr. – Meddr.Norsk inst.skogforsk.

Correspondence: English.

Trondheim

Institutt for Landbruksbebyggelse, Norges Tekniske Högskole (Institut of Farm Buildings, Norwegian Institute of Technology)
Arkitektavdelingen, Trondheim

Activities: Study courses leading to a degree in farm building.

Correspondence: Norwegian, English, German.

Vollebek

Avdeling for Mejerimaskiner, Norges Landburkshøgskole, Meieriinstituttet (Section of Dairy Engineering, The Agricultural College of Norway, Department of Dairy Industry)
Vollebek

Status: The Division of Dairy Engineering was established in 1953. Governmental institution.

Agricultural Engineering activities: (a) Teaching: Courses in milk plant layout, construction, building and building materials and courses in milk plant engineering –

Elements of mechanical and chemical engineering, boilers, steam, fuels and combustion, water supply, refrigeration, ventilation, insulation, pumping - Various types of machinery and equipment for handling of milk and milk products in milk plants. - (b) Research: To some extent testing of dairy machinery, and also some extension work.

Publications: Articles concerning dairy techniques and dairy engineering published in various journals. Mimeographed textbooks (compendiums) for the courses in dairy engineering.

Correspondence: Norwegian, English.

PAKISTAN

Faisalabad

Faculty of Agricultural Engineering and Technology, University of Agriculture
Faisalabad   Tel.: 25911-19
                                                         Telegr. Agrivarsity

Hanif Anwar Hussain   Ph.D.   Dean
                       M.Sc. (Agri)
                       B.Sc. (Agri)
                       P.G.H.P. ChemTech.
                       D.I.C.

Department of Basic Engineering
Department of Irrigation and Drainage
Department of Farm Machinery and Power

Islamabad

Agricultural Machinery Division, Pakistan Agricultural Research Council
P.O. National Institute of Health, Islamabad, Pakistan

Zia-ur-Rahman, Dr.   Dr.   Project Director
Javed Akhtar   Agricultural Engineer
Irfan Saleem Ahmad   Asst. Agri. Engineer
Hasan Ali Rizvi   Asst. Agri. Engineer
Munir Ahmad   "   "   "
Amjad Pervaiz   "   "   "
Muhammad Farooq   "   "   "
Saleem Ahmad   Research Assistant
M. Iqbal
Status: Established in 1979. Semi-Government Institute.

Agricultural Engineering activities: Design and development of appropriate agricultural machines - Field testing of agricultural machinery - Industrial extension of Division's designed machines - Mechanization research surveys.

Publications: On above activities.

Correspondence: English.

Lahore

Irrigation Research Institute
Lahore, Punjab

Multan

Agricultural Mechanization Research Institute
P.O. Box 416, Multan, Pakistan

Peshawar

Directorate of Agricultural Engineering
Tarnab, Peshawar, Pakistan

Agricultural Engineering Department, Faculty of Engineering, University of Peshawar
N.W.F.P., Pakistan

Arshad Aziz	M.Sc. B.E.(Agri.Engg)	Head of Department – Assoc.Prof.
Rafiq Ahmad	M.Sc. B.E.(Mech.Engg)	Assoc.Prof.
Badrud-Din	B.Sc.(Agri.Engg)	Assistant Professor
Mirajud-Din	B.Sc.	Assistant Professor
Syed-ul-Abrar	B.Sc.(Agri.Engg)	Lecturer
Mohammad Alamgit	B.Sc.(Agri.Engg)	Lecturer

Status: Date of establishment, 1960 – Semi-governmental.

Agricultural Engineering activities: a) Teaching: Four years B.Sc.Agri.Engineering. Students' enrolment each year 30. Masters Degree program in "Irrigation and Drainage" is in progress. – b) Research: The Department is engaged in completing the design and construction of hand/bullock drawn sugar-beet planter for small farm holdings in Pakistan and the study of consumptive use of water for sugar-beet. This project is being funded from PL-480 funds with the collaboration of USDA and Pakistan Agricultural Research Council –Research is also going on "The Farm Water Management, improvement of water measurement and control systems." This study has been funded from Cess Funds with the assistance of Pakistan Agri. Research Council – For the final year students, small research projects are essential towards their Degree requirements.

Correspondence: English.

Tandojam

Faculty of Agricultural Engineering, Sind Agriculture University
Tandojam, Pakistan   Tel.: 26881

Koondhar, I.D.M.	M.Sc.(Agri) Hons Agril.Engg.	Head, Faculty of Agricultural Engineering
Devrajani, B.T.		Senior Agricultural Research Engineer and Principal Investigator USDA Projects

Department of Farm Power and Machinery:

Bukhari, S.B.	Ph.D. Agril.Engg.	Asst.Professor
Malik, R.J.	M.Sc.(Agri)Hons Agril Engg.	Asst.Professor
Mughal, A.Q.	M.Sc. Agril.Engg.	Associate Professor
Soomro, M.S.	M.Sc. Agril.Engg.	Asst. Prof. and Chairman

Department of Irrigation and Drainage:

Chandio, B.A.	M.Sc. Agril.Engg.	Asst.Professor
Khattri, K.C.	M.Sc. Agril.Engg.	Asst.Professor
Koondhar, I.D.M.	M.Sc.(Agri)Hons Agril.Engg.	Associate Professor and Head of Faculty
Soomro, G.M.	M.Sc.(Agri)Hons Agril.Engg.	Asst.Professor

Department of Farm Buildings and Structures:

Kalwar, M.I.	M.E.	Asst.Professor

Department of Basic Engineering:

Chandio, S.M.	M.Sc.	Asst.Professor

Status: First established as the Department of Agril. Engineering of Agriculture College in 1940. The Department was upgraded to the status of the Faculty of Agril. Engineering in 1970. It is part of the Sind Agricultural University of Tandojam, and is financed by the Government of Pakistan and governed by the University syndicate and senate.

Agricultural Engineering activities: (a) Teaching: 4 year study program after intermediate science (Agril) to award B.E.(Agri) degree. - One year study program after graduation to award M.Sc. degree. For M.Sc. degree the student has to take up a research project of 10 credit hours - Short training courses like Tractor Operators course, Mechanics courses, Land Levelling courses, etc. - 120 students in four professional years of B.E.(Agri.) program - 4 post-graduate students. - (b) Research: is being conducted in the field of Farm Power and Machinery management, mechanization, Soil-water-plant relationships, irrigation and drainage, water resources. Staff members besides their normal academic activities are engaged in applied research projects.

Publications: Agricultural Mechanization in Asia; Sind Quarterly; J.Agril.Engg: ISAE, etc.

Correspondence: English.

PANAMA

Panamá

Instituto de Recursos Hidráulicos y Electrificación (IRHE) (Institute of Water Resources and Electrification)
Apartado 5285, Panamá 5, Panama

División de Desarrollo:

Algandona, C.	Ing.Eléctrico	Jefe de la División
Solva, C.	Ing.Electromec.	Jefe Dep.to de Planeamiento
Hinestroza, A.	Ing.Eléctrico	Jefe Sección de Electrificaión Rural

División de Operación y Mantenimiento:

Urrutia, V.	Ing.Eléctrico	Jefe de la División
Sánchez, N.	Ing.Eléctrico	Gerente Nacional de Producción y Comercialización
Segovia, E.	Ing.Eléctrico	Gerente Regional Prov. Centrales
Mihalitsianos, M.	Ing.Eléctrico	Gerente Regional Prov. Chiriquí
Berrocal, J.	Ing.Eléctrico	Gerente Regional Panamá Occidente

División de Ingeniería:

De GRacia, R.	Ing.Civil	Jefe de la División
Vásquez, R.	Ing.Eléctrico	Jefe Depto. de Diseño
Medina, A.	Ing.Eléctrico	Jefe Sección de Diseño Eléctrico

Dependencias en: Arraiján, Penonomé, Aguadulce, Chitré, Las Tablas, Santiago, San Lorenzo, David, Pto. Armuelles, Soná, San Blas, Yaviza, La Palma (Darién), Puerto Pilón, Colón y Bocas del Toro.

Caractér y estructura: El IRHE fue creado en 1961. Tiene carácter gubernamental.

Correspondencia: Español.

Santiago

Dirección Nacional de Ingeniería Rural, Ministerio de Desarrollo Agropecuario
Santiago, Provincia de Veraguas

Echevers, J. de. C.	Ing.Civil	Director
Escobar, F.E.	Ing.Agrícola	Sub-Director

Estudios y Diseños

Construcciones

Dependencias en otras localidades: En la Dirección existen seis Departamentos Regionales con sus respectivos coordinadores: Región Chiriquí, Veraguas, Herrera, Coclé, Capira, Chepo, Los Santos.

Caractér y estructura: La Dirección Nacional de Ingeniería Rural forma parte del Ministerio de Desarrollo Agropecuario, creada mediante Ley #25 de Enero de 1973. La Dirección Nacional de Ingeniería Rural, es una entidad de carácter gubernamental, ya que pertenece al MIDA y, está basada en un presupuesto de funcionamiento e inversión que proviene del Tesoro Público, o sea de los Fondos del Gobierno, los cuales son asignados anualmente. - Esta integrada por una Dirección, Sub-Dirección, Administración, Depto. de Estudio y Diseño y Depto. de Construcciones.

Actividades de Ingeniería Rural: La Dirección Nacional de Ingeniería Rural, como parte del Sector Agropecuario - MIDA, conciente del crecimiento de nuestra agricultura, brinda todo el apoyo a la producción agropecuaria del país. Es así como emanan muchas actividades como lo son: Asesoría y Entrenamiento de Personal, para manejos de aguas e instalaciones agrícolas, Supervición y seguimiento a los estudios de factibilidad de los diferentes proyectos, construcciones y mantenimientos de las infraestructuras para la producción (obras de riego y drenaje, silos, edificios, caminos, etc.). Habilitación y conservación de tierras en base a Riego y Drenaje, Asistencia Técnica a Grupos Organizados e Independientes para la implantación de sistemas de riego y drenaje, los cuales garanticen la producción e incremente la productividad de los productos agrícolas. - Capacitación: Dentro del ámbito agropecuario se capacita al personal profesional a través de Proyectos y Programas, con participación de personal Nacional e Internacional. En donde se asignan tareas específicos como lo son: Trabajos e Investigaciones.

Publicaciones: Documentos (no publicaciones)

Correspondencia: Español.

PERU

Cuzco

Departamento de Ingeniería Rural, Facultad de Agronomía, Universidad Nacional de San Antonio Abad del Cuzco (Department of Agricultural Engineering, Faculty of Agronomy, National University of San Antonio Abad del Cuzco)

Carácter: Institución gubernamental, establecida en 1956.

Actividades: Cursos de matemática, física, dibujo, hidráulica agrícola, topografía, irrigación y drenaje, moto-cultura, maquinaria agrícola, construcciones rurales - Al final de 5 años de estudio se concede el título de "Bachiller en Ciencias Agrícolas", y despues de la defensa de la tesis o mediante la aprobación del examen especial de la Facultad se otorga el título de ingeniero agrónomo - Se llevan a cabo también trabajos de investigacion sobre irrigación.

Correspondencia: Español.

Lima

Programa de Ingeniería Agrícola y Departamento de Construcciones Rurales - Universidad Nacional Agraria (UNA) (Agricultural Engineering Program and Department of Rural Constructions, National Agrarian University)
La Molina, Apartado 456 - Lima 100

Sub-Dirección de Conservación de Suelos - Dirección de Manejo de Cuencas, Dirección General de Aguas, Suelos e Irrigación, Ministerio de Agricultura (Soil Conservation

Sub-Department, Watershed Department, General Department of Water Soil and Irrigation, Ministry of Agriculture)
Jirón Washington '894, Lima

Servicio Nacional de Maquinaria Agrícola SENAMA, Instituto Nacional de Investigación y Promoción Agropecuario - INIPA (National Farm Machinery Service - SENAMA, National Research, Agriculture and Livestock Promotion Institute - INIPA)
Edificio El Regidor, Residencial San Felipe Jesus María, Lima

Personal Docente del Programa de Ingeniería Agricola (Universidad Nacional Agraria La Molina); Apartado n. 456, Lima   Tél.: 352035

Departamento de Mecanización Agricola (DMA):
Lecca Rodríguez, M.L.	Ing. M.Sc.
Gilardi Rodríguez, J.	Ing.Agr.
Maezono Yamashita, L.	Ing. M.Sc.
Araujo Rodríguez, H.	Ing. M.Sc.
Carrasco Rodríguez, J.	Ing.Agric.
Yamamoto Miyakawa, A.	Ing.Agric.
Lizárraga Mandujano, A.	Ing.Telec.
Rodríguez Flores, A.	Ing.Agr.Mg.
Cáceres Guerrero, F.	Ing.Agr.
Vidal Vidal, M.	Ing.Agr.
Salas Pinto, D.	Ing.Agr.
Garro Santillana, L.	Ing.Agr.

Departamento de Recursos de Agua y Tierra (DRAT)

Departamento de Construccciones Rurales

Correspondencia: Español, Ingles.

PHILIPPINES

Cagayan de Oro

Xavier University
Cagayan de Oro City, Misamis Oriental, Phil.

Boisier, B.V.	BSAE - M.Engg.	Department Head - Irrigation and Drainage Hydraulics Engineering
Branzuela, R.M.	BSAE - M.S.	Water Resources and Management Engineering specializ.: Water Resources
Noble, B.F.	BSME	Farm Power, Refrigeration and Biogas
Salimbangon, P.	BSAE	Farm Machineries and Soil Water
Villamor, A.	BSAE	Farm Structures, Farm Electrification and Crop Processing

Status: Established School Year 1977-1978. Governing bodies: private.

Agricultural Engineering activities: a) Teaching: Study courses in irrigation, drainage, soil and water conservation - Farm power, farm machineries and mechanization, waste utilization - Farm structures, electrification, refrigeration and crop processing - Agricultural economics (handled by Economics Division) - Type of degree awarded: Bachelor of Science in Agricultural Engineering. Number of students: 194 (1st – 5th year). – Research: Biogas (ongoing) – Crop consumption use (ongoing) – Irrigation structures (proposed)

Correspondence: Filipino and English.

Laguna

Department of Agrometeorology, Institute of Agricultural Engineering and Technology, U.P. at Los Baños, College, Laguna

Gayanilo, V.G.	Ph.D.	Head of Department – Processing (Solar Energy Utilization)
Dagaas, F.M.	BSAE	Agrometeorology
Ramirez, A.M.	BSAE	Agrometeorology
Zara, P.M.	BSAE	Agrometeorology

Status: INSAET established June, 1976. Governing bodies which direct and finance the institution: UPLB.

Activities: Teaching: The Department handles courses such as Hydrometeorology, Fundamentals of Agricultural Engineering, Plant Climate, Mechanics of Rigid Bodies in Motion in BSAE degree. In M.S. degree, courses are Tropical Agrometeorology, Special Topic/Problem, Micrometeorology and Master's Thesis. Research: Graduate student research (thesis).

Publications: Yearly weather data.

Correspondence: English.

Department of Land and Water Resources Engineering and Technology (LAWREAT), INSAET U.P. Los Baños, College, Laguna

David, W.P.	Ph.D.	Associate Professor and Chairman

Status: One of the four departments of the Institute of Agricultural Engineering and Technology (INSAET) – state-supported institution.

Department activities: Teaching: The Department handles courses on Soil and Water Engineering such as irrigation, drainage, soil conservation, water management, fluid mechanics, solid dynamics, environmental control engineering and surveying. It also offers an M.S. degree program on Soil and Water Engineering. The average number of students is about 20 per course. – Research: The Department is engaged in various research activities dealing with Land and Water Resources Engineering specifically modelling, water management and cropping systems.

Publications: Articles and Research Papers on Land and Water Resources Engineering and Management have been published since 1973.

Correspondence: English.

Department of Agricultural Machinery Engineering and Technology, INSAET
U.P. at Los Baños, College, Laguna

Del Rosario, C.R.   Ph.D.   Assisstant Professor and Chairman

Status: one of the four departments of the Institute of Agricultural Engineering and Technology (INSAET). State supported institution.

Activities: Teaching: The Department handles 11 undergraduate courses, one service course (AE 82), and one Rural High School subject (Agr.IV). The Department handles a total volume of 929 undergraduate students. Sixty-seven percent of them are AE students and the rest (35%) are non-AE students. It also offers 6 Graduate courses (200 level) and thesis (AE 300) a total volume of forty-five (45) graduate students in a year. - Research: The Department is engaged in various research activities that deal in Agricultural Machinery Engineering specifically in design, development, testing of agricultural machines, and implements and manufacturing of agricultural machines and implements. - Extension: The Department participated in three international workshops co-sponsored by RNAM. (One workshop on rice transplanters was organized by the department and held in Los Baños. The one on weeders was in Sri Lanka and the one on harvesters in India. It hosted the 2nd Subnetwork Workshop on Cereal Harvesters which was held on August 5-8, 1980 in Los Baños. It participated actively in the deliberation of the technical committee on standardization of machinery for agriculture and forestry of the Phil. Bureau of Standards.

Publications: RNAM Newsletter and PSAE Yearbook.

Correspondence: English.

Manila

Soil Conservation Division, Bureau of Soils, Ministry of Agriculture
Sanvesco Building, Taft Avenue, Manila   Tel.: 58-01-11

Martin, C.R.   B.S.M.E.   Head of Division

Land Use Section

Vegetative Erosion Control Section

Mechanical Erosion Control Section

Water Conservation Section

Soil Conservation Section

Soil Conservation Project Stations

Soils Regional/Provincial Offices

Status: Established 1951. Governing body: Ministry of Agriculture.

Activities: Planning and policy formulation on proper land use and soil and water conservation - Supervision of soil and water conservation works, engineering and other erosion control measures. Research on soil and water.

Correspondence: English.

Bureau of Plant Industry - Agricultural Engineering Division - Ministry of Agriculture
Malate, Manila

Gonzalo, B.C.	BSAE, MSAE	Chief Agril.Engg. Chief of Division Farm Mechanization, FP and M
Aldaba, P.B.	BSAE	Supervisor, Agril.Engg. Head of Section Farm Power and Machinery
Castañeda, T.C.	BSAE	Supervisor, Agril.Engg. Head of Section Irrigation and Drainage
Mercado, A.M.	BSAE	Supervisor, Agril.Engg., Head of Section Crop Processing and Storage
Santos, H.B.	BS Arch.	Architect II - Head of Section Building and Farm Structures
Fernandez, C.P.	BSAE	Agric. Engineer - Crop Processing
Vera Cruz, C.G.	BSAE	" " - Farm Power and Machinery
Icatlo, H.C.	BSAE	" " - Crop Processing
De Vera, C.G.	BSAE	Jr.Agr. " - Irrigation and Drainage
Ayson, F.A.	BSCE	Civil Engineer - Building and Farm Structures

Status: Government Institution, supervised by the Minister of Agriculture.

Research activities: Conduct basic studies on water, and plant relationships including irrigation and drainage; Develop survey plans for the installation and improvement of existing irrigation systems; Conduct studies on seed irrigation, drying, and storage techniques and facilities; Develop, improve and maintain existing laboratory and field equipment and other facilities of the Bureau; Conduct studies on the design, improvement, construction, testing and development of farm tools and implements and also on the economy and application of suitable equipment used in crop production and processing, plan, design and coordinate construction of farm structure, office buildings, and laboratories.

Publications on small and medium size machinery requirements; Rootcrop processing and storage; Farm mechanization, its technical aspects on the selection and proper use of machinery; Farm planning for irrigation and drainage; Irrigation guide for lowland rice in the Philippines; Mechanical land conditioning, etc.

Correspondence: English.

National Post-Harvest Institute for Research and Extension NGA Central Office
101 E. Rodriguez Sr. Avenue, Metro Manila, Philippines   Tel. 62-21-55
           Local telex: NGA/NAPHIRE 2021 -   Int.telex: NGA/NAPHIRE
                                                       Area Code 742 42007

De Padua, D.B.	Ph.D. Agric.Eng.	Executive Director
Cachuela, R.L.	Agricultural Engineer	

Status: Established in 1978 as an independent institution but for purposes of funding, policy and planning coordination it is considered a subsidiary of the National Grains Authority. A Board of Trustees acts as the institute's governing body.

Agricultural Engineering activities: (a) Teaching: The Institute organizes and implements short term non-degree training courses in Post-harvest Technology. - Research: Grain drying with solar energy collector - Grain quality deterioration in on-farm operations - Assessment and reduction of losses in commercial level of operations.

Correspondence: English.

Kagawaran ng Sakahaning Agsikapan, Dalubhasaan ng Sakahanin at Sigmuing Agsikapan (Department of Agricultural Engineering, Institute of Agricultural and Mechanical Engineering, Gregorio Araneta University Foundation), Victoneta Park, Malabon, Metro-Manila, Philippines

Acosta, J.A.	B.S.A.E. A.B.(Math) B.S.M.E. B.S.E.E.	Dean of the Institute and Head of Department Irrigation and Drainage Farm Structures
Blancaflor, A.B.	B.S.A.E. B.S.M.E.	Assistant Dean - Farm Machinery Farm Mechanization

Branches at other locations: Salikneta, the Agricultural Experimental Farm of G. Araneta University Foundation, 18 km away from the main campus.

Status: Non-profit, non-stock private foundation operating under a public law. Established in 1947.

Agricultural Engineering activities: Five-year course to an average of 1,500 students leading to the Degree of Bachelor of Science in Agricultural Engineering - Research: Farm Product Processing Equipment and Farm Machineries for major field crops - non-conventional sources of energy like solar heater, windmill, and bio-gas - Irrigation structures.

Publications: Araneta Research Journal.

Correspondence: English, Pilipino.

Marawi City

Department of Agricultural Engineering, College of Agriculture, MSU
Marawi City, Mindanao State

Octura, E.R.	B.S.A.E. M.Sc.	Department Head - Soil and Water Engineering (specialization)

Status: Established June, 1965 - Government financed institution

Agricultural Engineering activities: The Department is offering a General five year curriculum leading to a degree in Bachelor of Science in Agricultural Engineering. Average number of students enrolled - 200. - Programs/Activities/Projects undertaken: a) Farm Mechanization Assistance to about 50 farmers in the area. b) Research proposals: Soil erosion problems - study under local farm practices - Crop response to moisture stress level and fertilizer level of local crops - Fish drier study.

Correspondence: English.

Muñoz

Central Luzon State University
Muñoz, Nueva Ecija, Philippines

Angeles, H.L.　　　　　Ph.D.　　　　　　　　Chairman - Irrigation Engineering

Division of Soil and Water Management
Division of Farm Power and Machinery/Farm Mechanics
Division of Crop Processing/Farm Structures and Environmental Sanitation

Status: Established 1956. Government institution.

Agricultural Engineering activities: (a) Teaching: Bachelor of Science in Agricultural Engineering - Certificate in Farm Mechanics (one year) - Research: Design of labour saving devices for small farmers - Design of grain driers for small farmers - Experiment on indigenous sources of energy - Soil and Water Management researches - Post harvest researches.

Publications: on above activities.

Correspondence: English

Pasquin

Mariano Marcos State University - College of Agriculture
Pasquin, Ilocos Norte

Acob, J.V.　　　　　　B.Sc. Ag.Engg.
Caliva, E.A.

Correspondence: English

Mariano Marcos State University (Main University)
Batac, Ilocos Norte

Correspondence: English.

Quezon City

Pambansang Pangasiwaan ng Patubig (National Irrigation Administration - NIA)
Diliman, Quezon City, Philippines

Irrigators' Assistance Department:

Galvez, J.A.　　　　　Ph.D.　　　　　　　　Officer-in-Charge, Irrigators' Assistance
　　　　　　　　　　　　　　　　　　　　　　Department and Project Director, Watershed
　　　　　　　　　　　　　　　　　　　　　　Management and Erosion Control Project
Lazaro, R.C.　　　　　Ph.D.　　　　　　　　Director, Agriculture Department

Irrigators' Organization and Training Division

Irrigators' Assistance Division

Evaluation and Reports Division

Agricultural Coordination Division

Agricultural Development Division

Research and Development Department:

Research and Development Sections

On-Farm Facilities Study

Research and Development Department (Central Office)

Branches with other locations: Region 1: Urdaneta, Pangasinan - Region 2: Cauayan, Isabela - Region 3: Tambubong, San Rafael, Bulacan - Region 4: Pila, Laguna - Region 5: Naga City - Region 6: Iloilo City - Region 7 and 8: Tacloban City - Region 9: Zamboanga City - Region 10: Cagayan de Oro City - Region 11: Davao City - Region 12: Cotabato City.

Status: The Agricultural Engineering activities in the NIA have been undertaken with the creation of the office in 1964, but it was until the early part of the year 1973 when the Agency was reorganized to expand its activities by the integration of the Irrigation Service Unit (ISU) as called for in the Government reorganization that the agricultural engineering field was given more emphasis and significance. The NIA, at present (1980) is again reorganizing. The agency is a government-controlled and/or semi-governmental corporation.

Activities: Research: Water management - Water use - Transplanting methods - Drainage effects - Silting and desilting - Economics of irrigation (pump and gravity) - Effective rainfall - Cost studies of irrigation - Soil moisture stress - Production input cost - Irrigation association/cooperatives - On farm facilities - Benefit cost analysis.

Publications: Water management - Feasibility studies - Economics of pump irrigation - Groundwater exploration, development, utilization, and conservation - Production inputs cost studies - Water resources, inventory and development - Water and erosion control studies - Water conveyance, distribution and application - Agro-socio-economics of irrigation projects.

Correspondence: English and Pilipino.

National Post-Harvest Institute for Research and Extension
National Grains Authority Bldg., E. Rodriguez Avenue Ext., Quezon City

Department of Agricultural Engineering, University of the Philippines in the Visayas
c/o College of Fisheries, Diliman, Quezon City

Kawanihan Sa Pagpalaganap Ng Karunungang Pansakahan (Bureau of Agricultural Extension)
Diliman, Quezon City

Villorente, A.R.   B. S.A., M.P.A.   Chief, Agricultural Programs Division

Status: Governmental.

Activities: Extension services in Post-harvest handling – Irrigation and drainage – Biogas generation and waste recycling – Farm mechanization.

Correspondence: English.

POLAND

Kraków

Wydział Techniki i Energetyki Rolnictwa, Akademia Rolnicza (Faculty of Agricultural Engineering and Energetics, Agricultural Academy)
30-149 Kraków, ul. Balicka 104   Tel. 740-44, 740-47

Michałek, R.	Prof. dr hbl.	Dean of the Faculty
Pelc, K.	Doc. dr inz.	Deputy Dean of the Faculty

Status: Governmental institution, established 1923 and supervised by the Ministry of Scientific Research, Higher Learning and Technology.

Teaching activities: 4,5 years courses on agricultural mechanization and energetics (120 students per year). Mechanization as part of the 4,5 years course of other Faculties (Agronomy – 180 students, Livestock Farming – 90 students, Horticulture – 120 students, Land and Water Management – 70 students, Forestry – 75 students). The degree awarded after the 4,5 years course is Mgr., corresponding to M.Sc.

A. Instytut Mechanizacji i Energetyki Rolnictwa, Akademia Rolnicza (Institute of Agricultural Mechanization and Energetics, Agricultural Academy)

Michałek, R.	Prof. dr hbl	Director of the Institute – Farm mechanization management
Kloc, T.	Doc. dr inz.	Deputy Director of the Institute – Mechanization of horticulture
Zalewski, P.	Doc. dr hbl.	Deputy Director of the Institute – Machinery testing and Ergonomics
Broda, R.	Dr	Farm power and machinery
Borcz, J.	Dr inz.	Mechanization of horticulture
Burkiewicz, B.	Dr inz.	Machine design
Czernik, Z.	Dr inz.	Mechanization of forestry
Dabkowski, J.	Dr	Applied mathematics
Debski, J.	Dr inz.	Farm power and machinery
Dyduch, Cz.	Dr inz.	Mechanization of forestry
Gerard, V.	Doc. dr inz.	Energetics
Gedzior, J.	Dr inz.	Mechanization of forestry
Kogut, St.	Dr inz.	Farm power and machinery
Kokoszka, St.	Dr inz.	Transport in agriculture
Kolowca, J.	Dr hbl inz.	Physical properties of plant materials
Kowalski, J.	Dr inz.	Farm power and machinery
Marks, N.	Dr inz.	Farm power and machinery

Pelc, K.	Doc. dr inz.	Mechanization of animal husbandry
Ptaszek, K.	Dr inz.	Farm power and machinery
Ślipek, Zb.	Dr inz.	Physical properties of plant materials
Wachacki, B.	Doc. dr inz.	Mechanization of forestry
Walczyk, J.	Dr inz.	Farm power and machinery
Walczyk, M.(Mrs)	Dr inz.	Farm power and machinery
Wilkus, St.	Dr inz.	Machines for food industry
Winkler, M.	Dr inz.	Mechanization of horticulture.

Research activities: Experiment, testing and extension work. Programming of mechanization on large and small farms. Profitability of mechanization under various economic and climatic conditions. Constructional and exploitation parameters of various types of machines. New sources of energy for agriculture. Physical properties of plant materials. Ergonomics. Mechanization of forestry.

Publications: on above activities. Textbooks on: Mechanization of animal production - Ergonomics for agricultural engineers.

Correspondence: Polish, Czech, English, German, Russian.

B. Instytut Napraw i Organizacji Zaplecza Technicznego, Akademia Rolnicza (Institute of Repairs and Organization of Servicing, Agricultural Academy)

Bala, W.	Prof. dr hbl.	Director of the Institute - Organization of repairs and servicing.
Cieślikowski, B.	Dr inz.	Agricultural vehicles
Gruszczyński, J.	Doc. dr hbl.	Mechanization and organization of land and water works
Langman, J.	Dr. inz.	Organization of repairs
Lokas, M.	Dr inz.	Mechanization and organization of land and water works
Todorow, Ch.	Dr. inz.	Technical mechanics.

Research activities: Technology and organization of repairs. Infallibility of machines and installations. Organization of servicing. Technology and organization of land and water works.

Publications: on above activities. Textbook on planning and organization of land and water amelioration works.

Correspondence: Polish, German, English, Russian.

Institute of Agricultural and Forest Melioration, Academy of Agriculture
Kraków, Al Mickiewicza 24/28   Tel.: 390-98

Agrometeorology Section

Natural Basis of Melioration Section

Agricultural and Forest Melioration Section

Peat Science Section

Status: Date of establishment 1971. National institution supervised by the Ministry of Science, Higher Education and Technology.

Research activities: Water management in mountain areas. Drainage management. Groundwater drainage.

Publications on above activities.

Correspondence: Polish, English, German, Russian.

Lublin

Instytut Mechanizacji Rolnictwa, Akademia Rolnicza (Institute of Agricultural Mechanization, Agricultural University of Lublin)
Aleja P.K.W.N. 28, 20-612 Lublin, Poland

Orzechowski, J.	Professor Dr habil inz.	Director Equipment management
Bichta, H.	Dr inz.	Fruit production mechanization
Bieluga, B.	Assoc. Prof. Dr habil inz.	Agricultural mechanization
Burski, Z.	Dr inz.	Tractors and power
Bzowska-Bakalarz, M.	Dr inz.	Agricultural mechanization
Cichocki, A.	Dr inz.	Husbandry mechanization
Furtak, J.	Assoc. Prof. Dr inz.	Fruit production mechanization
Gawda, H.	Dr inz.	Physical properties of agricultural materials
Gieroba, J.	Professor Dr habil inz.	Agricultural mechanization
Gilewicz, K.	Dr inz.	Agricultural machinery
Gruszczynski, L.	Dr inz.	Agricultural mechanization
Karczewski, T.	Dr inz.	Tractors and power
Kliszczewski, W.	Dr inz.	Husbandry mechanization
Koper, R.	Dr	Physical properties of agricultural materials
Koszel, T.	Dr inz.	Equipment management
Koziej, J.	Assoc. Prof. Dr habil inz.	Husbandry mechanization
Kowalczuk, J.	Dr inz.	Fruit production mechanization
Krasowski, E.	Assoc. Prof. Dr habil inz.	Tractors and power
Kurzyp, T.	Dr	Soil mechanics
Kusz, A.	Dr inz.	Agricultural machinery repair
Kwiecien, S.	Dr inz.	Agricultural mechanization
Kwiecinski, A.	Assoc. Prof. Dr habil inz.	Husbandry mechanization
Lizut-Skwarek, M.	Dr	Physical properties of agricultural materials
Marciniak, A.L.	Dr inz.	Agricultural machinery
Marciniak, A.W.	Dr inz.	Agricultural machinery repair
Marciniak, J.	Dr	Agricultural machinery repair
Miczynski, J.	Dr	Soil mechanics
Niedziołka, I.	Dr inz.	Agricultural mechanization
Otmianowski, T.	Professor Dr habil inz.	Agricultural machinery repair

Piekarski, W.	Dr inz.	Tractors and power
Pietrzyk, W.	Dr inz.	Farm electrification
Pietruszewski, S.	Dr	Soil mechanics
Przesmycki, J.	Dr inz.	Materials technology
Sawicki, Cz.	Dr	Mechanics of liquids
Siwiło, R.	Dr inz.	Equipment management
Skorzyńska, Z.	Dr	Physical properties of agricultural materials
Skwarek, W.	Dr inz.	Tractors and power
Staszczak, Z.	Dr	Metrology in agricultural engineering
Szpryngiel, M.	Dr inz.	Equipment management
Szymański, W.	Dr inz.	Equipment management
Szwed, G.	Dr inz.	Materials technology
Tomaszewski, K.	Assoc. Prof. Dr habil inz.	Equipment management
Wrona, T.	Dr inz.	Equipment management

Status: Governmental Institution supervised by the Ministry of Education

Activities: Teaching on University level/M.S. degree/, additional studies for Ph.D., agricultural mechanization, research on tractors and agricultural machinery, soil mechanics, agricultural machinery repair, equipment management, physical properties of agricultural materials, tractors and power, fruit production mechanization, husbandry mechanization, farm mechanization, materials technology, mechanics, mechanics of liquids, metrology in agricultural engineering, soil mechanics.

Publications: Several books on agricultural mechanization and agricultural machinery construction. Papers are published in Roczniki Nauk Rolniczych, series C and D.

Correspondence: Polish, English, French, German, Russian.

Katedra Melioracji i Budownictwa Rolniczego, Akademia Rolnicza (Chair of Land Reclamation and Animal Buildings, Agricultural University)
ul. Leszczyńskiego 7, 20-069 Lublin.   Tel.: 20644

Mazur, Z.	Prof. Dr hab.	Head, Soil conservation
Głuski, T.	Mgr inz.	Farm buildings
Kaczyński, J.	Dr inz.	Farm buildings
Mazurek, T.	Dr inz.	Land reclamation
Orlik, T.	Doc. Dr hab.	Land reclamation
Pałys, S.	Dr inz.	Soil conservation
Przemycka, E.	Mgr inz.	Farm buildings
Wegorek, T.	Mgr inz.	Forestry

Experimental Station in Elizówka, Lublin province.

Status: Governmental institution supervised by the Ministry of Science, Higher Education and Technology.

Activities: 4,5 years study courses, attended by about 300 students, leading to Mgr/M.Sc. - Research: Soil erosion control in Poland. Animal buildings.

Publications: Papers are published in Zeszyty Problemowe Postepów Nauk Rolniczych and Roczniki Gelboznawcze.

Correspondence: Polish, English, German, Russian, French.

Olsztyn

Instytut Maszyn i Urzadzeń Rolniczych Akademii Rolniczo-Technicznej (Institute of Agricultural Machines and Equipment, Academy of Agriculture and Technology)
10736 Olsztyn, Kortowo 50   Tel.: 28436, 28920

Motorization and Transportation:

Kossowski, E.	Doc. dr inz.	Director of the Institute, Head of the Group, Engines and Tractors

Farm Mechanization:

Wierzbicki, K.	Dr inz.	Head of the Group, Construction of agricultural machinery

Machinery Exploitation and Repair:

Malewski, T.	Dr inz.	Head of the Group, Mechanical technology and repair

Electrification and Automation:

Orliński, J.	Dr inz.	Head of the Group, Rural electrification

Food Industry Machines Exploitation:

Katewicz, Z.	Doc. dr inz.	Head of the Group, Drying of agricultural products, Exploitation of food industry machines.

Status: Governmental institution supervised by Ministry of Science, Universities and Technology.

Activities: Study courses, $4\frac{1}{2}$ years leading to Mgr inz./M.Sc./in Agricultural Engineering, 45 students in each course, non resident study courses, 5 years leading to Inz./B.Sc./in Agricultural Engineering, 40 students in each course, 3 semesters courses on agricultural mechanization as part of the 5 years courses on general agriculture/200 students/, livestock breeding/180 students/, inland fisheries/150 students). - Research: Movement condition and energy consumption of agricultural machines in hilly terrain - Seed cleaning-procedures and machines, intendent cylinders - Plastic elements as members of agricultural machines - Technical equipment on farms - Fluid power in agricultural machines - Lubrication and greasing - Machines part regeneration - Repair shops for agr. machines - Reliability of technical systems - Rural electrification, electrical power demand and energy consumption in animal production - Chemical conservation of cereals, drying and storage of agricultural products - Mechanization of animal production - Seals in dairy machines - Mechanical vibration measurement in food industry machines.

Publications: in ART-Olsztyn and papers on subjects listed in item "Research" printed in:'Roczniki Nauk Rolniczych', 'Zeszyty Problemowe Postepów Nauk Rolniczych PAN', 'Eksploatacja Maszyn', 'Maszyny i Ciagniki Rolnicze', Przemysł Spozywczy','Nowe Rolnictwo', Przemysł Mleczarski', 'Mechanizacja Rolnictwa', 'Zeszyty Naukowe Akademii Rolniczo-Technicznej Olsztyn'.

Correspondence: Polish, English, German, Russian.

<u>Wydział Technologii Żywności</u> (Faculty of Food Technology)

Kisza, J. Prof. Dr. Dean of Faculty

<u>Instytut Inzynierii i Biotechnologii Żywności</u> (Institute of Food Engineering and Biotechnology) Olsztyn-Kortowo bl. 31

Jakubowski, J. Prof. Dr. Head of Institute

<u>General Food Technology</u>

<u>Dairy Technology</u>

<u>Plant Processing</u>

<u>Meat Technology</u>

<u>Food Microbiology</u>

<u>Food Engineering</u>

<u>Human Nutrition</u>

Status: Established in 1970. Governmental institution.

Activities: a) Teaching: Courses leading to the degree "magister inzynier" and "doktor nauk". Average number of students attending for all courses: 1300.
b) Research: Biotechnology and bioengineering of food production – Research on the biophysical and biochemical properties of food in connection with quality, storage and processing – Maximal utilization of food by-products – Factors affecting the nutritive value of food – Biosynthesis of food and food composites – Cleaning and hygiene during food production, storage and processing.

Publications: List available on request.

Correspondence: Polish, English, German.

<u>Katedra Ekonomiki i Organizacji Przemysłu</u> (Department of Economics and Food Industry Organization)
Olsztyn, Bl. 43 Tel.: 28-514

Stachowski, T. Doc. dr hab. Head

Status: Government institution.

Correspondence: Polish, English.

P o z n a ń

<u>Przemyslowy Instytut Maszyn Rolniczych/PIMR</u> (Institute of Agricultural Machinery)
ul. Starolęcka 31, 60-963 Poznań 11

Mielec, K.	Prof. Dr inz.	Director
Machowiak, W.	Doc. Mgr inz.	Deputy Director, Research
Timm, Z.	Dr inz.	Deputy Director, Technical activities
Baloniak, M.	Mgr inz.	Cultivating machinery and potato and beet harvesters
Cywiński, M.	Doc. Dr inz.	Measuring apparatus and measurements
Górski, A.	Doc. Dr inz.	Scientific and technical documentation and information.

Hajnowski, J.	Dr inz.	Equipment for animal breeding
Kiełpiński, Z.	Dr inz.	Technological research
Koczorowski, B.	Mgr inz.	Power and Transport
Liska, M.	Mgr inz.	Corn and forage harvesters and drying equipment
Lacki, J.	Mgr inz.	Machines for sowing, fertilizing and plant protection
Wize, R.	Doc. Dr	Quality testing and economics

Status: Governmental institution, established in 1946, supervised by the Ministry of Machinery Industry.

Activities: Research on the construction of new agricultural machines and implements of control, testing and measuring apparatus, of plant protection implements, domestic and breeding equipment - Research on technological problems, standardization, documentation and information on industries of agricultural machines - Construction of prototypes - Research, experiment and testing work on construction and utilization of agricultural machines - Exchange of experience and the development of agricultural engineering.

Publications: "Prace Przemysłowego Instytutu Maszyn Rolniczych", quarterly - "Przeglad Dokumentacyjny Maszyn Rolniczych", monthly.

Correspondence: Polish, English, French, German, Italian, Russian, Swedish.

Instytut Mechanizacji Rolnictwa, Akademia Rolnicza (Institute of Agriculture Mechanization, Academy of Agriculture)
ul. Wojska Polskiego 50, 60-627 Poznań   Tel.: 224-581

Woyke, W.	Doc. Dr inz.	Head of the Institute
Cegielski, B.	Dr inz.	Vice-Head of the Institute

Mechanization of Agriculture: Didactic Group
Kozicz, J.	Dr inz.	Head of the Group

Mechanization of Animal Production: Didactic Group
Woyke, W.	Doc. Dr inz.	Head of the Group

Theory of the Agricultural Machines: Didactic Group
Weres, S.	Assoc. Prof. Mgr	Head of the Group

Motorization of Agriculture: Didactic Group
Krugiełka, W.	Dr inz.	Head of the Group

Organization of the Repair Engineering: Didactic Group
Cegielski, B.	Dr inz.	Head of the Group

Exploitation of Farm Machinery: Didactic Group
Sęk, T.	Doc. Dr hab.	Head of the Group

Electrification of Agriculture: Didactic Group
Kalińska, H.	Dr inz.	Head of the Group

Status: Governmental institution supervised by the Ministry of Science, Universities and Technology.

Activities: Resident and non-resident studies. Average number of students: 1,500, awarded degree: engineer, master.

Research and Publications: Utilization of tractor power in typical field work - Influence of roots of some cultivated plants on the ploughing resistance - The effect of humus content in the resistance of ploughing soil - The straw harvesting by the selfloading trailers - The exploitation of milking machines - The influence of tractor wheels on the soil.

Correspondence: Polish, English, German, Russian.

Instytut Melioracji Rolnych i Leśnych Akademii Rolniczej w Poznaniu (Institute for Land and Forest Reclamation, Agricultural Academy in Poznań)
ul. Wojska Polskiego str. 71E, 60-625 Poznań

Kosturkiewicz, A.	Doc. Dr inz.	Director
Przybyła, C.	Dr inz.	Deputy Director

Agro-Meteorology

Plant Ecology

Hydrogeology and Hydromorphic Soils

Land and Water Management Research

Status: Established in 1962. Governmental institution, supervised by the Ministry of Education, Science and Technics.

Activities: Teaching: graduate study courses at the Faculty of Land Reclamation and Improvement, each year by 15 to 30 students, final degree "magister inzynier", equivalent to M.Sc.Agr. with specialization soil and water conservation engineering. Research is done on all problems connected with land and water management. Research is carried out on agrometeorology, agro-forest-hydrology, drainage, sprinkling and subirrigation, transport of water in soil, physical behaviour of soil-water system, soil moisture control, applied radioisotopic method for soil water control, interpretation of aerial photographs.

Correspondence: Polish, English, German, Russian.

Instytut Budownictwa Wodno-Melioracyjnego, Akademia Rolnicza w Poznaniu (Institute of Water Engineering, Agricultural University Poznań)
ul. Mazowiecka 26, 60-623 Poznań

Lewandowski, J.B.	Doc. Dr inz.	Director
Przedwojski, B.	Dr inz.	Deputy Director

Geotechnics

Surveying

Theory of structures and farm building

Water Engineering

Status: Established in October 1970, Governmental institution, supervised by the Ministry of Science, Higher Education and Technics.

Activities: a) Teaching: graduate study courses at the Faculty of Land Reclamation and Improvement, each year attended by 30 students - Final degree "magister inzynier wodnych melioracji", equivalent to M.Sc. with specialization water engineering for land reclamation. - b) Research: research is carried out on soil mechanics, foundation engineering, road engineering, cartography, detailed surveying, land surveying, photogrammetry, theory of structure, farm buildings, corrosion of building materials, hydraulics, control structures, river training, numerical methods in engineering.

Publications: on above subjects.

Correspondence: Polish, English, Russian, French, German.

Szczecin

Zakład Eksploatacji Sprzętu Rolniczego i Agrolotnictwa, Akademia Rolnicza w Szczecinie (The cathedral of farm machines exploitation and agricultural aviation operations, Agricultural University of Szczecin)
ul. Akademicka 1, 71-442 Szczecin
Tel.: 712-71
Telex : 0425494 ar pl

Michalski, M.	Asst.Prof.Ph.D., M. Eng.	Head
Gajewski, B.	Ph.D., M.Eng.	Ag-aviation expert
Sienkiewicz, J.	Ph.D., M.Eng.	Ag-aviation expert

and next 5 persons.

Status: Established December 1, 1974. Governmental institution supervised by the Ministry of Research, Universities and Technique.

Activities: (a) Teaching: Four and half years study courses leading to M.Sc., M.Eng. No pilot training. - (b) Research: Research in area of ag.aviation and especially; 1) designs of loading equipment both for dry and liquid materials (for fixed-wing and rotary-wing aircrafts); 2) optimalization of parameters of ag-aviation mission and ground works in the time of operational cycle; 3) adaptation of the farms to ag-aviation requirements; 4) designs of ag-aircraft home landing fields organization and operational landing fields localization; 5) different economical analysis for ag-aviation.

Publications: Among the more important, since 1977: Dispersal of low-rates of chemicals by helicopters - The operational problems in ag-aviation - Development of agricultural aviation - Development of agricultural aviation - Organization of labour during mineral fertilization by aid of helicopters - Influence of swath speed changes on preciseness of airborne agricultural application - The ag-copter basis for agricultural enterprises - Economical aspects of utilization of loading equipment for agricultural aircrafts - Guidelines for designing of loading equipment and choice of the most optimal loading equipment - Establishing of the turn time of the Mi2 helicopter in airborne agricultural operations - New mathematical formula for agricultural aviation operational cycle.

Correspondence: English, German.

Akademia Rolnicza - Instytut Gleboznawstwa i Godpodarki Wodnej Zakład Melioracji Rolnych (Agriculture Academy - Institute of Pedology and Water Management - Land Reclaiming Branch)
ul. Slowackiego 17   71434 Szczecin, Poland

Duda, L.	Prof.Dr habil inz. melioracje rolne (in Ag.Engg)
Zygas, M.	Dr inz. (in Ag.Engg)
Winkler, L.	Dr inz. " "

Status: Established 1956; Ministerstwo Nauki, Techniki, Szkolnictwa Wyzszego (Ministry of Science, Technics, Higher Schooling.

Activities: a) Lectures and training in land reclamation and in service and maintenance of land reclaiming installations. Number of students: 300. - b) Research on Economy of water resources, hydrology of small rivers, studies of the influence of agriculture intensification on charges of river flows and contamination of waters.

Publications: Prognostics on the flow of medium level waters in rivers. Hydrology of small rivers of the ungaged areas. Retention of lakes. Losses of soil chemical compounds in river basin.

Correspondence: Polish, Russian, German, English.

W a r s z a w a

Instytut Budownictwa Mechanizacji i Elektryfikacji Rolnictwa - IBMER (Institute for Buildings Mechanization and Electrification in Agriculture)
ul. Rakowiecka 32 -   02-532 Warszawa   Tel.: 49 32 31

Tymiński, J.	Assist.Prof.Dr	Director - Mechanization of animal production
Kowaliszyn, J.	Dr inz.	Deputy Director - Research, Assessing of farm equipment quality
Krzemiński, J.	Assist.Prof.Dr	Deputy Director - Research, Mechanization of crop production
Witebski, Z.	Assist.Prof.Dr	Deputy Director - Research, Farm building
Zaremba, W.	Prof.Dr	Deputy Director - Research, Operation of tractors and farm machinery
Antolak, T.	Dr inz.	Head of Department - Mechanization and organization of farm building
Bilowicki, J.	Assist.Prof.Dr	Head of Department - Agricultural progress extension
Bojańczych, J.	Assist.Prof.Dr	Head of Department - Organization of research
Fąfara, A.	Mgr inz.	Head of Department - Land reclamation, water and sewage management
Karwowski, T.	Prof.Dr	Head of Department - Mechanization of root crop cultivation and harvesting
Nieborowski, H.	Assist.Prof.Dr	Head of Department - Material engineering

Olszewski, T.	Dr inz.	Head of Department	– Machine operation
Pabis, J.	Assist.Prof.Dr	Head of Department	Mechanization of horticulture
Pabis, S.	Prof.Dr	Head of Department	– Cybernetics
Pawlik, A.	Assist.Prof.Dr	Head of Department	– Development of quality and reliability
Rauszer, Z.	Dr inz.	Head of Department	– Mechanization of livestock raising systems
Tomczych, S.	Assist.Prof.Dr	Head of Department	– Designing foundations for farm buildings
Wejher, A.	Dr inz.	Head of Department	– Electrification of agriculture
Wójcicki, Z.	Prof.Dr	Head of Department	– Management and economics of agricultural engineering

Branch Division of IBMER in Kludzienko:

Tabiszewski, A.	Dr inz.	Manager	
Kamiński, E.	Dr inz.	Head of Department	– Mechanization of tillage and fertilizing
Kijowska, W.	Dr inz.	Head of Department	– Central staff training service
Marszalek, T.	Dr inz.	Head of Department	– Measurement techniques
Mazur, J.	Dr inz.	Head of Department	– Motorization and transport
Roszkowski, A.	Dr inz.	Head of Department	– Mechanization of cultivation and harvesting of grain, green and technical crops
Szyszło, J.	Dr inz.	Head of Department	– Processing and storage of agricultural products

Branch Division of IBMER in Poznań-Strzeszyn: 60-463 Poznań ul. Biskupińska 67

Borowski, M.	Dr inz.	Manager	
Przygórzewski, S.	Assist.Prof.Dr	V ce Manager	
Eicke, W.	Mgr inz.	Head of Department	– Farm buildings mechanization
Kowalewski, Z.	Mgr inz.	Head of Department	– Feedlot mechanization for cattle
Pankowski, Z.	Dr inz.	Head of Department	– Farm devices (in agric.)
Pater, Z.	Dr inz.	Head of Department	– Feedlot mechanization for sheep and fur-bearing animals
Waligóra, T.	Dr inz.	Head of Department	– Feedlot mechanization for poultry

In all, 47 other scientific personnel work in the field of technical and agricultural sciences.

Regional status: 11 Machine Assessment Stations in the following places: Bonin, Grodków, Kąty Wrocławskie, Kraków, Krynica-Bradowiec, Lublin, Plock, Sanok, Stalowka Środa Wielkoposka, Tczew. A Machine Assessment Project in Brałystok.

Regional Centres: 9 Demonstration Stations for Farm Buildings placed in the following towns: Byałistok, Bydgoszcz, Kielce, Kraków, Łodz, Olsztyn, Szczecin, Wrocław, Zielona Góra. Demonstration projects for Farm Buildings in Koszalin and Lublin.

Status: Governmental institution established in 1950, supervised by the Ministry of Agriculture.

Activities: a) Professional training of the management and technical staff in agricultural engineering; Films as a research and extension method; Agricultural progress extension; recording, processing and spreading scientific, technical and economic information for the need of farm buildings, mechanization and electrification in agriculture; Contribution to patent protection; Contribution to initiate achieved findings and to provide instructions as well as technical assistance for initiating new material, Construction and production solutions. - b) New technical and technological developments for buildings and mechanization of agriculture; optimum ways for operation of power machinery for cultivation, transport and building purposes in agriculture; technical and operation trials in farm buildings sites and their involved designing; assessment of technical advance and potential for agriculture machinery equipment, materials, building units and constructions, produced domestically or imported; development of standards and unification or typification of projects for farm buildings, mechanization and electrification in agriculture.

Publications: Since 1973 there were issued 800 articles on research in the scope of farm building, crop production mechanization, animal production mechanization, operation of tractors and farm machinery, assessing of farm equipment quality and electricity applying in agriculture. - Publications in: CINTE Warszawa, PWRiL Warszawa, PWN Warszawa, ZW CRS Warszawa, IBMER Warszawa, plus several textbooks.

Correspondence: Polish, English, French, German, Russian.

Wydzial Techniki Rolniczej i Leśnej, Szkola Glówna Gospodarstwa Wiejskiego w Warszawie (Department of Agriculture and Forestry Engineering, Warsaw Agricultural University)
Instytut Mechanizacji Rolnictwa i Leśnictwa (Institute for Mechanization of Agriculture and Forestry)
ul. Nowoursynowska 166,   02-975 Warszawa   tel. 431876

Haman, J.S.	Prof.Dr	Dean of Faculty, Head of Institute - Physical properties of soil and plants
Bernacki, H.	Prof.Dr	Mechanization of horticulture
Bocheński, C.	Dr	Farm machinery maintenance
Dabkowski, W.	Dr	Tractors and power
Dmitrewski, J.	Doc.Dr	Mechanization of animal production
Fabirkiewicz, A.	Dr	Mechanization of animal production
Jakubiak, R.	Dr	Farm machinery construction
Kuczewski, J.	Prof.Dr	Farm machinery construction
Majewski, Z.	Dr	Soil cultivation
Majka, K.	Doc.Dr	Electrification of agriculture
Nowacki, T.	Prof. Dr	Mechanization of agriculture
Piotrowski, S.	Dr	Transport in agriculture
Skrobacki, A.	Dr	Tractors and power
Sosnowski, A.	Dr	Mechanization of horticulture
Waskiewicz, C.	Dr	Drying of Farm products
Zdun, K.	Doc.Dr	Mechanization of animal production

Status: Governmental Institution established in 1976, supervised by the Ministry of Science, High Education and Technology.

Activities: 4,5 year study leading to Master in Agriculture and Forestry Engineering - Postgraduate studies for Ph.D. in Agriculture and Forestry Engineering Research - Tractors and Power - Construction of agricultural and forestry machines - Economic aspects and forecasting of farm mechanization - Mechanization of field production - Mechanization of horticulture - Mechanization of animal production - Physical properties of soil and plant materials.

Publications: Several handbooks on farm mechanization - Scientific reports published in "Roczniki Nauk Rolniczych".

Correspondence: Polish, English, German, Russian.

Instytut Użytkowania Lasu i Inżynierii Leśnej, Skoła Główna Gospodarstwa Wiejskiego - Akademia Rolnicza (Institute of Forest Utilization and Forest Engineering, Agricultural University of Warsaw)
02-528 Warszawa, 26/30 Rakowiecka str.

Kamiński, E.	Prof.Dr hab.	Director, harvesting, logging, tapping
Laurow, Z.	Doc.Dr hab.	V-Director, harvesting, logging, grading of wood

Department of Logging and Wood Transport:

Laurow, Z.	Doc.Dr hab.	Head, logging, grading of wood
Głowacki, S.	Dr Eng.	Adiunct, minor forest products
Kosicki, K.	Dr Eng.	Adiunct, skidding, wood transport
Kowalski, J.	Dr Eng.	Adiunct, logging
Lenart, E.	Dr Eng.	Adiunct, skidding
Makała, S.	M.Sc., Eng.	Specialist, charcoal production
Paschalis, P.	Dr Eng.	Adiunct, wood science, grading of wood
Porter, B.	M.Sc., Eng.	Senior assistant, harvesting

Department of Forest Engineering and Protection of Work:

Józefaciuk, J.	Doc. Dr	Head, work protection
Kaczor, Z.	Dr Eng.	Adiunct, forest roads construction and projects
Pieńkos, K.	Dr Eng.	Adiunct, forest roads construction and projects
Schenke, P.	M.Sc.Eng.	Senior assistant, work protection
Szewczyk, J.	Dr Eng.	Adiunct, forest roads construction and projects

Laboratory of Forest Use for Recreation:

Giorliński, T.	Dr Eng.	Head, forest management
Rutkowska, J.	Eng.	Assistant, forestry

Branches with other location:
Laboratory of Forest Products Analysis at Rogów.
Road Laboratory at Rogów.
Center of Charcoal Production Analysis at Janów Lubelski.
Center of Charcoal Production Analysis at Moczarne.

Status: Government institution, established in 1969 (earlier two independent departments: "Forest Utilization" and "Forest Engineering" established 1920), supervised by the Ministry of Science, Higher Education and Technology.

Agricultural Engineering activities: a) Teaching: Normal study courses of forest utilization (harvesting, skidding, transporting of wood, wood science, grading of wood, minor forest products, forest engineering), construction of forest roads, small bridges, small buildings, forest amelioration (and Work Science), protection, physiology of work, ergonomy, sociology of work. Postgradual Technological Forest Study: technology of harvesting, skidding and transporting of wood, exploitation of forest roads. Number of students of intramural study (one course) ca 80-90, extra mural study ca 30, specializing for Master of Science Degree in Forest utilization ca 10 students and for Forest Engineering ca 10 students. Number of students of postgradual technological study ca 40. - b) Research: Properties of wood raw material, timber quality harvested from different site conditions, grading of logs, technology of harvesting, skidding, transporting and yarding of wood, tapping, quality of minor forest products, forest roads system, its characteristic and significance, forest road constructions and stabilization, organization of recreation in forest with special reference to the engineering equipment (small hats, parkings, etc.) protection of work in forestry, charcoal production in iron stove in forest conditions.

Publications: Standards of wood, grading, quality of wood, technology of wood exploitation in Poland and other countries (with special reference to the undeveloping countries), tapping, charcoal production in forest conditions, construction and stabilization of forest roads, work protection in forestry, forest as recreation object, importance of forests and proper system of wood harvesting and extracting in tropical and subtropical forests.

Correspondence: Polish, English, Russian, French.

Instytut Melioracji i Uzytków Zielonych (Institute for Land Reclamation and Grassland Farming)
Falenty n. Warsaw P.O. Raszyn 05-550

Land Reclamation

Hydraulic Constructions

Water Supply of Rural Settlements and Agriculture

Natural Conditions of Land Reclamation

Agricultural Utilization of Waste Waters

Grassland Farming

Land Reclamation Economics

Branches with other location: The Institute has 6 Branch Divisions in: Bydgpszcz – Elblag – Kraków – Lublin – Szczecin – Wroclaw. And 6 Experimental Stations at: Biebrza – Falenty – Jaworki – Kamieniec – Wroclawski – Leszkowice.

Status: State institution established in 1963, supervised by the Ministry of Agriculture.

Activities: Teaching: At the headquarters of the Institute a training centre exists for specializing staff in the construction, operation and design of water development and land reclamation. The research workers of the Institute take active part in lecturing and training. They also take part in teaching agricultural staff at various training centres all over the country. Some of the research workers with greater theoretical and practical experience are lecturers at various agricultural colleges. – Research: The Institute is carrying out research on land reclamation, water economy, hydrotechnical constructions for land reclamation purposes, grassland farming methods, management of reclaimed grasslands, agricultural utilization of peatland and peat. The research is being carried out both in field and laboratory.

Publications on above activities.

Correspondence: English, German, Russian, Polish.

Instytut Melioracji Rolnych i Leśnych, Akademia Rolnicza (Institute of Land and Forest Reclamation, Agricultural University)
02-975 Warszawa ul. Nowoursynowska 166

Branches: Agricultural Experimental Station at Puczniew, district Łódź.

Status: Established 1948, reorganized 1970. Governmental institution supervised by the Ministry of Science.

Activities: Study courses on land surveying, hydrology, river training, land and forest reclamation, agricultural use of sewage water, irrigation systems and water management as part of the regular 5-year course for Master Engineer degree. Postgraduate special courses for engineers on land drainage and irrigation, and additional studies, leading to Dr. degree. - Research: Methods of measurement and adjustment of geodetic networks. Water balances, especially groundwater discharge in river flow and runoff. Flood control. River training. Soil water relationships. Drainage of arable lands. Surface irrigation of grasslands. Agricultural use of sewage water in surface and sprinkling irrigation.

Publications: on above activities. Textbooks (in Polish): Drainage of grasslands - Irrigation principles - Hydrological basis for designing the land reclamation structures.

Correspondence: Polish, English, Russian, French.

Instytut Budownictwa Melioracyjnego i Rolniczego, Akademia Rolnicza (Department of Water and Agricultural Engineering, Agricultural University of Warsaw)
02-975 Warszawa, ul. Nowoursynowska 166

Status: Governmental institution, established in 1970, supervised by the Ministry of Science, Education and Technic. The Institute was created by joining: Chair of Soil Mechanics and Earth Structures, Chair of Hydraulic Engineering, Chair of Engineering Structures and Chair of Farm Building.

Activities: a) Teaching: Study courses on hydrology, hydraulics, soil mechanics, foundation engineering, earth structures, mechanization and organization of land reclamation works, hydraulic structures, small hydro-power plants, rural water supply and sewage systems, as part of the regular 5-year courses for Mgr. and 4-year courses for Inz. Additional postgraduate study. - b) Research: Goundwater movements influenced by water reservoirs, seepage of earth dams and dikes, quality control of earth works, river mechanics, channel and river sedimentation, reservoir sedimentation, prefabricated constructions in hydraulic structures, construction of waste lagoons, physical and mechanical properties of ashes, construction of earth dams, the use of finite elements method for calculating of pressure and strain in earth dams.

Publications on above activities. Textbooks in Polish.

Correspondence: Polish, English, French, German, Russian.

Wrocław

Instytut Mechanizacji Rolnictwa, Akademia Rolnicza (Institute of Farm Mechanization)
ul. Chelmońskiego 37/41   51-630 Wrocław

Bogdanowicz, J.	Prof.Dr hab inz.	Director
Drozd M.	Doc. Dr inz.	Deputy Director
Kamiński, E.	Doc. Dr hab inz.	Deputy Director

Agricultural Mechanization

Farm Machinery

Mechanization of Animal Production

Materials Engineering

Heat Technique and Drying of Agricultural Products

Farm Electrification

Motors and Tractors

Technology of Repairs

Branches with other location: None

Status: Governmental institution, established in 1950, supervised by the Ministry of Science, Schools of Academic Rank and Technics.

Agricultural Engineering activities: a) Teaching: Study courses of 4.5 years for Mgr inz. of Agricultural Mechanization with an average of 100 students per year. - Courses lasting one semester each year for students of Agronomy Faculty and Animal Production Faculty (500 per year). b) Research: Drying of agricultural crops - Bio-physical properties of agricultural materials - Materials Ingeneering (Tribology) - Agricultural transport - Technical diagnostics of tractors and machines - Agricultural economy - Theory and construction of machines - Reology of agriculture materials - Farm electrification - Dynamics of engines and tractors - Hydraulic systems of farm machines - Production economics.

Publications: Papers are published in: Zeszyty Problemowe Postępów Nauk Rolniczych - Biuletyn Informacyjny Instytutu Hodowli i Aklimatyzacji Roślin - Biuletyn Informacyjny Instytutu Budownictwa, Mechanizacji i Elektryfikacji Rolnictwa - Międzynarodowe Czasopismo Rolnicze - Mechanizacja Rolnictwa - Maszyny i Ciagniki Rolnicze - Nowe Rolnictwo - Przemysł Zbozowo Młynarski - Biuletyn Informacyjny Przemysłu Paszowego - Rolnictwo na Swiecie - Rolnik Dolnośląski- Zeszyty Naukowe Akademii Rolniczej we Wrocławiu - Roczniki Nauk Rolniczych, Seria C - Zagadnienia Eksploatacji Maszyn Polskiej Akademii Nauk.
And about 180 papers on maize, green forage and forest cones drying - grain drying theory - production economics - hydraulic systems of farm machines - dynamics of engins and tractors - farm electrification - reology of agriculture materials - theory and construction of machines - Agricultural economy - technical diagnostics of tractors and machines - agricultural transport - materials engineering (tribology) - bio-physical properties of agricultural materials. - Textbooks on: Heat technique for students of Agricultural Mechanization - Theory and Technology of Corn Drying - Exercises of Agricultural mechanization and electrification - Agricultural technology of machines works.

Correspondence: Polish, Russian, English, German, French.

Instytut Melioracji Rolnych i Leśnych, Wydział Melioracji Wodnych, Akademia Rolnicza
(Institute of Agricultural and Forest Improvement, Faculty of Land Reclamation and Improvement, Agricultural Academy)
ul. Plac Grundwaldzki 24, 50-363 Wrocław

Marcilonek, St.	Prof. Dr habil.	Director - Operation of land reclamation structures and systems
Szymański, J.	Prof. Dr habil.	Deputy Director - Land Reclamation
Bočko, J.	Prof. Dr habil.	Purification and agricultural utilization of sewage
Dejas, D.	Doc. Dr	Organization and technology of land reclamation works, pump stations
Kostrzewa, St.	Doc. Dr habil.	Drainage, land improvement of mountain areas

Status: Governmental institution, established in 1946, supervised by the Ministry of Science, Academy Education and Technics.

Activities: Teaching: $4\frac{1}{2}$ year courses for Mgr Eng. with about 50 students, and five-year courses (Extramural studies) for Mgr Eng. with about 40 students, one-year additional specialistic studies with about 50 students, on: Drainage and irrigation of agricultural lands - Soil conservation - Forest reclamation - Management of irrigation and Drainage systems - Agricultural utilization of sewage - Organization technology and mechanization of land reclamation works. - Research activities: Water balance of small basins - Water needs of agricultural crops - Evaporation of peat-soil - Sprinkler irrigation of grassland and arable land - Drainage of agricultural lands - Management of sprinkler irrigation systems - Groundwater flow of sewage irrigated land - Sewage purification in the soil - Organization and technology of drainage works - Land improvement of mountain areas.

Publications: Reports on the above in: "Zeszyty Naukowe Akademii Rolniczej we Wrocławiu - Melioracja; Wiadomości Melioracyjne i Łąkarskie". And textbooks in Polish.

Correspondence: Polish, English, Russian, German.

PORTUGAL

Lisboa

Comissão Nacional Portuguesa de Irrigação e Drenagem (Portuguese National Committee on Irrigation and Drainage)
Rua de S. Mamede ao Caldas, 23 Lisboa Codex (Portugal)

Ferreira, J.F. Faria President	Civil Engineer	General Director of Hydraulic Resources and Reclamation
Members:		
Oliveira, E.M. Lamas de	Civil Engineer	General Inspector of Public Works and Transports
Pereira, Luis Santos	Professor Civil Engineer	General Director of Hydraulic and Agricultural Engineering
Gonçalves, Adolfo	Civil Engineer	General Subdirector of Hydraulic Resources and Reclamation

Martins, C.C. de Azambuja	Agricultural Engineer	Adjunct General Director of Hydraulic Resources and Reclamation
Peça, C. de Oliveira	Civil Engineer	Superior Inspector of Public Works
Mendonça, P.V.M.	Professor, Agricultural Engineer	University of Agricultural Engineering
Rego, Z. de Castro	Professor, Agricultural Engineer	University of Agricultural Engineering
Manzanares, A.V.V.J.	Professor, Civil Engineer	University of Agricultural Engineering
Abecasis, F.M.A. Manzanares	Civil Engineer	Head of Hydraulic Service National Laboratory of Civil Engineering
Folque, J. de Brito	Civil Engineer	Researcher, National Laboratory of Civil Engineering
Raposo, J. Rasquilho	Professor, Agricultural Engineer	University of Agricultural Engineering
Figueira, M.	Agricultural Engineer	Caixa Geral de Depósitos
Curado, A. Martins	Agricultural Engineer	
Direito, F.J.T. Teixeira	Civil Engineer	Head of Projects Department, General Direction of Hydraulic Resources and Reclamation
Pote, F. Coelho	Civil Engineer	General Direction of Hydraulic Resources and Reclamation
Ferreira, A. César	Agricultural Engineer	General Direction of Hydraulic Resources and Reclamation
Santos, A. Lousada dos	Civil and Agricultural Engineer	Head of Hydrology Department of Hydraulic Resources and Reclamation
Lencastre, A.M.S. Coutinho de	Civil Engineer	Director of "Hidroprojecto" President of Association of Engineers
Vieira, A. Barrancos	Agricultural Engineer	Director of "APAGEL" Hydraulic Studies Cabinet
Franco, M. Macedo	Agricultural Engineer	"COBA"
Júnior, A.G. Santos	Professor, Agricultural Engineer	Academical Institute of Evora

Status: Established in 1954. The annual subscription for the International Commission on Irrigation and Drainage is paid by the Ministry of Foreign Afdairs. The President of the Committee is always the General Director of Hydraulic Resources and Reclamation, dependent upon the Ministry of Housing and Public Works.

Activities: Development and Implementation of Irrigation, Drainage, River Training and Flood Control Techniques.

Publications: The Committee cooperates in the publications of the International Commission on Irrigation and Drainage.

Correspondence: French.

Conselho Nacional da Agua (C.N.A.) (National Council for Water)
Rua das Pedras Negras, 16   1200 Lisboa

Estevão Mendonça Lamas de Oliveira   President - General Inspector

Members: 3

Main activities: Steering Agency for surface and underground water uses, either within planning or maintenance and exploitation. Under installation.

Status: affiliated to Ministry of Housing and Public Works.

Direcção Geral dos Recursos e Approveitamentos Hidraulicos (D.G.R.A.H.)(General Directorate of Hydraulic Resources and Reclamation)
Rua de S. Mamede (ao Caldas), 23 - 1196 Lisboa Codex

Faria Ferreira, J.F. General Director

Principal Members: Sub-General Director; Deputy General Director; Head of the Planning Office, Head of the Department for Hydraulic Works, Head of the Department of Hydrology, Head of the Department for Monitoring of Pollution, Heads of the Regional Departments.

Status: affiliated to Ministry of Housing and Public Works.

Main activities: Water management, Hydraulic plants planning, Study, Design and Works on Irrigation, Drainage, River Regulating and Flood Control, Monitoring of Pollution.

Direcção Geral de Hidráulica e Engenharia Agrícola (D.G.H.E.A.) (General Directorate of Hydraulics and Agricultural Engineering)
Rua Artilharia Um, 101 - 6º. Lisboa

Head: General Director

Principal Members: Two Sub-General Directors; Head of the Department of Projects and Works; Head of the Agricultural Hydraulics Department; Head of the Soils and Agro-Ecology Department; Head of the Department for Mechanization; Head of the Department of Construction and Rural Infrastructures; Director of the Support Office to Irrigated Perimeters; Director of the Office for Machinery Parks Management.

Status: affiliated to Ministry of Agriculture and Fishery.

Main activities: Management of Hydroagricultural projects; Appraisal and studies on projects of irrigation, drainage, defence, development and soil maintenance; Testing of mechanical equipment and legislation on agricultural mechanization; Studies leading into a better development of the agro-ecological field and soil maintenance; Management of irrigated perimeters and hydroagricultural management; Development, study and execution of equipment for agricultural enterprises within building, agricultural ways, rural electrification and collective equipment of rural communities; Management of industrial machinery for rural engineering.

Associação Portuguesa dos Recursos Hidricos - APRH (Portuguese Water Resources Association - National Laboratory of Civil Engineering)
Av. do Brasil, 101 - 1799 Lisboa Codex/Portugal

Head: President of the Executive Board

Principal Members: President of the General Assembly and General Council;
                               President of the Financial Council; President of the
                               Executive Board.

Status: affiliated to International Water Resources Association

Main activities: To further progress of knowledge on the National level, as well as studies and discussion of water resources problems, namely in the fields of water management, planning, development, administration, science, technology, research and education - To promote and give support to activities in order to obtain co-operation from the interested individuals and entities in creating the suitable institutional framework and means for solving national water resources problems - To support and participate in actions intended for the dissemination of the basic concepts of a suitable policy for water resources management in Portugal - To collaborate with similar foreign organizations and to promote Portuguese participation in International water resources programmes that may be of interest for Portugal.

Instituto Superior de Agronomia, Universidad Tecnica de Lisboa (College of Agriculture, Technical University of Lisbon)
Tapada de Ajuda, Lisboa

Activities: Study courses and research in soil conservation engineering, water resources development, irrigation and drainage, farm power and machinery, farm mechanization, rural electrification, farm buildings and work organization in agriculture.

Publications since 1962: on above activities.

Estaçao de Cultura Mecánica, Instituto Superior de Agronomia (Institute of Agricultural Engineering, College of Agriculture)
Tapada da Ajuda, Lisboa

Status: Governmental institution, established in 1911, reorganized in 1927, 1931 and 1968.

Activities: Experiments, testing and extension work.

Publications: on above activities.

Correspondence: Portuguese, Spanish, French, English.

Ordem dos Engenheiros   (Association of Engineers)
Av. António Augusto de Aguiar, 3 - D   1097 Lisboa Codex

Head: President

National Directing Council

Regional Directing Council from: North(Head Office in Oporto) - Centre (Head Office in Coimbra) - South (Head Office in Lisbon).

Status: The Association is a Free Institution.

Main activities: To promote Engineering Science and Technology that can further collective interests, while improving the standard of efficiency of its Members and dignifying their Professional ethics.

Evora

Instituto Universitário de Evora - I.U.E.   (Academical Institute of Evora)
Largo dos Colegiais - 7001 Evora Codex

Head: Rector
Principal Members: Head of the Department of Agricultural Engineering; Head of the Department of Agronomy; Head of the Department of Rural Architecture and Engineering; Head of the Department of Natural Sciences; Head of the Department of Hidrobionomics.

Status: affiliated to Ministry of Education

Main activities: Research on drainage, irrigation, soil conservation and agricultural systems; Land use planning; Survey and study of natural resources (Soil and Water); Biological Utilization of Waters.

Oeiras

Estaçao Agronomica Nacional - Oeiras   (National Agronomic Station)

Head
Principal Members: Head of the Soil Physics Section; Head of the Soil Technology Section; Head of the Land Drainage Laboratory.

Status: affiliated to Ministry of Agriculture and Fishery.

Main activities: Study of the water economy and efficiency in irrigation of important crops such as maize, rice, sugar beet, tomato - Mapping and reclamation of saline soils, namely marine soils (Tiosols, etc.) - Hydropedological characterization of soils needed for drawing up drainage projects of agricultural land.

Comissão Nacional Portuguesa de Engenharia Rural C.N.P.E.R.) (Portuguese National Commission for Rural Engineering)
Nova Oeiras, Bloco D 50, E   Oeiras

Head: President

Status: affiliated to Ministry of Economy.

Main activities: Liaison with the International Commission for Rural Engineering.

PUERTO RICO

Mayaguez

Departamento de Ingeniería Agrícola, Universidad de Puerto Rico - Recinto de Mayaguez (Agricultural Engineering Department, University of Puerto Rico - Mayaguez Campus)
College Station, Mayaguez, Puerto Rico   00708

Allison, W.F.	B.Sc. A.E., M.Sc. A.E.,	Professor
Collazo, A.	B.Sc. Agr., B.Sc. A.E., M.Sc. A.E.	Director, Services and Engineering Department
Colom-Lecároz, A.	B.Sc. A.E.	Drainage and Irrigation Extension Specialist
Colón-Martínez, V.A.	B.Sc. A.E., ?.Sc.	Farm Machinery Extension Specialist
Dávila, R.F.	B.S.A.E., M.S.A.E.	Instructor
Febre, A.C.	M.S.	Associate Professor
González-Corretjer, C.	B.Sc. C.E., B.Sc. A.E.	Farm Structures Extension Specialist
Lastra, J.	B.S. Agr.	Farm Machinery Extension Specialist
Otero-Dávila, R.	B.Sc. Agr., B.S.A.E.	Assistant Agricultural Engineer
Ravalo, E.J.	B.S.A.E., M.S.A.E.	Assistant Agricultural Engineer
Rivera-Negrón, F.	B.Sc. Agr., B.S.A.E.,	Assistant Agricultural Engineer
Rodríguez-Arias, J.H.	B.Sc..Agr., B.S.A.E., M.S.A.E., Ph.D.A.E.	Prof. Ad Honorem
Santaella-Pons, J.A.	B.Sc. Agr., M.S. Agr.	Irrigation Extension Specialist
Goyal, Megh	B.S.A.E., M.S.A.E., Ph. D.A.E.	Assistant Agricultural Engineer

Status: The University of Puerto Rico at Mayaguez is a government Land-Grant institution founded in 1911 under the name of College of Agriculture and Mechanic Arts, and according to a new University Law approved in 1966, presently organized as an autonomous campus comprising the Colleges of Agriculture, Engineering, and Arts and Sciences. The Department of Agricultural Engineering is part of the College of Agriculture. The University is governed by the Council of Higher Education consisting of the Commonwealth Secretary of Public Instruction (ex-officio member) and eight additional members appointed by the Governor with the advice and consent of the Senate of Puerto Rico. With the exception of the Secretary of Public Instruction they serve six-year terms.

Activities: Teaching: Study courses to agricultural students enrolled in 4-year curricula leading to a Bachelor of Science degree in Agriculture - Average number of students attending these courses vary from 180 per year for required courses (i.e.

Agricultural Power and Agricultural Machinery) to 20 for elective courses (i.e. Farm Drainage and Irrigation) - Administers a curricular option in Mechanical Technology in Agriculture with an average enrollment of 25 students - This latter programme emphasizes the practical application of engineering principles to the problems encountered in modern farming. A proposed 5-year professional curriculum leading to a Bachelor of Science degree in Agricultural Engineering, currently under consideration, is expected to be in force by academic year 1981-82. Short courses are also carried out aimed at giving soecialized practical training to non-degree students in pertinent phases of agricultural engineering. - Research: Development of rainfall deficiency index for agricultural regions in Puerto Rico - Monitoring ground water movement and salinity in the Lajas Valley of Puerto Rico - Study of mechanical harvesting efficiency of sugarcane - Systems analysis in mechanization of sugarcane production with emphasis in harvesting and transport operations - Evaluation of the bioclimate of Puerto Rico - Drip irrigation of tropical fruits and vegetables - Effect of irrigation flooding regime on rice yields - Tillage and cultural system for mechanized rice production - Test and evaluation of machinery for agricultural production in Puerto Rico especially on small hilly farms - Mechanization of tropical grasses for biomass production. - Extension: Power and machinery - Drainage and irrigation - Farm structures and environmental control - Farm electrification - Agricultural products processing.

Publications since 1965: Processing Arabica coffee by the wet method - Evaluation of performance of coffee harvesters - Effect of leaf loss during harvest on subsequent yield of coffee - Land engineering for growing and harvesting of sugar cane in the Central Coloso area of Puerto Rico - Water requirements of sugarcane under irrigation in Lajas Valley, Puerto Rico - A computer programme for selecting optimum systems for harvesting, loading and transportation of sugarcane - A psychometric approach to schedule irrigation - The effect of irrigation, harvest interval and nitrogen on the yield and nutrient composition of Napier grass - Two proposed rainfall characteristic indices suitable for agricultural planning in Puerto Rico - A differential psychro test of plant-soil-weather relationships and water use - The effect of meteorological conditions on the yield and nutrient composition of Napier grass - On the climate of Puerto Rico and related crop-weather research - Seasonal changes in soil and air temperature at three locations in Puerto Rico. Mechanical harvesting of tropical tubers, Equipment for mechanical harvest of tropical grasses for biomass production.

Correspondence: Spanish, English.

ROMANIA

București

Institutul de Cercetări pentru Mecanizarea Agriculturii, I.C.M.A. (Institute of Research, Design and Technological Engineering for the Mechanization of Agriculture) Bd Ion Inonescu de la Brad nr. 6, Bucahrest, Sector I - Băneasa

Abrihan, T.	Mech.Eng.	Director of the Institute Specialization: machine tools
Toma, D.	Prof. Dr.doc. Mech.Eng.	Deputy Director Specialization: agricultural machines
Petrescu, A.	Mech.Eng.	Deputy Director - Specialization: internal combustion engines

Vulpe, J.	Dr.Eng.	Deputy Director - agricultural machines
Stănilă, T.	Dr.Eng.	Director, Experimental Station for the mechanization of agricultural works on slopes - Cluj
Tomescu, D.	Dr.Eng.	Agricultural machines
Mariş, I.	Mech.Eng.	Agricultural machines
Popescu, G.	Mech.Eng.	Agricultural machines
Mitrescu, C.	Mech.Eng.	Agricultural machines
Săplăcan, L.	Dr.Eng.	Agricultural machines
Raileanu, G.	Mech.Eng.	Agricultural machines
Raba, T.	Mech.Eng.	Agricultural machines
Niţu, C.	Mech.Eng.	Agricultural machines
Stoicescu, G.	Dr.Eng.	Agricultural machines

Branches: Research groups in the field of mechanization of agriculture which are carrying out their activity in the research institutes belonging to the Academy of Agricultural and Forestry Sciences. The experimental station for the mechanization of agricultural work on slopes Cluj.

Structure: The institute was established in 1979 as a result of the unification of the Research Institute for the Mechanization of Agriculture (established 1952) and the Research and Design Institute for Agricultural Engineering (established 1960). The institute is subordinated to the General Economic Direction for Mechanization of Agriculture in the Ministry for Agriculture and Food.

Activities: Research, design and technological engineering institute for the mechanization of agriculture is concerned with the following main activities: technical-economic studies and research with a view of establishing tractor and machines systems needed for new technologies applying to crop cultivation and animal husbandry - technological engineering studies in order to elaborate new mechanization technologies for crop cultivation and animal husbandry - optimization studies concerning efficient use of power means, tractor and agricultural machines repair, maintenance technologies - research in view of elaborating work rate and energy, fuel and spare parts consumption norms - studies aiming at improving technical indices of tractors and agricultural machines from serial production - studies on future development in the field of mechanization of agriculture in accordance with improved technologies in field crops and animal husbandry - elaborating technical projects and machines, experimental models in order to verify research results in operation conditions - assure, together with other Romanian firms the delivery, assembling and technical assistance for cereal silos and cereal mills - elaboration of designs for prototypes: execution, testing and proving of prototypes in view of their finishing off and introduction into serial production.

Publications: Scientific work of the Institute containing the results of research and experiment studies in the field of agricultural mechanization. Key words: study, research work, experiment, machine approval, tractor, agricultural machine, experimental model, mechanization technology, repairs, maintenance, reliability, operation, work rate, consumption rate.

Correspondence: Romanian, Russian, French, English, German.

Institutul de cercetari hidrotehnice (ICH) (Hydraulic Engineering Research Institute)
Spl. Independenţei 294, 77749 Bucureşti

Hâncu, S.	Prof.Dr.doc.Eng.	Directeur de l'Institut
Jeler, V.	Dr.Eng.	Directeur adjoint scientifique
Bally, R.J.	Dr.Eng.	Chercheur principal
Celan, B.	Dr.Eng.	" "
Comşa, R.S.	Dr.Eng.	" "
Danchiv, A.	Dr.Eng.	" "
Ivan, C.	Dr.Eng.	" "
Kellner, L-M.	Dr.Eng.	" "
Moroianu, A.	Dr.Eng.	" "
Oncescu, V.	Dr.Eng.	" "
Perlea, V.	Dr.Eng.	" "
Pietraru, J.	Dr.Eng.	" "
Platagea, G.	Dr.Eng.	" "
Popescu, M.	Dr.Eng.	" "
Spătaru, A.	Dr.Eng.	" "
Zaharescu, E.	Dr.Eng.	" "

Structure: Institution fondée en 1957, autofinancement, coordonnée par l'Institut Central de Recherches, Projection et Direction en Constructions.

Activités: Etudes et recherches dans le domaine de l'hydraulique des systèmes, des travaux et des installations hydrauliques pour différents usages, des structures hydrauliques et de la mécanique des sols, ainsi que de la technique d'exécution des travaux hydrauliques.

Publications: "Etudes d'hydraulique" et "Etudes et recherches de constructions hydrotechniques, mécanique des sols et des roches", édités par l'Institut.

Correspondance: Français, anglais, italien.

Institut de studii si projectári i agricole (ISPA) (Institut d'Etudes et de Projections Agricoles)
B-dul Expozitiei nr. 1, raionul 30 Decembrie, Bucureşti

Station affiliée à Braila dans la région Galati.

Structure: Institution fondée en 1950, contrôlée et financée par le Département des entreprises agricoles de l'Etat du Conseil Supérieur de l'Agriculture.

Activités: ISPA est un institut de recherches et projections, qui élabore les documentations techniques en vue d'organiser et d'aménager le territoire agricole, ainsi que les constructions du secteur agricole.

Correspondance: Roumain, anglais, français, russe.

Institutul de Geodezie, fotogrammetrie, cartografie şi organizarea territoriului
(The Institute for Geodesy, Photogrammetry, Cartography and Land Management)
Bd. Expozitiei nr. 1 A, Sector 1, Bucharest

Status: Established in 1954, as "The Institute for Studies and Design of Land Management". Since 1971 as the Institute for Geodesy, Photogrammetry, Cartography and Land Management. National Institute co-ordinated by the Ministry of Agriculture, Food Industry and Waters.

Activities: Researches, studies, projects as regards: land management for each district and commune, natural-economic zones (hydromeliorative systems, hydrographic basis, etc.), land consolidation (for each district, hydrographic basin, hydromeliorative system, etc.); studies of land management and agricultural output development for each natural-economic zone, hydromeliorative system, state farm and collective farm; location studies; feasibility studies; execution studies for organizing vineyard orchard, natural and cultivated pasture lands, a.s.o. Studies and researches for geodesy, photogrammetry, topography and making plans and maps, necessary to the land reclamation, planning and management works etc.

Publications: on above activities.

Correspondence: Romanian, French.

Instituttul de cercetări şi proiectări pentru gospodărirea apelor (Institut de Recherches et Projections d'Aménagement des Eaux)
Splaiul Independenţei nr. 294, Sector 1, Bucureşti, România

Institutul de meteorologie şi hidrogeologie (Institut de Météorologie et Hydrogéologie)
Şoseaua Bucureşti - Ploeşti nr. 97, Sector I, Bucureşti

Staţiunea Centrală de cercetări pentru ameliorarea solurilor săraturate (Station Centrale de Recherches pour l'Amélioration des Sols Salins)
Şoseaua Vizirului Km 9, Brăila, Judeţul Brăila, România

Staţiunea centrală de cercetări pentru combaterea eroziunii solului-Perieni
(Station Centrale de Recherches pour la lutte contre l'Erosion du sol-Perieni)
Comuna Perieni, Judeţul Vaslui, România

Directia generală economică de îmbunătăţri funciare şi construcţii agricole
(Direction Générale Economique des Améliorations Foncières et Constructions Agricoles)
Şoseaua Olteniţei nr. 35-37, Sector 4, Bucureşti

Institutul de cercetări şi inginerie tehnologică pentru irgaţii şi drenaje
(Institut de Recherches et Génie Technologique pour Irrigations et Drainages)
Comuna Băneasa Giurgiu, Judeţul Ilfov, România

RWANDA

Kigali

Service du Génie Rural-Hydrologie (Service of Agricultural Engineering and Hydrology)
B.P. 323, Kigali

Division Hydrologique
Bureau d'Etudes Aménagement Hydroagricole
Section Topographique
Bureau des Technologies Agricoles

Annexes: Dans les différentes Préfectures il existe des observateurs du Génie Rural chargés des mesures hydrologiques.

Structure: Le Service est une direction du Ministère de l'Agriculture et de l'Elevage dépendant de la Direction Générale de l'Agriculture - Le financement du Service est national avec l'aide en personnel et pour certains moyens des assistances techniques étrangères.

Activités: Recherches hydrologiques - Réseau national totalement reconstitué en 1972 - analyse des ruissellements et des modules - Aménagements des marais - Alimentations en eau - Industries Agricoles et Alimentaires.

Publications: Annuaire hydrologique de la République Rwandaise - Projets d'aménagement hydro-agricole ou industriel - Etudes et réalisation d'usines - Nombreuses notes hydro-climatologiques.

Université Nationale du Rwanda (National University of Rwanda)
B.P. 117 Butane   Tél.: 271/2/3

SAUDI ARABIA

Riyadh

Agricultural Engineering Division, Agricultural Research and Development Department, Ministry of Agriculture and Water
Riyadh

Status: Governmental institution directed and financed by the Government.

Activities: Facilitating the introduction and use of machinery by farmers - Technical supervision of mechanized farms - Development and modification of equipment to suit local needs - Programming special mechanization projects for new crop production - Studying the possibility of initiating test facilities for new machinery and implements - Studying the possibility of modifying presently used irrigation methods and means of water control – Co-operating with the Agricultural Bank in the farm machinery subsidy programme – Offering technical consultations regarding training and maintenance programmes for other departments – Control of the importation and pricing of agricultural machinery and spare parts.

Correspondence: Arabic, English.

Faculty of Agriculture – Faculty of Engineering – University of Riyadh
P.O.B. 2454
Tel. 4769345
Telex 201019 R UNIV SJ
Correspondence: Arabic, English

Faculty of Engineering – King Abdulaziz University
P.O.B. 1540 Jeddah
Tel.: 6879033
Correspondence: Arabic, English

Faculty of Agriculture – King Faisal University
Damman – P.O.B. 1982

SENEGAL

Bambari

Station de Recherches de l'Institut de Recherches du Coton et des Textiles Exotiques (I.R.C.T.) (Research Station of the Research Institute for Cotton and Tropical Textiles)
B.P. 17, Bambari

Structure: Organisme privé de coopération technique internationale – Direction générale: 34 rue Des Renaudes, Paris 17e.

Activité en matière de génie rural: Mécanisation de la culture cotonnière – Conservation des sols en fonction de la culture cotonnière.

Publications: Divers rapports et notes dans la revue "Coton et Fibres Tropicales", IRCT, 34 rue Des Renaudes, Paris 17e.

Correspondance: Français, anglais, espagnol.

Bambey

Centre National de la Recherche Agronomique de Bambey (CNRA), Institut de Recherches Agronomiques et des Cultures Vivrières du Sénégal (IRAT-Sénégal) (Agricultural Research Centre Bambey of the Institute for Tropical Agriculture and Food Crops)
Bambey

Division du Machinisme Agricole

Station Régionale de Richard-Toll à Fleuve

Station Régionale de Séfa à Casamance

Structure: L'IRAT-Sénégal dépend du Ministère de l'Economie Rurale et est géré par l'IRAT-France, 110 rue de l'Université, Paris 7e.

Activité en matière de machinisme agricole: Culture attelée et motorisée - Essais de matériels agricoles - Adaptation de prototypes: définition de matériels répondant aux besoins propres des différentes régions - Interactions machinisme-milieu, en particulier avec le sol - Interaction de la machine dans les techniques culturales.

Correspondance: Français, anglais.

Dakar

Direction du Génie Rural et de l'Hydraulique Rurale (Agricultural Engineering and Hydrology Service)
Route des Mères Maristes, Hann, Dakar

Gaye, M.D.   Ingénieur   Directeur

Annexes: Huit. Inspections du Génie Rural, c'est-à-dire: du Cap-Vert à Dakar - de la Casamance à Ziguinohor - de Dioubel à Diourbel - du Fleuve à Saint-Luis - du Sénégal Oriental à Tambacounda - du Sine-Salou à Kaolack - de Thiés à Thiés - de Louga à Louga.

Structure: Le Service du Génie Rural du Sénégal est une institution gouvernementale qui a été créée en 1957 - Il est placé sous l'autorité du Ministère de l'Equipement du Sénégal.

Activités: Le service du Génie Rural est chargé de l'ensemble des actions d'équipement ayant trait au milieu rural - Il contrôle l'exécution des travaux ruraux d'aménagement et d'infrastructure confiés à l'entreprise.

Publications: Rapports sur les projets d'aménagement locaux.

Section Génie Rural, Ecole Nationale des Travaux Publics et du Bâtiment (Agricultural Engineering Section, National School of Public Works and Buildings)
B.P. 4004, Dakar

Structure: La section est complètement intégrée à l'Ecole Nationale des Travaux Publics et du Bâtiment qui est un établissement d'enseignement supérieur professionnel court.

Activités concernant le Génie Rural: L'Ecole recrute sur titre des bacheliers - Formation en 2 années scolaires normales à raison de 32 H. de cours par semaine plus de 8 H. de visites, manipulation, exercises - Examen de sortie en fin de 2e année et stage - Nombre d'élèves par classe: 10 en moyenne - L'Ecole décerne un diplôme de fin d'études, et les anciens élèves sont en principe intégrés dans l'Administration avec le titre d'Ingénieur des Travaux du Génie Rural.

Correspondance: Français.

Institut Sénégalais de Recherches Agricoles (ISRA) (Senegalese Institute for Agricultural Research)
B.P. 3120, Dakar

Université de Dakar (University of Dakar)
Dakar-Fann
Correspondence: French

Tel.: 250530
Telex: 262

SIERRA LEONE

Freetown

University of Sierra Leone
Private Mail Bag, Freetown

Fourah Bay College
P.O.B. 87   Freetown

Tel.: 27260

Correspondence: English

SOMALIA

Mogadishu

Agricultural Engineering Section, Department of Agriculture, Ministry of Agriculture and Animal Husbandry
Mogadishu

Activities: This Section is responsible for farm equipment and the Ministry's transports.

Faculty of Agriculture - Faculty of Engineering - Somali National University
P.O.B. 15 Mogadishu

Tel.: 25035/40042

SOUTH AFRICA

Cape Province

Afdeling Landbou-ingenieurswese (Division of Agricultural Engineering)
Privaatsak X515 (Private Bag X515) Silverton -)127 Republiek van Suid-Afrika
(Republic of South Africa)

Bruwer, J.J.	Pr.Eng., MSc Eng	Director - Agricultural energy and mechanization, soil and water
Boshoff, B.v.D.	Pr.Eng., Ph D	Chief Eng. - Agricultural mechanization, tillage and planting equipment
Crosby, C.T.	Pr.Eng., BSc (Agric. Eng)	Chief Eng. - Agricultural machinery, irrigation and drainage, soil conservation
De Witt, P.G.	BSc Eng. (Agric.) Engineer	Engineer - Mechanized irrigation, fertilizer application, pumping systems

Name	Qualifications	Position and Field
Du Plessis, J.B.	Pr.Eng., BSc (Agric. Eng)	Engineer – Tractor power testing, tractor safety frame testing, liaison tractor manufacturers
Fuls, J.	Pr.Eng., BSc Eng. (Mech)	Principal Eng. – Agricultural energy
Glatthaar, J.O.	Pr.Eng., BSc Eng. (Civil)	Principal Eng. – Grainsilos, technical advisor to Grainsilo Committee
Grobler, J.H.	Pr.Eng., BSc Eng.(Hons) (Agric)	Engineer – Mechanization planning, tillage equipment, tractor performance
Hawkins, C.S.	Pr.Eng., BSc Eng. (Mech)	Principal Eng. – Agricultural energy
Henning, P.A.	Pr.Eng., BSc Eng. (Agric)	Principal Eng. – Farm buildings, animal housing, agricultural structures
Heyns, A.J.	Pr.Eng., BSc Eng.(Hons) (Agric)	Principal Eng. – Mechanization planning, tillage equipment
Horvath, J.A.	Pr.Eng., Dipl Eng. (Mech) (Ctuj)	Principal Eng. – Feed processing, producer gas as substitute energy, animal housing equipment
Hugo, F.J.C.	Pr.Eng., MSc (Agric. Eng)	Asst. Chief Eng. – Agricultural energy, agricultural sprayers
Louw, A.A.	BSc Eng. (Agric)	Engineer – Irrigation, agricultural energy
McPhee, P.J.	Pr.Eng., BSc Eng. (Agric)	Principal Eng. – Soil conservation research, rainfall simulation, irrigation
Reid, N.W.	B. Eng., (Agric)	Engineer – Animal housing, animal waste management
Reinders, F.B.	BSc Eng. (Civil)	Engineer – Irrigation and drainage, glasshouses
Struck, H.W.P.W.	Pr.Eng., (Agric Eng)	Asst. Chief Eng. – Soil conservation
Van Niekerk, A.S.	BSc Eng., (Agric)	Engineer – Irrigation, filtering of irrigation water
Van Schalkwyk, C.J.	Pr. Eng., BSc (Agric Eng)	Asst.Chief Eng. – Agricultural Engineering extension, liaison with agricultural regional offices
Van Staden, J.F.	Pr.Eng., BSc (Agric. Eng)	Asst. Chief Eng. – Irrigation and drainage, glasshouses
Veenstra, J.	Pr.Eng., MSc Eng (Agric)	Asst.Chief Eng. – Agricultural mechanization, milking machines, farm dairy buildings
Vorster, A.	BSc Eng. (Agric)	Engineer – Agricultural structures, glasshouses, poultry housing

Winter Rainfall Complex (Stellenbosch): Private Bag X5013, Stellenbosch 7600

De Weerdt, S.J.P.	Pr.Eng., BSc.B Eng (Mech)	Principal Eng. - Agricultural mechanization, spraying equipment
Meyer, A.	Pr.Eng., B Eng. (Civil)	Engineer - Irrigation and drainage
Scheepers, I.	Pr. Eng., MSc Eng. (Civil)	Asst. Chief Eng. - Soil conservation, irrigation and drainage
Snyders, H.J.	Pr.Eng., BSc. B Eng. (Mech)	Principal Eng. - Wine cellar design and wine making machinery
Venter, J.C.	BSc.B Eng. (Civil)	Engineer - Animal housing and waste management

Natal Regional Office (Cedara, Pietermaritzburg): P/Bag X9059, Natal

Scott, J.J.	BSc.(Agric Eng)	Engineer - Soil conservation
Smithen, A.A.	MSc.(Agric Eng)	Engineer - Soil conservation
Whittal, W.A.C.	Pr.Eng., BSc.(Agric Eng)	Engineer - Agricultural mechanization

Subdivision Soil Conservation Works (Aliwal North):

Geertsema, F. de K.	Pr.Eng., BSc (Agric Eng)	Principal Eng. - Soil conservation
Jacobsz, S.W.	Pr.Eng., BSc (Agric Eng)	Principal Eng. - Soil conservation
Pienaar, P.J.	Pr.Eng., BSc (Agric Eng)	Asst. Chief Eng. - Soil conservation
Smit, J.C.L.	B Eng. (Civil)	Engineer - Soil conservation
Strümpher, P.J.	Pr.Eng., BSc. B Eng (Civil)	Engineer - Soil conservation

Highveld Regional Office (Potchefstroom):

Greeff, M.D.J.	Pr.Eng., BsC Eng	Principal Eng. - Agricultural tillage machinery
Kleinhans, J.F.	Pr.Eng., BSc (Agric Eng)	Principal Eng. - Soil conservation
Van Staden, H.J.	Pr.Eng., BSc (Agric Eng)	Principal Eng. - Soil conservation

Orange Free State Regional Office (Glen, Bloemfontein): P.O. Glen, 9360

Du Rand, D.J.	Pr.Eng., BSc Eng. (Agric)	Principal Eng. - Soil conservation
Kruger, G.H.J.	Eng BSc Eng (Agric)	Engineer - Irrigation and drainage

A total number of seventy one engineering posts exist within the entire Division (including all branch offices). Twelve of these posts are presently vacant. Engineers are assisted by engineering technicians for whom sixty two posts are available of which twenty one are presently vacant.

Branch offices: Eight Regional Branch Offices of the Head Office at Silverton, near Pretoria, (Transvaal Province) exist as follows:

(i) Subdivision - Winter Rainfall Complex (Western and Southern Cape Province) at Stellenbosch near Cape Town.

(ii) Subdivision - Soil Conservation Works (North Eastern Cape Province) at Aliwal North.

(iii) Natal Regional Office (Natal Province) at Cedara near Pietermaritzburg.

(iv) Highveld Regional Office (Transvaal Province) at Potchefstroom, Western Transvaal.

(v) Karoo Regional Office (Cape Province) at Middelburg.

(vi) Transvaal Regional Office (Transvaal Province) at Pretoria.

(vii) East Cape Regional Office (Eastern Cape Province) at Dohne near Queenstown.

(viii) Free State Regional Office (Orange Free State Province) at Glen near Bloemfontein.

Status: The Division of Agricultural Engineering is a governmental institution under the State Department of Agriculture and Fisheries and is directed and financed by state funds provided by that Department. The Division was established in 1961. It is the main co-ordinating body for all agricultural engineering research, development, testing and advisory work of the various other institutes, divisions and agricultural regional offices of the Department of Agriculture and Fisheries. Eight Regional Branch Offices exist, some of them specialising in certain fields of agricultural engineering important in those regions.

Agricultural Engineering activities: (a) Teaching: The Division of Agricultural Engineering is not a teaching institution as its main task is research, development, testing and advisory work in the field of agricultural engineering. It does however provide opportunities for practical training of agricultural engineering technicians in addition to their theoretical training at various Technikons in South Africa.
(b) Research: Research, development, testing and advisory work is conducted in the following fields: (i) Agricultural energy: Alternative fuels, sunflower oil as replacement or extender of diesel fuel, solar power for agricultural use, electricity, biogas. (ii) Agricultural mechanization: Planning of mechanization systems, official testing of tractors and machinery for animal and plant production, tractor traction studies, processing of agricultural products, grain drying, hay and silage machinery, feed processing, machinery for wine and fruit production. (iii) Irrigation and

drainage: Design of irrigation systems, flood, furrow, drip, sprinkler and micro irrigation and related equipment, official testing of irrigation pumps, drippers, sprinklers; planning and design of drainage systems. (iv) Agricultural structures and related facilities: Design and planning of farm buildings, animal housing, handling facilities, feed structures, animal waste handling systems, glasshouses, storage facilities for agricultural products, engineering control of grain silo design and construction, erection of game proof fencing. (v) Soil conservation: Design, planning and construction of soil conservation works, studies on soil erosion, rainfall intensity, silt gradients, rainfall simulator and wind erosion, engineering measures to prevent erosion.

Publications: Since 1973 a wide range of test reports and technical bulletins were published on most of the subjects listed under points (i) to (v). This included a series of 26 bulletins on fuel conservation (1979). – Since 1977 the Division also started publishing Comparative Reviews of Technical Specifications of agricultural machinery.

Correspondence: Afrikaans, English.

Natal

Department of Agricultural Engineering, University of Natal
P.O. Box 375, Pietermaritzburg, 3200   Tel.: 42851

Meiring, P.	Pr.Eng., B.Sc.Agric.Eng., M.Sc.Eng., F.S.A.I.A.E.	Professor and Head of Department, Power and Machinery, Mechanization, Farm Structures and Buildings
Lyne, P.W.L.	Pr.Eng., B.Sc.Eng(Agric), M.Sc.Eng., F.S.A.I.A.E.	Senior Lecturer, Power and Machinery, Mechanization, Farm Structures and Buildings
Dent, M.C.	Pr.Eng., B.Sc.Eng(Agric), M.Sc.Eng., M.S.A.I.A.E.	Lecturer, Mechanization planning, Soil and Water Engineering, Farm Structures and Buildings
Green, J.E.P.	Pr.Eng., B.Sc.Agric.Eng., F.S.A.I.A.E.	Lecturer, Soil and Water Engineering, Farm Structures and Buildings
Schulze, R.E.	M.Sc., Ph.D., U.E.D.	Senior Research Fellow, Hydrology, Soil and Water Engineering

Status: University department financed by University of Natal. Research projects supported by the Department of Agriculture and Fisheries, Council for Scientific and Industrial Research, Water Research Commission and numerous private concerns such as tractor, type and fuel companies.

Activities: Teaching: Study courses in basic agricultural engineering are provided for students studying for the degree of B.Sc.Agric. as well as courses for students taking the professional degree of B.Sc.Eng.(Agric.). An average number of 80 students attend the B.Sc.Agric. courses, with an average number of 40 students studying for the B.Sc. Eng.(Agric.) degree – Postgraduate degrees of M.Sc.Eng., Ph.D. and D.Sc. are offered. – Research: Research has been concentrated on tractor engine and traction performance; diesel fuel injection systems, diesel fuel extenders; small agricultural catchments hydrology in particular runoff modelling; aspects of the hydrology and climatology of Natal; establishment of a data bank of digitized rainfall and runoff data for South

Africa; soil loss estimation techniques and prediction of irrigation requirements; design criteria for design of multiple outlet low pressure irrigation.

Publications: Mechanization - tractor engine power and performance, tractor traction performance, type performance, performance transducers, research instrumentation, diesel injection systems, mechanization planning. - Hydrology: Land use planning, climatology, flood modelling, water resources planning, irrigation requirements estimation, soil loss modelling, data bank of digitized autographic records, flood studies, veld burning and soil moisture studies - Structures: silos, plastic silo covers, water tight jointing in precast concrete, mechanized silage production.

Correspondence: English, Afrikaans.

Orange Free State

Department of Agricultural Engineering, University of the O.F.S.
P.O. Box 339, Bloemfontein

Status: The Faculty of Agriculture was established in 1958 with a one-man department of Agricultural Engineering. The status of the Faculty of Agriculture is on par with all other faculties at this university which is a semi-governmental institution.

Activities: The main task of this department is teaching of ancillary subjects to agricultural students for the B.Sc.Agric. (4 years) course and B. Agric. (3 years) course.

Transvaal Province

University of Pretoria - Department of Agricultural Engineering, Faculty of Engineering and Agriculture
Hillcrest, Pretoria, Transvaal

Grobler, B.J.G.W.	Pr.Eng. D.Sc. (Agric.) Eng.	Professor - Head of Department Farm Mecahnization, Structures, Machine design and development
Du Plessis, H.L.M.	Pr.Eng. M.Sc.Eng. (Agric.)	Senior Lecturer - Farm Mechanization, Power Units, Soil dynamics
Smit, C.J.	Pr.Eng. B.Sc. (Agric.)	Senior Lecturer - Soil conservation, Hydrology, Farm Buildings, Irrigation
De Beer, R.W.	B.Sc.Eng(Agric.)	Lecturer - Thermodynamics, Cold storage fruit, Heat transfer, Fans and drying
De Kock, L.D.N.	B.Eng.(Agric)	Lecturer - Farm machinery (Processing), Farm mechanization, Irrigation, Farm Buildings
Wasserman, W.	Senior Tech. Gr I)	Instrument and model making, Testing,
Wasserman, J.	" " Gr II)	Practical demonstration
	)	Research Assistants

Status: Established 1934. Under semi-governmental direction and financed by Governmental and Private Institutions.

Agricultural Engineering activities: (a) Teaching: Four year course leading to the degree of B.Eng.(Agric.) - further study lead to: B.Eng.(Honours)(Agric.Eng.); M.Eng.(Agric.Eng.); D.Eng.(Agric.Eng.) - Average number of students: 80 to 90. - (b) Research Activities: Research in various fields of agricultural engineering viz. Research and development of: Alfalfa high protein - Low protein separator - Soil tillage machines - Optimum liter-per-hectare indicator for tractors - Utilization of sun energy - Various other projects in connection with the energy problem - Minimum tillage - Development of an electronic grading facility for orchid fruits.

Publications: Pneumatic planter development - Direct heat from wind energy - Head balance in drumdryers - Roller hay wafering - Gypson application in sprinkler irrigation.

Correspondence: English, Afrikaans, Dutch, German.

S P A I N

M a d r i d

Dirección General de la Producción Agraria, Ministerio de Agricultura, Sección de Maquinaria y Medios Auxiliares (General Bureau of Agricultural Production, Section Machinery and Auxiliary Means)
Paseo de Infanta Isabel, 1 - Madrid

Miguel Díez, A.	Dr.Ing.Agr.	Jefe del Departamento
Arenillas Asín, A.	Dr.Ing.Agr.	
Moro Serrano, R.	Dr.Ing.Agr.	
Enebral Casares, M.	Dr.Ing.Agr.	
Matamoros Sánchez-Capuchinos, E.	Dr.Ing.Agr.	

Actividades en ingeniería agrícola: Concursos internacionales de maquinaria agrícola (azúcar de remolacha, recolección de aceitunas, leguminosas), estadísticas de la maquinaria agrícola nacional, distribución de combustible agrícola, asistencia a los labradores.

Publicaciones: Mercados de maquinaria agrícola, estadísticas, homologación de la potencia de los tractores, desarrollo de la mecanización agrícola en el extranjero.

Correspondencia: Español, francés, inglés.

Sección de Conservación de Suelos del Instituto Nacional para la Conservación de la Naturaleza (ICONA) (Soil Conservation Section, National Institute for Nature Conservation)
Gran Vía de San Francisco, 35 - Madrid 5

Aguiló Bonnin, J.	Dr.Ing.Agrónomo	Ingeniero Jefe
Botella Fuster, E.	Dr.Ing.Agr.	
Aparicio Santos, J.	Dr.Ing.Agr.	
Magister Hafner, M.	Dr.Ing.Agr.	

Carácter: Brigadas, compuestas de un Ingeniero Agrónomo uno o dos Peritos Agrícolas en las capitales de provincias en Madrid, - Zaragoza, Valladolid, Badajoz, Huelva, Sevilla, Cadiz, Granada, Córdoba, Jaén, Almeria, Albacete, Castellón, Cuenca, Málaga, Murcia y Tarragona. - Por la Ley de Conservación de Suelos y Mejora de su fertilidad, de 20 de julio de 1955, fue costituido el Servicio de Conservación de Suelos. En 1971 pasa a depender del nuevo organismo autónomo ICONA, dependiente del Ministerio de Agricultura.

Actividades: Erosión hídrica. Experimentación en parcelas para determinar la pérdida del suelo y del agua en función de los distintos cultivos y labores. Ensayos de estructuras de retención y defensa.

Correspondencia: español, francés, inglés.

Escuela Técnica Superior de Ingenieros Agrónomos - Sección Ingenieria Rural - Ciudad Universitaria - Madrid

García Lozano, F.	Prof.Ing.Agr.	Jefe Depto. -	Hidráulica Agrícola
Lujan García, F.J.	Prof.Ing.Agr.		" "
Castanon Lión, G.	Prof.Ing.Agr.		" "
Camacho Matilla, J.	Prof.Ing.Agr.	Jefe Depto. -	Electrotecnia
Lago Lago, J.	Prof.Ing.Ind.		"
Sancho Llerandi, J.A.	Prof.Ing.Agr.		"
Maseda Eimil, F.	Prof.Ing.Agr.		"
Marcet Roig, J.R.	Prof.Ing.Agr.	Jefe Depto. -	Construcción I
Maléndez Falkowski, J.L.	Prof.Ing.Agr.		"
Garcimartin Molina, M.A.	Prof.Ing.Agr.		"
Dal-Re Teneiro, R.	Prof.Ing.Agr.	Jefe Depto. -	Construcción II
García-Vaquero Vaquero, E.	Prof.Ing.Agr.		"
Arrue Bengoa, A.	Prof.Ing.Agr.		"
Ortiz-Canavate, J.	Prof.Ing.Agr.	Jefe Depto. -	Motores y Maq.Agr. I
Hernanz Martos, J.L.	Prof.Ing.Agr.		" "
Munoz Valero, J.A.	Prof.Ing.Agr.		" "
Ruiz Altisent, M.	Prof.Ing.Agr.		" "
García de Diego, J.	Prof.Ing.Agr.	Jefe Depto. -	Motores y Maq.Agr. II
Enebral Casares, M.	Prof.Ing.Agr.		" "
Marquez Delgado, L.	Prof.Ing.Agr.		" "
Trueba Jainaga, I.	Prof.Ing.Agr.	Jefe Depto. -	Proyectos y Planificación Rural
Gómez Orea, D.	Prof.Ing.Agr.		" "
Goldaracena Saralegui, J.	Prof.Dr.Agr.		" "
Levenfeld González, G.	Prof.Ing.Agr.		" "
Marco Gutierrez, J.L.	Prof.Ing.Agr.		" "

Carácter: Gubernamental, su institución se remonta al año 1961.

Actividades: Enseñanza: Título Ingeniero Agrónomo, Especialidad Ingeniería Rural. Promedio 30 alumnos. - Investigación: Hidráulica de conducciones en tubería y canales. Análisis de diversos sistemas de riego y drenajes. Medición y Tarificación de la Energía Eléctrica en el ámbito rural. Organización y control de instalaciones electricas rurales. Utilización de fuentes de energía para el aumento de la producción vegetal. Técnicas de Laboreo, Laboreo minimo. Digestores de Biogas. Recolección mecánica y propiedades físicas de productos hortofrutícolas. Mecánica del suelo arable. Rodadura y resbalamiento, Ensayo de tractores toma de fuerza y en pista. Ensayos de Pulverizadores y abonadoras.

Ensayo de cabinas. Servomecanismos. Estudio de maquinaria de recolección. Proyectos de Mecanización. Ordenación del territorio en áreas rurales. Evaluación de Proyectos Agrarios.

Publicaciones: Maquinaria Agrícola: Motores y máquinas agrícolas. Ensayo de máquinas. Cálculo y proyecto de máquinas agrícolas. Mecánica del terreno y de labor. Hidráulica agricola: Riegos y saneamientos. Construcción rural: Instalaciones ganaderias industrias agrícolas y caminos vecinales.

Correspondencia: Español, Inglés, Francés, Alemán, Italiano.

Instituto Nacional de Reforma y Desarrollo Agrario (IRYDA) (Agriculture Development and Reform National Institute)
Paseo de la Castellana 112, Madrid-16   Telef.: 2611200

Mardones Sevilla, L.	Dr. Veterinario	Presidencia
Baquero de la Cruz, G.	Dr.Ing.Agr.	Secretaría General
Gómez Benita, V.	Dr.Ing.Agr.	Administr. General
Fernandez del Moral, J.M.	Dr.Ing.Agr.	Dirección Técnica

Direcciones:

Sanchez de Miguel, P.	Dr.Ing.Agr.	Director, Estudios y Planificación
Mene Lampré, A.	Letrado	Director, Personal
Alcat Salvoch, J.	Dr.Ing.Agr.	Director, Recursos Económicos
Canovas y Cobo del Prado, R.	Dr.Ing.Agr.	Director, Mejora del Medio Rural
Gonzales Ferrando, S.	Dr.Ing.Agr.	Director, Estructuras Agrarias
Villanueva Echevarria, T.	Dr.Ing.Agr.	Director, Obras y Mejoras Territoriales
Pérez Quintano, J.	Dr.Ing.Agr.	Director Adjunto " "

Servicios:

Bueno Gómez, M.	Dr.Ing.Agr.	Jefe, (Ing. geografo) Gabinete Técnico
Maqueda Valbuena, A.	Dr.Ing.Agr.	Jefe Adjunto,(Economista) " "
Gomez y Gomez Jordana, F.	Abogado del Estado	Jefe, Gabinete de Derecho Agrario
Sanchez Gil, J.L.	Letrado	Jefe Adjunto " '"
Hernandez Cano, J.	Dr.Ing.Agr.	Jefe, Revision
Corsanego Ulloa, A.	Dr.Ing.Agr.	Jefe, Planificación
Santa Cruz Tobalina, G.	Dr.Ing.Agr.	Jefe, Proceso de datos
Quintana Ibañez, J.A.	Dr.Ing.Agr.	Jefe, Asistencia Económica
Serrano Coca, J.	Dr.Ing.Agr.	Jefe, Oficina Técnica Relaciones con TRAGSA
Martín Liñan, L.	Dr.Ing.Agr.	Jefe, Concentración Parcelaria
Muñoz Durán, A.	Dr.Ing.Agr.	Jefe, Planes Especiales

Secciones:

Romero Sanz, E.	Técnico de gestión	Jefe, Contabilidad General
Vila Rodriguez, B.	Técnico de gestión	Jefe, Contabilidad Analítica
Navarro Menendez, J.	Contador del Estado	Jefe, Fiscalización de ingresos y gastos
Del Barrio Martín-Gamero, E.	Dr.Ing.Agr.	Jefe, Ingeniería Rural
Torroja Menendez, L.	Dr.Ing.Agr.	Jefe, Asuntos Generales
Pazos Diaz-Pimienta, G.	Dr.Ing.Agr.	Jefe, Informes Técnicos

Gutierrez Valdeón, A.	Letrado	Jefe, Revisión Procedimiento Administrativo
Bardaji Cando, J.	Dr.Ing.Agr.	Jefe, Suelos
Murcia Viudas, A.	Dr.Ing.Agr.	Jefe, Aguas Subterraneas y Residuales
Garrote Balmaseda, A.	Dr.Ing. de Caminos	Jefe, Estudios previos
Jordan de Urries, J.M.	Dr.Ing.Agr.	Jefe, Planes
Pou Tejero, J.	Dr.Ing.Agr.	Jefe, Programas
Gomez Jover, F.	Dr.Ing.Agr.	Jefe, Analisis y Explotación
Ruiz Izquierdo, G.	Dr.Ing.Agr.	Jefe, Información Estadistica
De la Cruz Fernandez, C.	Dr.Ing.Agr.	Jefe, Organización y Servicios
Alvarez y Ruiz de Vinaspre, R.	Letrado	Jefe, Régimen y Gestión de Personal
Campos Juez, A.	Técnico de Gestión	Jefe, Asuntos Economicos y sociales de Personal
Gonzales Niño, J.	Dr.Ing.Agr.	Jefe, Gestión del Patrimonio
Cubillo de Merlo, A.	Dr.Ing.Agr.	Jefe, Presupuestos y Contratación
García Viana Caro, A.	Dr.Ing.Agr.	Jefe, Recaudación
Gimenez Cacho, E.	Dr.Ing.Agr.	Jefe, Empresas individuales
Rabinal del Val, J.	Dr.Ing.Agr.	Jefe, Agrupaciones y Entidades
Leal Noguera, J.	Dr.Ing.Agr.	Jefe, Coordinación y Contratación de obras
Gonzales Hernandez, A.	Dr.Ing.Agr.	Jefe, Mejora de infraestructura y equipamiento comarcal
Medina Cubillo, F.	Dr.Ing.Agr.	Jefe, Desarrollo Social
Del Agua Tuero, A.	Dr.Ing.Agr.	Jefe, Colaboración con Entidades
Alonso García, C.	Letrado	Jefe, Regimen juridico de la tierra
García Pérez, J.M.	Dr.Ing.Agr.	Jefe, Adquisición de tierras
Gonzales Egido, M.	Dr.Ing.Agr.	Jefe, Ordenación de la Propiedad
Excribano Blesa, J.A.	Dr.Ing.Agr.	Jefe, Planes de Explotación y Concesionarios
Alabart Miranda, R.	Dr.Ing.Agr.	Jefe, Planes de Mejora
Andreu Peon, J.	Dr.Ing.Agr.	Jefe, Proyectos-Oficina Supervisora
Herrero Urgel, M.	Dr. Arquitecto	Jefe, Técnicas Especiales
Valcarcel Juan, J.M.	Dr.Ing.Agr.	Jefe, Ejecución de Obras
Sanchez Guardamino, J.A.	Dr.Ing.Agr.	Jefe, Liquidación y conservación de obras

<u>Delegaciones en las siguientes ciudades</u>: Albacete, Alicante, Almería, Avila, Badajoz, Barcelona, Bilbao (Vizcaya), Burgos, Caceres, Castellón, Ciudad Real, Cordoba, Coruna (La), Cuenca, Gerona, Granada, Granada, Guadalajara, Huelva, Huesca, Jaen, Jerez de la Frontera (Cadiz), Leon, Lérida, Logroño, Lugo, Madrid, Málaga, Murcia, Orense, Oviedo, Palencia, Palmas (Las), Palma de Mallorca (Baleares), Pamplona (Navarra), Pontevedra, Salamanca, Santander, Segovia, Sevilla, Soria, Tarragona, Tenerife, Teruel, Toledo, Valencia, Valladolid, Vitoria (Alava), Zamora, Zaragoza.

<u>Carácter y estructura</u>: Organismo Autónomo del Ministerio de Agricultura creado a 1971 por fusión del Instituto Nacional de Colonización y el Servicio Nacional de Concentración Parcelaria y Ordenación Rural.

<u>Actividades en Ingeniería Agrícola</u>: Ejecución de las obras para el mejor aprovechamiento de los recursos naturales. Transformaciones en regadio. Saneamiento de marismas y terrenos pantanosos. Mejora del regadio y del secano. Concentración Parcelaria, caminos rurales. Asesoramiento técnico y ayuda financiera a los agricultores para mejoras rurales. Estudio mecanico de suelos. Cartografía y máquinas. Economía Agraria.

Publicaciones desde 1973: Estudios monográficos: Evaluación del Programa de Concentración Parcelaria y Ordenación Rural en España - Evaluación de la Acción del IRYDA en apoyo de las agrupaciones para la explotación en común de la tierra - Comentarios a la Ley de Reforma y Desarrollo Agrario. - Serie de Información: La lucha por el agua en Aragon - Jornadas sobre modernos sistemas de riego - Comentarios sobre la comarca "Sierra Norte de Sevilla" y su plan de mejora - Jornadas sobre riego por goteo - El Estuario del Guadalquivir y su problemática agrosocial - Industrias Agrarias de las Vegas del Guadiana. - Serie Técnica: Investigación sobre los datos acumulados en dos piaras experimentales.

Instituto Nacional de Investigaciones Agrarias - Servicios Centrales
c/ José Abascal, 56

Dirección Técnica de Relaciones Científicas: Director: D. Manuel Vidal Hospital

Relaciones Internacionales: D. Eduardo Prieto Heraud

Coordinadores de temas relacionadas con Ingeniería Rural:
- Centro de Andalucía - Córdoba:

Berengena Herrero, J.	Dr.Ing.Agrón.	Riegos
Herruzo Sotomayor, B.	Dr.Ing.Agrón.	Mecanización en Olivicultura

Numerosos investigadores de los demás Centros Agrarios, están relacionados más o menos directamente con el tema Riegos y Drenajes, y Mecanización.

Nueve centros regionales de investigación: Galicia: Pastos y Forrajes, Ganado de carne, Maizicultura - Aragón: Arboricultura frutas, Cereales, Horticultura, Ganado ovino - Cataluña: Plantas ornamentales, Viticultura - Castilla la Vieja: Cereales y leguminosas, Plantas industriales - Levante: Citricultura, Horticultura, Arboricultura frutal - Extremadura: Pastos y forrajes de zonas áridas, Ganado vacuno y ovino - Andalucía: Olivicultura, Cereales, Vid, Plantas industriales - Canarias: Cultivos tropicales y subtropicales.

Estructura: Establecida la nueva estructura del Instituto en 1972 - tiene carácter gubernamental.

Actividades en Ingeniería Rural: Enseñanza: Cursos de riego por goteo - Diploma de asistencia. No. de alumnos: 40. - Investigación: Numerosos proyectos aplicados a diversos cultivos.

Publicaciones: Anales del INIA, Monografías, etc.

Correspondencia: Español.

SRI LANKA

Amparai

Hardy Senior Technical Institute
Amparai   Telegr.: HARDYTEC

Ismail, S.M.   B.Sc.(Agric)   Head

Kulatunge, D.V.P.Y.	M.Sc.(Agric)	
Saverimuthu, C.E.	I.D.D., Postgraduate Trg. in Extn.Education	
Shanmugam, N.	Dipl. Agric.Sci.	
Kuruppu, W.	" " "	
Wijesinghe, G.D.M.O.	Dipl.(Agric)	
Sandirasegaram, J.	B.Sc., A.M.I.R.E.	
Kumaratunge, M.K.N.	B.Sc.	
Amaradeva Appuhamy, L.D.A.	Dipl.(Commerce)	
Balasubramaniam, C.	B.Sc.	
Gunasekera, R.M.K.G.	Trained Teacher	
Weerasekera, W.M.N.B.	N.D.T.(Mech.Eng.)	
Werahera, A.E.	N.D.T.(Mech.Eng.)	
Sivaloganathan, P.	B.Sc.	
Patkunarajah, K.	Dipl.Electr.Eng.	
Jayawardane, R.T.R.		
Jayawardane, K.A.D.C.		
de Lima, R.A.L.	Dipl.Mech.Eng.	
Bandara, K.S.M.P.		
Samaratunga, N.T.		
Jagath Kumara	N.C.T.(Mech.Eng.)	
Methsiri Piyananda, H.	Teacher Trg.(Trade Cert.)	
Gunawardana, P.H.G.	B.A.	

Status: Established 1967 – Ministry of Higher Education.

Activities: Teaching: in Crop husbandry – Animal husbandry – Agricultural chemistry – Agriculture botany – Plant protection – Principles of Agriculture – Horticulture – Farm machinery – Agric. extension – Agric. economics, Farm management and accounts – Surveying and levelling and irrigation – Building construction and Farm electrification – Workshop technology – Engineering drawing. National Diploma in Technology (Agriculture). Average number of students, 30 per year. – Field work, Motor vehicle technology and Workshop practice, Fitting, Machining, Carpentry etc. – Extension work in the Second Year of study on Farm, Farm family and their problems.

A n u r a d h a p u r a

Farm Mechanization Training Centre (F.M.T.C.)
Anuradhapura, Sri Lanka

de S. Abeysuriya, A.T.S.	B.Sc.Eng(Hons.)	Project Manager
Mahindapala, R.P.	B.Sc.Agri.	Deputy Project Manager
Arampola, B.S.	Dipl.Agric.	Instructor in Farm Machinery
Balachandran, A.C.	Dipl.Mech.Eng.	" " "
Ekanayake, K.B.	Dipl.Agric.	" " "
Ekanayake, Y.N.		" " "
Fernando, M.N.R.F.	Dipl.Motor Industry	" " "
Fernando, P.M.C.A.	Dipl.Mech.Eng.	
	Dipl.Workshop Management"	" " "
Gooneratne, L.M.C.	Dipl.Agric.	" " "
Ismahoon, T.M.	City & Gilds Motor Mechanics	" " "
Jothipala, P.A.	Dipl.Mech.Eng.	" " "
Nanayakkara, G.L.G.	" " "	" " "

Samaranayake, G.A.	Dipl.Agric.	Instructor in Farm Machinery
Thilakaratne, H.M.	Dipl.Mech.Eng.	" " "
	" Elect.Eng.	
Velauthapillai, P.	" Auto Eng.	" " "

Status: Project under West German Aid, established in 1971. Directed and financed by the Department of Agriculture of Sri Lanka and West German Ministry for Economic Cooperation.

Agricultural Engineering activities: Teaching: Operation and Maintenance of all farm machinery - Workshop technology and Repairs of Farm Machinery - Special Courses in Economical aspects of farm mechanization - Workshop establishment and management - Extension Work: Advisory services to the training centres involved in the field of agricultural engineering - Advisory services to the public and private institutions in establishment and management of workshops.

Publications: Training programme and syllabus of the Farm Mechanization Training Centre, Anuradhapura. - Lesson Notes in Farm Engineering for the Agriculture Schools in the Department of Agriculture - Lesson Notes for courses in all subjects conducted at F.M.T.C. - Booklets published on Operation and Maintenance of Farm Machinery, in 3 languages: Sinhala, Tamil, English.

Correspondence: English.

Colombo

Irrigation Department - Ministry of Lands and Land Development
Colombo-7

Fernando, R.U.	Director of Irrigation
Chelvarajah, K.	Senior Deputy Director (Eastern Zone)
Ponrajah, A.J.P.	Senior Deputy Director (Designs)
Sivasubramaniam, S.	Senior Deputy Director (Mahaweli)
Gunawardana, G.J.P.	Senior Deputy Director (Western Zone)
Perera, K.D.P.	Senior Deputy Director (Southern Zone)
Gunawardana, O.A.	Senior Deputy Director (Specialized Services) and Research and Training
De Silva, S.H.C.	Senior Deputy Director (Machinery and Supplies)
Kanagaratnam, K.	Senior Deputy Director (Northern Zone)
Gunasekara, J.	Consultant (Feasibility Studies)

Branches with their location: Zonal Directors offices in Colombo - Range Directors Offices in: Colombo, Galle, Hambantota, Monaragala, Polonnaruwa, Bandarawela, Ampara, Batticaloa, Kekirawa, Kandy, Kurunegala, Puttalam, Trincomalee, Killinochchi, Vavuniya and Anuradhapura. With Divisional Offices manned by Irrigation Engineers under each of these Deputy Directors.

Status: Date of establishment 1900 - Government of Sri Lanka.

Agricultural Engineering activities: (Irrigation Engineers activity) - Research: Experiments and tests in Hydraulic and Irrigation Structures and River Outlets - Tests on soils and materials for construction of Irrigation and Hydraulic Structures.

Correspondence: Singalese, English.

The River Valleys Development Board
Colombo 7, P.O. Box 524, 415 Bauddhaloka Mawatha, Sri Lanka

Tillak Palamakumbura	Chairman
Mahes Danansuriya	Member
Pujitha Manawadu	Member

Status: Corporate body established under the Ministry of Land, Irrigation and Power in 1949.

Activities: Development of the Gal Oya and Walawe areas - Promotion and operation of irrigation, water supply and drainage schemes - Generation, transmission and supply of electrical energy and flood control - Promotion and control of irrigation and fisheries - Promotion of afforestation - Control of soil erosion - Promotion of Public health - Prevention and control of plant and animal diseases - General promotion of agricultural and industrial development, and economic and cultural progress.

Correspondence: Sinhalese, English.

Maha Illuppallama

Farm Machinery Research Centre - Department of Agriculture
Maha Illuppallama, Sri Lanka

Pillainayagam, M.G.	B.Sc., M.Sc. (Ag.Eng), C.Eng. M.I. Mech.E., A.M.I.E.E.	Head
Nithiananthan, T.	B.Sc.(Eng), M.Sc. (Ag.Eng.)	Agricultural Engineer (Designs)
Wijesiri, W.S.	B.Sc.(Eng.	Agricultural Engineer (Testing)

Status: Fully financed and controlled by Department of Agriculture, Government of Sri Lanka.

Activities: Testing and evaluation of all types of agricultural machinery under local field conditions - Designing and developing simple machinery, tools and equipment for rice cultivation, multi cropping system and agrobased industrial crops - Field experiments on mechanization system research - Advising project personnel on appropriate mechanized technological system for varying agricultural environment.

Publications: on Mechanization of rice cultivation in Sri Lanka - Cost of tractor power in agricultural production - Energy for agricultural use and future development - Rice production machinery development for Sri Lanka.

Correspondence: English.

Peradeniya

Department of Agricultural Engineering – Faculty of Agriculture – University of Peradeniya
Peradeniya, Sri Lanka   Tel.: 8301-05 ext 320

Ilangantileke, S.G.	B.Sc. Agric., M.Sc. and Ph.D.	Head of Department – Specialization: Agricultural Processing and Post-harvest Technology
Alvappillai, P.	B.Sc. Agric.	
Ariyaratne, A.R.	B.Sc. Agric.	Farm Machinery
Basnayake, B.F.A.	B.Sc. Agric.	Energy
Elias, S.	B.Sc. Agric.	Mechanization
Gunasekara, K.	B.Sc. Agric., M.Sc. Agric.	Hydrology
Gunawardena, E.R.N.	B.Sc. Agric.	Soil and Water
Jayasekara, A.A.(Mrs.)	B.Sc. Agric.	Water Management
Jayatissa, D.N.	B.Sc. Agric.	
Jeganayagam, S.S.	B.Sc. Agric.	Waste Management
Mills, R.C.	B.Sc. Agric.Eng. M.Sc.	Machinery
Prathapar, S.A.	B.Sc. Agric.	Soil Conservation
Punidadas, P.	B.Sc. Agric.	
Ranbanda, M.	B.Sc. Agric.	Mechanization
Sivapalan, S.	B.Sc. Agric.	

Branches with other locations: University Unit, Maha Illupallama, Sri Lanka

Status: Date of establishment of the Department of Agricultural Engineering – 1975. Semi government.

Agricultural Engineering activities: a) Teaching: Agricultural soil and water Engineering. Agricultural Machinery and Management. Post harvest technology. Agricultural waste management. Energy for Agricultural development. Degree awarded: B.Sc. Agric., (450 students) M.Sc. Agric. and M.Phil. Agric. (10 students), Ph.D. Agric. (1 student) – b) Research, Experiment testing or Extension work under progress: Soil and water – Comparative study of furrow irrigation system with locally fabricated trickle irrigation unit. Evaluation of performance of two trickle irrigation systems. Estimating channel seepage losses of Kalawewa basin 'H' area. Influence of land use on erodibility. Soil loss from small Agricultural watershed for selected storms. Monitoring the physical environment at Mahaweli 'H' area – Climate and hydrology, financed by I.M.F. The optimum drainage spacing in low humic clayey soils in Kalawewa basin, Mahaweli system. Water balance study of 'H' Walagambahuwa reservoir. – Energy: Technical evaluation of existing biogas plants in Kandy area. Solar grain drier – Design and evaluation of a Proto type. – Machinery: Conduct tests to determine and compare draught requirements of animal drawn implements. Row seeder for gingelly or kurakkan – Design and evaluation of seeders for small grain crops.

Publications: Numerous publications by staff members on field machinery, soil-water engineering, post harvest technology.

Correspondence: English.

Department of Agriculture - Engineering Research and Development Division
Peradeniya, Sri Lanka　　　　　　　　　　　　　Telegr.: AGRIENG Peradeniya
　　　　　　　　　　　　　　　　　　　　　　　Tel.:　　8155

Mechanical Engineering Research and Development Division:

Kathirkamathamby, S.	B.Sc.(Eng), F.I.Agr.E, F.I.E.	Superintending Engineer - Research and Development
Kanesu, N.	B.Sc.(Eng), M.I.Mech.E	Chief Engineer Repair Workshop
Vamadevan, A.	B.Sc.(Eng)	Agricultural Tender and Machine Supplies
Pillainayagam, M.G.	B.Sc.(Eng), M.I.Mech.E	Machinery Design and Development
Paskaran, S.	B.Sc.(Eng)	Implement Factory
Kanesha, E.	B.Sc.(Eng), M.I.E.	Repair Workshop, Colombo
Hulangamuwa, S.	B.Sc.(Eng)	Repair Workshop, Kandy
Jebarajah, E.M.	B.Sc.(Eng)	Electrical Works
Perera, J.	B.Sc.(Eng)	Seed Processing
Nithianandan, T.	B.Sc.(Eng)	Mechanical Testing
Jayasena, R.G.	B.Sc.(Eng)	Repair Workshop, Anuradhapura
A.T. de S. Abeyasuriyá	B.Sc.(Eng)	Farm Machinery Training
Amirthanesan, A.	B.Sc.(Eng)	Machinery Extension
Wijesiri, W.R.	B.Sc.(Eng)	Machinery Testing
Thangarajah, N.	B.Sc.(Eng)	Repair Workshop, Polonnaruwa
Gunawardena, D.A.	B.Sc.(Eng)	Repair Workshop, Ambalantota

Activities: The repair and maintenance of farm machinery including transport vehicles. The production of agricultural implements. The testing and development of agricultural implements and machinery for use in local agriculture.

Correspondence: Sinhalese, Tamil, English.

Civil Engineering Division:

Jayasundera, L.S.S.	B.Sc.(Eng), M.I.C.E.	Superintending Engineer (Civil)
Weerasinghe, A.V.S.	B.Sc.(Eng)	Construction Engineer
Jayawardena, P.H.	B.Sc.(Eng)	" "
Perera, K.A.R.C.	B.Sc.(Eng)	" "
Atukorala, P.S.W.	B.Sc.(Eng)	" "
Abeykoon, J.B.	B.Sc.(Eng)	" "

Activities: The design and construction of farm buildings, roads, bridges, etc. for government farms and research stations.

Correspondence: Sinhalese, Tamil, English

Status: 1960. Governmental.

S U D A N

K h a r t o u m

Department of Agricultural Engineering - Ministry of Agriculture Food and Natural Resources
Khartoum

Department of Agricultural Engineering, Faculty of Agriculture, University of Khartoum
P.O. Box 321 Khartoum   Tel.: 75100

Status: Semi-governmental institution.

Study courses covering tractors, field machinery, stationary equipment, farm buildings, irrigation and soil conservation – These courses are taken by all students in their first three years reading for the B.Sc.(Hons) degree in agriculture – During the 4th year specialization in farm mechanization – The Department also offers courses leading to B.Sc.A.E. – Students spend 2 years in the Faculty of Engineering followed by 2 years in the Faculty of Agriculture – Postgraduate students reading in agricultural engineering can register for M.Sc. and Ph.D. degrees by research – Research and experimental work on mechanization of field crops, irrigation, minimum tillage, soil-machine relationships, solar energy use.

Publications: on above subjects.

Correspondence: Arabic, English.

Faculty of Agriculture – University of Gezira
P.O.B. 20 Wadi Medani

SURINAM

Paramaribo

Centre for Agricultural Research in Surinam – Branch of University of Surinam
P.O.B. 1914, Paramaribo

Status: Established in 1965.

Agricultural Experiment Station – Department of Agriculture, Animal Husbandry and Fisheries
P.O.B. 160 Paramaribo

SWAZILAND

Luyengo

Faculty of Agriculture – University of Botswana and Swaziland
Luyengo Campus, Post Office Luyengo

Agricultural Engineering activities: a) Teaching: Buildings and farm mechanization for Diploma in Agriculture – Land use and irrigation for Diploma in Agriculture – General Diploma in Agriculture – The Agricultural Engineering section is an integral part of the Department of Land Use and Mechanization. b) Research on: Irrigation water management under Swaziland climatic conditions – Assessment and documentation of irrigation potential and conservation development possibilities – Watershed (field sited) studies to investigate soil loss from various regimes of land management – Development of low cost tractor – Testing and modification of ox-drawn equipment, particularly planters.

Publications on various agricultural engineering activities.

Correspondence: English.

Malkerns Research Station - Ministry of Agriculture and Cooperatives
P.O. Box 4 - Malkerns

S W E D E N

Alnarp

Alnarpsinstitutet, institutionen för maskinteknik (Alnarp Institute of Agriculture, Farm Machinery Department)
S-230-53, Alnarp

Status: Governmental institution belonging to the Royal Agricultural College, Ultuna, Uppsala. Financed fully by the Government.

Activities: Study courses for the "Lantmästarexamen" (leading to degree in Agriculture) and "Trädgårdsteknikerexamen" (leading to degree in horticulture), within the general curricula. - Research in harvesting and drying of field beans, projects carried out by students in various subjects.

Publications: Articles on various farm machinery subjects.

Correspondence: Swedish, English, German, Russian.

Lund

Institutionen för lantbrukets byggnadsteknik (LBT) (Department of Farm Buildings -
P.O. Box 624, S-220 06 Lund 6   Tel.: 046-117510   Swedish University of Agric.Sciences)

Henriksson, R.	Professor	Head of Department

Teaching Division   (Box 7032, 750 07 Uppsala 7):

Samuelsson, S.	Techn.Dr.	Teaching
Palmgren, H.	Arch.	Teaching
Nordström, A.	Civil Eng.	Teaching
Bengtsson, L.	Agr.	Teaching

Division of Farm Building Constructions:

Jansson, I.	Docent, MS	Head of Division

Division of Planning and Environment:

Gustafsson, B.	Ph.D. Docent	Head of Division

Division of Horticultural Buildings:

Landgren, B.	Civil Eng.	Head of Division - Building construction - Covering materials

Status: The Department belongs to the Swedish University of Agricultural Sciences and is financed by the Government.

Activities: a) <u>Teaching</u>: Study courses on structures, planning, environmental control and cost estimation of farm buildings attended by 20-30 students. Courses on horticultural building construction for landscape architects attended by 15-20 students and horticultural buildings and climate control for horticulturists by 5-10 students. b) <u>Research</u>: The farmer's living house - Systems for farm building planning and design - Silo constructions - Corrosion in animal buildings - Heat and moisture problems in roof structures - Strength and durability of materials and fittings - Air contamination in animal buildings - Air inlets and control of climatization devices in pig houses - Natural ventilation - Climatic control in poultry houses - Animal reactions for thermal factors in buildings - Energy from straw combustion - Pre heating of ventilation air for animal buildings with solar collectors - Fire hazards in farm buildings - Heat recovery with heat pump - Structural problems in connection with the building process - Wear tests on materials for green-houses - Green-house structures adapted to low energy system - Energy storage for green-gouses - Long time test of wind protection - Sun collectors for green-houses - Insulation systems for green-houses - Heating system for pot-tables - Technical analyses for green-house enterprises - Moisture control in storage rooms - The use of waste heat for unprotected crops - Watering systems for dairy cows - Systems for decreasing the rate of pig mortality - Close surroundings in dairy buildings - Prevention of accidents among workers - Narrow escape of accidents in green-houses - Work and machine data in animal production - Loose housing system for small dairy buildings - Data planning of buildings - Statistical economical investigations - Dealing with greenhouses and equipment.

<u>Publications</u>: Platform construction - Low cost buildings - Building information service - Animals demand on buildings - Cubicle buildings - Remodelling animal buildings - Concrete floors - Liquid manure storage tanks - Improving the thermal insulation - Insulated platforms for cows - Sheep buildings - Climate in animal buildings - Building materials - Silage quality with respect to wall tightness - Planning the coarse fodder storage - Controlling the heating and ventilating system - Heating in animal shelters - Cooling - Stretchmetal and concrete slatted floors - Heat flow in platforms for cows - Prefabricated elements - New building materials - Building techniques for big enterprises - The building process - Ventilation in potato stores - Air inlets - Temperatures in liquid manure tanks - Pressure in silos - Animal behaviour - Organizing the erection of buildings - Temperature and air velocity around cattle - Ventilation breakdown in animal shelters - Liquid manure gases - Wind damage - Weighing the coarse fodder - Active coal filter - Ventilation versus animal health - Foundation - Roof trusses - Manure scraping system - Plywood buildings - Outside walls - Water absorption - Safety factor - Risk of failure - Swine housing - Animal health - Time studies - Building cost index - Farm building investments - Labour requirement in green-houses - Green-house construction - Heat distribution in benches - Plastic houses - Light measurements - Heating - Covering materials - Ventilation - Temperature and humidity in plastic houses - Storage building construction - Noise and dust in animal buildings - Wind shelter for green-houses - Protective coatings in silos - Water distribution in green-houses - Warm air heating in green-houses - Low energy technique and energy saving in green-houses - Ventilation system with air inlet through the ceiling - Building system with simple wood frames - Planning a building for pig production - Energy saving with heat pump using the exhaust air as heat source - Building for pig production - Floor studies in animal buildings - Silos - Sheet materials for walls and ceiling - Fittings for hogs - The building for 30 cows.

<u>Correspondence</u>: Swedish, English, French, German.

Uppsala

Institutionen för markvetenskap, Avdelningen för lantbrukets hydroteknik, Sveriges Lantbruksuniversitet (Department of Soil Sciences, Division for Agricultural Hydrotechnics, Swedish University of Agricultural Sciences)
S-750 07 Uppsala

Hakansson, A.	Agr.Dr. Prof.	Head of Division
Berglund, G.	Agr.Lic.	Land drainage
Bjerketorp, A.	Agr.Lic.	Hydrology
Brink, N.E.	Agr.Dr.	Environmental water research
Ericson, A.	M.Sc.	Land drainage
Eriksson, J.	Agr.Lic.	Soil conservation
Gustafson, A.	M.Sc.	Environmental water research
Johansson, W.	Agr.Dr. Prof.	Soil, water and plant
Linnér, H.	M.Sc.	Irrigation
Sandsborg, J.	Agr.Dr.	Water and energy balance for crops

Status: The Division belongs to the University of Agricultural Sciences and is financed by the Government.

Activities: a) Teaching: Study courses (soil physics, hydromechanics, agrohydrology, drainage and irrigation) for Master of Science in Agriculture (40 students) and in Landscape Architecture (30 students). b) Research: Studies on physical properties of cultivated soils - texture - structure - waterholding capacity - permeability - capillarity. Soil physical conditions and plant growth. Water- and energy balance for crops. Drainage as an important hydrotechnical measure under different soil and climatic conditions - drain spacing and drain depths - runoff from small watersheds. Effect of sprinkler irrigation on different crops - irrigation scheduling - water quality and water supply - sewage and salt water for irrigation. Environmental water research - leaching from cultivated and forest land.

Publications: to be found in the Swedish Journal of Agricultural Research, in the periodical publication "Grundförbättring" as far as to 1976, and in separate Reports from the Division for Agricultural Hydrotechnics (mainly in Swedish with English summaries).

Correspondence: Swedish, English, German.

Institutionen för landbrukets arbetsmetodik och teknik, Lantbruksvetenskapliga fakulteten, Sveriges lantbruksuniversitet (Department of Agricultural Engineering, Swedish University of Agricultural Sciences)
S-750 07   Uppsala 7   Tel.: 018/10 20 00

Möller, N.	Agr.Dr. Prof.	Chairman and Head of Department Tractors and soil tillage
Bergström, T.	Agr. Asst.Prof.	Power and machinery
Erikson, G.	B.For. Asst. Prof.	Working environment
Nilsson, B.	Agr.Dr. Asst.Prof.	Optimization of Farm Mechanization
Olsen, H.J.	Agr.Lic. Asst.Prof.	Mechanical properties of soil
Svensson, K.	Agr. Lic. Prof.	Harvesting and storage

Division of field experiment techniques:
Hallerström, B.	Dipl.Agr.

Division of Horticultural Engineering, S-230 53 Alnarp, Tel. 040/41 50 00
Vacant

Status: The Department was established in 1945 within the Swedish Agricultural College. After the re-organization of the College in 1962 the Department changed name and its tasks were extended. The Department is part of the Swedish University of Agricultural Sciences which is subordinated directly to the Ministry of Agriculture.

Agricultural engineering activities: a) Teaching: Study courses in power and machinery, farm machinery and its application. Work studies and optimization of farm mechanization. Ergonomics and working environment. The courses are attended by 15-25 students a year. Four and a half years' study leading to the "Agronom" degree (M.Sc.) in Agricultural Engineering. Graduate program for Agr. Dr. degree. b) Research: Mechanical properties of the soil - tractor tires - tilling and drilling machines. Hitching and control of tractor implements. Automatic recording of tractor and field machinery operation data - optimization of farm mechanization. Spraying and spraying equipment. Uneven distribution of fertilizers. Harvesting and drying of green and grain crops. Measurement and control systems - adaptive process control - system control and automation. Agricultural engineering instrumentation and data acquisition. Studies of accidents in agriculture. Ergonomic studies on farm machinery.

Publications: Scandinavian languages, English, German.

Jordbrukstekniska Institutet (Swedish Institute of Agricultural Engineering)
S-750 07 Uppsala   Tel.: 018/30 19 30

Skarp, S.U.	Agr.Eng.	Director
Norén, O.	Mech.Eng.	Vice-Director, Senior Principal Research Officer
Danfors, B.	Agr.Eng.	Principal Research Officer - Soil
Ekström, N.	Agr.Eng.	Principal Research Officer - Grain, Straw
Aas, M.	M.Sc.Eng.	Principal Research Officer - Tractors
Larsson, K.	Agr.Eng.	Principal Research Officer - Potatoes, concentrates
Nilsson, E.	Agr.Eng.	Principal Research Officer - Ley crops, fertilizers
Sörlin, S.	Electr.Eng.	Principal Research Officer - Electricity
Thyselius, L.	Agr.Eng.	Principal Research Officer - Manure

Status: Semi-state research institute, established in 1954, supervised and financed by the Swedish Government and the Foundation for Swedish Agricultural Engineering Research.

Activities: Basic and applied research for developing new and improving older harvesting, handling and preservation methods and machines of different kinds - Soil managements - Harvesting, conservation and storage of cereals, straw and pulses - Hay and silage making and feeding - Planting, harvesting and handling of potatoes - Handling and feeding of concentrates - Handling of fertilizers - Handling, storage and utilization of manure - Rational utilization of electricity on the farm - Alternative energy sources from agricultural crops and residues - Tractor ergonomics.

Publications: Five bulletins per year, one of which is the annual report with a complete English translation including a list of available publications.

Correspondence: Swedish, English, German.

Jordbrukstekniska Föreningen (Swedish Association of Agricultural Engineering)
S-750 07 Uppsala 7   Tel.: 018/30 19 30

Berg, M.   Chairman
Skarp, S.U.   Secretary

Activities: Promotion of agricultural engineering studies - Exchange of experience by annual meetings and excursions.

Publications: Annual reports.

Correspondence: Swedish, English, German.

Statens maskinprovningar (National Testing Institute for Agricultural Machinery)
S-750 07 Uppsala 7   Tel.: 018/325100

Department of Ultuna (Central Sweden)

Department at Alnarp (South Sweden)

Department at Röbäcksdalen (North Sweden)

Department of Dairy Equipment

Status: Governmental institution, established in 1897.

Activities: Testing and extension work - Testing machines for agriculture, forestry, commercial gardening and dairy industry.

Publications: "Meddelande f. Statens maskinprovningar", official testing reports.

Correspondence: Swedish, English, German.

S W I T Z E R L A N D

Bern

Eidgenössisches Meliorationsamt, Bundesamt für Landwirtschaft (Service fédéral des améliorations foncières, Office fédéral de l'agriculture)
Mattenhofstrasse 5, 3003 Bern

Helbling, F.   Dipl.Kult.Ing. ETH   Abt. Chef
Flury, W.   Dipl.Kult.Ing. ETH   Sektionschef
Schläpfer, H.P.   Dipl.Kult.Ing. ETH   Sektionschef

Services cantonaux d'améliorations foncières à: 5001 Aarau - 6460 Altdorf - 9050 Appenzell - 4001 Basel - 6501 Bellinzona - 3001 Bern - 7001 Chur - 2800 Delémont - 8500 Frauenfeld - 1700 Friburg - 1204 Genève - 8750 Glarus - 1000 Lausanne - 4410 Liestal - 6002 Luzern - 2001 Neuchâtel - 6060 Sarnen - 8201 Schaffhausen - 6430 Schwyz - 1951 Sion - 4500 Solothurn - 6370 Stans - 9001 St. Gallen - 9053 Teufen - 6301 Zug - 8090 Zürich.

Structure: Subventionnement d'améliorations foncières et de constructions rurales.

Correspondance: Allemand, français.

Eidg. Forschungsanstalt für Milchwirtschaft (Station fédérale de recherches laitières)
CH-3097 Liebefeld-Bern

Blanc, B.	Prof.Dr.	Directeur
Flückiger, E.	Dr.	Technologie
Stettler, H.		Centrale fédérale du service d'inspection et de consultation en matière d'économie laitière

Annexes situées dans d'autres localités: Uettlingen: fromagerie d'essai pour l'emmental
Moudon: fromagerie d'essai pour le gruyère et les pâtes molles.

Structure: Date de la fondation: 1901. Institution gouvernementale placée sous le contrôle de la Division de l'Agriculture du Département fédéral de l'économie publique.

Activité concernant le génie rural: Formation: Notre section de technologie ne délivre pas de diplômes cependant elle donne des cours qui s'étendent sur plusieurs jours à l'intention des inspecteurs de machines à titre. Ces cours ont lieu plusieurs fois par année et réunissent chaque fois 40 à 50 participants. - Recherches: Recherches techniques des procédés de fabrication des produits laitiers. - Essais et contrôles d'équipements laitiers et de machines à traire - Contrôle des détergents et produits de nettoyage en laiterie.

Publications: "12 Tätigkeitsbericht 1968-70" de la station.

Correspondance: Français, Allemand.

B r u g g

Landwirtschaftliches Bau- und Architektbüro LBA des Schweizerischen Bauernverbandes in Brugg (Office de constructions agricoles OCA de l'Union suisse des paysans)
LBA Geschäftsleitung: Laurstrasse 10   5200 Brugg   Tel.: 056 4167 55

Indergand, R.   Dipl.Ing.agr. ETH   Mitglied SIA (Member SIA)

Offices à: Bern, Chur, Lausanne, Küsnacht, Neuchâtel, St. Gallen, Weinfelden etc Winterthur.

Structure: L'Institution, fondée en 1916, dépend de l'Union Suisse des paysans, mais se finance elle-même.

Activités: développement, planification et exécution des constructions agricoles.

Correspondance: Allemand, français.

M a r c e l i n - s u r - M o r g e s

Station d'Essais de Machines Agricoles
CH-1110 Marcelin-sur-Morges   Tél.(021) 711451 - (021) 711455

Gobalet, R.   Ing.mec.   Directeur

<u>Activité</u>: Etudes et expérimentation de machines agricoles, notamment pour la moto-viticulture - Vulgarisation.

<u>Correspondance</u>: Français, allemand.

Nyon

<u>Station fédérale de recherches agronomiques de Changins</u> (Federal Agricultural Research Station)
Route - 1260 Nyon
Tél.: 022 61 54 51
Télex: 22 785

Vez, A.	Dr.Ing.agronome	Directeur - Travail du sol et techniques culturales
Rod, P.	Dr.Ing.agronome	Propriétés physiques du sol et lutte contre l'érosion
Neyroud J-A.	Dr.Ing.agronome	Physique du sol
Calame, F.	Ing.agronome	Agroclimatologie et irrigation

<u>Stations</u>: Station fédérale de recherches agronomiques de Changins, Groupe Viticulture, Domaine de Caudoz, CH-1009 Pully: Murisier, F., Ing.agronome - Station fédérale de recherches agronomiques de Changins, Groupe Arboriculture-Horticulture, Centre des Fougères, CH-1964 Conthey: Perraudin, G., Dr.agronome, Lutte contre le gel par aspersion, chauffage, etc.

<u>Structure</u>: Institution gouvernementale placée sous le contrôle de l'Office fédéral de l'agriculture du Département fédéral de l'économie publique, CH 3003 Berne.

<u>Activités concernant le génie rural</u>: Etude des techniques de culture en relation avec le machinisme - Travaux du sol et entretien des cultures - Lutte contre le gel - Erosion: couverture du sol, utilisation des gadoues - Certains problèmes d'irrigation.

<u>Publications</u>: Revue suisse d'agriculture (paraît tous les deux mois), Revue suisse de viticulture, d'arboriculture et d'horticulture (paraît tous les deux mois).

<u>Correspondance</u>: français, allemand, italien, anglais.

Riniken

<u>Schweizerischer Verband für Landtechnik SVLT</u> (Association suisse pour l'équipement technique de l'agriculture ASETA)
Postfach 53, CH 5222 Riniken AG

Bächler, H.	Ing.Agr.	Président de l'Association
Bühler, W.		Directeur

<u>Sections</u>: 22 sections dans 22 cantons différents.

<u>Structure</u>: Organisation privée, fondée en 1924 - Financement par cotisations des membres.

Activité: Promotion de l'emploi rationnel des tracteurs et machines agricoles ainsi que du développement technique en agriculture.

Publication: "Schweizer Landtechnik" (Technique agricole), publiée 15 fois par an en français et en allemand.

Correspondance: Allemand, français, italien, anglais.

Tänikon

Eidg. Forschungsanstalt für Betriebswirtschaft und Landtechnik (Swiss Federal Research Station for Farm Management and Agricultural Engineering)
CH-8355 Tänikon

Meier, W.	Dr.	Director
Bergmann, F.	Ing.Agr.ETH	Scientific Assistant
Bisang, M.	Dipl.Ing.Agr.ETH	Forage production technique
Fankhauser, J.	Maschinening.ETH	Farm power and machinery
Göbel, W.	Bauing.	Farm structures
Glöcke, G.	Elektrotechniker HTL	Electronic measurements
Hilty, R.	Architekt HTL	Farm structures
Jakob, P.	Dipl.Ing.Agr.ETH	Animal production technology
Irla, E.	Dipl.Ing.Agr.	Crop production technology
Kramer, E.	Dipl.Ing.Agr.ETH	Farm power and machinery
Luder, W.	Dipl.Ing.Agr.ETH	Rationalization
Nosal, D.	Dipl.Ing.Agr.	Animal production technology
Ott, A.	Dipl.Ing.Agr.ETH	Forage production technology
Schönenberger, A.	Dipl.Ing.Agr.ETH	Rationalization
Stuber, A.	Architekt HTL/SIA	Farm structures
Studer, R.	Dipl.Ing.Agr.ETH	Farm power, machinery and Energy problems
Uenala, N.	Dipl.Maschinening.ETH	Farm power
Zihlmann, F.	Dipl.Ing.Agr.ETH	Special advisor
Zumbach, W.	Dipl.Ing.Agr.	Crop production technology

Status: The Institute was established in 1969 and is responsible to the Eidgenössisches Vorkswirtschafts-Department.

Activities: Teaching: The Institute organizes in co-operation with the "Schweizerische Betriebsberatung", training courses for extension workers in the field of agricultural engineering and farm structures - Training courses are attended by approximately 50-60 students - No formal certificates or diplomas are given.

Research: Research is carried out on various problems dealing with farm power, machinery and energy problems, farm structures, animal production, crop production, etc.

Publications: Numerous publications (in German and French) are issued, dealing with various research projects.

Correspondence: German, French, Italian, English.

Institut d'économie rurale - Institut für Agrarwirtschaft
Universitätsstrasse 2, CH-8006 Zürich   Tél.: 01/326211

Heusser, H.   Prof.Dr.   Directeur
Rist, M.   Dr.agr. Dipl.Ing.
Schoch, R.   Dipl.Arch.
Studer, R.
Fässler, P.   Dr.

Abteilung für Kulturtechnik und Vermessung, Eidgenössische Technische Hochschule Zürich (ETHZ) - Division de génie rural et de Géodésie, Ecole Polytechnique Fédérale, Zürich (EPFZ)
Institut für Kulturtechnik (Institut de génie rural) CH-8093 Zürich-Hönggerberg

Grubinger, H.   Prof.Dr.techn. Dr.rer.nat.   Hydraulique agricole, hydrogéologie
Flury, U.   Prof.Dr.techn.   Remembrement agricole, colonisation agricole.
Schmid, W.   Prof.Dr.techn.   Aménagement du territoire

Structure: L'EPFZ est une Institution fédérale depuis 1855. Fondation de la chaire de Génie Rural en 1889, transformation en Institut en 1953. L'Institut est divisé en deux sections: sol et eau, remembrement et planification; Laboratoire et champs d'essai.

Activités concernant le Génie Rural: a) Formation normale des ingénieurs du Génie Rural (30-50 étudiants par an), diplôme d'Ingénieur de Génie Rural EPFZ. Cour de perfectionnement, troisième cycle. - Recherches: expérimentation et vulgarisation: hydrologie de petits bassins versants en montagne; hydraulique agricole (assainissement, drainage, irrigation, correction des cours d'eau et des torrents, mise en valeur des terres), économie d'eau. Amélioration alpestre; chemins ruraux; constructions et colonisations rurales. Remembrement agricole du point de vue technique, législatif, économique et social; préservation du paysage; aménagement dans l'espace rural, spécialement dans des régions peu développées.

Publications: L'écoulement dans des petits bassins versants alpins, assainissement sélectif en montagne, consolidations de glissements, levé de zones en danger, stratégie et méthodologie concernant l'amélioration des structures agraires.

Correspondance: allemand, français, anglais.

Eidg. Forschungsanstalt für landwirtschaftlichen Pflanzenbau Zürich-Reckenholz (Station Fédérale de Recherches Agronomiques - Swiss Federal Research Station for Agronomy)
8046 Zürich, Reckenholzstrasse 191/211   Tél.: (01) 578800

Brönnimann, A.   Dr.sc.techn.   Directeur
                   dipl. Ing.agr.
Jäggli, F.   Dr.sc.techn.   Travail du sol, laboratoire de Physique
                   dipl.Ing.agr.   du sol
Peyer, K.   Dr.sc.techn.   Irrigation, drainage.
                   dipl.Ing.agr.

Structure: Fondation 1878. Institution gouvernementale, placée sous le contrôle de la Division de l'Agriculture du Département Fédéral de l'Economie Publique.

Activités concernant le Génie Rural: Recherches et expérimentation sur: Besoin d'irrigation – Tassement des sols organiques – Sous-solage et profondeur du travail du sol – Besoin de drainage.

Correspondance: Allemand, français, anglais, italien.

Schweizerische Vereinigung Industrie und Landwirtschaft (SVIL) (Association Suisse Industrie et Agriculture)
Schützengasse 30, CH-8001 Zürich   Tél.: 01/234630

Annexes: Bienne, Olten, Frauenfeld, Landquart.

Statut juridique: Association privée d'utilité publique en 1918.

Activités: Préparation et exécution de projets de construction – Acquisition de terrain pour l'industrie et le trafic routier ainsi que pour le réseau suisse à gaz naturel – Amélioration de la culture du sol par des cercles non-agricoles – Préparation et exécution de projets de colonisation, d'assainissement de constructions rurales, de constructions rurales nouvelles; adaptation de constructions existantes – Service de consultation en matière de constructions rurales – Assistance aux paysan dans les régions de la montagne.

Publications sur les activités sus-mentionnées.

Correspondance: Allemand, français, italien.

SYRIA

Aleppo

Faculty of Agriculture – University of Aleppo, Aleppo
Syrian Arab Republic

Fayes Elyassin	Dr.	
Ghadri M. Ghassan	Dr.	Associate Dean for Scientific Affairs
Alashram Mahmoud	Dr.	Associate Dean for Official Affairs

Department of Food Technology:
Kayali Ali Ziad   Dr.   Head of Department

Department of Plant Protection:
Bayaa Bassam   Dr.   Head of Department

Department of Agricultural Economics:
Najjar K. Elsabeh   Dr.   Head of Department

Department of Animal Production:
Mizyad Mihye   Dr.   Head of Department

Department of Field Crops:
Kaff Elghazal Rami   Dr.   Head of Department

Department of Soil:
Durmosh M. Khaldun   Dr.   Head of Department

Department of Rural Engineering:
Barbara Suheil   Dr.   Head of Department

Department of Forestry and Horticulture:
Wereh Hassan   Dr.   Head of Department

Agricultural Research Centre:
Hafez, A.K.   B.Sc.   Head of Centre

Status: The Faculty of Agriculture was established in 1960. It is a Government financed Institution.

Agricultural Engineering Activities: Teaching: Scientific research, Agricultural consultations. Degree awarded: B.Sc. and M.Sc. in Agricultural Sciences.

Correspondence: Arabic, English, French.

Cotton Bureau - Ministry of Agriculture and Agrarian Reform
Aleppo, Syrian Arab Republic   Cables: COTTONBUREAU   Tel.: 12813/14/15 - 17500 - 20230

Deiri, M.A.   Dr.   Director
                                            Head of Cotton classification and control of export section
Ammouney, W.   B.Sc.   Head of Breeding section
Hakim, A.   B.Sc.   "   Agricultural section
Eade, M.   B.Sc.   "   Seed and Mechanization section
Jaara, A.   B.Sc.   "   Varieties preservation section
Syraice, D.   B.Sc.   "   Yarn and Fiber Lab.

Annexes: Cotton Research Experimental Station in: Tal-Hadia (Aleppo); Draikila (Aleppo); Deir-ez-zor; Farhania (Homs); Hassakat; Hama. - Cotton Sub-section in each of cotton producing region - Cotton Sub-section in ports of cotton export.

Status: The Cotton Bureau was established in 1952 and it is a Governmental Organization financed by the Government.

Activities in Agricultural Engineering: a) Training: Various training courses attended by approximately 20 students every year. b) Research: Progeny, strain and variety trials - Hybdridization - Development of farming practices and production - Irrigation frequencies and quantities - Mechanical techniques - Varieties preservation - Classification - Defoliants - Herbicides.

Publications: Annual report of experimental results in Arabic - Half-monthly report about Syrian cotton in Arabic and English and various publications.

Correspondence: Arabic, English and French.

D a m a s c u s

Rural Engineering Department, Ministry of Agriculture and Agrarian Reform
Damascus   Tel.: 22700-13613

Subdivisions in all Syrian districts

Activities: The Department's major task is to run all agricultural and rural engineering works of the Ministry of Agriculture and Agrarian Reform - Preparation and execution of agricultural and rural engineering projects - Extension and field services to the farmers - Participation in governmental committees related to the field of agricultural and irrigation engineering and planning of activities along this line.

Publications: Reports on agricultural engineering activities.

Correspondence: Arabic, English, French.

Faculty of Agriculture - University of Damascus
Damascus

Department of Agricultural Engineering:

Hasan, Abdul Hamid S.	Ph.D.	Head of Department - Farm Farm Machinery
Shawa, Farouk M.	Ph.D.	Irrigation and Drainage
Smetah, Ahmed		Agricultural Engineer
Soubeh, Ghassan		Agricultural Engineer

Status: Established 1976. Governmental institution.

Activities: Farm mechanization (tractors, automobiles, machines, exploitation); Agricultural irrigation and drainage - Surveying: 400 students.

Publications: on Farm mechanization and Agricultural irrigation and drainage, and textbooks on this subject.

Correspondence: Russian, Bulgarian, English.

TANZANIA

Arusha

Tanzania Agricultural Machinery Testing Unit (T.A.M.T.U.)
P.O. Box 1389, Arusha   Tel.: 3666

Mujemula, F.K.		Head of Department
Kaaya, A.N.	Dip.Agric.Eng.	
Kajagu, C.S.R.	Dip.Agric.Eng.	
Kyaruzi, E.K.M.	Dip.Mech.Eng.	
Tulapona, D.	M.Sc.Agric.	
Wawa, N.M.	Cert.Mech.Eng.	

Branches: Nzega in Tabora Region fully under TAMTU. Others in Musoma - Mara Region. Malya - Mwanza Region. Wigehe - Shinyanga Region. Kigwe - Dodoma Region and Songea in Ruvuma Region are administered by the regions and only get technical advice from TAMTU.

Status: Established in 1955, a government institution under the Ministry of Industries.

Activities: Designing: Various agricultural machinery and implements are designed which are usually not available on the local market. A few prototypes of a particular design are constructed and tested at TAMTU and in the villages before a final prototype is released for production. - Production: The approved prototypes are produced and sold to the farmers. Demand on some of the products is quite high at the moment and this calls for large-scale production. – *Testing*: Locally manufactured and imported agricultural machinery and implements are tested for their suitability to local conditions and an official test report published. – *Training*: Training of Rural Craft Workshop (RCW) personnel. These craftsmen, at the completion of their training at TAMTU go back to their respective RCWs and continue producing implements approved for manufacture by TAMTU. – *Intermediate technology/or village technology work*: Involves the making of simple tools, devices and implements for the villages and in the villages by the village people themselves using locally available materials. – *Publicity*: Demonstrations are carried out in the villages on machinery and implements considered useful for the rural areas but not available there. Information on this equipment is published in pamphlets or booklets for circulation to the public.

Publications: Test reports – Development reports – Monthly and annual reports – also occasional articles in Ukulima wa Kisasa.

Correspondence: Swahili, English.

Tropical Pesticides Research Institute
P.O. Bos 3024, Arusha, Tanzania

Mushi, C.S.K.   B.Sc.   Officer in charge, Physics Section

Status: The Institute was established in 1949 and administered by the East African Community from 1967 up to June 1977. Thereafter it was only administered by the Tanzania Government.

Agricultural Engineering activities: The Physics and Engineering Sections are concerned with research into best methods of pesticides application in the tropics. Spray machinery is developed, modified, calibrated and tested.

Publications: The Institute has published numerous papers on spraying machinery and related subjects. The recent titles include: physical properties of knapsack mistblowers, tractor powered sprayers and aerial application of pesticides.

Correspondence: English, Swahili.

D a r e s S a l a a m

Division of Agriculture, Ministry of Agriculture and Co-operatives
P.O. Box 9071, Dar-es-Salaam

Agricultural Engineering activities: Promotion of farm mechanization – Farm machinery testing, both mechanical and oxdrawn equipment – Study of tractors and farm implements suitable for local conditions – Advisory services – Research on minimal cultivation techniques – Studies on the economics of mechanization as applied to tropical agriculture.

Correspondence: English.

Ministry of Agriculture Training Institute (MATI) Mlingano
P.O. Box 5051, Tanga

Offering diploma courses in Agro Mechanization

Ministry of Agriculture Training Institute (MATI) - Nyegezi
Box 1400, Mwanza

Offering diplomas in Irrigation and Land Planning

Ministry of Agriculture Training Institute (MATI) - Ukiriguru
Box 1434, Mwanza

Offering diplomas in Agro-Mechanization

Water Development and Irrigation Division, Ministry of Lands, Settlement and Water Development - P.O.B. 9153   Dar-es-Salaam   Tel. 23247

Engineering Technical Assistants Training School in Ubungo

Activities of the School: Two six months periods of practical inservice training as part of the Dar-es-Salaam Technical College programme with training in basic principles of civil engineering, surveying, hydraulics, building construction, engineering drawing, mathematics and irrigation techniques.

Correspondence: English.

Morogoro

Department of Agricultural Engineering and Land Planning - Faculty of Agriculture, Forestry and Veterinary Science - University of Dar-es-Salaam
P.O. Box 643, Morogoro, Tanzania

Mrema, G.C.	Ph.D.	Head of Department - Processing
Beck, A.	Dipl.Ing.	Lecturer - Workshop Tech.
Boldholt, O.	Ph.D.	Senior Lecturer - Structures
Dihenga, H.O.	M.Sc.	Lecturer - Mechanization
Dujardin, R.	Dipl.Ing.	Lecturer - Structures
Dumelow, J.	B.Sc.	Lecturer - Machinery
Gidey, R.L.	B.Sc.	Asst. Lecturer - Mechanization
Kayombo, B.	M.Sc.	Lecturer - Soil and Water
Kihupi, N.I.	M.Sc.	Assist.Lecturer - Irrigation
Kimboka, T.	B.Sc.	Assist.Lecturer - Hydrology
Massawe, J.B.M.	B.Sc.	Assist.Lecturer - Processing
Tiwari, H.C.	M.Sc.	Senior Lecturer - Irrigation

Status: A Department of the University of Dar-es-Salaam established in 1969. Governed by Faculty Board and University of Dar-es-Salaam Senate and Council. Financed through grants from Tanzania Government.

Agricultural Engineering activities: a) Teaching: Undergraduate courses for B.Sc. Agriculture students - general introduction courses in Agricultural Engineering (90) students. Some students specialize in their final year in Agricultural Engineering (12 students). A full four year programme leading to the award of B.Sc. degree in Agricultural Engineering will start, in collaboration with the Faculty of Engineering, University of Dar-es-Salaam. Postgraduate courses - 1 year course work, and 1 year research are available in Agricultural Processing and Structures, Soil and Water Engineering, Agricultural Mechanization and Machinery, leading to the award of M.Sc. degree. b) Research: activities in minimum tillage, soil and water conservation, appropriate mechanization, crop storage and processing.

Publications: A number of papers have been published in international and local journals on oilseed processing, appropriate technology, agricultural mechanization, soil and water conservation.

Correspondence: English, Swahili.

THAILAND

Bangkok

Agricultural Engineering Division, Department of Agriculture, Ministry of Agriculture and Cooperatives
Bangkhen, Bangkok 9

Rugtrakul, S.	B.Sc.	Director
Ketupanya, W.	B.Sc.	Assistant Director
Chakkaphak, C.	B.Sc.	Cert. in Farm mechanization - Head of Research and Testing Section
Krisnaseranee, S.	M.Sc.	Agricultural Engineer
Suksri, C.	B.Sc.	Agricultural Engineer
Rojanasaroj, C.	M.Sc.	Agricultural Engineer
Singhakachein, S.	B.Sc.	Head of Storage and Processing Section
Thongsawang, M.	M.Sc.	Agricultural Engineer
Mongkontanatat, J.	M.Sc.	Head of Workshop and Service Section
Pasabutra, R.	B.Sc.	Head of Farm Machinery Training Section
Suwannarak.	B.Sc.	Agricultural Engineer
Busapawet, V.	B.Sc.	Agricultural Engineer
Wanprakop, T.	B.Sc.	Head of Machinery Repair and Maintenance Sect.
Boon-it, A.	B.Sc.	Head of Farm Mechanization Section
Singhasuk, N.	B.Sc.	Cert. in Farm Mechanization, Agric. Engineer
Tangkosakul, V.	B.Sc.	Agricultural Engineer

Status: Governmental institution, established in 1954, directed and financed by the Department of Agriculture, Ministry of Agriculture and Co-operatives.

Agricultural Engineering activities: Training: Training young farmers in the application of farm machinery and equipment - $\frac{1}{2}$ month and 3 months duration courses. - Research projects: Agricultural engineering survey - Native and small tools development - Water lifting devices - Farm machinery and equipment - Agricultural products industry - Crop processing and storage -

Publications on above activities.

Correspondence: Thai, English.

Co-operatives Promotion Department
Bangkok

Department of Agricultural Engineering - Faculty of Engineering - Kasetsart University
Kampangsaen Campus, Kampangsaen, Nakornpathom

Suwachirat, B.	M.Eng.	Dean of Faculty (Structural Engineering)
Kapilakan, V.	M.S.	Secretary of Faculty (Mechanical Engineering)
Sermsatanasvudsi, P.	M.Eng.	Head of Department Hydraulics and Hydrology (Applied Hydraulics)
Bhaholyotin, B.	M.S.	Head of Department Agricultural Engineering (Agricultural Eng. - Power and Machinery)
Pahniputt, W.	Ph.D.	Head of Department Mechanical Engineering (Mechanical Engineering)
Komarakul Na Nakorn, Ch.	M.Eng.	Head of Department Irrigation Engineering (Irrigation Engineering)
Intrasuwan, K.	M.S.	Head of Department of Electrical Engineering (Electrical Engineering)
Chaichanavong, T.	Ph.D.	Head of Department of Civil Engineering (Civil Engineering)

Status: Established in 1966. Governmental Institution.

Activities: Degree awarded: Bachelor of Engineering. Average number of students: 21. Curriculum of Department of Engineering: Tractor and Farm machinery operation - Agric. Engineering instrumentation - Agricultural engines and tractors - Agricultural machine design - Heat and mass transfer in Agricultural Engineering - Farm structures and equipment - Spraying and dusting methods - Agricultural machinery production engineering - Farm operation engineering - Hydraulic systems and machines - Power and farm machinery management - Theory of soil-machine systems - Agricultural process engineering - Thermal processing and cold storage - Agricultural product handling equipment - Physical properties of agricultural products - Drying of farm crops - Farm production planning and control - Pumps and distribution systems - Agricultural waste water - Farm electrification - Application of solar energy in agriculture - Energy use and control in agricultural systems - Pumps and fans - Agricultural environmental engineering. - Research: Laboratory and field tests.

Publications: in Thai Society of Agricultural Engineers.

Correspondence: Thai and English.

TOGO

Lomé

Direction du Génie Rural - Ministère de l'Aménagement Rural
B.P. 1463, Lomé, Togo   Tél.: 21-02-92

Emoe, K.	Ingénieur de l'Equipement Rural	Directeur du Service

Division de la Topographie:

Samlan, K.	Techn.Sup.Hydraulique et Equipement Rural	Chef de Division

Division de la Programmation des Etudes

Division de l'Aménagement Rural

Division de l'Hydraulique

Division du Machinisme:
Pedanou, K.   B.Sc.Ag.   Chef de Division

Division des Industries Agricoles et Alimentaires:
Kodjovi-Numado, A.H.   Techn.Sup. H E R   Chef de Division

Section Pistes Rurales

Annexes situées dans les localités: 5 divisions régionales à: Aneho, Atakpame, Sokode, Kara, Dapaong

Structure: Date de création: 1969. Organe Directeur: Gouvernement (Etat Togolais)

Activités concernant le Génie Rural: a) Formation: Cours de formation à la Section du Génie Rural au Centre de Formation Professionnelle Agricole de Tove (Kpalime) – Diplôme d'Ingénieur Adjoint du Génie Rural. – b) Recherche et Expérimentation: Création de 5 Centres d'appui technique aux irrigations pour initiation et vulgarisation des méthodes d'irrigation.

Correspondance: Français.

Institut National de Formation Professionnelle Agricole – Section du Génie Rural
Ministère de l'Enseignement du 3ème et 4ème degrés et de la Recherche Scientifique
B.P. 3221, Lomé, Togo

TRINIDAD AND TOBAGO

Trinidad

The University of the West Indies, Faculty of Engineering
St. Augustine, Trinidad, West Indies   Tel. 662-7171

Phelps, H.O.	B.Sc., Ph.D. DIC, FICE, CEng, FAPE	Hydraulics, Water Engineering Temporary Coordinator of Teaching and Research Programme Professor of Civil Engineering
Ahmad, N.	M.Sc., Ph.D., AICTA	Soil Physics – Faculty of Agriculture
Griffith, S.	M.Sc., Ph.D.	Soil Physics – Faculty of Agriculture
Narayan, C.V.	B.Sc.A., M.Sc., Ph.D., MASAE	Agricultural Mechanization
Sankat, K.S.	B.Sc., M.Sc., Ph.D.	Agricultural Product Processing

Note: Other engineering subjects serviced in common with other engineering discipines. Subjects in Agriculture are taught by the Faculty of Agriculture.

Status: Agricultural Engineering is a unit in the Faculty of Engineering. This programme is financed by the Government of Trinidad and Tobago.

Agricultural Engineering activities: (a) Teaching: Bachelors Degree Programme in Agricultural Engineering with inputs from the Faculty of Agriculture; teaching commenced in October, 1979. Ten students 1979/80, the first year of the programme.
(b) Research: Priority areas of research are: Agricultural Mechanization, Agricultural Product Processing, Soil and Water Engineering.

Correspondence: English

TUNISIA

Medjez-El-Bab

<u>Ecole Supérieure des Ingénieurs de l'Equipement Rural de Medjez-El-Bab (ESIER)</u>
Medjez-El-Bab, Tunisie

Aouina, M.S.	Ingénieur Principal	Directeur de l'Ecole - Hydraulicien et conservateur des eaux et des sols
Batigne, M.	Ingénieur Principal	Topographie et Photogrammétrie
Ben Salem, M.	Assistant	Topographie
Daghari, H.	Assistant	Résistance des matériaux
Gharbi, A.	Assistant	Hydraulique et Mécanique des fluides
N'sir, A.	Assistant	Technologie de construction
Sellami, A.	Assistant	Mécanique moteur
Essid, M.	Ingénieur T.E.	Machinisme agricole
Abrougui, A.	Ingénieur T.E.	Machinisme agricole
Souissi, A.	Ingénieur en Chef	Pédologie
Hadri, H.	Ingénieur Principal	Conservation des eaux
Kallel, R.	Ingénieur Principal	Hydrologie
Haouet, H.	Ingénieur Principal	Machinisme agricole
Kammoun, M.	Administrateur en chef	Organisation administrative et financière

<u>Structure</u>: Etablissement gouvernemental; fondé en 1964 en collaboration avec la F.A.O. sous l'étiquette "Collège Nord Africain de Génie Rural", il dépend du Ministère de l'Agriculture et financé par l'état Tunisien.

<u>Activités</u>: Cours dans une des branches: Génie Rural et Equipement Rural - topographie, hydraulique agricole, conservation des eaux et des sols, construction rurale, informatique génie civil. Machinisme agricole: électromécanique, mécanique, mécanique moteur, machines agricoles, gestion du matériel de fabrication mécanique, informatique.
Les deux spécialisations sus-citées comprennent chacune d'elle deux cycles: a) cycle court - Bacc. + 2 ans, b) cycle moyen - Bacc. + 4 ans. Le diplôme décerné est celui de: l'ingénieur adjoint pour les étudiants du cycle court; l'ingénieur de Travaux de l'Etat pour les étudiants du cycle moyen. Le nombre moyen des étudiants par promotion est 180.

<u>Publications</u>: Cours illustrés en Français sur: Machinisme agricole - Hydraulique agricole - Hydraulique générale - Conservation des eaux et des sols - Topographie - Dessin - Gestion du matériel agricole - Résistance des matériaux - Photogrammétrie - Mécanique des sols - Informatique - Constructions Rurales et Génie Civil.

<u>Recherches</u>: Les thèmes traités touchent les domaines suivants: Bilan hydrique, quantification de l'érosion, relation travaux de sol et érosion; énergies renouvelables, retrait et gonflement des argiles.

<u>Correspondance</u>: Arabe, Français, Anglais.

Pont du Fahs

<u>Centre National Mécanique Agricole</u>
Pont du Fahs

Tunis

Direction du Génie Rural (Direction of Rural Engineering)
30, Rue Alain Savary, Tunis

Alouini, K.	Ingénieur en Chef	Directeur, Ingénieur du Génie Rural

Sous-Direction des Aménagements Hydro-agricoles:

Djebali, A.	Ingénieur en Chef	Sous-directeur - Ingénieur du Génie Rural
Beguith, N.	Ingénieur Principal	Chef de Service
Ghattassi, A.	Ingénieur en Chef	Chef de Service

Sous-Direction du Développement Rural:

Jaoua, M.	Ingénieur en Chef	Sous-directeur
Baccar, F.	Ingénieur Principal	Chef de Service
Baccar, M.	Ingénieur Principal	Chef de Service
Ben Taleb, H.	Ingénieur des Travaux de l'Etat - Chef de Service P.I.	

Sous-Direction de l'Equipement Rural:

Chaouachi, A.	Ingénieur Principal	Sous-directeur
Attia, M.	Ingénieur Principal	Chef de Service
Amara, M.	Ingénieur Principal	Chef de Service

Annexes situées dans d'autres localités: Vingt Arrondissements Régionaux du Génie Rural dans les gouvernorats de Tunis: Zaghouan - Nabuel - Bizerte - Beja - Jendouba - Le Kef - Kasserine - Gafsa - Gabès - Médenine - Sfax - Kairouan - Sousse - Nabeul - Sidi Bouzid - Tozeur - Siliana - Mahdia - Tataouine.

Structure: Création en 1936 du Service du Paysan. 1972: Création d'une Direction du Génie Rural chargée des activités relevant du domaine du Génie Rural. 1977: décret n. 77-647 du 5 Août 1977 portant attribution du Ministère de l'Agriculture (Article 22 - Attribution de la Direction du Génie Rural).

Activités concernant le Génie Rural: La Direction du Génie Rural est chargée: de l'étude, l'exécution ou le contrôle d'exécution en dehors des grands aménagements des infrastructures d'irrigation et de drainage d'assainissement agricole ainsi que des routes et des pistes agricoles y afférentes; - du contrôle de l'utilisation des eaux dans les périmètres irrigués; - de l'étude, la réalisation ou le contrôle de réalisation des programmes de constructions rurales et d'habitat rural conçus dans le cadre de projets intégrés de la mise en valeur hydro-agricole; - de la programmation, l'étude et le contrôle de l'exécution des points d'eau publics destinés à l'alimentation en eau humaine et animale dans les zones rurales; - de l'étude et la promotion du machinisme agricole; - de l'étude et la promotion des industries agricoles et des équipements du froid en relation avec les départements intéressés; - du contrôle de la pollution des eaux et du signalement aux services chargés de l'assainissement urbain des risques de pollution afin de remédier à ces risques.

Publications: Elaboration d'études sous forme de projets individualisés qui trouvent leur application sur le terrain aux fins de développer l'infrastructure dans l'espace rural tant dans le secteur de la production agricole que dans celui de la promotion de l'homme.

Correspondance: Arabe, Français.

Section Génie Rural de l'Institut National Agronomique de Tunis
43, Avenue Charles Nicolle, Tunis

Jarraya, A.	Professeur (entomologie)	Directeur de l'I.N.A.T.
Ennabli, N.	Ing.Agronome	Maître de Conférences (Génie Rural)

Annexes situées dans d'autres localités: aucune annexe dépendant directement de l'INAT mais la section Génie Rural s'appuie sur les différents arrondissements de Génie Rural couvrant l'ensemble du pays et dépendant de la Direction de Génie Rural du Ministère de l'Agriculture.

Structure: Date de fondation de la section Génie Rural de l'INAT: octobre 1975. Organe controlant et finansant l'institution: Ministère de l'Agriculture.

Activités concernant le Génie Rural: Formation: 2 types de formation: a) Ingénieur des Travaux de l'Etat de Génie Rural (Bac + 4 ans d'études). b) Ingénieur Principal de Génie Rural avec mention de l'option (Bac + 4 ans + 2 ans d'options). Les options actuellement en place sont: hydraulique, hydraulique agricole, hydraulique et aménagement rural, hydrologie, hydrogéologie, géophysique, machinisme agricole, industries alimentaires, génie sanitaire. Nombre d'étudiants suivant les cours: en moyenne 20. - Recherches: fondamentales et appliquées dans les domaines suivants: Aménagements hydro-agricoles, en particulier en irrigation - drainage; hydrologie urbaine; utilisation des eaux usées en agriculture; dessalement des eaux salées par énergie solaire; hydrodynamique: milieu non saturé; recharge artificielle des nappes; gestion intégrée de la ressource en eau; utilisation des grandes puissances en agriculture.

Publications: Differents cours publiés - différentes publications scientifiques.

Correspondance: Arabe, Français, Anglais.

Direction de l'Enseignement, de la Recherche et de la Vulgarisation (DERV)
3, rue de Hollande, Tunis

Activité de génie rural: Cours appliqués de génie rural avec détails d'aménagement agricole et des pratiques de conservation des eaux et du sol pour la formation des élèves - Cours de formation mécanique pour la conduite et l'entretien de tracteurs et machines agricoles - Recherches et expérimentations dans les fermes domaniales pour l'amélioration de la céréaliculture dans les zones semi-arides de la Tunisie.

La Direction des Etudes des Grands Travaux Hydrauliques - Ministère de l'Agriculture
rue Alain Savary, Tunis   Tél.: 280456
Correspondance: Français.

TURKEY

Ankara

Ankara Universitesi Ziraat Fakültesi, Kültürteknik Bölümü (University of Ankara, Faculty of Agriculture, Department of Agricultural Engineering (Irrigation-Drainage and Farm Structure Department) Ankara

Sönmez, N.	Prof.Dr.	Head of Department
Aküzüm, T.	Dr.	
Ayyildiz, M.	Prof.Dr.	
Balaban, A.	Prof.Dr.	
Benli, E.	Doc.Dr.	
Erözel, Z.	Dr.	
Evsahibioğlu, N.		Asst.
Girgin, I.	Dr.	
Güngör, Y.	Prof.Dr.	
Kodal, S.		Asst.
Korukçu, A.	Doc.Dr.	
Okman, C.	Doc.Dr.	
Lgun, M.		Asst.
Ones, A.	Dr.	
Öztürk, F.		Asst.
Selenay, F.		Asst.
Tokgöz, A.		Asst.
Yildirim, O.	Dr.	

Status: Ankara University is an autonomous institution. It was established in 1933.

Agricultural Engineering activities: (a) Teaching: The Faculty of Agriculture consists of 8 semester (4 years). The courses given at the Faculty can be divided into four major groups. First group is the Basic Science Courses. Second group is the Basic Courses for Agricultural Engineering. The third one consists of courses given in order to prepare the students for the department of Agricultural Engineering. The fourth group are major courses of Agricultural Engineering department. Students attending: approximately 15 in each semester. B.Sc. Degree in Agricultural Engineering is awarded to those students who have successfully completed their studies. The Department also has graduate students at the graduate school of the Agricultural Faculty, and M.Sc. and Ph.D. degrees are also awarded to graduate students in Agricultural Engineering. - (b) Research: testing of home-made irrigation pipes, couplers, sprinklers and other equipment. Experiments on uniformity of sprinkling, optimization of water use, hydraulic characteristics of irrigation pipes, irrigation scheduling, consumptive-use studies, hydrology, farm structures, comparison selection of filter materials for agricultural drainage of different drain spacing formulas.

Publications: Numerous publications about different irrigation techniques of several crops, drainage studies of agricultural lands, irrigation efficiencies, use of computer techniques on utilization of water resources, etc. Textbooks.

Correspondence: Turkish, English, German.

Department of Agricultural Engineering, University of Çukurova
Adana, Turkey

Research Department of TOPRAKSU General Directorate - Ministry of Village Affairs and Co-operatives
Ankara

Sarikatipoğlu, S.	Head
Yilmaz, Y.	Assist. Head

Çuhadaroglu, D.   Director of Application and Evaluation
                                           Section
Taşkin, S.   Director of Program and Project Section
Seyisoğlu, N.   Chief Engineer

Activities: Research.

Publications: Guide to water consumption of irrigated crops in Turkey - Guide to operational evaluation of agricultural machinery and equipment in Turkey - Guide to soil cultivation and planting equipment to be used in fallow grains farming system etc.

Correspondence: Turkish, English.

Directorate of Soil and Fertilizer Research Institute
Ankara

Ulgen, N.   Director
Hindistan, M.   Asst. Director
Yurtsever, N.   Specialist
Inceoğlu, I.   "
Ulgen, H.   "
Alemdar, N.   Chief Engineer
Kurucu, N.   "
Alkan, B.B.   "
Göksel, H.   "
Uzunoğlu, S.   "
Diğdiğoğlu, A.   "
Tüzüner, A.   "
Börekçi, M.   Chemical Engineer
Gürbüzer, E.   Agricultural Engineer
Sungur, M.   "   "
Şahin, A.   "   "
Yüncüoğlu, H.   "   "
Şencan, N.   "   "
Yörük, M.   "   "
Haciomeroğlu, H.   "   "
Gençer, S.H.   "   "
Tinay, E.   "   "
Türkoğlu, F.   "   "

Activities: Research.

Publications: Methods of estimating field capacity and others on research activities.

Correspondence: Turkish, English.

Merkez TOPRAKSU Araştirma Enstitüsü (Central TOPRAKSU Research Institute)
P.O. Box 253, Bakanliklar, Ankara

Dinçer, D.   Director of the Institute
Doğan, O.   Asst. Director

Division of Soil and Water Conservation:

Küçkçakar, N.	Agric.Engineer	Chief Engineer of Soil and Water Conserv.
Unlü, K.	" "	Engineer of Soil and Water Conservation

Division of Hydrology:

Soykan, I.	(Dr) Agric.Eng.	Specialist Engineer of Hydrology and Hydraulics (leader of the division)
Çelebi, D.	Agric.Engineer	Chief Engineer of Hydro-Meteorology
Ünal, R.	" "	Chief Engineer of Hydrology and Hydraulics
Rona, H.	" "	Chief Engineer of Sedimentation

Division of Irrigation and Drainage:

Madanoğlu, K.	Agric.Engineer	Specialist Engineer of Irrigation (leader of the division)
Aylâ, Ç.	" "	Chief Engineer of Irrigation
Becer, T.	" "	Chief Engineer of Drainage
Ustün, H.	" "	Engineer of Irrigation

Division of Agronomy:

Neftçi, A.	Agric.Engineer	Specialist Engineer of Agronomy (leader of the division)
Çelebi, F.	" "	Chief Engineer of Water Products
Güçer, A.	" "	Chief Engineer of Cultural Plants
Yildiz, I.	MS " "	Chief Engineer of Soil Fertility

Division of Economy and Management:

Bilir, M.	Agric.Engineer	Engineer of Farm Machinery and Equipment
Dernek, Z.	" "	Chief Engineer of Statistics
Eryilmaz, H.	" "	Chief Engineer of Farm Machinery and Equipm.
Özbayram, K.	" "	Chief Engineer of Economy and Management

Branches with other location: Bolu (Sub-station of Demonstrations and Researches) - Kesikköprü (Sub-station of Demonstration and Researches).

Status: Established in 1962. Institute directed and financed by the Government (Ministry of Village Affairs).

Activities: Researches are going on hydrology, water and soil conservation, water erosion, irrigation, drainage, soil fertility, adaptation of cultural crops, economics and farm management, machinery and equipment, water products.

Publications: Central TOPRAKSU Research Institute annual research reports.

Correspondence: Turkish, English, French.

TOPRAKSU Genel Müdürlüğü (General Directorate of Soil Conservation and Small Irrigation) Ankara

Regional Directorates in: Izmir, Mersin, Istanbul, Ankara, Antalya, Konya, Diyarbakir, Kayseri, Erzurum, Malatya, Samsun, Van, Sivas, Trabzon, Eskişehir, Bursa.

Regional Planning offices in: Adana, Manisa

Training Centres in: Ankara, Izmir, Tarsus

Research Institutes in: Ankara, Menemen, Eskişehir, Tarsus, Tokat, Samsun, Konya

Cartographic Laboratory in: Ankara

Machinery and Supply Directorate : Ankara

Status: The TOPRAKSU General Directorate was established in 1960 and is directed and financed by the Government.

Activities: Planning and implementation of field projects on soil conservation, irrigation, land levelling, land consolidation and land reclamation – Soil survey and land classification – Inventory surveys – Soil and water analysis for the farmers – Research and extension work on soil conservation and irrigation – Training courses on soil conservation and irrigation for staff members (engineers, technicians, foremen) and for farmers – Supervised credit to farmers for buildings, soil conservation and irrigation structures – Courses and seminars for professionals, sub-professionals, as well as farmers, are organized on subjects covered by the General Directorate of TOPRAKSU. Soil conservation and farm irrigation. A certificate is given to each trainee at the end of the course.

Correspondence: Turkish, English.

Bölge TOPRAKSU Araştirma Enstitüsü (Regional Soil and Water Research Institute)
P.K. 35, Eskişehir, Türkiye

Oylukan, S.	Agric.Eng.	Director
Güngör, H.	" "	Assistant Director
Akbay, S.	" "	Chief Engineer of Hydrology and Meteorology
Altinel, B.	" "	Engineer of Crop Adaptation
Ayday, E.	" "	Chief Engineer Soil and Water Conservation
Bolu, A.	" "	Chief Engineer of Crop Adaptation
Büker, M.	" "	Chief Engineer of Laboratory
Büyükacar, K.	" "	Chief Engineer
Çelik, Y.	" "	Chief Engineer of Agricultural Equipment and Machinery
Imrek, Ç.	Agric.Technician	Technician of Crop Production
Kalelioglu, G.	Agric.Engineer	Technical Adviser
Kidan, O.	Chemical Engineer	Engineer of Laboratory
Oruç, S.	Agric.Engineer M.S.,Ph.D.	Chief Engineer of Soil Fertility .
Ögretir, K.	Agric.Engineer	Engineer of Irrigation
Sefa, S.	" "	Agronomic Research Specialist
Sunay, O.	" "	Chief Engineer
Yalçin, M.	" "	Chief Engineer of Economics
Yildirim, B.	" "	Chief Engineer of Drainage and Land Reclamation

Branches with other location: Two experimental and demonstrational stations in M. Kemâlpaşa (Bursa) and Selevir (Afyon).

Status: Established in 1952 and Ministry of Village Affairs and Cooperatives is directing and financing it.

Agricultural Engineering activities: Research on: Hydrology and meteorology – Soil and Water conservation – Irrigation–Drainage and land reclamation – Agricultural equipment and machinery – Soil fertility – Adaptation of cultural plants – Economics and management.

Publications: Research reports of the Institute.

Correspondence: Turkish, English.

Erzurum

Directorate of Regional TOPRAKSU Research Institute
ERzurum

Kamburoglu, S.	Director
Avşar, F.	Asst. Director
Yalçinkaya, A.	Agricultural Engineer
Sevim, Z.	" "
Doğan, O.	" "
Akan, F.	" "
Demirtas, A.	" "

University of Atatürk, Department of Agricultural Engineering
Erzurum, Turkey

Istanbul

Ziraat Makinalari Kürsüsü Makina Fakultesi - Istanbul Teknik Universitesi
(Department of Agricultural Machinery, Mechanical Engineering Faculty, Technical University of Istanbul)   Tel.: 790632-790567-433100/2498

Öz, I.H.	Prof.Dr.	Head of Department - Farm Power and Machinery
Arpaci, M.	Dr.M.Sc.M.E.	Asst. - Agricultural Machines
Kantarci, M.S.	M.Sc.M.E.	Asst. - " "
Kurtay, T.	Assoc.Prof.Dr. M.Sc.M.E.	Agricultural Machines
Kut, T.	Dr.M.Sc.Agric.E.	Agricultural Machines
Saygili, I.	Prof.Dr.M.Sc.M.E.	" "

Branch: Institute of Agricultural Machinery, Florya, Istanbul

Status: Governmental institution. Department was established in 1954.

Teaching activities: Optional courses are provided for undergraduate and graduate engineering students in Mechanical Engineering Faculty. Further studies lead to award of M.Sc. and Ph.D. degrees in Mechanical engineering. Number of students at undergraduate level about 50, graduate level 20. **Research:** The Institute of Agricultural Machinery is mainly concerned with research and testing work. Research on Tractor field performance - Tractor tyre performance under different soil conditions - Agricultural Mechanization in Turkey - Orchard sprayers - Ultra-Low-volume sprayers - Mechanical harvesting of potatoes - Types and sources of seed loss in combines - The wear of metal shares in agricultural soils.

Publications: Agricultural machineries (Turkish) - The agricultural mechanization in Turkey (Turkish) - Important factors effected spraying characteristics in swirl chamber atomizers (Turkish) - Predicition of stress and strain in the soil by finite element method.

Correspondence: Turkish, English, German, French.

Izmir

University of Ege, Department of Agricultural Engineering
Izmir, Turkey

Konya

Directorate of Regional TOPRAKSU Research Institute
Konya

Özdoğan, N.	Director
Topçu, H.	Asst. Director
Tetik, Ş.	Consultant
Ertaş, R.	Chief Engineer
Alptürk, C.	" "
Yilmaz, A.	Agr. Engineer
Kayizmazbatir, N.	Chief Engineer
Uçar, I.	" "
Öktem, O.M.	Agr. Engineer
Abali, I.	" "
Oflaz, M.	" "
Altuğ, A.	" "
Tamyel, C.	Mechanical Engineer

Activities: Research.

Publications: Crop water consumption as measured in Lysimeters in Konya Plain conditions - Wind erosion control works in Karapinar, and other research papers.

Menemen

Regional TOPRAKSU Research Institute
Menemen

Batur, K.	Director
Sorgun, O.	Assistant Director
Altuğlu, B.	Chief Engineer of Hydrology and Meteorology
Aksoy, G.	Farm Economics and Management Engineer
Akbay, F.	Laboratory Engineer
Bilgin, A.E.	Agronomy Specialist Engineer
Biçer, A.	Agricultural Mechanical Engineer
Ersin, B.	Engineer of Soil Fertility
Gürkök, K.	Adviser
Güngör, O.	Chief Agricultural Mechanical Engineer
Isliel, E.	Adviser
Kirkalioglu, A.	Chief Laboratory Engineer
Kocabaylioğlü, H.	Chief Engineer of Cultural Plants
Omay, E.	Specialist Irrigation Engineer
Özkara, M.M.	Chief Irrigation Engineer
Sarper, M.	Adviser

Saatçiler, M.	Ph.D.	Drainage and Land Reclamation Chief Engineer
Şener, S.	Ph.D.	Specialist Engineer of Agricultural Engineering and Irrigation
Uner, K.		Soil Productivity Engineer
Ünver, N.		Irrigation Engineer
Yakar, M.		Soil and Water Conservation Chief Engineer
Yalçuk, A.		Chief Engineer of Farm Economics and Management
Yalçuk, H.		Irrigation Engineer
Yemışcioğlu, U.		Chief Soil Productivity Engineer
Yeşilyurt, G.	M.S.	Chemical Engineer
Zaloğlu, S.		Engineer of Cultural Plants

Status: Established 1949.

Activities: Research on: Hydrology - Meteorology - Soil and water conservation - Irrigation - Drainage and land reclamation - Farm machinery and equipment - Soil fertility - Adaptation of cultural plants - Economics and farm management.

Publications: on activities of the Institute.

Correspondence: Turkish, English.

Samsun

Directorate of Regional ROPRAKSU Research Institute
Samsun

Döner, S.	Director
Mavi, A.	Asst. Director
Bayrak, F.	Chief Engineer
Selçuk, F.S.	" "
Bahçeci, I.	Agr. Engineer
Özdemir, O.	" "
Dinler,, A.	" "
Ilkyaz, H.	" "
Güner, S.	" "
Uzun, Z.	" "
Aksoy, N.	" "
Ocaktan, A.	" "

Status: Established 1970.

Activities: Research.

Publications: Non-disturbed soil sampler and other research reports of the Institute.

Tarsus

Regional TOPRAKSU Research Institute
Tarsus

Aybas, M.C.	Director
Kamber, R.	Asst. Director
Ozus, T.	Specialist
Biçer, Y.	"
Mete, C.	Chief Engineer
Dogan, D.	Agricultural Engineer
Yüksek, G.	Chief Engineer
Yön, A.	Chief Engineer
Eplem, M.	Agricultural Engineer
Ozel, M.	" "
Şahin, A.	" "
Derviş, O.	" "

Status: established 1947.

Activities: Research.

Publications: Irrigation and Fertilization of cotton crop in Çukurova - Water consumption of clover in Çukurova - A Lysimeter Research on effects of irrigation applied at various levels of available moisture of some soil series on yield and water consumption of cotton crop in Çukurova conditions and other research reports.

Correspondence: Turkish, English.

Tokat

Regional TOPRAKSU Research Institute
Tokat

Sayin, S.	Director
Ozgün, A.R.	Assistant Director
Firat, B.	Engineer of Soil Fertility
Günbatili, F.	Irrigation Engineer
Güney, D.	Engineer of Agricultural Economics
Ozmen, M.	Soil and Water Conservation Engineer
Oztürk, O.	Engineer of Soil Fertility
Ozyurt, E.	Adaptation Engineer
Sipahi, M.	Hydrology Engineer
Sipahi, N.(Mrs.)	Engineer in Laboratory

Status: Established 1963 (governmental)

Activities: Research on hydrology and meteorology - Soil and water conservation - Irrigation -Drainage and land reclamation - Machinery and agricultural equipment - Soil fertility - Plant adaptation - Agricultural economics - Soil testing (Lab.).

Publications: Wheat irrigation in Tokat (Farmers' bulletin No.4) - Determination of water requirement of sunflower in Tokat - Effect of contouring on soil and water conservation in Tokat Region - Irrigation Directory of Kazova Plain - Early vegetable growing under plastic cover in Tokat Region - Hard Plan (Farmers' bulletin No.8) and other research reports.

Correspondence: Turkish, English.

Urfa

Directorate of Regional TOPRAKSU Research Institute
Urfa

Demirören, T.	Director
Köse, C.	Assistant Director
Karaada, H.	Chief Engineer
Deniz, Y.	" "
Deniz, N.	Agric. Engineer
Unal, O.F.	" "
Ferhatoglu, H.I.	" "
Özer, S.M.S.	" "
Orhan, A.	" "

UGANDA

Entebbe

Water Development Department
P.O. Box 19, Entebbe

Activities: Assessment of the total water resources of Uganda – Regional distribution of the water resources and the water development potentiality over the whole country – Provision and control of water for government irrigation schemes – Study of suitable methods of irrigation water application under various conditions found in the country – Design and construction of pilot schemes for encouraging swamp reclamation and advice on the most suitable techniques – Provision of water in rural areas for cattle and general purposes by the construction of dams, valley storage tanks, river intakes, channel systems, pipelines and other structures necessary to divert and control surface water – Technical advice to government departments and the general public on all water supply and drainage problems.

Correspondence: English.

Kampala

Agricultural Mechanization Division, Namalere, Department of Agriculture
P.O. Box 7144, Kampala

Engineering Unit

Land Planning Unit

Status: Government institution for the development of, promotion and research into mechanized agriculture; Division was established in 1949.

Activities: Trial on minimum land preparation techniques – Intensification of use of ox cultivation in areas most suited – Land clearing techniques and equipment – Simple mechanical aids to assist small-scale farmers to process their produce – Physical and soil survey – Farm plan layout – Water management and conservation – under Uganda conditions and farming practices.

Correspondence: English.

Department of Agricultural Engineering, Faculty of Agriculture, Makerere University
P.O. Box 7062, Kampala   Tel.: 42471 and 56661

Status: Para-statal institution.

Activities: Three-year course leading to the B.Sc.(Agric) degree - Postgraduate course leading to the M.Sc.(Agric) in agricultural engineering - Study of watershed runoff, soil infiltration and percolation - Design of ox-drawn tool frames with plough, weeder and seeder - Design of intermediate technology rural aids for small farmers - Design of low cost threshers - Solar energy for crop drying - Development and assembly of a small tractor designed for tropical conditions - Time and energy analysis of agricultural tasks.

Publications: on Department of Agricultural Engineering activities.

Correspondence: English.

Soroti

Serere Research Station, Department of Agriculture
P.O. Soroti

Status: Governmental institution established in 1960, supervised by the Chief Research Officer and the Chief Agricultural Mechanization Officer, Department of Agriculture, Uganda.

Activities: Short courses ranging from a week up to three months to train middle-class agricultural extension staff and junior field staff in basic farm mechanization, using ox-drawn implements - 100 students each year - Improving and testing ox-drawn implements, such as ploughs, ridgers, seeders, groundnut lifter and sprayers - The recommended implements are demonstrated to farmers.

Correspondence: English.

UNION OF THE SOVIET SOCIALIST REPUBLICS

Armeniya

Armyanskii Nauchno - Issledovatelski Institut Ghidrotekhniki i Melioratzii (Armenian Research Institute for the Development of Land and Water Resources)
Prospekt Ordzhonikidze 3, Erevan

Activities: Research on hydraulics - Land improvement through irrigation.

Armyanskii Selskokhozyaistvennyi Institut (Armenian Agricultural Institute)
Ul. Teryana 74, Erevan

Activities: Mechanization of agricultural production - Land improvement through irrigation, motorized transports.

Azerbaydzhan

<u>Azerbaidzhanskii Nauchno-Issledovatelskii Institut Mekhanizatzii i Elektrifikatzii Selskogo Khozyaistva</u> (Azerbaidjan Research Institute for Farm Mechanization and Rural Electrification)
Kirovabad

<u>Activities</u>: Advanced studies to improve the technology of mechanization in farming operations, crop and stock farming, under the conditions of Azerbaydzhan - Postgraduate training.

<u>Azerbaidzhanskii Selskokhozyainstvennyi Institut</u> (Azerbadjan Agricultural Institute)
Ul. Azizbekova 222, Kirovabad

<u>Activities</u>: Mechanization and electrification of agricultural production.

Byelorussiya

<u>Byelorusskii Nauncho-Issledovatelskii Institut Melioratzii i Vodnogo Khozyaistva</u> (Byelorussian Research Institute for Land Improvement and Water Management)
Prospekt Stalina 68, Minsk

<u>Activities</u>: Improvement of agriculture and mechanization of irrigation work - Postgraduate training courses.

<u>Byelorusski Nauchno-Issledovatelskii Institut Mekhanizatzii i Elektrifikatzii Selskogo Khozyaistva</u> Byelorussian Research Institute for Farm Mechanization and Rural Electrification)
Vyistavka 15, Minsk

<u>Activities</u>: Advanced studies in technology of mechanization of farming operations, in crop and stock farming under the conditions of Byelorussiya - Post-graduate training courses.

<u>Byelorusskii Institut Mekhanizatzii i Elektrifikatzii Selskogo Khozyaistva</u> (Byelorussian Institute for Farm Mechanization and Rural Electrification)
Prospekt Stalina 129, Minsk

<u>Activities</u>: Farm mechanization and rural electrification.

<u>Byelorusskii Lyesotekhnicheskii Institut imeni S.M. Kirova</u> (Byelorussian Institute of Forest Technology named after Kirov)
Ul. Sverdlova 85, Minsk

<u>Forest engineering activities</u>: Mechanization of forest work - Technology of wood and wood industry.

Byelorusskaya ordena Trudovogo Krasnogo Znameni Selskokhozyaistvennyi Institut (Byelo-Russian Agricultural Institute of the Order of the Red Flag of Labour)
Gorki, Mogilevskoya oblast

Activities: Farm management - Farm mechanization - Development of land and water resources - Land improvement through irrigation.

E s t o n i y a

Estonskaya Selskokhozyaistvennaya Akademya (Estonian Agricultural Academy)
Pl. Lenina 1, Tartu

Activities: Farm Mechanization - Land improvement through irrigation - Forest engineering - Land management.

G r u z i y a (G e o r g i a)

Kutaisskii Selskokhozyaistvennyi Institut (Kutaisi Agricultural Institute)
Ul. Voroshilova 67, Kutaisi

Activities: Farm mechanization.

Vsesoyuznyi Nauchno-Issledovatelskii Institut Chaya i Subtropicheskikh Kultur (All-Union Research Institute for Tea and Sub-Tropical Crops)
P/O Anaseuli, Makharadze

Activities: Studies of equipment for sub-tropical crops - Mechanization in tea and sub-tropical crop production - Post-graduate study courses.

Gruzinskii Institut Subtropiccheskoyo Khozyaistva (Georgia Institute of Subtropical Agriculture)
Kelasuri, Sukhumi

Activities: Mechanization of agricultural production.

Gruzinskii ordena Trudovogo Krasnogo Znameni Selskokhozyaistvennyi Institut (Georgian Agricultural Institute of the Order of the Red Flag of Labour)
Prospekt Chavchavadze Avenue 33, Tbilisi (Tiflis)

Activities: Mechanization of farm labour - Development of equipment for complete electrification of agriculture - Machinery for mountain farming - Post-graduate training courses.

K a z a k h s t a n

Kazakhskii Nauchno-Issledovatelski Institut Mekhanizatzii i Elektrifikatzii Selskogo Khozyaistva (Kazakh Research Institute for the Farm Mechanization and Rural Electrification) Pos. Tashkenskaya 3-12, Alma-Ata

<u>Activities</u>: Mechanization of farm work – Development of machinery for complete mechanization and electrification of operations in crop and stock farming under the conditions of Kazakhstan – Post-graduate training.

<u>Kazakhskii Selskokhozyaistvennyi Institut</u> (Kazakh Agricultural Institute)
Ul. Krasina 123, Alma Ata

<u>Activities</u>: Farm mechanization – Rural electrification – Development of land and water resources – Irrigation.

<u>Tzelinogradskii Selskokhozyaistvennyi Institut</u> (Tzelinograd Agricultural Institute)
Tzelinograd, Tzelinniy Kray

<u>Activities</u>: Mechanization of agricultural production – Land management.

Kirghiziya

<u>Kirghizskii Selskokhozyaistvennyi Institut imeni K.I. Skriabina</u> (Kirghiz Agricultural Institute n.a. Skriabin)
Aktiubinskaya ul. 34, Frunze

<u>Activities</u>: Farm mechanization.

Latviya

<u>Latviiskaya Selskokhozyaistvennaya Akademiya</u> (Latvian Agricultural Academy)
Dvorstovii Ostrob, Yelgava, Riga

<u>Activities</u>: Farm mechanization – Land improvement through irrigation – Forest engineering – Mechanization wood technology – Farm management – Motorized transport.

<u>Latviiskii Nauchno-Issledovatelski Institut Ghidrotechniki i Melioratzii</u> (Latvian Research Institute for the Development of Land and Water Resources)
Bulvar Kommunarov, Riga

<u>Activities</u>: Mechanization of irrigation works under conditions of Latviya – Post-graduate training courses.

Litva (Lithuania)

<u>Litovskaya Selskokhozyaistvennya Akademiya</u> (Lithuanian Agricultural Academy)
Ul. Kestutchio 15, Kaunas

<u>Activities</u>: Farm mechanization – Development of land through irrigation – Mechanization in forestry – Land management.

Litovskii Nauchno-Issledovatelskii Institut Mekhanizatzii i Elektrifikazii Selskogo Khozyaistva (Lithuanian Research Institute for the Farm Mechanization and Rural Electrification)
Lithuanian SSR, Kaunas district, Raudondvaris

Status: Governmental institution, supervised by the Ministry of Agriculture Lithuanian SSR. Established in 1956.

Activities: Research and experiment work under the conditions of Lithuania - The drying of fodder by active ventilation and in high temperature dryers, the harvesting and storage of corn, hay, grassy seed, plants and roots, the mechanization of cattle farms and pigsties, the utilization of dung, the application of electric power. Post-graduate course (5 post-graduates studies to receive the candidate degree of technical science).

Publications: Institute scientific works on the problems of agricultural mechanization, materials of annual scientific conferences, articles in periodicals.

Correspondence: Lithuanian, Russian, English, German.

Moldaviya

Kishinevskii Selskokhozyaistvennyi Institut imeni M.V. Frunze (Kishenev Agricultural Institute n.a. Frunze)
Sadovaya ul. 121, Kishenev

Activities: Farm mechanization.

Moldavskii Nauchno-Issledovatelskii Institut Oroshaemogo Zemledelya i Ovoshchevodstvá (Moldavian Research Institute for Irrigated Crops and Market Gardening)
Ul. Mira 34, Tiraspol

Activities: Selection and breeding of market garden crops - Cultivation techniques for irrigated market garden crops - Development of mechanized techniques - Post-graduate training.

Russian Federation (Asia)

RSFSR/A: Barnaul

Altaiskii Selskokhozyaistvennyi Institut (Altai Agricultural Institute)
Pushkinskaya ul. 82, Barnaul

Activities: Mechanization of agricultural production.

RSFSR/A: Blagoveshchensk

Blagoveshchenskii Selskokhoyyaistvennyi Institut (Blagoveshchensk Agricultural Institute)
Politekhnicheskaya ul. 50, Blagoveshchensk, Amurskaya oblast.

Activities: Farm mechanization.

RSFSR/A: Irkutsk

<u>Irkutskii Selskokhozyaistvennyi Institut</u> (Irkutsk Agricultural Institute)
Ul. Timiryazeva 59, Irkutsk

<u>Activities</u>: Farm mechanization.

RSFSR/A: Krasnoyarsk

<u>Krasnoyarskii Selskokhozyaistvennyi Institut</u> (Krasnoyarsk Agricultural Institute)
Prospekt Stalina 82, Krasnoyarsk

<u>Activities</u>: Farm mechanization.

<u>Sibirskii Teknologicheskii Institut</u> (Siberian Institute of Technology)
Prospekt Stalina 82, Krasnoyarsk

<u>Forest engineering activities</u>: Machinery and equipment used in the lumber industry and forest management – Mechanical technology of wood – Chemical wood technology and apparatus.

RSFSR/A: Novosibirsk

<u>Novosibirskii Selskokhozyaistvennyi Institut</u> (Novosibirsk Agricultural Institute)
Bolshevistskaya ul. 172, Novosibirsk

<u>Activities</u>: Farm mechanization.

RSFSR/A: Omsk

<u>Omaskii Selskokhozyaistvennyi Institut</u> (Omsk Agricultural Institute)
Zagorodnaya Roshcha, Omsk

<u>Activities</u>: Development of land and water resources – Geodetic engineering – Farm mechanization

RSFSR/A: Primorskiy Kray

<u>Primorskii Selskokhozyaistvennyi Institut</u> (Agricultural Institute for the Maritime Province)
Zeleznodorozhnii Prospekt 42, Ussurisk Primorskiy Kray

<u>Activities</u>: Mechanization of agricultural production.

RSFSR/A: Ulan-Ude

<u>Buryatskii Selskokhizyaistvennyi Institut</u> (Buryat Agricultural Institute)
Ul. Kalandarashvili 18, Ulan-Ude

<u>Activities</u>: Farm mechanization – Farm building construction.

Russian Federation (Europe)

RSFSR/E: Arkhangelsk

<u>Arkhangelskii ordena Trudovogo Krasnogo Znameni Lyesotekhinicheskii Institut iemni V.V. Kuibysheva</u> (The Arkhangelsk Institute of Forest Technology named after V.V. Kuibyshev)
Naberezhnaya im. Stalina 17, Arkhangelsk

<u>Forest engineering activities</u>: Machinery and equipment used in the lumber industry and forest management – Mechanical and chemical technology of wood – Technology of board making – Building of factories and dwellings.

RSFSR/E: Balashikha

<u>Vsesoyuznyi Selskokhozyaistvennyi Institut Zaochnogo Obrazoyaniya</u> (All-Union Agricultural Institute for Education)
P/O Leontovo, Balashikha, Moskovskaya oblast.

<u>Activities</u>: Farm mechanization – Rural electrification – Development of land an water resources – Geodetic engineering.

RSFSR/E: Bryansk

<u>Bryanskii Lyesokhozyaistvennyi Institut</u> (Bryansk Forest Management Institute)
Sovietskaya ul. 18, Bryansk

<u>Forest engineering activities</u>: Forest machinery – Mechanization of forest work – Wood technology.

RSFSR/E: Chelyabinsk

<u>Chelyabinsk Institut Mekhanizatzii i Elektrifikatzii Selskogo Khozyaistva</u> (Chelyabinsk Institute for Farm Mechanization and Rural Elektrification)
Krasnaya ul. 38, Chelyabinsk

<u>Activities</u>: Farm mechanization and rural electrification.

RSFSR/E: Gorkiy

<u>Gorkovskiy Selskokhozyaistvennyi Institut</u> (Gorki Agricultural Institute)
Pl. Minina i Pozharskogo 7/1, Gorkiy

<u>Activities</u>: Farm mechanization.

RSFSR/E: Ioshkar-Ola

Povolzhskii Lyesotekhnicheskii Institut imeni A.M. Gorkogo (Volga Region Institute of Forest Technology n.a. Gorki)
Sovietskaya ul. 152, Ioshkar-Ola, Mariiskaya ASSR

Forest Engineering activities: Forest industries – Lumber machines and equipment – Wood technology.

RSFSR/E: I z h e v s k

Izhevskii Selskokhozyaistvennyi Institut (Izhevsk Agricultural Institute)
Ul. Kirova 18, Izhevsk

Activities: Farm mechanization.

RSFSR/E: K a z a n

Kazanskii Selskokhozyaistvennyi Institut (Kazan Agricultural Institute)
Ul. Karla Marksa 65, Kazan

Activities: Farm mechanization.

RSFSR/E: K i n e l

Kuibyshevskii Selskokhozyaistvennyi Institut (Kuibyshev Agricultural Institute)
Kinel, Kuibyshevskaya oblast

Activities: Farm mechanization.

RSFSR/E: K i r o v

Kirovskii Selskokhozyaistvennyi Institut (Kirov Agricultural Institute)
Oktyabrskaya ul. 73, Kirov

Activities: Farm mechanization.

RSFSR/E: K o s t r o m a

Kostromskii Selskokhozyaistvennyi Institut (Kostroma Agricultural Institute)
Ul. 1 Maya 14, Kostroma

Activities: Farm mechanization.

RSFSR/E: K r a s n o d a r

Kubanskii Selskokhozyaistvennyi Institut (Kuban Agricultural Institute)
Ul. Druzhby 107, Krasnodar

Activities: Farm mechanization.

Vsesoyuznyi Nauchno-Issledovatelskii Institut Maslichnykh i Efiro-Maslichnykh Kultur (All-Union Research Institute for Oleaginous and Essential Oil-bearing Plants)
P/Ya 50, Krasnodar

Activities: Mechanization of the cultivation, harvesting and processing of oleaginous and essential oil-bearing crops and their processing - Post-graduate study courses.

Vsesoyuznyi Nauchno-Issledovatelskii Institut Ispitanya Traktorov i Selskokhozyaistvennikh Mashin (Kuban Research Institute for Tractor and Machinery Testing)
P/O Mayak Revolyutzii, Novokubanskii Rayon, Krasnodarskiy Kray

Activities: Development of techniques and methods for testing of tractors and agricultural machinery - Tractor and machine testing for complete mechanization of growing of staple crops - Devising of testing equipment.

Vsesoyuznyi Nauchno-Issledovatelskii Institut Tabaka i Makhorki imeni A.I. Mikoyana (All-Union Research Institute for Tobacco and Shag n.a. Mikoyan)
P/Ya 55, Krasnodar

Activities: Design of machinery for planting, harvesting and curing tobacco of all grades.

RSFSR/E: Leningrad

Vsesoyuznyi Nauchno-Issledovatelskii Institut Zashchity Rastenii (All-Union Research Institute for Plant Protection)
Ul. Gertzena 42, Leningrad

Activities: Design of equipment for controlling plant pests and diseases taking into account various soil-climatic conditions and the agricultural practices in the cultivation of various farm crops - Study courses.

Vsesoyuznyi Zaochnyi Lyesotekhnicheskii Institut (All-Union Institute for Forest Technology)
Institutskii per. 3, Leningrad 18

Activities: Machinery and equipment used in the lumber industry and forest management - Mechanical and chemical technology of wood - Automation and remote control technique - Technology of board making.

Leningradskaya ordena Lenina Lyesotekhnicheskaya Akademyia imeni S.M. Kirova (Leningrad Academy for Forest Technology of the order of Lenin n.a. Kirov)
Institutskii per. 5, Leningrad 18

Forest engineering activities: Automation and remote control techniques - Machinery and equipment used in the lumber industry and forest management - Wood technology and pulp and paper making industry.

RSFSR/E: Michurinsk

Nauchno-Issledovatelskii Institut Sadovadstva imeni I.V. Michurina (Research Institute for Horticulture n.a. Michurin)
Michurinsk

Activities: Mechanization of work in the cultivation and care of vegetable and berry crops - Post-graduate training courses.

RSFSR/E: Moskva

Moskovskaya ordenya Lenina Selskokhozyaistvennaya Akademiya K.A. Timiryazeva (Moscow K.A. Timiryazev Agricultural Academy of the Order of Lenin)
Novoie Chaussée 51, Moskva

Activities: Mechanization and electrification of agricultural production - Land improvement through irrigation - River hydraulics works - Hydro-electric power stations.

Gosudarstvennyi Soyuznyi Nauchno-Issledovatelskii Traktornyi Institut (NATI) (State Institute for Tractor Research)
Ul. Verkhnaya 34, Moskva 40

Activities: Formulation and development of theories of tractor engine design - Manufacture and production of machinery based on new principles or new materials - Determination of technical requirements of tractor fuels and oils - Improvement in the main features of existing tractor types - Development of new tractor testing methods - Collection of documents for determining tractor trends - Assessment and technical analysis of tractor production in the world - Standardization - Circulation of up-to-date information from tractor factories and institutes.

Nauchno-Issledovatelskii Institut Technologii Traktorov i Selskokhozyaistvennogo Machinostroeniya (Research Institute for Tractor and Agricultural Machinery Construction)
Marksistaskaya ul. 20, Moskva

Branches: at Rostov-on-Don and Barnaul

Activities: Study and development of new machinery and construction methods - Theory and practice of manufacturing - Development of means for mechanizing technical supervision - Automation of mass production of spare parts of tractors and agricultural machines - Technical aid to factories in the production of new models - Organization of improved systems of production and determination of engineering and economic production standards on the basis of progress achieved.

Vsesoyuznyi Nauchno-Issledovatelskii Institut Selskogkaozyaistvennogo Mashinostroeniya (VISKHOM) (All-Union Research Institute for Agricultural Machinery Construction)
Pochtovoe otdelenie Wagonoremont, Moskovskaya oblast

Activities: Construction of new agricultural machinery in conjunction with factories on the basis of research and prototype testing - Development of new methods of machine design - Standardization of machine elements and parts - Preparation of scientific publications.

Vsesoyuznyi Nauchno-Issledovatelskii Institut Mekhanizatzii Selskogo Khozyaistva (All-Union Research Institute for Farm Mechanization)
Moskva, J-389

Activities: Advanced studies of the principles of farm mechanization - Technological processes in farm production - Extension work - Post-graduate training courses.

Vsesoyuznyi Nauchno-Issledovatelskii Institut Elektrifikatzii Selskogo Khozyaistva (All-Union Research Institute for Farm Electrification)
Moskva, J-389

Activities: Study of the principles of electric power applied to farm production - Elaboration of schemes for electric powered machinery - Rational utilization of power resources for rural electrification - Extension work - Post-graduate training courses.

Vsesoyuznyi Nauchno-Issledovatelskii Institut Chidrotekhniki i Melioratzii (All-Union Research Institute for the Development of Land and Water Resources)
Ul. Prianishnikove 19, Moskva A-8

Activities: Research in the field of agricultural hydrology, irrigation and drainage, water supply and the mechanization of land improvement works - Post-graduate study courses.

Vsesouznyi Selskokhozyaistvennyi Institut Zaotchnogo Obrazovanya (All-Union Agricultural School by Correspondence)
P/O Leonovo, Balashikha, Moskovskaya oblast

Activities: Mechanization and electrification of forest industries - Land improvement through irrigation.

Vsesoyuznyi Nauchno-Issledovatelskii Institut Udobrenii i Agropochvovodenya (All-Union Research Institute for Fertilizers and Soil Science)
Ul. Prianishnikova 31, Moskva, A-8

Activities: Mechanization of fertilizer application - Chemical weed control - Post-graduate training.

Vsesoyuznyi Nauchno-Issledovatelskii Institut Kormov imeni V.R. Willamsa (All-Union Research Institute for Fodder Crops n.a. V.R. Williams)
P/O Lugovaya, Moskovskaya oblast

Activities: Study of problems of harvesting fodder crops - Technology of haymaking - Theory and technology of ensilage making and feed mixtures - Post-graduate training.

Nauchno-Issledovatelskii Institut Ovoshchnogo Khozyaistva (Research Institute for Vegetable Production)
Perlovskaya, Moskva

Activities: Study of complete mechanization of vegetable growing in open fields and under cover - Post-graduate training.

Nauchno-Issledovatelskii Institut Kartofelnogo Khozyaistva (Research Institute for Potato Growing)
P/O Malakhovka, Raminskii Rayon, Moskovskoya oblast

Activities: Mechanization of potato planting and harvesting - Design of potato growing and harvesting equipment - Potato storage - Post-graduate training courses.

Moskovskii Institut Inzhenerov Zemleustroistva (Moscow Institute of Agrarian Engineering)
Ul. Kazakoba 15, Moskva

Activities: Land management - Geodetic engineering.

RSFSR/E: Mytishchi

<u>Moskovskii Lyesotekhnicheskii Institut</u> (Moscow Institute for Forest Technology)
Moselok Stroitel, Mytishchi, Moskovskaya oblast

<u>Forest engineering activities</u>: Machinery and equipment used in the lumber industry and forest management – Mechanical technology of wood – Technology of pulp and paper making – Automation and remote control techniques.

RSFSR/E: Novocherkassk

<u>Novocherkasskii Inzhenerno-Meliorativnyi Institut</u> (Novocherkassk Institute for Reclamation Engineering)
Pushkinskaya ul. 101, Novocherkassk

<u>Activities</u>: Development of land and water resources – Land improvement through irrigation – Forest management.

RSFSR/E: Ordzhonikidze

<u>Severo-Osetinskii Selskokhozyaistvennyi Institut</u> (North-Ossetian Agricultural Institute)
Timiryazevskii per. 3-5, Ordzhonikidze

<u>Activities</u>: Farm mechanization.

RSFSR/E: Orenburg

<u>Orenburgskii Selskokhozyaistvennyi Institut</u> (The Orenburg Agricultural Institute)
Ul. Cheluikintzev 18, Orenburg

<u>Activities</u>: Farm mechanization.

RSFSR/E: Penza

<u>Pensenzkii Selskohozyaistvennyi Institut</u> (Penza Agricultural Institute)
Posielok Akhuny, Penza

<u>Activities</u>: Farm mechanization.

RSFSR/E: Perm

<u>Permskii Selskokhozyaistvennyi Institut imeni D.N. Prianishnikov</u> (Perm Agricultural Institute n.a. D.N. Prianishnikov)
Kommunisticheskaya ul. 23, Perm

<u>Activities</u>: Mechanization of agricultural production.

RSFSR/E: P u s h k i n

<u>Leningradskii Selskokhozyaistvennyi Institut</u> (Leningrad Agricultural Institute)
Komsomolskaya ul. 14, Pushkin, Leningradskoi oblasti

<u>Activities</u>:   Farm mechanization and rural electrification.

RSFSR/E: P u s h k i n o

<u>Vsesoyuznyi Nauchno-Issledovatelskii Institut Lyesovodstva i Mekhanizatzii Lyesnogo Khozyaistva</u> (All-Union Research Institute for Forestry and Mechanization of Forest Management)
Pisareveskaya ul. 12, Pushkino, Moskovskaya obl.

<u>Activities</u>:   Mechanization of work in forest management - Post-graduate study courses.

RSFSR/E: R y a z a n

<u>Ryazanskii Selskokhozyaistvennyi Institut imeni Prof. P.A. Kostycheva</u> (Ryazan Agricultural Institute n.a. Kostychev)
Ul. Sverdlova 26, Ryazan

<u>Activities</u>: Farm mechanization.

RSFSR/E: S a r a t o v

<u>Saratovskii Institut Mekhanizatzii Selskogo Khozyaistva imeni M.I. Kalinina</u> (Saratov Institute for Mechanization of Agriculture n.a. Kalinin)
Sovietskaya ul. 60, Saratov

<u>Activities</u>: Farm mechanization and rural electrification - Development of water resources.

<u>Saratovskii Selskokhozyaistvennyi Institut</u> (Saratov Agricultural Institute)
Pl. Revolutzii 1, Saratov

<u>Forest engineering activities</u>:   Forest mechanization.

RSFSR/E: S t a v r o p o l

<u>Stavropolskii Selskokhozyaistvennyi Institut</u> (Stavropol Agricultural Institute)
Zootekhnicheskii per. 10, Stavropol

<u>Activities</u>:   Farm mechanization.

RSFSR/E: S v e r d l o v s k

Sverdlovskii Selskokhozyaistvennyi Institut (Sverdlovsk Agricultural Institute)
Ul. Klara Liebknechta 42,   Sverdlovsk

Activities: Farm mechanization.

Uralskii Lyesotekhnicheski Institut (Ural Institute for Forest Technology)
Sibirskii trakt, 5-i kilometr, Sverdlovsk

Forest Engineering activities: Machinery and equipment used in the lumber industry and forest management – Mechanical technology of wood – Chemical wood technology – Board manufacture.

RSFSR/E: T o r j h o k

Vsesoyuznyi Nauchno-Issledovatelskii Institut Lna (All-Union Research Institute for Flax)
Lunarcharskogo 35, Torjhok, Kalininskaya oblast.

Activities: Technology in the processing of flax – Mechanization and economics of flax cultivation in various zones of the USSR – Post-graduate training courses.

RSFSR/E: U f a

Bashkirskii Selskokhozyaistvennyi Institut (Bashkir Agricultural Institute)
Ul. Karla Marksa 12, Ufa

Agricultural Engineering activities: Farm mechanization.

RSFSR/E: U l y a n o v s k

Ulyanovskii Selskokhozyaistvennyi Institut (Ulyanovsk Agricultural Institute)
Novyi Veneyz 1, Ulyanovsk

Activities: Farm mechanization.

RSFSR/E: V o l g o g r a d

Vsesoyuznyi Nauchno-Issledovatelskii Institut Agrolyesomelioratzii (All-Union Research Institute for Land and Forest Improvement)
Volgograd 21 (former Stalingrad)

Activities: Investigation on wind and water erosion of soils and their control through mechanized shelter belt and shrub plantations – Post-graduate training courses.

RSFSR/E: V o l o g d a

Vologodskii Molotchnyi Institut (Vologda Dairy Institute)
Prospekt Shmidta 2, Poselok Molotchnoy, Vologda

Activities: Mechanization of agricultural production.

RSFSR/E: V o r o n e z h

<u>Voronezhskii Selskokhozyaistvennyi Institut</u> (Voronezh Agricultural Institute)
Ul. Mitchurina 1, Voronezh

<u>Activities</u>: Farm mechanization.

<u>Voronezhskii Lyesoteknicheskii Institut</u> (Voronezh Institute for Forest Technology)
Ul. Timiryazeva 8, Voronezh

<u>Forest Engineering activities</u>: Machinery and equipment used in the lumber industry and forest management – Road construction – Motorized transports.

RSFSR/E: Z e r n o v o y

<u>Vserossiskii Nauchno-Issledovatelskii Institut Mekhanizatzii i Elektrifikatzii Selskogo Khozyaistva</u> (All-Russian Research Institute for Farm Mechanization and Rural Electrification)
Zernovoy, Rostovskaya oblast

<u>Activities</u>: Advanced technological studies of machinery design and mechanization of farm work with special emphasis on grain and hay crops – Post-graduate training courses.

<u>Azovo-Chernomosrskii Institut Mekhanizatzii i Elektrifikatzii Selskogo Khozyaistva</u>
(Azov-Black Sea Institute for Mechanization and Electrification of Agriculture)
Zerzovoy, Rostovskaya oblast

<u>Activities</u>: Farm mechanization and rural electrification.

T a d z h i k i s t a n

<u>Tadzhikskii Selskokhozyaistvennyi Institut</u> (Tadzhik Agricultural Institute)
Ul. Shevchenko 17, Duchambe (former Stalinabad)

<u>Activities</u>: Farm Mechanization – Development of land and water resources.

T u r k m e n i y a

<u>Turkmenskii Selskokhozyaistvennyi Institut imeni V.V. Kalinina</u> (Turkmen Agricultural Institute n.a. Kalinin)
Sad Keshi, Asnhkhabad

<u>Activities</u>: Farm mechanization.

U k r a i n a

Ukr: D n i e p o p e t r o v s k

Dniepopetrovskii Selskokhozyaistvennyi Institut (Dniepopetrovsk Agricultural Institute)
Ul. Voroshilov 25, Dniepopetrovsk

Activities: Farm mechanization.

Vsesoyuznyi Nauchno-Issledovatelskii Institut Kukuruzy (All-Union Scientific Research Institute for Maize)
Ul. Dzerjhinskogo 14, Dniepopetrovsk

Activities: Along with selective seed growing work the institute conducts research on the agricultural engineering in maize cultivation and the generalization of advanced experimentation in the production of abundant steady crops of maize in the various soil and climatic zones of the U.S.S.R. - Post-graduate study courses.

Ukr: Kharkov

Kharkovskii Institut Mekhanizatzii Selskogo Khozyaistva (Kharkov Institute for Mechanization of Agriculture)
Prospekt Stalina 45, Kharkov

Activities: Farm mechanization and rural electrification.

Kharkovskii Ordnya Trudovovo Krasnovo Znameni Selskokhozyaistvennyi Institut imeni V.V. Dokuchaevea (Kharkov V.V. Dokuchaev Agricultural Institute of the Order of the Red Flag of Labour)
Ul. Artema 44, Kharkov

Activities: Farm mechanization - Work organization.

Ukr: Kiev

Ukrainskii Nauchno-Issledovatelskii Institut Mekhanizatzii i Elektrifikatzii Selskogo Khozyaistva (UNDIM)(Ukrainian Research Institute for Farm Mechanization and Rural Electrification)
Goloseyevo 41, Kiev

Activities: Study of the principles of farm mechanization and rural electrification under the conditions of the Ukrainian S.S.R. - Research on new techniques for mechanization of staple crop growing - Post-graduate study courses.

Ukrainskaya ordena Trudovogo Krasnogo Znameni Selskokhozyaistennya (Ukrainian Agricultural Academy of the Order of the Red Flag of Labour)
Goloseyevo 41, Kiev

Activities: Farm mechanization - Rural Electrification - Forest engineering - Machinery and equipment used in the timber industry.

Vsesoyusnyi Nauchno-Issledovatelskii Institut Sakharnoi Sviokli (All-Union Research Institute for Sugar Beets)
Klinichevskaya 23, Kiev

Activities: Agricultural engineering problems and mechanization of sugar beet cultivation, harvesting and processing in the various zones of the USSR - Post-graduate study courses.

Ukrainskaya Akademiya Selskokhozyaistvennich Neuk (Ukrainian Academy of Agricultural Sciences)
Goloseyevo, Kiev

Activities: Mechanization and electrification of agricultural production – Mechanization in forestry.

Ukr: L v o v

Lvovski Selskokhzyaistvennyi Institut (Lvov Agricultural Institute)
Nesterovskogo Raisona Duelyany, Lvovskaya oblast

Activities: Farm mechanization – Farm management.

Lvovskii Lyesotekhnicheskii Institut (Lvov Institute for Forest Technology)
Pushkinskaya ul. 103, Lvov

Forest engineering activities: Mechanization of wood technology.

Ukr: Y a l t a

Vsesoyuznyi Nauchno-Issledovatelskii Institut Vinodyelya i Vinogradstva "Magarach"
(All-Union Magarach Scientific Research Institute for Viti-Viniculture)
Ul. Kirova 25, Yalta, Crimskaya oblast

Filipovitz, W.	Director
Walvjko, G.	Deputy Director
Svjatinov, I.	Deputy Director

Status: One Branch office in Moscow, 4 research farms – Founded 1928 – Directed by Ministry of Food Economy of the USSR.

Activities: University teachers' training (improving of the vine quality, agrotechnics, production of landing vine material, protection of vineyards against pests and diseases, mechanization at the vineyards, fertilization systems, irrigation and mastering of useful areas, vineyard economics, winemaking technology, environmental protection, complex mechanization and automatization of the winemaking processes, winemaking microbiology). – Research works: improvement of the vine quality, agrotechnics, production of landing vine material, protection, fertilization, irrigation, economics, winemaking technology, mechanization.

Publications: Biochemistry and colour wines technology – Mechanization of the auxiliary processes in winemaking – Hermetic tanks and protective coverings in winemaking – Question about chemistry of vines and wine – Complex processing of the secondary products of winemaking – Increasing of vine yields and improvement of the quality in wine production - Scientific and technical achievements in the field of winegrowing and winemaking - Processing line of vine into the white wines - Microbiology of winemaking - Grape wines - Solar radiation and productivity of vineyards.

Correspondence: English, French, German, Spanish, Italian.

U z b e k i s t a n

Tashkentskii Selskokhozyaistvennyi Institut (Tashkent Agricultural Institute)
Ul. Kirova 32, Tashkent

Activities: Farm and forest mechanization.

Tashkentskii Institut Irrigatzii i Mekhanizatzii Selskogo Khozyaistva (Tashkent Institute for Irrigation and Mechanization of Agriculture)
Uchitelskaya ul. 29, Tashkent

Activities: Farm mechanization - Development of water resources - Land improvement through irrigation.

Sredne-Aziatskii Nauchno-Issledovatelskii Institut Mekhanizatzii i Elektrifikatzii Selskogo Khozyaistva (Central Asian Research Institute for Mechanization and Electrification of Agriculture)
Yanghi-Yul, Tashkentskoi oblast

Activities: Mechanization and electrification of crops under irrigation - New technique of cotton cultivation and harvesting - Technical studies for complete mechanization of basic farming operations - Post-graduate training courses.

UNITED KINGDOM

Aberdeen

Agricultural Engineering/Farm Buildings Group, School of Agriculture, Craibstone, Bucksburn
Aberdeen AB2 9TR Tel.: Aberdeen 713741 (0224)

Shiach, J.G.	BSc(Eng) BSc(Agr) FIAgrE	Chairman of Group
Gerrie, W.A.G.	MC ARICS	Head of Farm Buildings Division

Status: Engineering Division was established in 1947. School of Agriculture consists of North of Scotland College of Agriculture in association with Agriculture Department of Aberdeen University. College is financed from Government Sources through a Board of Governors.

Activities: Teaching: The Group provides courses for students studying for BSc(Agric), OND and HND in Agriculture. - Extension: The Group provides a Mechanization and Farm Buildings Advisory Service throughout the North of Scotland. - Research and Development: Mechanization systems, drainage, grain handling and processing, building monitoring, anaerobic digestion of slurries, energy saving in farm dairies, whole-crop cereals harvesting.

Publications: Back-up publications for Advisory work, waste treatment, ventilation, fencing.

Correspondence: English, French.

Ashford

Wye College, University of London
Wye, near Ashford, Kent, England

Warboys, I.B.	M.Sc., F.I.Agr.E.	Soil tillage – in charge of Agricultural Engineering
Wilkes, J.M.	M.Phil., A.M.I.Agr.E.	Tractor performance.

Status: University: Financed by University Grants Committee.

Activities: a) Teaching: Courses in agricultural engineering and farm mechanization for B.Sc. agriculture students, attended by 45 students; also horticultural mechanization for B.Sc. horticulture students, attended by 20 students. - b) Research: Development of implements for deep soil tillage. Analysis of tractor performance in the field.

Publications: Hay storage preservatives. Development of Double Digger.

Correspondence: English.

Ayr

Department of Agricultural Engineering and Surveying, The West of Scotland Agricultural College, Donald S. Hendrie Building
Auchincruive, by Ayr, KA6 5HW

Mouat, G.C.	B.Sc.Agr., N.D.Agr.E., F.I.Agr.E.	Head of Department
Hair, I.	B.SC.Agr.(Hons.)	
Jones, A.	B.Sc.A.E.(Hons.)	
Pickering, J.	N.D.Agr.E., Cert.Ed.	
Veitch, A.	B.Sc.M.E.(Hons., M.I.Agr.E.	Senior Lecturer
Walker, A.E.	N.D.A., N.D. Agr.E., H.N.C.M.E., M.I.Agr.E.	
Wallington, R.S.	B.Sc.Agr., M.Sc.A.E., F.I.Agr.E.	Senior Lecturer

Status: Independent institution, but relying on funds provided by Government and supervised by a Board of Governors.

Activities: Agricultural engineering instruction as part of the study courses in all diploma courses within the College. Specialized teaching for the courses leading to the College Diploma in Agricultural Engineering and to a Higher National Diploma in Engineering (Agriculture) mounted in conjunction with the Bell College of Technology, Hamilton.

Correspondence: English.

Biggleswade

Shuttleworth Agricultural College
Old Warden Park, Biggleswade, Beds. SG18 9DX

Kennedy, P.D.	N.D.Agric. E. Cert, Ed.	Lecturer in Farm Mechanization
Costley, A.G.	B.Sc.(Agric) Cert. Ed.	Lecturer in Farm Mechanization
Davey, A.P.	Dip. Agric. Eng. A.I. Agr. E.	Lecturer in Farm Mechanization

Status: Established 1947. Direct Grant College.

Activities: Teaching: Higher National Diploma (Agriculture) - Higher National Diploma (Arable Option 1981) - Ordinary National Diploma - Farmers Course - H.N.D. 40 students - O.N.D. 40 students - F.C. 30 students. Research: Arable trials and demonstration plots in collaboration with the Crops Department.

Correspondence: English.

Cambridge

Department of Applied Biology, University of Cambridge
Pembroke Street, Cambridge   Tel.: 358381

Barker, M.G.	M.A.	Lecturer, Farm implements and machinery

Activities: Some problems of food supply.

Correspondence: English.

Edinburgh

Department of Agricultural Engineering and Mechanization, The Edinburgh School of Agriculture
West Mains Road, Edinburgh, EH9 3JG   Tel.: 031-667-1041

Witney, B.D.	M.Sc., Ph.D., NDA, CEng., MIMechE, FIAgr.E.	Head of Department Soil Mechanics; Mechanization Systems Analysis
Buckingham, J.F.	B.Sc.	Lecturer – Environmental Control of Livestock Buildings
Jack, D.A.	M.Sc., NDA, NDAgr.E. MIAgr.E.	Lecturer – Field and Crop Protection Machinery; Seed Technology
Parkes, M.E.	M.Sc., Ph.D., MIAgr.E.	Lecturer – Water Resource Management
Morrison, R.R.	B.Sc., MIAgr.E.	Senior Mechanization Adviser – Forage Crop Conservation
Elrick, J.D.	B.Sc., M.Phil.	Mechanization Systems Economist Machinery Investment
Jeffrey, W.A.	B.Sc.	Mechanization Adviser – Farm Power
Langley, A.	M.Sc., NDAgr.E.	Mechanization Adviser – Cultivations.

Status: Semi-governmental institution, directed and financed by the Department of Agriculture and Fisheries for Scotland and the University of Edinburgh.

Activities: Courses in Farm Mechanization for: postgraduate Degree in Seed Technology with 10 students; undergraduate Degrees in Agriculture and Agricultural Sciences with 40-50 students; Higher Diploma in Agriculture with 40-50 students; Diploma in Agriculture with 20-30 students. Postgraduate research programme. Provision of advisory, development, demonstration and information services for the East of Scotland on all aspects of Agricultural Engineering. Research investigations on Soil Machine Mechanics, Stone Windrowing for Potato Production, Mechanization System Analysis, Environmental Control of Agricultural Buildings, Grass Conservation.

Publications: Bulletins, reports, advisory leaflets, popular press articles and research papers on machinery systems for silage; hay and silage in big roll bales; stone windrowing for potatoes; combine harvester selection; tractor power selection for primary tillage; cultivations; grain drying; environmental control of livestock buildings; and water resource management.

Correspondence: English.

Kenilworth

Farm Buildings Information Centre, National Agricultural Centre
Kenilworth, Warwicks CV8 2LG

Blockway, B.N.	N.D.A., D.M.S.	Manager – Agriculturist
Loynes, I.J.	B.Sc.(Hons.), Ag.Eng.	Agricultural Engineer
Cowin, A.J.	A.R.I.C.S.	Surveyor (Buildings)

Status: Private

Activities: Collects, collates and disseminates information on farm buildings, materials and components, plus all fixed and mechanical equipment used in buildings. Surveys and case studies. Trade, technical and scientific reference library. Worldwide abstracting.

Publications: Farm Buildings Digest (quarterly).

Correspondence: English.

London

Agricultural Development and Advisory Service, Ministry of Agriculture, Fisheries and Food
Great Westminster House, Horseferry Road, London SW 1P 2AE   Tel.: 01-216-6311

Evans, D.P.   B.Sc., M.Sc.(Agr.E.)   Senior Mechanization Advisory Officer
ADAS, Liaison Unit,
NIAE, Wrest Park,
Silsoe, Bedford MK 45 4HS   Tel.: Silsoe (0526) 60077

Regional Mechanization Advisory Officers: **Region**

Finney, J.B.   B.SC. Dip(Agric.) NDAgrE.   Eastern
Government Buildings
Brooklands Avenue
Cambridge CB2 2DR

Jamieson, M.   B.Sc., M.Sc.(Agr.E.)   Northern
Government Buildings
Kenton Bar
Newcastle upon Tyne, NE1 2YA

Nicholson, R.J.   B.Sc.(Agr.) Hons.   South-Western
Government Buildings
Burghill Road
Westbury on Trym
Bristol BS10 6NJ

Shipway, G.P.   B.Sc., M.Sc.(Agr.E.)   West Midlands
Woodthorn
Wolverhampton WV6 8TQ

Smith, G.   NDA, NDD   East Midlands
Shardlow Hall
Shardlow, Derby DE7 2GN

Spear, G.B.H.   NDAgr.E.   Wales
Trawscoed
Aberystwyth SY23 4HT

Grundey, J.K.	NDA, NDAgr.E.	Yorks and Lancs

Government Buildings
Lawnswood
Leeds LS16 5PY

Willows, D.E.	B.Sc.	South Eastern

Government Buildings
Coley Park
Reading RG1 6DT

<u>Activities</u>: Technical advice to farmers and growers on all agricultural engineering matters. The work is carried out by Mechanization Advisory Officers grouped into regional teams.

<u>Publications</u>: Bulletins and leaflets on: Mowers - Field crop sprayers - Machinery for swath of hay - Grain driers - Drills - Farmyard manure handling and waste disposal - Sugar beet harvesting - Pick-up balers - Farm seed cleaning and grading machinery - Forage harvesters - Fertilizers handling and broadcasting - Combine harvesters - Farm grinding - Barn hay drying - Food mixers - Livestock feeding of forage and concentrates - Potato harvesting - Mechanical thinning and gapping of row crops - Glasshouse heating and ventilation - Irrigation equipment for horticultural crops - Environmental control in glasshouses, cold stores and stockbuildings - Handling unit loads.
Reports of studies: combine monitors, sugar beet harvesters, carrot damage, bale handling, feeding silage, slurry stores, handling liquid fertilizer.

<u>Correspondence</u>: English.

Royal Agricultural Society of England, National Agricultural Centre
Stoneleigh, Kenilworth, Warwickshire, England

Jackson, G.H.	Agricultural Director
Culpin, C.	Honorary Consulting Engineer

<u>Activities</u>: Organization of the Annual Royal Show with exhibits of all major agricultural machines and implements manufactured by British firms or under licence in U.K. - Silver Medal Awards for new implements at Royal Show - Development of new National Agricultural Centre, Kenilworth, Warwickshire, which includes a comprehensive electro-agricultural centre.

<u>Publications</u>: Annual Journal - NAC News, bi-monthly review of the Society's activities - Agricultural education leaflets.

<u>Correspondence</u>: English.

British Standards Institution
2 Park Street, London W1A 2BS

Tel.: 01 629 9000
Telex: 266933 (BSI LONDON)
Head Office
23218 (STNADARDS LDN)
Sales and Accounts

Feilden, G.B.R.	CBE FRS	Director General
Robinson, L.T.		Agricultural Machinery standards
Kenyon, P.		Farm building and fences standards
		Landscape work standards and technical information services

Meredith, C.   Landscape work standards
                                          Dairying engineering standards
                                          Agricultural food products standards

Branches: Manchester Office, 115 Portland Street, Manchester M1 6EB – Textiles Dept.

Status: Autonomous institution, established in 1901, comprising representatives of government, industry, engineering, professional institutions and scientific organizations.

Agricultural engineering activities: Standardization of agricultural machinery and equipment, farm buildings and fixtures, landscape work nomenclature, materials and operations.

Publications: Fences, farm gates, cattle grids, general purpose farm buildings, glossary of terms for landscape work, topsoil, turf, peat, tree work, transplanting semi-mature trees, farm dairy buildings and fixed equipment for cow houses, agricultural machinery, nomenclature, tractors and attachments, safety requirements, pesticide names, milking installations, refrigerated farm tanks, milk cans, slat conveyors for milk bottles, packaging (cartons) for dairy products, milk tankers, cleaning-in-place of dairy equipment, use of detergents in dairy industry, chemical analysis and microbiological examination of dairy products, analysis of oil seeds, oilseed residues, oils and fats, cereal and pulses, meat and meat products, spices and condiments, tea, coffee, cocoa, starch, tobacco and tobacco products, storage of cereals.

Correspondence: English, French, German, Russian, Spanish, Italian, Swedish.

The Electricity Council
30 Milbank, London SW1P 4RD

Agricultural Section, located at: Farm-electric Centre, National Agricultural Centre, Stoneleigh, Kenilworth, Warwickshire CV8 2LS

Wakeford, P.	CEng, MIEE, FIAgrE	Head of Section
Clark, E.M.C.	CEng, MIEE, FIMH	Markets Officer
Mitchell, C.D.	BSc, MAgSc, PhD, MIAgrE	Product Officer, Livestock
Paterson, H.	BSc(Agric), NDA, FIAgrE	Product Officer, Crops
Pavey, S.R.	BSc(Hons)	Assistant Product Officer, Monitoring and Development
Southall, M.C.	BSc(Agric), NDA	Assistant Product Officer, Maintenance and Monitoring
Weir, J.A.C.W.	NDA, MIAgrE	Product Officer, Horticulture

Status: Re-organized and nationalized 1 April 1948. Semi-governmental with direct finance. Agricultural Marketing and Development Section established 1970.

Activities: Information: Maintaining technical and commercial information on all aspects of electricity utilization for agriculture and horticulture. – Demonstration: By exhibitions and full scale practical methods. – Teaching: By utilization appreciation courses and conferences. – Marketing: Dissemination of information and close collaboration with all leading agricultural and horticultural authorities in the United Kingdom. – Research: Promoting and assisting in utilization research development projects by either official and/or direct financed University and other research centres.

Publications: Safe use of electricity on the farm and in horticulture – Farm-electric handbooks: grain drying and storage, hay drying, vegetable storage, essentials of farm lighting, pumping and irrigation. – Grow-electric handbooks: growing rooms, lighting in greenhouses, ventilation for greenhouses. - Technical Information Sheets are also available.

Correspondence: English.

National Farmers' Union
Agriculture House, Knightsbridge, London, SW1X 7NJ

Brutey, C.V.   Tech. CEI, A.I.Agr.E., M.A.S.A.E.   Assistant Director, Technical and Commercial and Machinery Officer.

Branches: The Union has 49 County Branches with Offices in most County Towns but no technical staff are employed at Branch Offices.

Status: The Technical and Commercial Division is a section of National Farmers' Union Headquarters working under the oversight of the Technical and Machinery Committee of the Union. The Union was formed in 1909.

Agricultural Engineering activities: Liaison between farmers and Research Organizations - advice to Research Organizations and Development Departments of Machinery Manufacturing Companies on agricultural machinery matters - surveys on machine life, performance etc. - representations of farmers on various Technical Committees of the British Standards Institution formulating standards for Agricultural Machines and components - Advice to farmers on safety matters with regard to machinery and representation of farmers on Government and National Organization Committees concerned with farm machinery safety.

Publications: No specialist publications relating to Farm Machinery are issued.

Correspondence: English, French, Dutch, German.

Muckamore

Agricultural Engineering Department, Greenmount Agricultural College
Antrim, Co. Antrim, N. Ireland   Tel.: Antrim 2114

Martin, W.   B.Agr., NDAg.E.   Head of Department
McCracken, W.R.   NDA, NDAg.E.   Lecturer

Status: Department established within the College in June 1961.

Activities: Teaching consists of mechanization as a main subject in the one-year certificate agricultural course, Advanced certificate course, Three year ordinary national diploma course in agriculture (O.N.D.)

Correspondence: English.

Newcastle-upon-Tyne

Department of Agricultural Engineering, The University of Newcastle-upon-Tyne
NE1 7RU   Tel.: Newcastle (0632) 28511 ext.2865

Greig, D.J.   Head of Department

Feeney, J.J.		Demonstrator Agricultural Mechanization	
Hettiaratchi, D.	Dr.	Senior Lecturer Soil Mechanics	
McCarthy, T.T.		Lecturer – Electronics, control and Dynamics	
O'Callaghan, J.R.		Professor – Thermodynamics	
Reece, A.R.	Dr.	Reader – Soil Machine Mechanics	
Wills, B.M.D.	Dr.	Senior Lecturer – Design	
Woods, J.L.	Dr.	Lecturer – Thermodynamics	

Status: Established as a Department within the Faculty of Agriculture in 1947. Undergraduate and post-graduate teaching financed by the University. Research financed, in general, from contracts outside the University.

Agricultural Engineering activities: a) Teaching: Two undergraduate courses leading to the B.Sc.(Hons) Degree. The first established in 1964 in Agricultural Engineering with an intake of 20 students per year and the second established in 1975 in Agricultural Mechanization with an intake of 15 students per year. – M.Sc. degrees in Agricultural Engineering and Mechanization by 1 year course for suitably qualified candidates (otherwise a qualifying year also required) or by research over 1 year. – Post-graduate research leading to Ph.D. degree. – b) Research: Activities in: soil-machine mechanics; crop production; mechanization and conservation; grain drying and storage; mechanization systems; energy in agriculture, including solar wind; farm waste disposal and recycling; systems analysis and mathematical modelling; design and development of machines including low cost tractors; electronics in agriculture for control and monitoring; pipe laying in saturated soils.

Publications: Soil mechanics; grain drying; pipe laying; farm waste recycling; modelling of mechanization systems; economics of grain drying and storage; losses during storage.

Correspondence: English, French, German.

Newport

Agricultural Engineering Department, Harper Adams Agricultural College
Newport, Shropshire, TF10 8NB   Tel.: Newport, Shropshire (0952) 811280

Gedye, I.D.	NDA, NDAg.E., FIAg.E., MASAE, Cert.Ed. Head of Department
Bradley, R.F.	NDA, CDAg.E., AIAg.E., FTC, Cert.Ed.
Giles, D.F.	FTC, NDAg.E.
Hemingway, P.F.	B.Sc.
Hunt, P.J.	B.Sc., MIAg.E.
Thirtle, P.J.	B.Sc., MIAg.E.

Branches with other location: Agricultural Engineering course run jointly with the Mechanical Engineering Department, The Polytechnic, Wolverhampton.

Status: College established in 1901. The Department of Education and Science direct and finance the College through the Governing Body of the College.

Agricultural Engineering activities: (a) Teaching: Higher National Diploma in Agricultural Engineering, three year, full time sandwich course. Annual intake, 23-30 students. – Application techniques and equipment as part of a one-year post diploma course in Crop Protection (20-25 students). – Farm Mechanization as part of the following courses: B.Sc. Sandwich Degree in Agricultural Technology (30 students); H.N.D. Agriculture (80-100 students); H.N.D. Agricultural Marketing and Business Administration (36 students). - (b) Research: Developmental research in controlled environment in pig and poultry housing, farm effluent disposal, crop storage, crop mechanization systems, crop protection application systems.

Correspondence: English.

Newton-Abbot

Engineering Section, Seale-Hayne College
Newton Abbot, Devon, TQ12 6NQ, England

Tel.: Newton Abbot 2323
Telegr.: Seale-Hayne, Newton Abbot

Jewett, D.W.	M.Sc., MA, FIAg.E.	Head, Engineering Section Agricultural Engineering
Bomford, P.H.	M.Sc., CEng., MIAg.E.	Agricultural Engineering
Carpenter, J.L.	M.Sc., NDAg.E., CEng., FIAg.E.	Farm Buildings
Heath, R.P.	NDAg.E., NDA, MIAg.E., T.Eng(CEI)	Agricultural Engineering
Morris, D.	B.Sc., MIAgE., T.Eng.(CEI)	Agricultural Engineering
Vranch, A.T.	B.Sc.	Chemical Engineering

Status: The College was established in 1901 - It has own Governing Body, but is directed and financed by the Government Department of Education and Science.

Activities: a) Teaching: Agricultural Engineering/Farm Mechanization is part of the Agriculture, Natural Resources and Rural Economy and Farm Management Courses. Additional courses provided at the College are in Food Technology, Applied Biology and Diploma in Management Studies. Agriculture courses are taught at Higher National Diploma and Degree level, other courses at Higher National Diploma level. - Student numbers: Agriculture (350), Natural Resources and Rural Economy (80), Food Technology (80), Applied Biology (40), Diploma in Management Studies (10), Diploma in Farm Management (20). b) Research: Optimisation of water use in farm dairies. Forage equipment testing for manufacturers.

Publications in Farm Business, Farm Mechanization and other journals.

Correspondence: English.

Oxfordshire

Rycotewood College
Priest End, Thame OX9 2AF, Oxfordshire, England

Tel.: Thame 2501

Teaching arrangements are not rigidly structured and although individual members have certain specializations they also teach other subjects. Any enquiries should be addressed to the Principal.

Status: Governmental.

Activities: Teaching: Technician and Higher Technician level. 200 plus students.
Research: Various projects for manufacturers.

Correspondence: English, French, German.

Penicuik

Scottish Institute of Agricultural Engineering
Bush Estate, Penicuik, Midlothian, Scotland

Tel.: 031-445-2147

Blight, D.P.	B.Sc., M.Sc.(Agr.Eng), Ph.D., CEng., FIMech.E.	Director
Agricultural Section:		
Soane, B.D.	B.Sc., MS, Ph.D., FIAE	Head of Department Senior Principal Scientific Officer
Campbell, D.J.	B.Sc., M.Phil.	Senior Scientific Officer - Acting Leader, Soil Mechanics Section
McRae, D.C.	B.Sc., M.Sc.(Agr.Eng) FIAgr.E.	Principal Scientific Officer - Potato Crop Mechanization
Pascal, J.A.	B.A., NDA, TEng.(CEI) MIAgr.E.	Principal Scientific Officer - Basic Cultivations and Liaison Section
Engineering Department:		
Gilfillan, G.	B.Sc., M.Sc., ARCST, CEng., FIMech.E.	Head of Department - Senior Principal Scientific Officer
Bailey, P.H.	B.Sc.(Eng), M.Sc.(Agr.Eng) AKC, FIAgr.E.	Principal Scientific Officer - Crop drying and storage
Palmer, J.	B.Sc., M.Sc.(Agr.Eng), MA, FIAgr.E.	Principal Scientific Officer - Machinery controls
Ramsay, A.M.	HND (Mech.Eng.), C.Eng.	Senior Scientific Officer - Raspberry harvesting
Spencer, H.B.	M.Sc., HND (Mech.Eng.), MFRAeS., C.Eng.	Principal Scientific Officer - Behaviour of implements and machines on sloping ground
Instrumentation Department:		
Parks, R.	B.Sc., Dip. AD, Ph.D., C.Eng., FIERE	Head of Department - Principal Scientific Officer
Porteous, R.L.	B.Sc.	Principal Scientific Officer - Spectral properties of agricultural materials

Status: Established in 1946. Funded by the Department of Agriculture and Fisheries for Scotland; Governed by the British Society for Research in Agricultural Engineering.

Activities: Research into cultivations, direct drilling and reduced cultivations for cereals - Potato crop mechanization - Automatic controls - Crop drying - Solar energy in crop drying - Behaviour of tractors, machines and implements on sloping ground - Soil compaction - Operational research and modeling - Hill land improvement - Utilization of farm machines - Raspberry harvesting machines.

Publications:since 1973: Tillage - Soil physical properties - Direct drilling of cereals - Instruments for measuring soil bulk density - Soil compaction by machines - Vibratory soil cutting - Potato planting - Potato damage - Combine harvesters - Soil clods in potato production - Rotary cultivators - Hill land improvement - Cultivations for potatoes - Drying large round hay bales - Grain drying - Potato harvesters - Automatic control of field machinery - Tractor stability and control on sloping land - Raspberry harvesting.

Correspondence: English, German, French.

Reading

Department of Agriculture and Horticulture, University of Reading
Earley Gate, Reading, Berkshire, RG6 2AT

Agricultural Engineering and Building Section:

Gibb, J.A.C.	OBE, MA, M.Sc., CEng., Hon.FIAgr.E., Mem.ASAE, FRAgS.	Senior Lecturer - Head of Section
Brooke, D.W.I.	BA, CEng., MIEE	Lecturer - Electronics and instrumentation
Morgan, K.E.	B.Sc., NDAgr.E., FIAgr.E. Mem.ASAE	Lecturer - Farm Power and Machinery
Owen, J.E.	M.Sc., FIAgr.E.	Lecturer - Agricultural structures environment control.

Horticultural Section:

Canham, A.E.	M.Sc.(Eng), CEng., MIEE, MCIBS	Lecturer - Electrical Engineering, Environmental control, greenhouse technology

Experimental Staff:

Constantine, T.R.	CEng., MIEE

Status: Governmental institution established in 1926, financed by endowments and annual grant from government sources, and governed by a Court and Council which are largely autonomous.

Activities: (a) Teaching: Agricultural and horticultural engineering courses forming part of 3 year B.Sc. degree courses in agriculture (250 students) and horticulture (75 students). Research higher degrees: M.Phil. 2 years, and Ph.D. 3 years (2 graduate students) - (b) Research: Automatic control systems for farm machinery - Data-logging techniques for agricultural and biological research - Instrumentation of agricultural processes. Energy conversion and conservation. Data acquisition and data processing for agricultural enterprises. Evaluation of light sources for horticulture.

Publications: Papers on: Electronics and automation in agriculture. Recirculatory ventilation systems for pigs and poultry - control of environmental factors in animal housing - Refrigerative dehumidification - Alternative energy sources and storage - Electronic systems of data collection - Data processing systems in agriculture - A simplified tubular solarimeter. Effects of light, temperature and carbon dioxide on the growth of young tomato plants.

Correspondence: English.

Shinfield

National Institute for Research in Dairying (N.I.R.D.)
Shinfield, Reading, Berkshire, RG2 9AT   Tel. Reading 883103

Porter, J.W.G.	MA, Ph.D., FIBiol.	Director

Dairy Husbandry Department:

Dodd, F.H.	B.Sc., Ph.D., FIBiol.	Head of Department
Akam, D.N.		Machine milking
Griffin, T.K.		Machine milking and mastitis
Grindal, R.J.,	NDA	Machine milking and mastitis
Pain, B.F., B.Sc., Ph.D.		Anaerobic fermentation and farm waste handling
Phipps, R.H.	B.Sc., MPhil., Ph.D., DipTropAgric.	Automatic cattle feed dispensing
Shearn, M.F.H.		Automation of teat disinfection

Process Technology Department:

Burton, H.	BEng., M.Sc., D.Sc., CEng., MIEE, DIFST	Head of Department
Belcher, J.R.		Milking cooling on farms
Dawkins, J.	MIMH	Cleaning of farm milk vats

Status: A research institute of the University of Reading, financed and supervised by the Agricultural Research Council.

Activities: Investigation of the role of the milking machine in relation to mastitis and development of control techniques - Automatic test disinfection - Automation of feed dispensing to cattle - Problems of slurry disposal on land - Methods of slurry disposal including anaerobic fermentation - Cost reduction and energy saving in milk cooling - Specification testing of farm milk vats - Cleaning and disinfection of farm milk vats.

Publications since 1975: Machine milking - Mechanization and automation of milking parlours - Rotary milking parlours - Milking machine design and mastitis - Mastitis detection - Udder washing and teat disinfection - Refrigeration on the farm - Heat recovery from milk cooling - Slurry disposal and utilization.

Correspondence: English, French, German.

Silsoe

National Institute of Agricultural Engineering
Wrest Park, Silsoe, Bedford MK45 4HS   Tel: Silsoe (0525) 60000
                                              Telex: 825808

Bell, R.L.	B.Sc., Ph.D. CEng. FIM FInst.P. Prof.	Director
Cox, S.W.R.	B.Sc., FInstP., FIAgrE.	Deputy Director

Crop Engineering Division:

Boyce, D.S.	B.Sc., Ph.D., MemASAE, FIAgr.E.	Head of Division
Audsley, E.	BA, M.Sc.	Operational research

Hughes, M.	FIST	Central laboratory
Marchant, J.A.	B.Sc.,Ph.D., CEng, MIEE	Spraying
Nellist, M.E.	B.Sc., M.Sc.,Ph.D., CEng., MIAgrE.	Crop drying and ventilation

Engineering Design and Construction Division:

Maher, L.E.	B.Sc.,CEng., MIMechE.	Head of Division
Cowell, J.	CEng., MIMechE.	Drawing office
Dyer, M.J.	CEng., MIMechE.	Design
Osborne, L.E.	TEng.,MIAgrE.	Engineering services

Farm Buildings Division:

Messer, H.J.M.	B.Sc.ARICS	Head of Division
Carpenter, G.A.	B.Sc.,CEng., MIChemE.,MIAgrE.	Environmental engineering
Cumby, T.R.	B.Sc.	Animal feeding
Hepherd, R.Q.	B.Sc., MIAgrE.	Waste engineering
Hoxey, R.P.		Structures

Glasshouse and Scientific Information Division:

Cox, S.W.R.	B.Sc.,FInstP., FIAgrE.	Head of Division
Bailey, B.J.	B.Sc.,Ph.D.,MInstP.	Greenhouse engineering
Field, Mrs.E.M.	BA, M.Sc.,AIInfSc.	Scientific information

Instrumentation and Control Division:

Moncaster, M.E.	MA	Head of Division
Bowman, G.E.	B.Sc.	Instrumentation research
Turner, M.J.B.	B.Sc.,M.Sc., CEng.MIEE,MInstP.	Livestock control
Wignall, W.R.	FInstMC.	Instrumentation services

Machine Division:

Manby, T.C.D.	OBE M.Sc.,CEng. FIMechE, FIAgrE.	Head of Division
Armold, R.E.	MA, MIAgrE.	Forage conservation
Brown, F.R.	B.Sc.	Field vegetables
Chisholm, C.J.	B.Sc.,Ph.D., CEng.,MIMechE.	Machine dynamics and reliability
Holt, J.B.	B.Sc.,M.Sc.,CEng. MIMechE.,MIAgrE.	Materials handling
Klinner, W.E.	FIAgrE.,MemASAE	Forage machinery
Lindsay, R.T.	B.Sc., TEng.,MIAgrE.	Product evaluation and testing

Overseas Division:

Bell, R.D.	B.Sc.,CEng., FIMechE.,FIAgrE.	Head of Division

Tractor and Cultivation Division:

Matthews, J.	B.Sc.,MInstP., FIAgrE.	Head of Division
Dwyer, M.J.	B.Sc.,M.Sc.,Ph.D. CEng.,MIMechE.,MIAgrE.	Tractors
Patterson, D.E.	B.Sc.,M.Sc.,MIAgrE.	Cultivations
Stayner, R.M.	B.Sc., M.Sc.	Ergonomics

Status: Governmental institution, established 1924 and controlled by the Agricultural Research Council.

Research and Development activities: Tyres, wheels and traction - Ride vibration - Tractor driver monitoring and control: steering - Tractor cab climatic environment - Tractor and machinery noise - Tractor monitoring and control - Design of tractors and vehicles - Farm materials transport - Machine dynamic performance and overload protection - Soil dynamics and the design of cultivation implements - Cultivation implements and seed drills for cereals - Soil compaction - Wear - Spray physics: electrostatics - Field sprayer design: spray boom stability - Air carried sprays - Performance of crop mowing and conditioning equipment - Simulation of forage conservation - Forage chopping - Application of preservatives during hay harvesting - Moisture measurement - Drying of agricultural crops - Methods and machines for compacting and transporting straw - Simulation of cereal grain drying and harvesting costs - Separation of grain from straw and other materials at harvest - Rowcrops establishment: drilling seeds and transplanting - Irrigation - Sugar beet topping mechanisms - Harvesting and preparation of vegetables in the packing shed - Efficient use of energy in greenhouses - Greenhouse heating - Monitoring and control of greenhouse environment - Materials handling and allied operations in glasshouses - Performance monitoring of livestock - Dairy parlour engineering - Statistical time series analysis - Animal feeding - Environmental engineering - Waste engineering - Wind loading - Retaining walls for agricultural materials - Optimum labour and machinery for arable farms - Annual cost of machinery - Reference library and internal abstracting and translation service - Mechanization in developing countries.

Publications: Annual reports - Programmes of research - Translations - Occasional reports - Test reports on tractors.

Correspondence: English, French, German, Russian, Dutch, Italian.

The Institution of Agricultural Engineers
West End Road, Silsoe, Bedford MK45 4DU   Tel.: Silsoe 61096

Norman, R.F.	B.Sc.(Agric), M.Sc.(AgrEng) CEng, FIAgrE, FBIM	President
Fryett, R.J.	Assoc.I Gas E, FInstPet., CIAgrE	Secretary

Status: The Institution is primarily financed by subscriptions and fees paid by individual members. It is not financed from governmental or semi-governmental sources.

Activities: National and local day Conferences, technical meetings, seminars, works visits - Three national and some six local conferences, 60 technical meetings, throughout the U.K. are organized each year.

Publications: "The Agricultural Engineer" the proceedings of the Institution of Agricultural Engineers.

Correspondence: English.

National College of Agricultural Engineering
Silsoe, Bedford MK45 4DT, England   Tel.: Silsoe (0525) 60428

May, B.A.	B.Sc.,NDAgrE., CEng., MIMechE.,FIAgrE.,	Dean of Faculty of Agricultural Engineering, Food Production and Rural Land Use, Professor of Environmental Control and Processing.

Design, Manufacture and Mechanization Group:

Radley, R.W.	B.Sc., Ph.D.	Professor in Agricultural Production Technology
Inns, F.M.	MA, M.Sc.,CEng., MIMechE, FIAgrE.	Professor of Agricultural Machinery Engineering
Cowell, P.A.	B.Sc., M.Sc., Ph.D., CEng, MIMechE, MIAgrE.	
Dyson, J.	B.Sc., CEng., MIMechE.	
Crossley, C.P.	B.Sc.,M.Sc.,CEng., MIMechE.,FIAgrE.	
Kilgour, K.	B.Sc.,M.Sc.,CEng,MIAgrE.	
Watt, C.D.	B.Sc.,CEng.,MIMechE, MIAgrE.	
Lewis, R.T.	B.Sc.,M.Sc.,MIAgrE.	
Taylor, J.C.	MS, Ph.D.	
Noble, D.H.	B.SC.,MA, FSS.	
Bascombe, M.L.A.	B.SC.,CEng.,MIMechE.	
Morris, J.	B.Sc.,M.Sc.	
Grant, A.	MA, DIC	
Morgan, D.D.V.	BA,M.Sc.,FSS, MIS.	
Larkin, S.	BA	
Howard, A.	B.Sc.	

Field Engineering Group:

Hudson, N.W.	B.Sc., M.Sc.,CEng., FICE, FIAgrE.	Professor of Field Engineering
Morgan, R.P.C.	BA, MA, Ph.D.	
Spoor, G.	B.Sc., M.Sc., MIAgrE.	
Keech, M.A.	B.Sc.,M.Sc.,FIAgrE.	
Godwin, R.J.	B.Sc.,MS,Ph.D., MIAgrE.,AMASAE	
Kay, M.G.	B.Sc.,M.Sc.,CEng, MICE, MASCE	
Carr, M.K.V.	B.Sc., Ph.D.	
Prins, R.J.	B.Sc.,M.Sc.,APEO, MIAHR.	
Leeds-Harrison, P.	B.Sc., Ph.D.	
Carter, R.	BA, MA, M.Sc.	
Weatherhead, K.	BA, MA, MICE.	

Buildings and Processing Group:

Tindall, H.D.	MBE, B.Sc., M.Sc., FLS, FIBiol.,CIAgrE,NDH	Professor of Tropical Agronomy
Strang, I.G.	B.Sc.,CEng., MIMechE., MCIBS, AFRAES	
Murfitt, R.F.A.	B.Sc.,DTA, MIBiol.	
Stenning, B.C.	B.Sc., CIAgrE.	
Douglass, M.P.	MS, TEng(CEI),MIAgrE.	
Clarke, B.	B.Sc., Ph.D., CEng., MIMechE.,ARTCS.	

Stone, G.T.   Dip.Tech., AMIEE

Industrial Development, Marketing and Management Group:
Hill, R.W.   BA, Ph.D.   Professor of Product Management
Austin, V.   B.Sc.(Econ), NDAgrE,
                    MIAgrE, MASAE
Crawford, I.M.   BA

Ministry of Agriculture Fisheries and Food
Field Drainage Experimental Unit Training Section:
Castle, D.A.   NDA, MRAC, DipAgrEng,
                    T Eng(CEI), MIAgrE
Gregory, J.E.   NDA, MRAC, DipAgrEng,
                    TEng(CEI), MIAgrE.

Design Engineer:
Le Flufy, M.J.   MA

Research Officers:
Fry, R.K.   B.Sc., M.Sc., GradIAgrE
Noble, Mrs. C.   B.Sc.
Monk, Miss A.S.   BA, MPhil

Senior Administrative Staff:
King, C.E.   MBIM   Head of Administration
Neville, J.H.   B.Sc., M.Sc., NDA, CEng,
                    FIAgrE   Faculty Registrar
Harding, P.G.   DMS(Ed), MBIM   Careers and Recruitment Officer
Morgan, B.A.   MLS, ALA   Librarian.

Status: The College was founded in 1960 with an independent governing body. In 1975 it became a School of the Cranfield Institute of Technology and became the Faculty of Agricultural Engineering, Food Production and Rural Land Use in 1977.

Activities: Advanced Education: B.Sc. Degree in Agricultural Engineering with honours: 3 years; approximately 140 undergraduates. M.Sc. in either one or 2 years. Taught courses in the following options: Agricultural Machinery Engineering, Agricultural Mechanization, Land Resource Management, Soil and Water Engineering, Tropical Crop Storage and Primary Processing and Product Management. - Postgraduate Diploma, one academic year in the following options: Soil and Water Engineering, Agricultural Mechanization and Machinery Engineering, Land Resource Planning, Marketing and Crop Processing. M.Sc. and Ph.D. Research Degrees. Approximately 180 postgraduates.
Short Courses: Short courses are undertaken in Aerial Photographic Interpretation, Surveying, Crop Storage and Processing, Electrical and Electronic Control of Farm Machinery, Irrigation and Drainage, Remote Sensing, Management for Agricultural Machinery Dealership Personnel. - Consultancy: Staff undertake consultancy work in the areas of irrigation, drainage, machinery development, product management, environmental control and rural industrial development. - Research and Development: Dynamics and control of tractor and implement - Physic-mechanical properties of crop and animal materials - Low-cost traction devices - Separation of granular materials - Ventilation of agricultural buildings - Ergonomic studies of capital intensive and labour intensive farm equipment - Plant response to environmental conditions - Cultivation of fine-textured soils - Resistance of soil to water erosion - Hydrology of small rural catchments - Treatment of farm wastes - Airphoto interpretation - Land resource planning and capability classification - Use of operational research techniques in development of agricultural systems - Influence on investment in field and fixed equipment.

Publications: Since 1965 several hundred articles, books and reports on many aspects of agricultural engineering have been published.

Correspondence: English, French, German, Spanish, Italian.

Slough

Tropical Stored Products Centre, Tropical Products Institute
Overseas Development Administration, Foreign and Commonwealth Office,
London Road, Slough, Berks, SL3 7HL   Tel.Slough 34626

Calverley, D.J.B.	B.Sc.,M.Sc., MS, FIAgriE	Head of Centre - Agricultural Engineering
Prevett, P.F.	B.Sc., Ph.D.,MIBiol.	Deputy Head of Centre - Storage Entomology; Training

Storage Engineering Section:

Hallam, J.A.	B.Sc.(Eng)	Head of Section
O'Dowd, E.T.	M.Sc., MIAgriE	
Gracey, A.D.	B.Sc.	
Gough, M.C.	B.Sc.,PhD.	Physical factors
Bisbrown, A.J.K.	B.Eng.	
Kenneford, Miss S.	B.Sc.	

Status: Department of the Tropical Products Institute and a Scientific Unit of the Overseas Development Administration.

Activities: a) Twice-yearly 'Course in the Storage of Durable Agricultural Products in the Tropics'- 3-6 months duration (20 participants) - covering all aspects of storage and handling of durable commodities). Collaboration with National College of Agricultural Engineering in 'Tropical Crop Storage and Primary Processing' M.Sc. course. Training courses in developing countries. - b) Storage Engineering Section: Moisture and temperature relations; design of warehouses for bagged produce; silos and other bulk storage systems including drying and handling equipment; farm and village level storage. Other Sections concerned with Pest Biology and Inspection; Chemical Control, Training and Information.

Publications: Tropical Stored Products Information, twice-yearly. Tropical Storage Abstracts, bi-monthly.

Correspondence: English.

Southampton

University of Southampton, Institute of Irrigation Studies
Southampton S09 5NH, Hampshire

Postgraduate Course in Irrigation Engineering:

Rydzewski, J.R.	B.Sc.,Ph.D.,CEng, FICE, F.ASCE	Director
Cooper, M.R.	B.Sc.,M.Sc., MICE	
Gordon, E.D.	MA, Dip.Ag.Econ.	
Hillman, P.F.	M.Sc.,CEng, MICE, MIWES,MIPHE	

Riley, M.J.	B.Sc., CEng, MICE
Rycroft, D.W.	B.Sc., Ph.D., CEng, MICE, MIWES
Smith, K.V.H.	M.Sc., Ph.D., CEng, MICE, M.ASCE, MIStructE, MIPHE
Svehlik, Z.	Dip.Ing., Ph.D., DHE Delft
Tanton, T.W.	B.Sc., Ph.D.
Ward, C.F. (Miss)	BA, M.Sc., FGS
Clarke, D.	B.Sc.
Stern, P.H.	MA, CEng, FICE (Visiting)
Sutcliffe, J.V.	MA, Ph.D. (Visiting)
Swynnerton, Sir Roger	CMG, OBE, MC, BA, DipAgric, AICTA (Visiting)
Clark, L.	B.Sc., Ph.D., FGS, MIGeo. (Visiting)

Status: Established in 1964. Financed by the University of Southampton and the Overseas Development Administration of the Foreign and Commonwealth Office of the U.K.

Activities: 12 month Postgraduate Course in Irrigation Engineering leading to M.Sc. Degree or Diploma – Syllabus includes: Surface Water Hydrology, Groundwater Hydrology, Soil Physics, Geotechnical Engineering, Open-Channel Flow, Hydraulics of Control Structures, Hydraulic Models, Dams, Fundamentals of Irrigated Agriculture, Design of Irrigation Systems, Drainage of Irrigated Lands, Project Planning, Appraisal and Monitoring, Management Economics, Statistics, Computation, Language and Communication, Visiting lecturers dealing with specialist topics related to the course, Seminars and Site Visits. Intake 35 students per year. – Postgraduate Research: leading to Ph.D. degree in: Irrigation Water Requirements, Effect of Dry Air on the Growth of Crops, Analyses of Irrigation Systems, Reclamation of Heavy Clay Soils, Drainage of Irrigated Lands, Irrigation Water Quality, Efficiency of Irrigation Water Management, Management and Operation of Irrigation Projects.

Publications: Numerous research publications.

Correspondence: English.

Sutton Bonington

University of Nottingham, Department of Agriculture and Horticulture, School of Agriculture, Sutton Bonington, Loughborough, Leicestershire

Wilton, B.   M.Sc., AgrEng.   Lecturer – Agricultural Engineering

Status: Governmental Institution, established in 1945. Governed through University Grants Committee.

Activities: (a) Degree courses in Agriculture (30 students), and Horticulture (10 students). – (b) Research on raising and transplanting seedlings – Crop drying – Processing animal feedstuffs – Whole crop cereal harvesting.

Publications since 1973: Whole crop cereal harvesting.

Correspondence: English.

Writtle

Writtle Agricultural College
Chelmsford CM1 3RR   Tel.: Chelmsford (0245) 420705

Bebb, D.L.	M.Sc., NDAgrE, NDA, Cert.Ed., MIAgrE	Head of Department
Oldacre, N.	NDAgrE, NDA, Cert.Ed., MIAgrE	
Carr-West, M.St.J.	B.Sc., MAgrSc., Cert.Ed., MIAgrE	
Keeble, B.M.	B.Sc., MIAgrE	
King, G.C.	B.Sc., M.Sc.	

Senior Lecturers:

Milne, L.	B.Sc., Cert.Ed., DipAgrE
Raggett, M.W.	DipAgrE

Activities: 1 year course NCA - 3 year sandwich courses:
Ordinary National Diploma and Higher National Diploma in Agriculture
Ordinary National Diploma and Higher National Diploma in Commercial Horticulture
Ordinary National Diploma in Amenitiy Horticulture
Higher National Diploma in Landscape and Recreation Provision and Management
(Amenity Horticulture)
Higher National Diploma in Agricultural Engineering (projected start 1981)

Correspondence: English.

UNITED STATES OF AMERICA

Alabama

Agricultural Engineering Department, Auburn University
Auburn, AL 36849

Turnquist, P.K.	Ph.D.	Professor and Head of Department
Busch, C.D.	Ph.D.	Associate Professor
Dumas, W.T.	M.Sc.	Associate Professor
Flood, C.A.	Ph.D.	Associate Professor
Hill, D.T.	Ph.D.	Associate Professor
Johnson, C.E.	Ph.D.	Professor
Koon, J.L.	Ph.D.	Associate Professor
Renoll, E.S.	M.Sc.	Professor
Rochester, E.W.	Ph.D.	Associate Professor

National Tillage Machinery Laboratory, Auburn, AL

Gill, W.R.	Ph.D.	Collaborator
Hendrick, J.G.	Ph.D.	Ag.Engineer
Schafer, R.L.	Ph.D.	Ag.Engineer, Interim Director
Taylor, A.C.	Ph.D.	Soil Scientist
Bailey, A.C.	Ph.D.	Ag.Engineer
Burt, E.C.	Ph.D.	Ag.Engineer
Pickering, W.D.	B.Sc.	Ag.Engineer

U.S. Forest Service, George W. Andrews Forest Research Laboratory, Auburn, AL

Sirois, D.	Forest Engineer
Steel, J.	Forest Engineer
Stokes, B.	Forest Engineer

Status: Established in 1919, Land Grant University (State and Federal)

Activities: Teaching: B.S., M.S., Ph.D. in Agricultural Engineering; B.S. in Forest Engineering. Average No. of Agricultural Engineering students – 80; Average No. of Forest Engineering students – 22. – Research: Power and Machinery, Soil and Water, Structures and Environment, Electric Power and Processing, Animal Waste Management.

Publications: Power and Machinery – Soil and Water – Structures and Environment – Electric Power and Processing, Animal Waste Management.

Correspondence: English.

Alaska

Agricultural Experiment Station
P.O. Box AE, Palmer, Alaska 99645

Allen, L.D.	B.Sc.(Ag.E.) M.Sc.(Ag.E.)	Head, Agricultural Engineering

Status: 1948, Experiment Station reorganized at this time to include Agricultural Engineering.

Activities: Teaching: No occasional short course or seminar given. Research: Full time research related to Northern Agriculture and Engineering.

Publications: University of Alaska: Cooperative Extension Service, Extension Bulletin; Alaska Agric. Experiment Station, Information leaflet, Progress reports, Circulars.

Correspondence: English.

Publications: University of Alaska: Cooperative Extension Service, Extension Bulletin; Alaska Agric. Experiment Station, Information leaflet, Progress reports, Circulars.

Correspondence: English.

Arizona

Department of Soils, Water and Engineering, The University of Arizona
401 Agricultural Sciences Bldg. No. 38 – Tucson, Arizona 85721

Gardner, W.R.	Ph.D.	Head of Department – Soil Physics
Cannon, M.S.	M.S.	Mechanization
Hinz, W.W.	B.S.	Mechanization
Fangmeier, D.D.	Ph.D.	Irrigation
Halderman, A.D.	Engineer	Irrigation
Roth, R.L.	M.S.	Irrigation
Flug, M.	Ph.D.	Water Resources
Matlock, W.G.	Ph.D.	Water Resources
Welchert, W.T.	M.S.	Livestock Housing and Equipment
Wiersma, F.	Ph.D.	Livestock Housing and Equipment
Larson, D.L.	Ph.D.	Agricultural Systems

Branches with other location: Yuma Experiment Station, The University of Arizona, 6425 W. 8th, Yuma, Arizona 85364 – Cotton Research Centre, The University of Arizona, 4201 E. Broadway, Phoenix, Arizona 85040.

Status: Established in 1890, the Land-Grant University of the State of Arizona.

Activities: Teaching: B.S. and M.S. in Agricultural Engineering; B.S. in Agri-Mechanics and Irrigation. Total enrollment in these programs – 50. – Research: Field crop mechanization, vegetable crop mechanization, irrigation, groundwater development, surface water development, arid land housing of livestock and poultry, range improvement, agricultural waste management, agricultural systems analysis and design.

Publications: Publications in all of the above areas.

Correspondence: English.

Arkansas

Agricultural Engineering Department, University of Arkansas
101 Agricultural Engineering Building, Fayetteville, Arkansas 72701

Bryan, B.B.	Head of Department	Soil and Water Engineering
Braker, C.R.	Ed.D.	Mechanization
Ferguson, J.A.	Ph.D.	Soil and Water Engineering
Griffis, C.L.	Ph.D.	Instrumentation
Harris, W.A.	M.S.	Structures and Environment
Magree, C.	Ph.D.	Structures and Environment
Matthews, E.J.	M.S.Ag.E.	Power and Machinery
Mote, C.R.	Ph.D.	Waste Management
Nelson, G.S.	M.S.Ag.E.	Crop and Food Engineering
Peralta, R.C.	Ph.D.	Soil and Water Engineering
Rokeby, T.R.C.	Ph.D.	Structural Design, Agricultural Power
Walker, J.T.	Ph.D.	Power and Machinery
Warnock, W.K.	Ph.D.	Electrical Power and Processing

Status: Established 1948; governing body: College of Engineering for academic purposes. College of Agriculture and Home Economics for funding and other programs.

Activities: Teaching: approximately 6 students per class. Research: Twenty active research projects encompassing approximately 80 percent of the total faculty time.

Publications: Arkansas Farm Research: Salts removed in winter runoff from rice fields - Center pivot sprinkler irrigation of rice - Laboratory size composting unit for agricultural waste materials - Dilute acid soluble carbon as a measure of available carbon in the composting process - Development of the bench scale composter - Alternate weed control systems for soybeans - Post-harvest quality of machine-harvested strawberries - A high-density brooding system for broilers using radiant heat - In-plant equipment for handling machine harvested strawberries - Quality of machine-harvested strawberries in relation to handling techniques - A rapid method of measuring moisture in litter used for broilers brooded at high density - A solar heated broiler house - The potential of broiler litter as fuel.

Arkansas Cooperative Extension Service, University of Arkansas
P.O.Box 391, Little Rock, AR 72203

Benz, R.C.	M.Sc.	Livestock, Poultry Structures and Environment; Grain Drying and Storage.
Huitink, G.W.	M.Sc.	Field Machinery.
Langston, J.M.	M.Sc.	Residential Housing; Electricity; Energy.

Status: Cooperative Extension Service (with U.S. Department of Agriculture), partly supported by the Federal Government and partly by the state.

Activities: Extension work only.

Publications: Water conditioning; Peach orchard irrigation; Rice drying on the Farm; Sprayer calibration; Tractor costs; Cotton harvesting; Flame cultivation; Insulate for comfort; Building plans for Arkansas residents; Production and use of ethanol.

Correspondence: English.

California

Department of Agricultural Engineering, University of California
Davis, California 95616

Garrett, R.E.	Ph.D.	Professor and Department Chairman
Akesson, N.B.	M.S.	Professor
Brune, D.E.	Ph.D.	Assistant Professor
Carroad, P.A.	M.B.A., Ph.D.	Assistant Professor Food Engineering
Carroll, J.J. III	Ph.D.	Associate Professor of Meteorology
Chancellor, W.J.	Ph.D.	Professor
Chen, P.P.	Ph.D.	Associate Professor
Dobie, J.B.	M.S.	Lecturer
Goss, J.R.	M.S.	Professor
Hills, D.J.	Ph.D.	Assistant Professor
Kaminaka, M.S.	Ph.D.	Assistant Professor
Kepner, R.A.	A.A., B.S.	Professor
Merson, R.L.	Ph.D.	Professor of Food Engineering

Miles, J.A.	Ph.D.	Assistant Professor
Morrison, S.R.	Ph.D.	Professor
O'Brien, M.	Ph.D.	Professor
Rumsey, T.R.	Ph.D.	Assistant Professor
Singh, R.P.	Ph.D.	Associate Professor
Studer, H.E.	M.S.	Associate Professor
Yates, W.E.	M.S.	Professor

Status: The Department of Agricultural Engineering was established on the Davis campus of the University of California in 1915. Currently it is a department in the College of Agricultural and Environmental Sciences where it teaches service courses to majors in the College of Agricultural and Environmental Sciences, but does not offer a major of its own. It is also a department in the College of Engineering where it offers a professional major in agricultural engineering and graduate studies. It is also a department in the Agricultural Experiment Station where research activities are funded and reported.

Activities: The Department of Agricultural Engineering offers a B.S. in agricultural Engineering, Master of Engineering, Master of Science in Agricultural Engineering, Doctor of Engineering, and Ph.D. in Agricultural Engineering. Teaching and research programs cover the areas of aquacultural engineering, food engineering, forest engineering, agricultural power and machinery, irrigation and drainage engineering, structures and environment, waste management, and alternate energy in agriculture. Agricultural Engineering Specialists in the Cooperative Extension Service are affiliated with the Department.

Research: Studies on movement of agricultural chemicals and potential damage to: agricultural chemicals and forest crops, human and domestic animal habitat, to wild life and the general environment, and to workers mixing, applying and working with pesticide materials. Analysis of the basic parameters controlling diffusion and transport of agricultural chemicals and other pollutants in the atmosphere and deposit on soils and in water systems. Development of systems to reduce pesticide losses and resultant potential ecosystem damages. Dispersal and control of contaminants from utilization and conversion of agricultural waste materials. - Aquaculture engineering; design, development, and optimization of systems for intensive culture of marine and freshwater organisms; water quality management and utilization of waste nutrients; growth kinetics and productivity of algal and bacterial mass cultures. - Food engineering: Food and biochemical engineering; innovations in food processing operations such as blanching, fluming, and thermal processing; bioconversion of food industry wastes; product oriented research on cottage cheese. - The structure of and mass and energy transfer within the atmospheric boundary layer, especially in polar regions. Air pollution transport and diffusion on local through regional scales. Instrumentation systems and measurement techniques applicable to the study of these phenomena. - Soil strength and dynamics relative to traction, compaction, and tillage (vibratory); agricultural machinery systems management and optimization; energy requirements and resources for agriculture; international agricultural development, particularly agricultural mechanization and processing in Southeast Asia. - Development and/or improvement of engineering systems for increasing production and utilization of foods and fiber. Specific areas include mechanical harvesting and handling systems for boysenberries, oranges, tree fruits, and fresh market tomatoes; physical properties of agricultural materials as related to the design of harvesting, handling, and sorting equipment. - Forage mechanization, processing, storage, and handling and related feeding and economic analysis; packaging of livestock feeds; harvesting, handling, processing, storage, and utilization of agricultural residues. Use of solar energy for crop and residue drying. Rural electric research program. - Design and development of equipment and systems for aquaculture and for production, harvesting, and handling of lettuce and other vegetable crops; studies of physical properties of agricultural materials; electronic and hydraulic controls; gamma radiation techniques for measuring density; economic feasibility and operations research. - Gasification of agricultural and forest residues. Harvesting and processing of certified seed crops. Performance and operation of the self-propelled combine, including automatic control. Forest Engineering education and research. - Agricultural wastes management including animal manures and crop residues; water and land pollution control; agricultural usage of treated domestic and industrial sewage effluents and sludges; rural water supply and sanitation. - Human factors engineering; noise abatement in canneries; machinery operator skills, appropriate technology, roadside marketing, microprocessor applications in agriculture, international agricultural development, bioengineering. - Farm machinery, including harvesting of forage crops, asparagus, seed crops, and cotton. - Application of chemical engineering to food processing. Thermal processing of foods (especially Steriflamme canning, the sterilization of canned foods by direct application of a gas flame), membrane processing (ultrafiltration, reverse osmosis), liquid chromatography, enzyme reactions, application of computers to food process control, freezing of foods, crystallization of sugars. - System design and analysis of forest culture, reforestation, harvesting and refuse management, with emphasis on conifer seed collection and processing, reforestation and thinning procedures and equipment, forest soil compaction, and improved cable yarding hardware and procedures. - Response of swine to environmental factors; environmental modification and manure management for beef animals in hot climates; environmental quality problems associated with beef and swine production; methane production from poultry manure. - Materials handling equipment; mechanical harvesting, handling, transportation, and storage of fruits and vegetables (cleaning, loading, transport, sampling, sorting, quality control grading, conveying, sizing, singulation, filling and containerization), cannery noise and waste management. - Food engineering; direct and indirect solar drying; computer simulation of solar energy collection systems; heat transfer in food materials; stress analysis of agricultural materials. - Food engineering; mathematical description of biological processes during processing and storage; energy efficiency improvements in food processing unit operations; heat and mass transfer in rice drying. - Mechanical harvesting and handling of wine, raisin and table grapes, mechanical pruning of vines, trellising of grape vines, mechanical harvesting, handling, and packing of fresh market tomatoes. - Agricultural chemical application technology. Design and development of aerial and ground application equipment with emphasis on control of particle size; transport and diffusion of spray; and assessment and evaluation of pesticide efficacy and environmental contamination.

Publications: Bulletins, Div. of Agric. Science, University of California; bulletins, Department of Agric. Engineering, University of California.
Correspondence: English

Land, Air and Water Resources (formerly Water Science and Engineering), University of California
Davis, CA 95616

Amorocho, J.	Ph.D.	Professor
Biggar, J.W.	Ph.D.	Professor
Burgy, R.H.	M.S.	Professor
Carlson, S.P.	M.S.	Coordinator (Cooperative Extension Service)
Davenport, D.C.	Ph.D.	Research Water Scientist
Fereres, E.	Ph.D.	Extension Irrigationist and Lecturer
Grimes, D.W.	Ph.D.	Lecturer
Hagan, R.M.	Ph.D.	Professor
Hanson, B.R.	Ph.D.	Extension Drainage and Groundwater Specialist
Henderson, D.W.	Ph.D.	Professor
Hsiao, T.C.	Ph.D.	Professor
Knight, A.W.	Ph.D.	Professor
Luthin, J.N.	Ph.D.	Professor
Malakoff, E.R.	J.D.	Lecturer
Marino, M.A.	Ph.D.	Professor

Miller, R.J.	Ph.D.	Lecturer
Nielsen, D.R.	Ph.D.	Professor
Pruitt, W.O.	M.S.	Lecturer
Robinson, F.E.	Ph.D.	Lecturer
Scott, V.H.	Ph.D.	Professor
Silk, M.W.K.	Ph.D.	Asst. Professor
Tanji, K.K.	M.S.	Professor

Activities: A wide variety of engineering research areas concerned with the supply, distribution, utilization and disposal of water resources is available for graduate study. These include hydrologic engineering relationships; groundwater yield and recharge; well problems; flow through porous media; hydrodynamics, including turbulence and potential-flow studies; hydraulics of measuring devices, structures and systems; mixing phenomena in streams and channels; water application and efficiency of surface and sprinkler systems; planning and utilization of surface and groundwaters; water pollution; disposal of agricultural, industrial and domestic waters; design of drainage systems; flow in tile lines; financial aspects of water developments; and optimization and analysis of water-resource systems.

Undergraduate programs available are: Atmospheric Science, Renewable Natural Resources, Soil and Water Science. Master's degree programs include: Water Science (hydrology, irrigation and drainage, water quality and pollution, and water resources management). Atmospheric Science (hydrometeorology). Ecology (water biology, hydroecology, pollution biology), Engineering (drainage engineering, water resources engineering, irrigation and drainage, hydraulics, hydrologic engineering, irrigation and drainage. Plant Physiology (plant-water relations, water stress and metabolism). Soil Science (physics of soil water, pollutant transport in soil, soil-water-plant relations). Doctor's degree programs are: Doctor of Engineering, Ph.D. in Engineering, Ph.D. in Ecology, Ph.D. in Plant Physiology, Ph.D. in Soil Science.

Publications: Bulletins and leaflets.

Correspondence: English.

Agricultural Engineering Department, California Polytechnic State University
San Luis Obispo, Calif. 93401

Status: State institution for undergraduate education in arts and sciences, engineering and agriculture.

Activities: Four-year undergraduate study courses leading to B.Sc. with major in agricultural engineering or mechanized agriculture. Two years' practical study programmes leading to technical certificate in mechanized agriculture. Special study programmes of 1-9 months duration for foreign trainees with major interest in mechanization of agriculture. Special courses in engineering support of fresh and salt water farming of fish and shellfish.

Correspondence: Engllish.

Colorado

Agricultural Engineering Department, Colorado State University
Fort Collins, Colorado 80523

Harper, J.M.	Ph.D.	Head of Department. Food Engineering
Bausch, W.	Ph.D.	Asst.Prof. Soil-Water Engineering
Clyma, W.	Ph.D.	Assoc.Prof. Soil Water Engineering
Corey, A.T.	Ph.D.	Prof. Soil-Water Engineering
Duke, H.R.	Ph.D.	Faculty Affiliate (USDA-SEA). Soil-Water Engineering
Evans, N.A.	Ph.D.	Prof. Irrigation
Hansen, R.W.	M.S.	Assoc.Prof. Extension
Heermann, D.F.	Ph.D.	Faculty Affiliate (USDA-SEA). Soil-Water Engineering
Hiller, R.L.	JD	Asst.Prof. Legal
Karmeli, D.	Ph.D.	Prof. Soil-Water Engineering
Kruse, E.G.	Ph.D.	Faculty Affiliate (USDA-SEA). Soil-Water Engineering
Loftis, J.C.	Ph.D.	Asst.Prof. Systems Engineering
McWhorter, D.B.	Ph.D.	Prof. Drainage
Podmore, T.	Ph.D.	Assoc.Prof. Soil-Water Engineering
Skogerboe, G.V.	M.S.	Prof. Soil-Water Engineering
Smith, J.L.	Ph.D.	Prof. Power and Machinery
Ward, R.C.	Ph.C.	Prof. Environmental Engineering

Status: State and Federal institution, established in 1959, supervised by State Board of Agriculture.

Activities: Graduate education leading to M.S. and Ph.D. with average attendance of 35. Basic and applied research is carried out, with special emphasis on irrigation and drainage engineering, food engineering, energy use in agriculture, and alternative fuels.

Publications: Salinity of return flow; drainage or irrigated lands; tile drainage design; conjunctive use; remote sensing; irrigation scheduling; sprinkler irrigation; ground water recharge; flow in porous media; water resources; water systems analysis; vibratory tillage; micro-plant environments; sewage sludge disposal; harvesting techniques; animal wastes; extrusion processing; food processing; alcohol fuels; energy use in agriculture.

Correspondence: English.

Connecticut

Agricultural Engineering Department, University of Connecticut
Storrs, Co.. 06268

Aldrich, R.A.	Ph.D.	Prof. and Head of Department - Agricultural Structures - Energy Conservation
Prince, R.R.	M.Sc.A.E.	Professor - Power machinery and environments
Bartok, J.B.	B.Sc.	Research Assoc.I. - Materials handling - Greenhouse mechanization
Greiner, J.B.	M.Sc.A.E.	Assoc.Prof. - Electrification - shop
Kolega, J.J.	Ph.D.	Assoc.Prof. - Conservation - waste disposal
Neyeloff, S.	Ph.D.	Assoc.Prof. - Agricultural structures - alternate energy sources.

Status: University of Connecticut established in 1881 by act of the State Legislature - The Agricultural Engineering Department was organized in 1918 as a part of the University of Connecticut.

Activities: Study courses are given in farm buildings, power and machinery, electrification, conservation and water resources - Approximately 10 major students attend the courses in each of the subject matter areas and other associated areas not listed - The study courses lead to a B.Sc. degree in agriculture - Proper selection of electives in engineering, mathematics and science prepare the student for a degree equivalent to a B.Sc. in agricultural engineering - Research projects are underway in poultry housing, individual waste disposal systems (septic tanks), mechanization of greenhouse operation, plant growth structures and equipment for humane handling of small animals for slaughter.

Publications: Ventilation of poultry and dairy structures - Special plans for farm buildings - Energy conservation in greenhouses - Greenhouse construction - Controlled environment plant growth - Waste disposal for rural homes - Solar energy for home and greenhouse heating.

Correspondence: English.

Delaware

Department of Agricultural Engineering, University of Delaware
Newark, Del. 19711

Status: State land grant institution.

Activities: Teaching: B.S. in agriculture - Major, Ag. Engineering Technology - 70 majors - average class size (including non-major course) - 30. Research: Agricultural Waste Disposal - Minimum Tillage - Poultry Housing - Closed System Aquaculture.

Publications: Poultry House Ventilation, Subsurface Asphalt Moisture Barrier, Closed System Catfish Production.

Correspondence: English.

District of Columbia

Engineering Division, Soil Conservation Service, U.S. Department of Agriculture
Washington, D.C. 20250

Activities: Primary activity of this office is to provide technical leadership and assistance to field staff of Soil Conservation Service. SCS provides assistance to over 3,000 local soil and water conservation districts in the development of plans and in the installation of soil and water conservation measures. In addition, SCS has leadership for the small watershed program. To date, over 1,000 projects have been approved for installation. The Engineering Division provides the technical direction for the planning, design, and installation of all engineering soil and water conservation measures.

Correspondence: English.

Florida

Department of Agricultural Engineering, University of Florida
Gainesville, FL 32601

Zachariah, G.L.   Ph.D.   Professor - Chairman of Department

Research and Teaching:

Bagnall, L.O.	Ph.D.	Assistant Professor – Processing and machinery; aquatic weed, forage and byproduct utilization; solar drying
Baird, C.D.	Ph.D.	Assoc. Professor – Solar energy applications in agriculture, food engineering, processing
Bottcher, A.B.	Ph.D.	Assistant Professor – Soil and water engineering non-point source pollution, agricultural water management.
Buffington, D.E.	Ph.D.	Associate Professor – Environmental modification for agricultural production and processing structures, environmental physiology, thermal analysis of structures
Campbell, K.L.	Ph.D.	Associate Professor – Watershed hydrology, agricultural water quality, irrigation and drainage, agricultural water management
Chau, K.V.	Ph.D.	Assistant Professor – Processing, energy utilization, solar energy, drying
Choate, R.E.	MSA	Professor and Assistant Chairman – Soil and water engineering
Fluck, R.C.	Ph.D.	Professor – Agricultural Energetics, instrumentation, systems analysis, physical properties of biological materials
Jones, J.W.	Ph.D.	Simulation of crop production systems, water and pest stresses
Lincoln, E.P.	Ph.D.	Production and processing of algae for livestock feed.
Mishoe, J.W.	Ph.D.	Associate Professor – Systems modelling and simulation, measurement and instrumentation of biological systems parameters
Myers, J.M.	MSA	Professor – Drip/trickle and sprinkler irrigation, evaporation losses, plant-soil-water relationships.
Nordstedt, R.A.	Ph.D.	Associate Professor – Agricultural water management, animal waste treatment, utilization and disposal.
Overman, A.R.	Ph.D.	Professor – Irrigation and drainage for production, groundwater recharge, wastewater renovation for agricultural and municipal systems, physics of soil water.
Shaw, L.N.	Ph.D.	Professor – Vegetable harvest mechanization, transplanting, hydraulic power, land locomotion.
Smajstrla, A.G.	Ph.D.	Assistant Professor – Evapotranspiration, irrigation water use efficiency, soil and water engineering.
Smerage, G.H.	Ph.D.	Associate Professor – Representation of biosystems, modelling of pest management and algal production systems.

Extension:

Baldwin, L.B.	MSA	Associate Professor – Extension waste management, pollution control
Cromwell, R.P.	ME	Associate Professor – Power and machinery, pesticide application
Harrison, D.S.	MSA	Professor – Irrigation and drainage, cold protection
Stanley, J.M.	MS	Vstg. Professor – Extension energy conservation.
Talbot, M.T.	MAE	Assistant Professor – Grain drying and energy.
Bowman, E.K.	BS	Associate Professor – Methods, equipment, facilities and systems for handling and packing fruits and vegetables
Gaffney, J.J.	MSAE	Associate Professor – Cooling of fruits and vegetables, optical quality detection

Haile, D.G.	Ph.D.	Assistant Professor – Insect control, mass rearing for sterile males, chemical application, population modelling
Henry, F.E.	BIE	Professor – Post-harvest handling and conditioning of fresh vegetables
Larson, L.W.	Ph.D.	Professor – Administration
Rogers, J.S.	Ph.D.	Associate Professor – Water management systems, drainage and irrigation
Webb, J.C.	MS	Assistant Professor – Equipment and techniques for non-chemical insect pest control.

Agricultural Research and Education Center, Lake Alfred:

Chruchill, D.B.	USDA	Mechanical harvesting and handling of citrus, pickup machines, cleaning equipment
Coppock, G.E.	MS	Professor, Department of Citrus – Citrus harvest mechanization, selective harvest of "Valencia", shaker-catch frame harvesting
Hedden, S.L.	MS	Associate Professor, USDA – Citrus harvest mechanization, shaker-pickup and foliage shaker systems
Miller, W.	Ph.D.	Professor – Citrus packinghouse handling methods, quality detection, energy utilization
Whitney, J.D.	Ph.D.	Associate Professor – Mechanical harvesting and production of citrus.

Agricultural Research and Education Center, Belle Glade:

Clayton, J.E.	MS	Associate Professor, USDA – Mechanization of the harvest of recumbent sugar cane, trash elimination, planting.
Eiland, B.R.	MD	Assistant Professor, USDA – Harvesting recumbent sugar cane, mechanical planting, trash elimination
Shih, T.S.F.	Ph.D.	Associate Professor – Water management, drainage and irrigation, modelling, hydrology.

Status: The Agricultural Engineering Department was established in 1923 – The University of Florida is a land-grant institution under the general supervision of a Board of Regents which reports to the cabinet of the state government acting as the Board of Education.

Activities: Curricula are offered for the degree of B.Sc. in agricultural engineering, a four year programme, and for the degree B.Sc. in agriculture, major in mechanized agriculture, a four-year programme. – A graduate programme is offered for the degree M.Sc. in engineering and Master in Engineering Research.

Publications: Progress reports on research are printed in the Annual Report of Florida Agricultural Experiment Stations.

Correspondence: English.

Food and Resource Economics Department – University of Florida
Gainesville, Florida 32611

Kiker, C.F.	Ph.D.	Agric.Econ. and Engr.

Status: Governmental institution supervised by Board of Regents of Florida.

Agricultural Engineering activities: Management and allocation of water resources. Production of energy from alternative sources.

Publications: Water allocation institutions – Operational management of water resource systems – Irrigation economics – Economic demand for water – Decentralized energy systems – Solar and biomass energy.

Correspondence: English.

Georgia

Agricultural Engineering Division, University of Georgia
Athens, Ga. 30601

Coastal Plain Experiment Station, Tifton

Georgia Agricultural Experiment Station, Griffin

Activities: Four-year courses leading to the degree B.Sc. in agricultural engineering and a graduate programme leading to the degree M.Sc.

Correspondence: English.

Hawaii

Agricultural Engineering Department, College of Tropical Agriculture and Human Resources, University of Hawaii
3050 Maile Way, Honolulu, Hawaii 96822

The Faculty:
Smith, M.R.	Ph.D.	Professor and Chairman – Agricultural equipment, mechanization
Gitlin, H.M.	MS	Professor – Post-harvest processing engineering, extension
Liang, T.	Ph.D.	Professor – Operations research, systems engineering
Singh, D.	Ph.D.	Assistant Professor – Livestock systems, machinery management
Wang, J-K	Ph.D.	Professor – Agricultural production systems analysis
Wu, I-P	Ph.D.	Professor – Irrigation engineering
Yang, K-P	Ph.D.	Assistant Professor – Systems engineering, crop processing
Yang, P-Y	Ph.D.	Oklahoma State, Associate Professor – Agricultural Waste Management.

Affiliate Graduate Faculty:
Gibson, W.O.	MA	Head, Field Engineering Department, Experiment Station, Hawaii Sugar Planters' Association – Sugar cane mechanization
Hundtoft, E.B.	Ph.D.	Professor – Agricultural equipment, statistical methods
Myers, A.L.	MS	Agricultural Engineer, USDA, ARS, WR – Harvesting equipment
Phillips, A.L.	Ph.D.	Research Associate, EW Food Institute, East-West Center, University of Hawaii – Agricultural mechanization, water management.

Emeritus Faculty:
Kinch, D.M.   Ph.D.   Emeritus Professor of Agricultural Engineering

Status: The Department of Agricultural Engineering offers the opportunity for graduate study leading to the degree of Master of Science in Agricultural Engineering. Current emphases of the department provide graduate study opportunities in tropical fruit, nut and flower production equipment, irrigation engineering, physical properties of tropical fruits, applied systems engineering, waste management, bio-energy systems, and aquacultural engineering. – The M.S. program requires a minimum of 21 credit hours of course work and nine credit hours of thesis research. The emphasis of the various programs is on the engineering application of the basic physical, mathematical and biological sciences to the problems of tropical agriculture. A candidate acquires his knowledge by conducting an original research investigation and by writing and defending a thesis. The department is also cooperating with the Department of Agricultural and Resource Economics to offer an option in Agricultural Systems Analysis in the Ph.D. program in Agricultural Economics. – Departmental projects currently focus on crop mechanization, development of production systems planning and management, improved packaging and transport methods, food crop delivery systems, trickle and drip irrigation technology, composting technology and microbial protein production in animal waste management, anaerobic bioconversion of organic wastes and modelling, simulation and probability theory in agricultural planning. – Hawaii's main crops are sugar cane, pineapple, coffee, macadamia nuts, papaya, and flowers which require different engineering techniques from those used in other parts of the country. Mechanized systems and equipment developed for mainland agriculture are frequently not directly applicable to tropical agricultural practice. Additionally, many tropical crops are not grown at all in temperate climates. Thus, the thrust of education and research programs of the Department centers on the application of engineering technology to the problems of tropical agriculture.

Publications: Aquacultural engineering – Mechanical harvesting – Tropical fruit processing – Physical properties – Pre-cooling of fresh vegetables and fruits – Systems engineering – Applied statistical methods – Waste management engineering – Water management and irrigation engineering.

Correspondence: English, Chinese, Japanese.

Idaho

Agricultural Engineering Department, University of Idaho
Moscow, Idaho

Fitzsimmons, D.W.	Ph.D.	Head of Department
Bloomsburg, G.L.	Ph.D.	Drainage
Brockway, C.E.	Ph.D.	Irrigation Water Management
Busch, J.R.	Ph.D.	Irrigation Water Management
Dixon, J.E.	Ph.D.	Structures and Environmental Systems
Dowding, E.A.	M.Sc.	Equipment Development
Everts, C.J.	M.Sc.	Nonpoint Source Control Extension
Guillerie, R.	M.Sc.	Nonpoint Source Control Extension
Halderson, J.L.	Ph.D.	Equipment Development
Karsky, T.J.	M.Sc.	Farm Safety Extension
Larsen, D.C.	B.Sc.	Irrigation Extension
Longley, T.S.	M.Sc.	Irrigation
McHargue, J.M.	M.Sc.	Agricultural Mechanics

McMaster, G.M.	M.Sc.	Irrigation
Moden, W.L.	M.Sc.	Equipment Development
Molnau, M.P.	Ph.D.	Hydrology
Peterson, C.L.	Ph.D.	Power and Machinery
Riesenberg, L.E.	Ph.D.	Agricultural Mechanics
Taylor, R.E.	M.Sc.	Agricultural Engineering Extension
Williams, L.G.	M.Sc.	Agricultural Mechanization
Worstell, J.R.	M.Sc.	Energy Extension

Status: Established in 1912; first accredited in 1950; financed by State Government of Idaho.

Activities: Study courses leading to B.Sc., M.Sc. and Ph.D. in agricultural engineering and B.Sc. with major in agricultural mechanization – Sixty-five undergraduate students and twelve graduate students – Research work dealing with crop water requirements, Irrigation water management, irrigation systems, trickle irrigation, water quality, runoff, erosion control, flow through porous media, modelling, crop storage and processing, crop harvesting and handling, and equipment development – Extension programs in irrigation water management, non-point source control, energy, machinery management, animal housing, waste management, and farm safety.

Publications: Irrigation, Irrigation systems, Water management, Water quality, Water resources, Flow into and through soils, Runoff, Erosion control, Crop storage, Harvesting equipment, Mechanization, Energy conservation, Energy sources.

Correspondence: English.

Illinois

Agricultural Engineering Department – University of Illinois Urbana-Champaign
202 Agricultural Engineering Building – Urbana, Illinois 61801

Yoerger, R.R.	Ph.D.	Department Head – Power and Machinery

Power and Machinery Area:
Hunt, D.R.	Ph.D.	Area Leader
Bode, L.E.	Ph.D.	
Butler, B.J.	MS	
Espenschied, R.F.	Ed.D.	
Goering, C.E.	Ph.D.	
Hoag, D.L.	Ph.D.	
Hummel, J.W.	Ph.D.	
Nave, W.R.	MS	
Siemens, J.C.	Ph.D.	

Electric Power and Processing Area:
Olver, E.F.	Ph.D.	Area Leader
Paulsen, M.R.	Ph.D.	
Peterson, W.H.	MS	
Puckett, H.B.	MS	
Rodda, E.D.	Ph.D.	
Shove, G.C.	Ph.D.	

Structures and Environment Area:
Curtis, J.O.	Ph.D.	Area Leader and Associate Department Head
Day, D.L.	Ph.D.	

Jedele, D.G.   MS
Muehling, A.J.   MS
Scarborough, J.N.   Ph.D.

Soil and Water Area:
Lembke, W.D.   Ph.
Drablos, C.J.W.   MS
Mitchell, J.K.   Ph.D.
Vanderholm, D.H.   Ph.D.
Walker, P.N.   Ph.D.

Status: The University of Illinois was founded in 1867 as a land grant institution of the State of Illinois.

Agricultural Engineering activities: a) Teaching: 4 year - B.S. Degree in Agricultural Engineering - 5 year combined - B.S. Degree Agriculture - B.S. Degree Agricultural Engineering - 4 year - B.S. Degree Agriculture - Agricultural Mechanization Option. Current enrollments total approxiamtely 160 in the first two programs and 75 in the third program. - b) Research: An extensive research program is carried out within the Agricultural Experiment Station with current focus on development, of new energy sources, energy conservation, field mechanization, grain drying and storage, electronic controls, livestock mechanization and waste handling, agricultural structures, drainage, hydrology, and erosion control. - c) Public Service: Six staff members serve as Agricultural Engineering subject matter specialists in interpreting and applying research results to production situations throughout the state of Illinois in cooperation with country extension advisors.

Publications: Technical papers are generally presented before meetings of the American Society of Agricultural Engineers and published in the Transactions of the ASAE.

Correspondence: English.

Department of Agricultural Economics, University of Illinois
Urbana, Ill. 61801

Activities in agricultural engineering: Lectures on the economic-engineering approach to farm operation - Investigation on the economics of farm size - Changing structure of agriculture.

Publications: Farm labour - Power - Machinery - Buildings.

Correspondence: English.

Indiana

Agricultural Engineering Department, Purdue University
West Lafayette, IN 47907

Isaacs, G.W.   Ph.D.   Professor and Head of Department.
                                 Grain drying and storage.

Barrett, J.R.	M.Sc.	Assist.Professor. Systems engineering
Beasley, D.	Ph.D.	Assist.Professor. Soil and water
Dale, A.C.	Ph.D.	Professor. Farm Buildings and solary energy
Field, W.E.	Ph.D.	Assist.Professor. Farm safety
Foster, G.H.	Ph.D.	Professor. Grain drying and storage
Foster, G.R.	Ph.D.	Assist.Professor. Soil and water
Friday, W.H.	M.Sc.	Assoc.Professor. Farm structures
Gibson, H.G.	M.Sc.	Assoc.Professor. Forest engineering
Hinkle, C.S.	Ph.D.	Professor. Farm structures
Huggins, L.F.	Ph.D.	Professor. Soil and water
Jones, D.D.	Ph.D.	Assoc.Professor. Farm structures and waste mgt.
Krutz, G.W.	Ph.D.	Assoc.Professor. Power and machinery
Ladisch, M.	Ph.D.	Assist.Professor. Biochemical conversion
Lien, R.M.	M.Sc. A.E.	Professor. Power and machinery
Liljedahl, J.B.	Ph.D.	Professor. Power and machinery
Marks, J.	Ph.D.	Assoc.Professor. Food engineering
McKenzie, B.A.	M.Sc. A.E.	Professor. Grain quality and energy
Monke, E.J.	Ph.D.	Professor. Soil and water
Nye, J.C.	Ph.D.	Assist.Professor. Waste management
Okos, M.R.	Ph.D.	Assoc.Professor. Food engineering and energy
Parsons, S.D.	Ph.D.	Assoc.Professor. Power and machinery
Peart, R.M.	Ph.D.	Professor. Energy and grain quality
Richey, C.B.	B.S.A.E. and M.E.	Assoc.Professor. Power and machinery
Strickland, R.M.	Ph.D.	Assist.Professor. Power and machinery
Strohshine, R.G.	Ph.D.	Assist.Professor. Grain drying and storage
Thompson, L.	Ph.D.	Assoc.Professor. Power and machinery
Holderly, L.G.	M.Sc.	Assist.Professor. Surveying
Tsao, G.C.	Ph.D.	Professor. Biochemical conversion

<u>Status</u>: Department was established in 1914 and is an integral part of Purdue University, a state-supported land grant institution.

<u>Activities</u>: Teaching: study courses leading to the Bachelor of Science, Master of Science, Ph.D. degree in Agricultural Engineering and the Bachelor of Science, and Master of Science degrees in Agricultural Mechanization. 1973 enrollments include 107 students working toward the Bachelor of Science degree in Agricultural Engineering and 113 students working toward the Bachelor of Science degree in Agricultural Mechanization. There are 43 students working toward the Master of Science and Ph.D. degrees. - Research: experimentation and extension work in farm machinery, soil and water, electrification, agricultural building, agricultural processing, physical properties, systems engineering and waste management.

<u>Publications</u>: Farm machinery, soil and water, electrification, farm buildings, agricultural processing, systems engineering, food engineering, waste management, forest engineering, biomass energy conversion, solar energy.

<u>Correspondence</u>: English.

Iowa

<u>Agricultural Engineering Department, Iowa State University</u>
Ames, Iowa 50011

Johnson, H.P.	Ph.D.	Professor. Acting Head of Department

Soil and Water:

Anderson, C.E.	Ph.D.	Assoc. Professor
Baker, J.L.	Ph.D.	Assoc. Professor
Beer, C.E.	Ph.D.	Professor
Hamlett, J.M.	B.Sc.	Res. Assoc.
Johnson, H.P.	Ph.D.	Professor
Laflen, J.M.	Ph.D.	USDA Professor

Power and Machinery:

Boyd, M.M.	M.Sc.	Instructor
Buchele, W.F.	Ph.D.	Professor
Chaplin, J.	M.Sc.	Res. Assoc.
Claar, P.W., II.	M.Sc.	Res. Assoc.
Colvin, T.S.	Ph.D.	USDA Collaborator
Erbach, D.C.	Ph.D.	USDA Collaborator
Lovely, W.G.	M.Sc.	Assoc. Professor
Marley, S.J.	Ph.D.	Professor
Smith, D.R.	M.Sc.	Graduate Asst.
Tevis, J.W.	B.Sc.	Instructor

Agricultural Construction and Maintenance:

Anderson, W.R.	M.Sc.	Asst. Professor
Bekkum, V.	Ph.D.	Asst. Professor
Everett, L.B.	M.Sc.	Adjunct Instructor
Hoerner, T.A.	Ph.D.	Professor
McCarthy, D.A.	M.Sc.	Adjunct Instructor
Morford, V.J.	M.Sc.	Prof. Emeritus
Shorter, G.S.	M.Sc.	Adjunct Instructor

Electric Power and Processing:

Bern, C.J.	Ph.D.	Assoc. Professor
Hurburgh, C.R.	M.Sc.	Instructor
Soderholm, L.H.	Ph.D.	USDA

Crop Conditioning and Storage:

Hukill, W.V.	B.Sc.	USDA Prof. (Retired)
Kline, G.L.	M.Sc.	USDA Assoc. Professor

Structures and Environment:

Bundy, D.S.	Ph.D.	Assoc. Professor
Hazen, T.E.	Ph.D.	Professor
Mangold, D.W.	Ph.D.	Assoc. Professor
Smith, R.J.	Ph.D.	Professor

Extension:

Glanville, T.D.	M.Sc.	Instructor
Greiner, T.H.	Ph.D.	Asst. Professor
Melvin, S.W.	Ph.D.	Professor
Meyer, V.M.	Ph.D.	Professor
Ozkan, H.E.	Ph.D.	Asst. Professor
Van Fossen, L.D.	M.Sc.	Assoc. Professor
Wilcke, W.F.	M.Sc.	Instructor
Williams, D.L.	M.Sc.	Instructor

Midwest Plan Service:

Pedersen, J.H.	Ph.D.	Professor. Director. Development and production of standard plans for agricultural buildings and production systems.

Agricultural Experiment Station, Engineering Service:
Hazen, T.D.   Ph.D.   Professor

Status: Federal land grant and State institution, supervised by a State Board of Regents – Department was established in 1907. The curriculum is fully accredited by the Engineering Council for professional development.

Agricultural Engineering activities: **Training activities** – Courses are provided for four-year B.Sc. degree, for graduate degrees of M.Sc. and Ph.D. – Major options are chosen in electric power and processing, structures and environment, power and machinery, and soil and water – and crop conditioning and storage. Service courses in agricultural engineering are provided for students in the College of Agriculture. A B.S. degree in Agricultural Mechanization is available for those who do not wish to pursue an engineering program. Approximately 130 majors plus 30 graduate students in the department. – **Research activities:** Field investigations of sub-surface and surface drainage – Erosion control and water conservation investigations in Iowa – Physical and economic analysis of watersheds as related to soil and water conservation – Utilization of electric service in rural areas of Iowa – Farm storage and conditioning of grain – Functional design requirements and performance of farm equipment – Water infiltration into soils – Response of swine to cycling physical factor variation – Efficiency of utilization of machines, power units and machine systems in crop production – Tillage requirements for crop establishment and growth for soil and water conservation – Development of specifications for improving machinery and methods for the control of weeds – Pasture and forage production management and utilization in relation to land use and agricultural adjustment in Iowa – Hydrological characterization of small watersheds – Atmospheric studies in confined animal housing – Animal waste management – Mechanical harvesting of small fruits – Evaluation of flood damage to corn (maize) from depth and frequency of flooding – Properties of tile drainage water – Corn (maize) harvester design – Maximizing the utilization of biological and physical resources in the development of superior soybean production systems – Study of the biotypes of the European corn borer. – **Extension activities:** Soil and water conservation – Power and machinery – Structures and environment – Farmstead planning – Farmstead mechanization – Utilization of electrical energy on farms – Crop conditioning and storage – Agricultural safety – Animal waste management.

Publications since 1973: 200 papers in the subject matter fields listed.

Correspondence: English, French, German.

Accident Prevention Section, Institute of Agricultural Medicine, College of Medicine, University of Iowa
Oakdale, Iowa 52319

Knapp, L.W., Jr.   M.Sc.A.E.   Director of International Program and Chief of Accident Prevention Laboratory
Monson, R.W.   B.A.   Accident Investigator

Status: State institution, established in 1959, supervised by Board of Regents.

Activities: Research in injury and occupational health problems in agriculture – product safety, epidemiological studies and laboratory analysis of agricultural injuries, investigation of low-back pain among Iowa farmers.

Publications: Epidemiology of farm tractor/motor vehicle accidents, Epidemiology of power take-off accidents, Epidemiology of rotary power lawn mower injuries, Burn injuries.

Correspondence: English.

Kansas

Department of Agricultural Engineering, Kansas State University
Manhattan, Kansas 66506

Johnson, W.H.	Ph.D.	Professor, Head of Department
Barnes, P.L.	Ph.D.	Asst. Professor, Irrigation
Baugher, E.E.	M.Sc.	Asst. Professor. Farm Mechanics and Teacher Training
Chung, D.S.	Ph.D.	Professor. Agricultural processing
Clark, S.J.	Ph.D.	Farm Power and Machinery
Fairbanks, G.E.	M.Sc.	Professor. Farm Machinery
Hay, D.R.	M.S.	Ext.Assoc.Professor. Irrigation
Hodges, T.O.	Ph.D.	Professor. Farm Structures
Holmes, E.S.	M.Sc.	Ext. Professor. Rural Electrification
Jepsen, R.L.	Ph.D.	Ext.Assoc.Professor. Agr'l safety
Larson, G.H.	Ph.D.	Professor. Farm Power Machinery
Lipper, R.I.	M.Sc.	Professor. Rural Electrification
Manges, H.L.	Ph.D.	Professor. Irrigation
Murphy, J.P.	M.S.	Ext.Assoc. Professor. Farm Buildings
Powell, D.M.	M.S.	Instructor. Irrigation
Powell, G.M.	Ph.D.	Ext.Assist.Professor. Soil and Water Conservation
Rogers, D.H.	Ph.D.	Ext.Asst.Professor. Irrigation
Schrock, M.	M.S.	Asst. Professor. Farm Power and Machinery
Spillman, C.K.	Ph.D.	Professor. Farm Buildings
Steichen, J.M.	Ph.D.	Assoc.Professor. Soil and Water Engineering
Stevenson, P.N.	M.Sc.	Assoc.Professor. Farm Mechanics and Teacher Training
TenEyck, G.R.	M.S.	Asst.Professor. Irrigation
Thomas, J.G.	M.S.	Ext.Asst. Professor. Irrigation
Wendling, L.T.	M.Sc.	Ext. Professor. Farm Buildings.

Status: State land-grant institution, supervised by State Board of Regents through the President of the institution.

Activities: Study courses leading to a B.Sc., M.Sc. and Ph.D. degree in agricultural engineering - Research on: Watersheds and water conservation - Irrigation engineering practices and equipment - Studies pertaining to operation of farm power units - Study of minimum tillage practices (soil dynamics) - Harvesting and conditioning of agricultural crops - Conditioning of grain in storage - Grain drying with solar heat - Building construction as it affects comfort of animals and poultry housing - Exposure tests of fence wire with various kinds and weight of coating - Mechanical and environmental grain damage studies - Animal waste disposal - Alternate fuels for agriculture, solar collectors for heating animal shelters.

Publications: Solar heating swine houses, solar swine house - Green house symbiotic system, alternate fuels for agriculture, solar grain drying, combine harvesting efficiency, equipment for conservation tillage, grain drying systems, exposure tests of fence wire with various kinds and weight of coating, animal waste disposal, automatic surface irrigation systems, water quality, irrigation for humid regions.

Correspondence: English.

Kentucky

Agricultural Engineering Department, University of Kentucky
Lexington, Ky. 40506 (Mail address: Room 126, 00751 - Agric. Engineering Bldg.)

Walker, J.N.	Chairman and Professor
Barfield, B.J.	Soil and Water, Plant and Animal Environment
Benock, G.T.	Crops Processing
Bridges, T.C.	Computer Modeling
Brooks, J.B.	Professor (Emeritus)
Casada, J.H.	Tobacco Mechanization
Colliver, D.	Structures and Environment
Duncan, G.A.	Tobacco and Horticulture Facilities
Fehr, R.L.	Animal Environment
Henson, W.H.	Tobacco Curing and Mechanization
Holland, S.	Energy
Loewer, O.J.	Crop Processing
McNeill, S.	Swine, Grain and Feed Facilities
Moore, I.	Soil and Water Hydrology
Nickell, W.T.	Tobacco Curing and Mechanization
Parker, B.F.	Structures and Environment
Payne, F.	Grain Energy
Piercy, L.	Agricultural Safety
Priddy, T.	Meteorology
Ross, I.J.	Agricultural Processing, Animal Waste Processing
Smith, E.M.	Agricultural Machinery, Soil Environment, Tobacco Mechanization
Tapp, J.	Soil and Water
Taraba, J.L.	Waste Management
Thompson, M.	Agricultural Meteorology
Turner, G.M.	4-H Programs
Turner, L.	Energy
Walton, L.R.	Tobacco Curing and Mech.
Ward, A.	Soil and Water Modeling
Wells, L.G.	Agricultural Machinery, Machinery/Soil Relationships
White, G.M.	Processing
Williams, R.	Soil and Water
Wilson, B.	Soil and Water
Yoder, E.E.	Tobacco Mechanization and Processing

Status: Department of Agricultural Engineering was established in 1956 as a part of the University of Kentucky which is the land-grant institution in the Commonwealth of Kentucky.

Activities: Educational programmes are provided leading to B.Sc., M.Sc. and Ph.D. in agricultural engineering - The major research area encompasses engineering for plant and animal life with several important projects in: Agricultural machines - Agricultural structures - Hydrology - Animal environment - Plant environment - Soil environment - Sedimentology - Micro-climate - Soil moisture - Thermal radiation - Tobacco mechanization - Tobacco curing - Physical properties of tobacco and potatoes - Waste utilization - Grain drying - Water pollution - Watershed management.

Publications: Burley tobacco curing - Mechanical handling of tobacco - Mechanical tobacco harvester (stalk cut) - Beef, swine, dairy buildings - Economy housing - Soil environment - Thermal radiation through plastics - Greenhouse environment - Animal ventilation - Frequency of drought in Kentucky - Water requirement for plant life - Grain handling, drying and storage - Water supply and treatment.

Correspondence: English.

Louisiana

Department of Agricultural Engineering, Louisiana State University and A&M College
Baton Rouge, Louisiana 70803

Agricultural Engineering Teaching and Research:

Brown, W.H.	Ph.D.	Prof. and Head,	Structures and Environment
Baldwin, J.D.C.	Ph.D.	Asst.Prof.	Structures and Environment
Baskin, G.R.	BS	Associate	Soybean Harvesting
Bengtson, R.L.	Ph.D.	Asst.Prof.	Soil and Water
Braud, H.J.	Ph.D.	Professor	Structures and Environment
Carter, C.E.	MS	USDA-AR	Subsurface Drainage
Cochran, B.J.	Ph.D.	Professor	Power and Machinery
Edling, R.J.	Ph.D.	Assoc.Prof.	Soil and Water
Mayeux, M.M.	MS	Professor	Power and Machinery
McDaniel, V.	MS	Instructor	Soil and Water
Newton, G.J.	BS	Associate	Sugarcane Systems
Pringle, H.C.	BS	Associate	Energy Systems
Rollason, S.H.	BS	Associate	Horticultural Crops
Sistler, F.E.	Ph.D.	Asst.Prof.	Electric Power and Processing
Smith, P.A.	BS	Associate	Computer Applications
Stipe, D.R.	MS	Professor	Electric Power and Processing
Thomas, C.H.	Ph.D.	Professor	Power and Machinery
Verma, L.R.	Ph.D.	Asst.Prof.	Electric Power and Processing
Wright, M.E.	Ph.D.	Professor	Power and Machinery

Agricultural Engineering Extension:

Baker, F.E.	MS	Assoc.Spec. and Leader,	Structures and Environment
Branch, J.W.	MS	Assoc.Spec.	Structures and Environment
Deason, D.L.	Ph.D.	Specialist	Electric Power and Processing
Hadden, W.A.	MS	Specialist	Soil and Water
McManus, R.V.	MS	Ext.Asst.	Farm Safety
Peavy, M.M.	MS	Assoc.Spec.	Residential Energy Systems
Rester, D.C.	MS	Assoc.Spec.	Pesticide App. Systems
Smilie, J.L.	MS	Specialist	Power and Machinery

Status: Louisiana State University and A&M College originated in 1877. The Agricultural Engineering Department was organized in 1911. The LSU System is governed by a Board of Supervisors. The Louisiana Agricultural Experiment Station administers research activities. Teaching, research and extension activities are jointly funded by the State of Louisiana, the U.S. Department of Agriculture, and by numerous public and private donors.

Teaching: B.S. in Agricultural Engineering - B.S. in Agricultural Mechanization - MS in Agricultural Engineering - M.E. in Engineering with Agricultural Engineering major - Ph.D. in Engineering with Agricultural Engineering major. - Research: Sweet potato mechanization; tabasco pepper harvesting; ammonia handling equipment; high-population sugarcane mechanization and harvesting; solar energy for grain drying; safety programs in a research institution; soybean harvest losses; soybean-wheat double-cropping; energy production from biomass; forage simulation systems; energy efficiency in cotton production; duckweed production for animal feeds; water source heat exchange for heat pumps; trickle irrigation; energy reduction in on-farm processing; wind energy applications; irrigation systems and water management; hay conditioning and storage; runoff and erosion from flat watersheds; subsurface drainage systems.

Publications: Several publications on above activities.

Correspondence: English.

Agricultural Engineering Department, Louisiana Tech University
Box 4535 Tech Station, Ruston, LA 71272

Robbins, J.W.	Ph.D.	Head of Department - Environmental Conservation
Albritton, J.V.	M.Sc.	Asst.Prof. - Electric Power and Processing
Nelson, J.D.	Ph.D.	Asst.Prof. - Soil and Water Conservation, Structures
Vidrine, C.G.	Ph.D.	Prof. - Machinery

Status: State institution, governed by State Board of Education.

Activities: Four-year study courses leading to B.Sc.A.E. degree, 35 students and one further year's course for M.Sc.Engr. and two further years course work for Doctor of Engineering. Research on: Field sprayers - Waste water for irrigation - Greenhouses for winter crops - Hog environment - Heat exchange - Waste management - Forestry Mechanization - Irrigation.

Publications: Municipal waste water for irrigation - Papermill wastewater for irrigation - Forestry Mechanization.

Correspondence: English.

Maine

Agricultural Engineering Department, University of Maine
Orono, Me. 04473

Smith, N.	Ph.D.	Head of Department - Farm power and machinery
Christensen, T.	M.Sc.	Assist.Prof. - Farm power and machinery
Hallee, N.D.	M.Sc.	Ext.Ag.Engr. - Farm power and machinery
Hedstrom, W.E.	Ph.D.	Assoc. Prof. - Soil and water engineering
Huff, E.R.	Ph.D.	Assoc.Prof. - Farm machinery materials
Hunter, J.H.	M.Sc.	Assoc.Prof. - Farm buildings and agricultural processing
Kittridge, C.W.	B.Sc.	Ext.Ag.Engr. - Farm buildings and equipment
Klinge, A.F.	Ph.D.	Prof. - Soil and Water Engineering
Rhoads, R.B.	M.Sc.	Prof. - Agricultural Processing
Rowe, E.J.	Ph.D.	Prof. - Farm machinery and instrumentation
Sides, S.E.	B.Sc.	Assoc.Prof. - Farm machinery
Soule, H.M. Jr.	M.Sc.	Assoc.Prof. - Farm machinery
Warner, M.R.	M.Sc.	Ext.Ag.Engr. - Farm buildings and agricultural processing
Riley, J.G.	Ph.D.	Assist.Prof. - Aquaculture, vehicle mechanics

Status: Governmental, supervised by Board of Trustees.

Training activities: 4 years courses - There are approximately 150 students in 4 year courses - The 4 year courses lead to a B.S. in agricultural or forest engineering or to a B.S. in agricultural mechanization. - Research: Mechanical harvesting of oysters - Combined agriculture/aquaculture systems - Forage harvesting systems - Energy conservation methods - Forest engineering systems - Potato handling methods - Mathematical model of physiological behaviour of potato tubers - Biomass harvesting - Biomass Combustion equipment - Solar assisted heat pumps.

Publications: Several publications on above activities.

Correspondence: English.

Maryland

Soil, Water, and Air Sciences - National Program Staff - Agricultural Research Science and Education Administration, U.S. Department of Agriculture
Washington, D.C. 20250

Barrows, H.L.	Ph.D.	Chief, Soil and Water Management
Farrell, D.A.	Ph.D.	National Research Program Leader, Hydrology
Jensen, M.E.	Ph.D.	National Research Program Leader, Water Management

Status: The U.S. Congress first established an appropriation for agricultural engineering in 1898. The first work was for irrigation information, drainage soon followed, and after that erosion control and hydrology. Several reorganizations followed until July 1972 when the current staff was organized. The reorganization of October 26, 1978, resulted in the formation of Science and Education Administration, the present structure.

Activities: The Soil, Water, and Air sciences staff is responsible for the quality and scope of all phases of the soil, water, and air research of the U.S. Department of Agriculture except those related to forestry. The staff consists of leading scientists and engineers, each having national responsibility for a given portion of the program. Research is in progress on the following: reduce salt damage to soils, crops, and water; improve irrigation and drainage of agricultural lands; develop tillage practices for improving soil properties and crop growth; manage and use precipitation and solar energy for crop production; reclaim and revegetate land areas disturbed by man; utilize, conserve, and manage agricultural water resources; agricultural application of remote sensing; and conserve energy in agricultural production.

Publications: Results of the research on the subjects listed above are published in technical journals, technical and farmers bulletins, agricultural handbooks, leaflets and articles. These average about 500 per year.

Correspondence: English.

Massachusetts

Food Engineering Department, University of Massachusetts, Amherst, Mass. 01003
Amherst, Mass. 01003

Clayton, J.T.	Ph.D.	Prof. Head of Department - Food Engineering
Feng, T.H.	Ph.D.	Prof. Environmental Engineering (waste proceeding)
Fitzgerald, G.A.	M.Sc.	Emeritus Prof. - Food Engineering (food processing and distribution)
Johnson, E.A.	M.Sc.	Prof. - Food Engineering (instrumentation, food processing plant design)

Light, R.G.	M.Sc.	Prof. - Agricultural Engineering (food production systems and environmental quality)
Mudgett, R.E.	Ph.D.	Assoc.Prof. - Fermentation engineering; dielectric properties, microwave processing
Norton, J.S.	M.Sc.	Assoc.Prof. - Food Engineering (cranberry production, harvesting and processing)
Peleg, M.	Ph.D.	Assoc.Prof. - Physical properties, rheology of foods, powder technology
Pira, E.S.	M.Sc.	Assoc.Prof. - Food Engineering (controlled environment food production)
Rosenau, J.R.	Ph.D.	Assoc.Prof. - Food Engineering (transport processes, dairy processing, waste utilization, solar energy)
Schwartzberg, H.G.	Ph.D.	Prof. - Food Engineering (food processin., organic waste utilization, solar energy)
Whitney, L.F.	Ph.D.	Prof. - Food Engineering (food machinery, food plant operations, systems analysis).

Status: State land-grant institution, established in 1965, governed by a Board of Trustees.

Activities: Food Engineering professional curriculum leading to B.Sc. F.E. degree - Graduate programmes leading to M.Sc. and Ph.D. degrees - 26 graduate students enrolled in 1979-80. - Research is carried out in the broad areas which relate macrophysical and microphysical environments to biological systems and the products derived from these systems. Special emphasis is placed on studies of the engineering properties of food materials, on the parameters of food process design, on the design of food processing machinery, on simulated and synthetic foods, on new food sources, on the storage and preservation of foods, on the packaging and protection of foods, and on the production and storage of foods.

Publications: Food process engineering, simulated and synthetic foods, engineering properties of foods, food packaging and distribution, food storage and preservation, processing and utilization of organic wastes.

Correspondence: English, French, German, Spanish.

Michigan

American Society of Agricultural Engineers (ASAE)
2950 Niles Road, St. Joseph, Mich. 49085 - P.O. Box 410

Butt, J.L.	M.Sc. A.E.	P.E. Executive Vice President
Basselman, J.A.	B.Sc. A.E.	Publications Manager
Castenson, R.R.	B.Sc. A.E.	Manager, Membership Activities
Hahn, R.H. Ms.A.E.	Ms.A.E.	Manager, Technical Operations
Purschwitz, M.A.	Ms.A.E.	Staff Engineer

Branches: ASAE has two major branches: (a) technical and (b) geographic. Technically, the Society is divided into five divisions - Power and Machinery, Soil and Water, Electric Power and Processing, Structures and Environment, and Food Engineering. Geographically ASAE has 51 geographic units throughout the United States and Canada - 10 regional, 39 state, and 2 local. The Society also has administrative departments concerned with education and research, professional development, finances, publications, and awards.

Status: ASAE is a private corporation, established in 1907, supported by annual dues of its members and income from sale of publications. It is governed by a Board of Directors of 26 engineers elected, staggered terms, by ASAE members throughout the world.

Agricultural Engineering activities: Teaching: ASAE brings together top educators from throughout the United States and other countries for the exchange of information on agricultural engineering courses, curriculum, and teaching techniques. This is done through regular ASAE meetings, through special conferences and seminars, and through the publication of ideas and concepts presented by individual educators. Most agricultural engineering teachers in the United States, and many from around the world, hold membership in ASAE and participate in these activities. ASAE also has a student branch organization made up of college students from all of the agricultural engineering departments in North America. They meet twice annually to conduct business meetings and to exchange ideas about student branch activities in their respective institutions. - Research: ASAE is the principal source of publication of research results by agricultural engineers at the colleges and universities throughout the United States, those with various agencies of government, and engineers in industry. These research results may be presented at one of ASAE's two national meetings each year, or at special ASAE in-depth conferences, or they may be submitted to ASAE for publication in one of its technical journals. ASAE may be considered to be the principal means whereby agricultural engineers may communicate with one another about matters of mutual interest.

Publications: ASAE has seven major categories of publications: (1) Agricultural Engineering, a monthly magazine describing agricultural engineering developments, news, trends, meetings, and lists of new bulletins, publications, and research reports. (2) Transactions of the ASAE is published six times a year and is made up entirely of manuscripts provided by individual researchers, teachers, extension workers, or industry engineers describing their significant findings for the benefit of others. (3) Agricultural Engineers Yearbook is published annually and contains the ASAE standards, recommendations, data, membership roster, and list of products and suppliers of agricultural machinery and components. (4) Proceedings of special conferences are published as a means of recording the technical manuscripts presented during the ASAE special conferences. (5) ASAE special publications are varied in nature and may include (a) collections of papers presented at ASAE meetings, (b) bibliographies, (c) summaries of research, (d) textbooks for use in agricultural engineering courses, and other material. (6) Technical papers - ASAE obtains technical papers presented at its two national meetings, about 1000 per year, and makes these available as individual reprints to interested persons. (7) Microfiche - ASAE technical papers, described in the preceding sentence, also are available on photographic negatives as microfiche for those who have microfiche reading equipment.

Correspondence: English.

Agricultural Engineering Department, Michigan State University
East Lansing, Michigan 48824

Edwards, D.M.	Ph.D.	Professor and Chairman - Head of Department
Bakker-Arkema, F.W.	Ph.D.	Professor - Processing of Agricultural Products, Storage and Drying; Computer Use
Bickert, W.G.	Ph.D.	Professor - Livestock Facilities and Environment; Extension Project Coordinator
Bralts, V.F.	MS	Instructor - Soil and Water; Irrigation Scheduling; Hydrology

Brook, R.C.	Ph.D.	Asst. Professor - Post Harvest Handling, Storage and Drying; Computer Use
Burkhardt, T.H.	Ph.D.	Assoc. Professor - Power and Machinery; Modelling of Crop Production Systems; Coordinator of Undergrad. Programs
Cargill, B.J.	Ph.D.	Professor - Fruit and Vegetable Mechanization and Storage
Cron, D.J.	MA	Specialist - Building Construction
Doss, H.J.	MS	Specialist - Agricultural Machinery Training and Farm Safety
Esmay, M.L.	Ph.D.	Professor - Structures and Environment; Solar Energy; International Programs; Coordinator of Graduate Programs
Galbavi, F.R.	BS	Specialist - Power and Machinery
Gerrish, J.B.	Ph.D.	Assoc. Professor - Agricultural Pollution Control; biochem. engineering; electrical power
Heldman, D.R.	Ph.D.	Professor - Food Engineering
Kampe, D.F.	MS	Specialist - Power and Machinery
Ledebuhr, R.L.	BS	Specialist - Fruit and Vegetable Mechanization
Lillmars, L.D.	MS	Instructor - Electrical Technology in Agriculture
Loudon, T.L.	Ph.D.	Assoc. Professor - Agricultural Waste Management; Livestock Facilities; Irrigation
Mack, L.A.	MS	Specialist - 4-H Youth; Home Weatherization; Family Housing
Merva, G.E.	Ph.D.	Professor - Agricultural Hydrology; Plant-Soil Relationships; Coordinator of Research Programs
Myers, C.A.	BS	Specialist - Energy for Agriculture
Person, H.F.	Ph.D.	Asst. Professor - Livestock Facilities and Environment
Pfister, R.G.	Ph.D.	Professor - Safety in Agricultural Systems, Human Factors
Rotz, C.A.	Ph.D.	Asst. Professor - Power and Machinery; Alcohol Fuels
Segerlind, L.J.	Ph.D.	Professor - Physical Properties; Finite Element Analysis
Srivastava, A.K.	Ph.D.	Asst. Professor - Power and Machinery; Hydraulics
Steffe, J.F.	Ph.D.	Asst. Professor - Food Engineering
Surbrook, T.C.	Ph.D.	Assoc. Professor - Electrical Power; Animal Environment and Instrumentation; Coordinator of Inst. of Ag. Techn.
Van Ee, G.R.	Ph.D.	Asst. Professor - Power and Machinery
Wilkinson, R.H.	Ph.D.	Professor - Power and Machinery; Human Factors Engineering

Status: Governmental Institution, supervised by an elected Board of Trustees

Training (Teaching) activities: Study courses in professional Agricultural Engineering; Four year undergraduate leading to B.Sc. degree, graduate programs for M.Sc. and Ph.D. Courses in Agricultural Engineering leading to B.Sc., M.Sc., and Ph.D. degrees, as well as B.Sc., M.Sc., and Ph.D. degrees in Agricultural Engineering Technology. Farm Equipment Service and Sales, Electrical Technology for Agriculture are special two-year courses for students in agriculture. Four year course leading to a B.Sc. degree in Residential Building Construction; M.Sc. degree also available in this area. Number of students - year: Agricultural Engineering 100; Agricultural Engineering Technology 60; Building Construction 160. Research: Processing of Biological Products; Engineering Aspects of Processing Biological Products; Animal Housing: Environment Improvement

Improvement with Solar Energy; Dynamics and Energetics of the Plant-Soil Atmosphere Continuum; Mechanical Harvesting, Handling of Fruits and Vegetables; Animal Waste Management System; Climatic Resources - North Central Region; Human Factors Engineering; Engineering Consideration in Food Processing; Energy in Agriculture; Climatology of Michigan; Engineering for Improved Cultural Practices; Mechanization in Developing Countries; Land Application of Animal Waste; Design Analysis and Optimization of Agricultural Machinery Systems; Development of Machinery Concepts for Mechanized Agriculture; Research Productivities for Michigan Crop Production Systems; Mechanical Harvesting and Handling Tree Shaping, and Spraying Peaches and Pears; Mechanical Harvesting and Handling of Vegetables; Post Rice Harvest Losses and Technology in Tropical Countries.

Publications on: energy, solar, wind, biomass, alcohol, gasifiers, energy management and conservation, food-fuel conflicts, anaerobic digestion, agricultural residues, milking parlours and livestock housing. Books on: Rice post-production technology in the tropics - Glimpses of agricultural mechanization in China - Applied finite element analysis - Food processors and engineers handbook - Engineering for dairy and food products - Dictionary of agricultural and food engineering - Food engineering systems 1 : operations - Food Engineering Systems 2 : Utilities - Food process engineering - Physioengineering principles - Buildings for small acreages.

Correspondence: English, Spanish, French, German.

Minnesota

Agricultural Engineering Department, University of Minnesota
St. Paul, MN 55108

Flikke, A.M.	Ph.D.	Head of Department - Environment and Processing
Aherin, R.A.	MS	Agricultural Safety
Allred, E.R.	MS	Soil and Water
Bates, D.W.	MS	Agricultural Structures
Bear, W.F.	Ph.D.	Farm Mechanics
Bergsrud, F.G.	MS	Soil and Water
Clanton, C.J.	MS	Waste Management
Cloud, H.A.	Ph.D.	Materials Handling
Goodrich, P.R.	Ph.D.	Waste Management
Gustafson, R.J.	Ph.D.	Power and Processing
Jacobson, L.D.	MS	Farm Structures
Janni, K.A.	Ph.D.	Farm Structures
Jordan, K.A.	Ph.D.	Environment and Processing
Larson, C.L.	Ph.D.	Soil and Water
Machmeier, R.E.	Ph.D.	Soil and Water
Morey, R.V.	Ph.D.	Systems Engineering
Onstad, C.A.	Ph.D.	Soil and Water
Pomroy, J.H.	MS	Agricultural Structures
Schertz, C.E.	Ph.D.	Power and Machinery
Schuler, R.T.	Ph.D.	Power and Machinery
Schultz, L.D.	MPH	Agricultural Safety
Slack, D.C.	Ph.D.	Soil and Water
Strait, J.	MS	Power and Machinery
Thompson, D.R.	Ph.D.	Food Engineering
True, J.A.	MS	Power and Machinery

Werner, H.D.	MS	Soil and Water
Wright, J.A.	MS	Soil and Water
Young, R.A.	Ph.D.	Soil and Water

Status: Established in 1913 as a Governmental institute.

Activities: Teaching: Agricultural Engineering Technology, B.S. Agriculture - Master of Agriculture - Ag.Eng.Tech. Agricultural Engineering, B.Agr.Engr. - Master of Agricultural Engineering, M.AgEn. - Master of Science, M.S. Ag. Engineering - Doctoral Program, Ph.D. Ag.Engineering. Research: Marketing and Delivery of Quality Cereals and Oilseeds - Reduction of Non-Renewable Energy Consumption for Grain Drying - Improved Soybean Harvest Methods to Reduce Harvest Losses and Improve Seed Quality - Engineering Studies related to the Production and Processing of Wild Rice - Reduced Tillage Studies emphasizing Energy, Soil, and Water Conservation on a Sandy Loam Soil - Durability of Concrete Drain Tile and Irrigation Pipe - Field Evaluation of Crop Water Stress - Irrigation Practices to Improve Resource Management in Minnesota - Hydrology of Small Watersheds - Farm Animal and Agricultural Waste Management - Transport Phenomena in Livestock Environments - Swine Production Facilities using no Fossil Fuels - Pathogenic Responses in Neonatal Farm Animals to Environmental Insults - Construction and operation of a Farm Scale Fuel Ethanol Plant - Extension Programs in Power and Machinery, Processing, Soil and Water Structures, Agricultural Wastes and Agricultural Safety.

Publications: Numerous publications on research activities.

Correspondence: English, French, Spanish.

Mississippi

Agricultural and Biological Engineering Department, Mississippi State University
Mississippi State, MS 39762

Status: The Department was established at Mississippi State University in 1914 - At the present, the Department is jointly administered by the Colleges of Agriculture and Engineering at Mississippi State University.

Teaching activities: Agricultural engineering, technology and business courses leading to B.Sc. degree from the College of Agriculture - Agricultural engineering and biological engineering courses leading to B.Sc. degree from the College of Engineering - Agricultural Engineering studies leading to M.Sc. and Ph.D. degrees from the Graduate School.
Research activities: Seed processing and storage - Soil-water-plant relationships - Machine and thermal systems - Crop production engineering - Animal and avian environment - Food process engineering - Forest engineering - Particle mechanics - Systems engineering - Instrumentation and control systems - Bio-physical properties of material.

Publications: Research publications dealing with areas of research work listed above.

Correspondence: English.

Missouri

Agricultural Engineering Department, University of Missouri
Columbia, Missouri 65211

Day, C.L.	Ph.D.	Head of Department
Anderson, M.E.	Ph.D.	Assoc.Prof. - Food Engineering
Baker, D.E.	M.Sc.	Farm safety
Brooker, D.B.	M.Sc.	Prof. - Electric Power and Processing
Constien, E.J.	B.Sc.	Asst.Prof. - Power and Machinery
Cromwell, C.F.	M.Sc.	Assoc.Prof.- Irrigation
Currence, H.D.	Ph.D.	Assoc.Prof. - Electric Power and Processing
Fischer, J.R.	Ph.D.	Res.Assoc. - Energy
Frisby, J.C.	Ph.D.	Prof. - Power and Machinery
Fulhage, C.D.	Ph.D.	Assoc.Prof. - Animal Waste Management
Gebhardt, M.R.	Ph.D.	Assoc.Prof. - Power and Machinery
George, R.M.	M.Sc.	Assoc.Prof. - Farmstead Mechanization
Gregory, J.M.	Ph.D.	Asst.Prof. - Erosion Control
Harris, F.D.	Ph.D.	Prof. - Power and Machinery
Hires, W.G.	Ph.D.	Asst.Prof. - Power and Machinery
Iannotti, E.L.	Ph.D.	Asst.Prof. - Microbiology
Kramer, L.A.	M.Sc.	Res.Assoc. - Watershed Hydrology
Linhardt, R.E.	Ph.D.	Assoc.Prof. - Farm Mechanics
Mayes, H.F.	M.Sc.	Res.Assoc. - Farm Structures
McCarty, T.R.	Ph.D.	Asst.Prof. - Soil and Water
McFate, K.L.	M.Sc.	Prof. - Rural Electrification
Meador, N.F.	Ph.D.	Prof. - Farm Structures
Phillips, R.E.	Ph.D.	Prof. - Farm Structures
Piest, R.F.	B.Sc.	Res.Assoc. - Hydrology and Hydraulics
Rausch, D.L.	M.Sc.	Res.Assoc. - Reservoir Sedimentation
Schottman, R.W.	Ph.D.	Asst.Prof. - Soil and Water
Shanklin, M.D.	Ph.D.	Prof. - Bio-engineering
Sievers, D.M.	Ph.D.	Assoc.Prof. - Waste Utilization
Smith, D.B.	Ph.D.	Assoc.Prof. - Power and Machinery

Status: The Department of Agricultural Engineering was established in 1916 within the University of Missouri which is the State University of Missouri.

Agricultural Engineering activities: Teaching: Study courses leading to the following degrees: B.Sc. in agricultural engineering (75 students); M.Sc. in agricultural engineering (5 students); Ph.D. in agricultural engineering (5 students); B.Sc. in agricultural mechanization (115 students); M.Sc. in agricultural mechanization (5 students). - Research: Controlling erosion - Grain conditioning - Forage harvesting - Pest control equipment - Utilization of electrical energy - Animal housing - Waste disposal - Anaerobic digestion - Alternate energy sources.

Publications: Grain drying - Farm safety - Forage management and drying - Machinery management - Animal waste management - Soil and water engineering - Pest control equipment - Energy utilization - Farm structures, and animal environment.

Montana

Agricultural Engineering Department, Montana State University
Bozeman, MT 59715

Erickson, L.R.	Ph.D.	Assist.Prof. - Mechanics
Hanson, T.L.	Ph.D.	Prof. - Soil and Water
Larsen, W.E.	Ph.D.	Head of Dept. - Machines
Linn, R.	MS	Assoc.Prof. - Agr. Safety and Energy Spec.

Milledge, R.K.	BS	Assist.Prof. - Shop Practices
Milne, C.M.	Ph.D.	Prof. - Structures and Environment
Westesen, G.	Ph.D.	Prof. - Irrigation

Status: State Institution.

Activities: Study courses leading to a BS and MS in Agricultural Engineering or a BS in Agricultural Mechanics. Research on: Irrigation, Farm structures, Tillage and Energy in Agriculture.

Publications: Tractor Use, Irrigation, Wood heating.

Correspondence: English.

Nebraska

Department of Agricultural Engineering, University of Nebraska
Lincoln, Nebraska 68583

Splinter, W.E.	Ph.D.	Prof. - Head of Department
Bashford, L.L.	Ph.D.	Assoc.Prof. - Power and Machinery
Berry, I.	Ph.D.	Res. Leader - USDA
Bodman, G.R.	MS	Assoc.Prof. - Farmstead Structures and Equipment
Chen, Y.R.	Ph.D.	Asst.Prof. - USDA
DeShazer, J.A.	Ph.D.	Prof. - Farmstead Strcutures and Equipment
Dickey, E.C.	Ph.D.	Asst.Prof. - Soil and Water
Eisenhauer, D.E.	MS	Asst.Prof. - Soil and Water
Fischbach, P.E.	MS	Prof. - Soil and Water
Gilbertson, C.B.	MS	Assoc.Prof. - Farmstead Structures and Equipment
Gilley, J.R.	Ph.D.	Assoc.Prof. - Soil and Water
Gooding, R. II	MS	Instructor - Youth Safety
Hahn, G.L.	Ph.D.	Prof. - USDA
Hanna, M.A.	Ph.D.	Assoc.Prof. Food Engineering
Hashimoto, A.G.	Ph.D.	Assoc.Prof. - USDA
Klocke, N.	MS	Asst. Prof. - Irrigation Research
Meyer, G.E.	Ph.D.	Asst.Prof. - Plant Growth Modelling
Mielke, L.N.	Ph.D.	Assoc.Prof. - USDA
Nienaber, J.A.	MS	Instructor - Farmstead Structures and Equipment
Schinstock, J.L.	EdD	Asst.Prof. - Power and Machinery
Schnieder, R.D.	MS	Prof. - Farm Safety
Schulte, D.D.	Ph.D.	Assoc.Prof. - Farmstead Structures and Equipment
Shelton, D.P.	MS	Asst.Prof. - Soil and Water
Schull, H.	MS	Prof. - Farmstead structures and Equipment
Silletto, T.	Ph.D.	Asst.Prof. - Shop Practices
Steinbruegge, G.W.	MS	Prof. - Power and Machinery
Stetson, L.E.	MS	Assoc.Prof. - Rural Electrification
Thompson, T.L.	Ph.D.	Prof. - Farmstead Structures and Equipment
Von Bargen, K.	Ph.D.	Prof. - Farmstead Structures and Equipment
Watts, D.G.	Ph.D.	Assoc. Prof. - Soil and Water
Wittmuss, H.D.	Ph.D.	Assoc.Prof. - Soil and Water

Nebraska Tractor Testing Laboratory

Leviticus, L.I.	Ph.D.	Prof. - In charge of Laboratory
Mumgaard, M.L.	B.Sc.	Assist. Prof. - Tractor Testing

Status: State institution, established in 1905.

Activities: Two courses of study: A technical engineering curriculum leading to B.Sc., M.Sc. and Ph.D. degrees in agricultural engineering, and an applied engineering curriculum leading to B.Sc. degree in mechanized agriculture. - Extensive tractor testing service and extensive research and extension programme in irrigation agricultural power and machinery, farmstead structures and equipment - Soil and water development and utilization programmes with emphasis on irrigation, soil management, and water pollution control.

Publications: Various publications on above activities.

Correspondence: English.

New Hampshire

Department of Soil and Water Science, College of Agriculture, University of New Hampshire
James Hall, N.H. 03824

New Jersey

Biological and Agricultural Engineering Department, Cook College, Rutgers University
New Brunswick, New Jersey 08903

Status: Department established in 1918 - Directed by State of New Jersey - Teaching activities financed by State of New Jersey - Research financed by State, U.S. Government and private sources - Extension work financed by State and U.S. Government sources.

Activities: Curriculum in professional agricultural engineering leading to B.Sc. degree - Graduate study leading to M.Sc. degree. - Research in: Self-feeding structures - Irrigation equipment performance - Drainage systems performance - Fruit and vegetable harvesting equipment - Rheological, electrical and thermodynamic properties of forage - Handling and disposal of poultry manure - Suburban hydrology.

Publications: Textbooks and various publications on above activities.

Correspondence: English.

New Mexico

Agricultural Engineering Department, New Mexico State University
P.O. Box 3268, Las Cruces, N. Mex. 88003

Abernathy, G.H.	Ph.D., Agr.Eng.	Prof. and Head of Department
Dean, J.W.	M.Ed.	Asst.Prof.
Freeburg, R.S.	Ph.D., Agr.Eng.	Assoc. Prof.
Hohn, C.M.	MS Agr.Eng.	Assoc.Prof., Extension
Hulsman, R.B.	Ph.D., Agr.Eng.	Asst.Prof.
Patterson, R.C.	BS, ME	Res.Engr. Agr.Eng.
Sammis, T.W.	Ph.D. Hydrol.	Asst.Prof.

Status: Agricultural Engineering Department was established in 1948 with a professional curriculum. The University is governmental.

Agricultural Engineering activities: (a) Teaching: Professional curriculum for agricultural engineering majors leading to a BS degree – Courses in Junior and Senior years have four to seven students and others not majoring in agricultural engineering – These courses have 9 to 43 students enrolled. – (b) Research: Energy Conservation: Improving the energy conversion efficiency of natural gas irrigation pumping plants – Utilization of dry agricultural by-products as a supplementary energy source for irrigation pumping – Irrigation water application efficiency improvement in New Mexico. Water Quality and Yield: Impact of recreational development on water quality and yield – Measurement of physical and institutional factors for managing environments on rural lands – Primary production model for selected forage species growing on mine-spoils – Trickle irrigation to improve crop production and management – Predicting consumptive use with climatological data – Irrigated agricultural decision strategies for variable weather conditions – Effects of decreased water on wheat yields, High Plains – Evapotranspiration and yield, cotton and alfalfa.

Publications (since 1973): Energy Conservation: Comparing energy costs for irrigation pumping – Pumping plant efficiency check for natural gas engines – Can sunshine power the irrigation pump? – Design and installation of solar-powered irrigation pump – Factors affecting the efficiency of natural gas-powered irrigation pumping plants – Alternate energy for irrigation pumping – Operational characteristics of a medium temperature solar-power system – Solar-powered agricultural irrigation pumping – Improving the efficiency of natural gas irrigation pumping plants – Coal-fired steam turbine power to drive irrigation pumps – Comparing energy costs for irrigation pumping – Irrigation pumping plant efficiency studies in various New Mexico counties – Efficiency testing of irrigation pumping systems – Power cost for irrigation pumping. Irrigation and plant response: An explanation for the growth advantage of drip irrigation – Application of modern irrigation technology in the Mesilla Valley of New Mexico – Demonstration of irrigation return flow salinity control in the Upper Rio Grande – Sizing plastic pipelines for water on the range – Physical modelling of plant microenvironments – Thermal environment of orchards as influenced by tree spacing – Water uptake by plants under desert conditions – Water infiltration under desert conditions – Plant growth and water transfer interactive processes under desert conditions – Measurement of evapotranspiration with monolith lysimeters – Crop production functions for alfalfa, cotton, grain corn, and grain sorghum – Consumptive use and yields of crops in New Mexico – Influence of irrigation methods on salt accumulation in row crops – Potato and lettuce response to irrigation methods and practices – Predicting consumptive use with climatological data – Vegetable production and water-use efficiency as influenced by drip, sprinkler, subsurface, and furrow irrigation methods. Miscellaneous: Tillage systems for cotton – a compairson in the U.S. Western Region – Evaluation of lignite shale from northeastern New Mexico as a prospective additive to cattle feeds – Impact of roads in recreational developments on forest environment – A bio-living home – Water disposition in ephemeral stream channels – I. Sampling soil changes; II. Channel transmission losses – Verification of an alfalfa model (Simed) under limited moisture condition.

Correspondence: English.

New York

Department of Agricultural Engineering, New York State College of Agriculture and Life Sciences, Cornell University
Riley-Robb Hall, Ithaca, NY 14853

Scott, N.R.		Department Chairman
Furry, R.B.		Coordinator of Research
Ludington, D.C.		Coordinator of Undergraduate Instruction
Rehkugler, G.E.		Coordinator of Graduate Instruction
Markwardt, E.D.		Department Extension Leader
Price, D.R.		Director, Office of Energy Programs
Stipanuk, D.M.		New York Food and Energy Council Leader
Albright, L.D.	Ph.D.	Assoc. Professor – Livestock Engineering; Thermal Systems
Baker, L.D.	MS	Sr. Extension Associate – Agricultural Safety
Bakker, U.B.	Ph.D.	Extension Associate – 4-H Energy Programs
Bartsch, J.A.	Ph.D.	Asst. Professor – Storage Structures and Facilities
Black, R.D.	Ph.D.	Assoc. Professor – Water Management Engineering
Campbell, J.K.	MS	Assoc. Professor – Power and Machinery
Cooke, J.R.	Ph.D.	Professor – Biological Engineering
Cukierski, G.	MS	Research Associate – Energy
Furry, R.B.	Ph.D.	Professor – Structures and Environment
Geohring, L.D.	MS	Research Associate – Water Management Engineering
Guest, R.W.	MS	Assoc. Professor – Materials Handling and Livestock Structures
Gunkel, W.W.	Ph.D.	Professor – Power and Machinery; Energy Utilization
Haith, D.A.	Ph.D.	Assoc. Professor – Environmental Systems Analysis
Hillman, P.E.	Ph.D.	Research Associate – Bioengineering
Irish, W.W.	MS	Assoc. Professor – Dairy Structures and Environment
Irwin, L.H.	Ph.D.	Associate Professor – Rural Transportation and Public Works
Jewell, W.J.	Ph.D.	Assoc. Professor – Agricultural Waste Management
Kabrick, R.M.	MS	Research Associate – Agricultural Waste Management
Koelsch, R.K.	MS	Extension Associate – Energy Conservation and Utilization
Lechner, F.G.	Ed.D.	Professor – Agricultural Mechanics
Levine, G.	Ph.D.	Professor – Water Management Engineering
Loehr, R.C.	Ph.D.	Professor – Agricultural Waste Management
Longhouse, H.A.	MS	Lecturer – Engineering Drawing, 4-H Programs
Lorenzen, R.T.	MS	Assoc. Professor – Structures and Environment
Ludington, D.C.	Ph.D.	Assoc. Professor – Agricultural Waste Management; Energy Utilization
Markwardt, E.D.	MS	Professor – Machinery and Irrigation
Martin, J.H.	MS	Research Associate – Agricultural Waste Management
Millier, W.F.	Ph.D.	Professor – Power and Machinery
Morris, J.W.	MS	Instructor – Agricultural Waste Management
Muck, R.E.	Ph.D.	Agr. Engr. – USDA/SEA – Agricultural Waste Management
Naylor, L.M.	Ph.D.	Research Associate – Agricultural Sludge Management
Parsons, R.A.	MS	Sr. Ext. Associate – Northeast Regional Agricultural Engineering Service
Pellerin, R.A.	MS	Research Associate – Power and Machinery
Pitt, R.E.	MS	Asst. Professor – Power and Machinery
Price, D.R.	Ph.D.	Professor – Electric Power and Processing; Energy Conservation and Utilization

Rehkugler, G.E.	Ph.D.	Professor – Power and Machinery
Sagi, R.	DS	Visiting Assoc. Prof. Structures and Environment
Scott, N.R.	Ph.D.	Professor – Structures and Environment, Bioengineering
Sobel, A.T.	MS	Research Associate – Agricultural Waste Management
Steenhuis, T.S.	Ph.D.	Asst. Professor – Water Management Engineering
Stipanuk, D.M.	MS	Res./Ext. Associate – Food and Energy Council
Updahyaya, S.	Ph.D.	Res. Associate – Biological Engineering
Walker, L.P.	Ph.D.	Asst. Professor – Energy
Walter, M.F.	Ph.D.	Asst. Professor – Water Management Engineering

Status: Cornell University, chartered in 1865, is the Land Grant University in New York State. It is privately endowed. The New York State College of Agriculture and Life Sciences is operated by Cornell University but obtains its major support from State and Federal funds. The first courses in Agricultural Engineering subjects were given in 1900-01, within the College of Agriculture, and the department has existed as a separate entity within the College since 1907, under various names, until 1930 when the present name was adopted.

Activities: Teaching: A professional agricultural engineering four-year program, offered jointly by the College of Agriculture and Life Sciences and the College of Engineering, leads to a B.S. degree from the College of Engineering and a professional Master of Engineering degree may be obtained following a fifth year. The department provides two programs of specialization leading to a B.S. degree from the College of Agriculture and Life Sciences: Agricultural Engineering Technology and Environmental Technology. An option in Agricultural Mechanization Teaching is also available. Enrollment in the professional A.E. program averages 150 students and an average of 70 students are enrolled in the B.S. agriculture program. Graduate programs leading to the M.S., Ph.D., M. Engr. and M.P.S. degrees are offered with an average enrollment of 35-40 per year. – Research: Basic and applied research and Extension programs include soil and water management engineering, soil mechanics, materials handling equipment, livestock housing systems, electric power applications, rural transportation systems, agricultural waste handling and management systems, environmental control in agricultural structures, rural structures, rural resource development, biological engineering.

Publications: Hydraulic structures – Precision planting – Seed pelleting – Poultry ventilation – Agricultural and rural waste handling – Milking parlors – Milk handling – Mechanical fruit harvesting – Onion storage – Forage handling – Dairy housing – Nursery crop mechanization – Irrigation – Fruit storage structures – Electric power applications – Mechanical harvesting of vegetables – Physiological measurements by telemetry – Rural road planning – Tillage principles – Bioengineering.

Correspondence: English.

North Carolina

Biological and Agricultural Engineering Department, School of Agriculture and Life Sciences, North Carolina State University
Raleigh, N.C. 27650   Tel.: 737-2011

Head of Department

Beasley, E.O.	MS	Biological and Agricultural Engineering
Blum, G.B.	MS	" " " "
Bowen, H.D.	Ph.D.	" " " "

Glover, J.W.	BS	Biological and Agricultural Engineering
Hassler, F.J.	Ph.D.	" " "
Huang, B.K.	Ph.D.	" " "
Humenik, F.J.	Ph.D.	" " "
Humphries, E.G.	Ph.D.	" " "
Johnson, W.H.	Ph.D.	" " "
Kriz, G.J.	Ph.D.	" " "
McClure, W.F.	Ph.D.	" " "
Rohrbach, R.P.	Ph.D.	" " "
Skaggs, R.W.	Ph.D.	" " "
Sneed, R.E.	Ph.D.	" " "
Sowell, R.S.	Ph.D.	" " "
Suggs, C.W.	Ph.D.	" " "
Watkins, R.W.	MS	" " "
Young, J.H.	Ph.D.	" " "

Status: State government institution - Department established in 1935.

Activities: Teaching: 2-year non-baccalaureate - B.Sc. in technology - B.Sc. in biological and agricultural engineering - M.Sc. - Ph.D. - Research in: Tobacco curing and mechanization - Peanut curing and mechanization - Cotton mechanization - Mechanization of horticultural crops - Soil-water-plant relationships - Electrostatic application of pesticides - Human engineering - Bioinstrumentation - Environmental engineering - Operations research - Watershed hydrology.

Publications: on above topics.

Correspondence: English.

North Dakota

Agricultural Engineering Department, North Dakota State University
Fargo, ND 58105

Pratt, G.L.		Department Chairman
Backer, L.F.	MS	Crop Storage and Environment
Borgen, V.	BS	Research Assistant
Disrud, L.A.	MS	Soil and Water Engineering
Kaufman, K.R.	MS	Instrumentation and Farm Power
Kucera, H.L.	MS	Farm Power and Machinery
Lindley, J.A.	Ph.D.	Waste Management and Structures
Moilanen, C.W.	MS	Farm Mechanics
Promersberger, W.J.	MS	Professor Emeritus
Roehl, L.	MS	Biomass Utilization
Sauvageau, G.		Drafting Technician
Solseng, E.	MS	Farm Mechanics
Stegman, E.C.	Ph.D.	Irrigation Engineering
Witz, R.L.	MS	Rural Electrification and Structures
Bartholomay, R.C.	MS	Extension - Irrigation Specialist
Cossette, R.H.	MS	" Artist Draftsman
Fanning, R.W.	BS	" Farm Safety Specialist
Hauck, D.D.	BS	" Energy Conservation Specialist
Hirning, H.J.	Ph.D.	" Crop Storage and Energy Conservation
Hofman, V.L.	MS	" Farm Power and Machinery

Johnson, D.W.	MS	Head, Ag.Engr.Ext., and Farm Structures
Lundstrom, D.R.	MS	Extension – Irrigation Engineering

Status: University established in 1890. Department of Agricultural Engineering established in 1924. State institution supervised by the State Board of Higher Education.

Agricultural Engineering activities: Study courses leading to B.S. and M.S. degrees in Agricultural Engineering and in Agricultural Mechanization. About 80 students normally enroll in Agricultural Engineering, and about 120 in Agricultural Mechanization. Research testing, and extension work is carried on: Energy conservation in agriculture, energy from biomass, waste management, livestock and crop environment studies, water quality, drainage and irrigation, grain drying, grain storage, feed handling, farm buildings, physical properties of agricultural products, grain losses in harvesting, irrigation scheduling by computer.

Publications: Textbook on: Modern Farmpower. Papers and bulletins on: Livestock Waste Management, Livestock and Crop Environment Studies, Irrigation, Irrigation Scheduling by Computer, Grain Drying, Feed Handling, Farm Buildings, Electric Power for Farm Use, Potato Storage, Water Quality, Weather Modification.

Correspondence: English.

Ohio

Agricultural Engineering Department, Ohio State University
Columbus, Ohio 43210

Blaisdell, J.L.	Ph.D.	Prof. – Food Engineering
Bondurant, B.L.	M.Sc.	Prof. – Soil and Water
Brazee, R.D.	Ph.D.	Adjunct Prof. – Aerosol Diffusion
Brugger, M.	Ph.D.	Asst. Prof. – Structures and Environment
Carpenter, T.G.	Ph.D.	Assoc. Prof. – Power and Machinery
Curry, R.B.	Ph.D.	Prof. – Plant Growth Simulation
Drew, L.O.	Ph.D.	Prof. – Power and Machinery
Dylla, A.	M.Sc.	Adjunct Instructor – Soil and Water
Fausey, N.R.	Ph.D.	Adjunct Assoc. Prof. – Soil Physics
Fox, R.D.	Ph.D.	Adjunct Prof. – Fine Particle Physics
Gliem, J.A.	Ph.D.	Asst. Prof. – Agricultural Mechanics
Hamdy, M.Y.	Ph.D.	Prof. – Computer Science and Simulation
Henry, J.E.	Ph.D.	Asst. Prof. – Power and Machinery
Herum, F.L.	Ph.D.	Prof. – Physical Properties
Holmes, R.G.	Ph.D.	Prof. – Power and Machinery
Huber, S.G.	M.Sc.	Prof. – Power and Machinery
Keener, H.M.	Ph.D.	Assoc. Prof. – Animal Growth and Production Dynamics
Miller, R.A.	M.Sc.	Assoc. Prof. – Rural Housing
Nelson, G.L.	Ph.D.	Prof. and Chairman – Structures and Processing
Nolte, B.H.	Ph.D.	Prof. – Watershed Development
Palmer, M.L.	M.Sc.	Prof. – Water Management
Papritan, J.C.	Ph.D.	Asst. Prof. – Agricultural Mechanics
Reeder, R.	M.Sc.	Asst. Prof. – Structures and Environment
Reichard, D.L.	M.Sc.	Adjunct Instructor – Pesticide Application Equipment
Roller, W.L.	Ph.D.	Prof. and Assoc. Chmn. – Structures and Environment
Schnug, W.R.	M.Sc.	Assoc. Prof. – Farm Electrification
Schwab, G.O.	Ph.D.	Prof. – Soil and Water Management

Short, T.H.	Ph.D.	Assoc. Prof. – Specialty Crop Mechanization
Stombaugh, D.P.	Ph.D.	Assoc. Prof. – Animal Environment and Growth Dynamics
White, R.K.	Ph.D.	Assoc. Prof. – Animal Waste Management
Hansen, C.H.	Ph.D.	Asst.Prof. – Animal Waste Management

Separate Branch: Ohio Agricultural Research and Development Center, Wooster, Ohio

Status: Institution of the State of Ohio established in 1870, supervised by a Board of Trustees and under control of a Board of Regents. Department of Agricultural Engineering established in 1914.

Activities: Study course of 4 years duration in Agricultural Engineering (options include: General Agr. Engineering, Food Engineering, Pre-Veterinary Medicine) attended by 70-90 students leading to B.Sc., A.E. Advanced studies in Agricultural Engineering lead to M.Sc. and Ph.D. degrees. An active Student Branch of the American Society of Agricultural Engineers and Student Mechanization Club have participated in extra curricular professional activities and have been recognized by several national awards. Service courses in mechanized agriculture are offered for students studying in various undergraduate engineering degrees. Research, extension and public service work is pursued in: livestock and crop environment studies – erosion control and water quality – livestock and human waste disposal – drainage and irrigation – tillage, cultivation and harvesting equipment – pest control procedures and equipment – grain storage and drying-feed and livestock handling – equipment management – advisement on foreign agricultural engineering institution development (India, Brazil, Somalia and Kenya) – agricultural structures systems engineering – energy conservation – alternative energy utilization – biomass fuels – solar energy, alcohol fuels – greenhouse engineering.

Publications, Textbooks, Papers and Bulletins on: House remodelling – tractor noise – drainage – watershed development – flood damage prevention – plastic tubing installation – grain shipments to developing countries – labour and machinery balance – Agricultural engineering institution development – food engineering curricula – controls of food science – food engineering developments – transient cooling of foods – water systems – tillage and emergence – maize and soybean harvest losses – grain storage, handling and drying – physical properties of fruits and vegetables – energy production from animal wastes – international methods of waste management – managing natural resources at the urban – rural interface – impact of livestock waste on environment – teaching pollution to agricultural engineers. – Textbooks on: Soil and Water Conservation Engineering; The Ferguson Foundation Agricultural Engineering Series, Elementary Soil and Water Engineering; Agriculture and Forest Hydrology.

Correspondence: English.

Department of Agricultural Engineering, Ohio Agricultural Research and Development Center (OARDC)
Wooster, Ohio 44691   Tel.: 216-264-1021

Nelson, G.L.	Ph.D.	Prof. and Chairman – Structures and Environment
Roller, W.L.	Ph.D.	Prof. and Assoc. Chairman – Structures, Environment and Energy
Blaidsdell, J.L.	Ph.D.	Prof. – Food Engineering
Bondurant, B.L.	MS	Prof. – Soil and Water
Brazee, R.D.	Ph.D.	Adjunct Prof. – Fine Particle Physics
Brugger, M.F.	Ph.D.	Asst.Prof. – Structures and Environment
Byg, D.M.	MS	Prof.Emer. – Power and Machinery
Carpenter, T.G.	Ph.D.	Assoc. Prof. – Tillage and Pesticide Application

Curry, R.B.	Ph.D.	Prof. – Plant and Growth Simulation
Drew, L.O.	Ph.D.	Prof. – Power and Machinery
Dylla, A.S.	MS	Adjunct Asst. Prof. – Drainage
Fausey, N.R.	MS	Adjunct Prof. – Drainage
Fox, R.D.	Ph.D.	Adjunct Asst. Prof. – Fine Particle Physics
Hamdy, M.Y.	Ph.D.	Prof. – Computer Science and Simulation
Hansen, C.L.	Ph.D.	Asst. Prof. – Waste Management
Harrold, L.L.	BS	Adjunct Prof. – Watershed Hydrology
Henry, J.E.	MS	Asst. Prof. – Power and Machinery
Herum, F.L.	Ph.D.	Prof. – Physical Properties
Holmes, R.G.	Ph.D.	Prof. – Specialty Crop Mechanization
Huber, S.G.	MS	Prof. – Power and Machinery
Keener, H.M.	MS	Assoc. Prof. – Animal Growth and Production Dynamics
Nolte, B.H.	Ph.D.	Prof. – Watershed Development
Palmer, M.L.	MS	Prof. – Water Management
Reeder, R.C.	MS	Asst. Prof. – Structures and Environment
Reichard, D.L.	MS	Adjunct Asst. Prof. – Pesticide Application Equipm.
Schnug, W.R.	MS	Assoc. Prof. – Farmstead Engineering
Schwab, G.O.	Ph.D.	Prof. – Drainage
Short, T.H.	Ph.D.	Assoc. Prof. – Greenhouse Energy Conservation
Stombaugh, D.P.	Ph.D.	Assoc. Prof. – Animal Environment and Growth Dynamics
White, R.K.	Ph.D.	Asst. Prof. – Animal Waste Management

Status: The Dept. of Agricultural Engineering was established in 1912 and is composed of faculty at OSU, Columbus and OARDC, Wooster, all integrated as one department – Governed and partially financed by the State of Ohio – Additional financing from USDA allocated and contract funds – Industrial Grants and Contracts.

Activities: Research and Development concerning: Implements and their function in tillage, planting, harvesting, pesticide application, drainage – Water drainage, pollution, filtration, purification – Specialty crop mechanization – Soil slips – Animal housing – Swine environment – Air conditioning – Physical properties of grain and fruits – Grain harvesting, drying, storage – Nature of drying – Analog models – Fine particle physics – Feed handling – Waste treatment – Modelling and simulation of plant growth systems, animal growth and regulatory systems – Energy conservation – Alternative Energy Utilization – Biomass fuels.

Publications: Aerosols–Air supported plastic greenhouses – Alfalfa harvesting – Animal environment – Animal housing – Atmospheric turbulence – Automatic animal feeding – Biomass fuels – Biological rheology–body temperature regulation – Cabbage harvester design – Centrifugal wheat separation–Colloid movement–Concurrent flow dryers – Dairy feeding systems – Deep bed drying – Diffusion in spheres – Drain spacing, subsurface drains – Drainage, crop yields, methods – Drainage design, subsurface drain openings – Dynamic modelling – Dynamic simulation of growth – Electrokinetic water purification – Energy utilization – Environment, simulation and control – Flood control, drainage channels, routing – Feedlot runoff – Greenhouse energy conservation – Greenhouse cooling – Hyperthermia, animal – Image processing – In storage grain drying – Land treatment of effluent – Laser beam automatic depth control – Low temperature drying – Mechanization of fruit and vegetable harvest – Milk production, product rheology – Drying of hay wafers – Meat smoking – Membrane processing – Maize harvesting systems – Maize losses and kernal damage – Maize storage in the semi-tropics – Natural resource management – Plastic drainage tubing – P ysical properties of tomatoes – Paste feeding systems – Particle analysis and measurement – Plant population statistics – Plant canopies, micrometeorology – Pesticides and equipment – Reject heat utilization – Root penetration – Receding water tables – Rainfall infiltration modelling – Soil and water conservation – Sediment movement – Subsurface drains – Soybean drying parameters – Seedling environment – Soil crust strength – Seedling thrust – Soybean harvest losses – Soybean viscoelastic – Soil anticrustants – Seed wafers – Solar grain drying - Solar ponds - Swine reproductive performance - Swine heat tolerance - Sugar beet emergence - Swine waste handling system - Spreading granular material - Sludge disposal on soil - Tillage, environmental effects - Water quality: drainage water, nutrient and sediment, loss, runoff, surface effects - Waste treatment: odors, energy, management, impact on environment, recycle system, aerosol sampling, gaseous sampling, odor effects on swine-aerobic composting - Watershed modelling - Water stress in plants - Whey processing.

Correspondence: English.

Oklahoma

Agricultural Engineering Department, Oklahoma State University
Stillwater, Oklahoma 74078 - 109 Ag Hall

Haan, C.T.	Ph.D.	Prof. and Head - Hydrology and Hydrologic Modelling
Barefoot, A.D.	MS	Assoc.Prof. - Irrigation; Soil and Water
Batchelder, D.G.	MS	Prof. - Farm Power and Machinery
Bloome, P.D.	Ph.D.	Prof. - Grain Drying, Feed Handling, Animal Environment
Brusewitz, G.H.	Ph.D.	Prof. - Processing; Environmental Engineering
Clary, B.L.	Ph.D.	Prof. - Process Engineering
Cook, G.E.	MS	Assoc. Prof. - Mechanized Agriculture Farm Mechanics
Crow, F.R.	MS	Prof. - Hydrology; Soil and Water
Downs, H.W.	MS	Asst. Prof. - Farm Power and Machinery
Garton, J.E.	Ph.D.	Prof. - Soil and Water, Irrigation
Gerling, J.F.	MPS	Asst. Prof. - Safety
Gwinn, W.R.	Ph.D.	Prof. - Hydraulic Structures
Huhnke, R.L.	Ph.D.	Asst. Prof. - Farm Structures
Jones, L.K.	MS	Asst. Prof. - Energy
Lewis, A.P.	MS	Assoc. Prof. - 4-H
Mahoney, G.W.A.	Ph.D.	Assoc. Prof. - Light Structures; Environment Systems
Porterfield, J.G.	MS	Prof. - Farm Power and Machinery
Rice, C.E.	Ph.D.	Assoc. Prof. - Hydraulics of Conservation Structures
Roth, L.O.	Ph.D.	Prof. - Farm Power and Machinery; Pesticide Application Equipment
Schwab, D.P.	MS	Prof. - Irrigation
Taylor, W.E.	MS	Assoc. Prof. - Farm Power and Machinery
Temple, D.M.	MS	Asst. Prof. - Erosion; Hydraulic Structures
Whitney, R.W.	Ph.D.	Assoc. Prof. - Farm Power and Machinery

Status: State Institution. Department established in 1920. Principal source of funds - State.

Agricultural Engineering activities: (a) Teaching: Average 85 students in undergraduate (B.S.), 20 students in graduate (M.S. and Ph.D.) programs in Agricultural Engineering, and 45 undergraduate (B.S.) students in Mechanized Agriculture program. (b) Research and Extension: Research and Extension - The Hydraulics of conservation channels - Animal waste management - Forage handling systems - Feedlot environment and management - Agricultural hydrology - Evaporation control studies - Engineering phases of irrigation - Farm water supply - Design and construction of farm equipment - Tillage practices and cropping systems for soil and moisture conservation and production of cotton, wheat, and grain sorghum - Development of improved machines and methods for Seedbed preparation, planting, and weed control - Machines and methods for the establishment and improved maintenance of pasture and ranges - Grain and crop drying and conditioning investigation - Development of farm building plans and specifications - Ventilation and temperature control in animal shelters - Air quality sampling - Aerial application of pesticides - Precision seed metering - Tractor performance monitoring - Crop residue studies - Destratification of surface water impoundments - Energy - Solar Energy - Biomass energy - Alcohol fuel - Greenhouses - Trickle irrigation - Agricultural safety - Energy audits - Energy conservation - Heating with wood.

Publications: Technical papers, reports, bulletins and fact sheets issued in all of the above areas. - Books: Engineering Applications in Agriculture - An Introduction to Agricultural Engineering.

Correspondence: English.

Oregon

Department of Agricultural Engineering, Oregon State University
Corvallis, Oregon 97331

Miner, J.R.	Ph.D.	Head - Animal Waste Management
Booster, D.E.	MS	Power and Machinery
Brooks, R.H.	Ph.D.	Drainage and Hydrology
Corey, A.T.	Ph.D.	Drainage and Hydrology
Cuenca, R.H.	Ph.D.	Irrigation and Drainage
English, M.J.	Ph.D.	Irrigation Optimization and Systems Analysis
Hansen, H.E.	Ph.D.	Agricultural Mechanics
Hansen, H.J.	MS	Extension Agricultural Engineer
Hellickson, M.L.	Ph.D.	Structures and Environment
Kirk, D.E.	MS	Food Process Engineering
Kolbe, E.R.	Ph.D.	Fisheries Engineering
Long, D.R.	MS	Power and Machinery
Matson, W.E.	MS	Extension Agricultural Engineer
Moore, J.A.	Ph.D.	Extension Agricultural Engineer
Shearer, M.N.	MS	Extension Agricultural Engineer
Willrich, T.L.	Ph.D.	Extension Agricultural Engineer
Wolfe, J.W.	Ph.D.	Irrigation and Pumping

Status: The Department of Agricultural Engineering was established in 1916. The Oregon Board of Higher Education governs Oregon State University; apppropriations for the institution are made by the Oregon legislature in biennial sessions.

Agricultural Engineering activities: Teaching: Offers B.S. and M.S. degrees in Agricultural Engineering and B.S. in Agricultural Engineering Technology. The Department graduates approximately 20 students per year. - Research: Mechanical harvesting, Animal waste management, Agricultural hydrology, Food process engineering, Drainage of agricultural soils, Irrigation systems, Farm safety, Crop storage, Animal housing and handling systems, Aquaculture, Fisheries engineering, Solar energy applications.

Publications: Technical articles and reports are issued on all of the above subjects.

Correspondence: English.

Pennsylvania

Agricultural Engineering Department, The Pennsylvania State University
University Park, PA 16802

Walton, H.V.	Ph.D.	Prof. - Head of Department - Food Engineering
Anderson, P.M.	M.Sc.	Assoc. Prof. - Electric Power and Processing
Bartlett, H.D.	M.Sc.	Prof. - Structures and Environment
Beppler, D.C.	M.Sc.	Assoc. Prof. - Power and Machinery

Daum, D.R.	M.Sc.	Prof. – Ext. – Power and Machinery
DeTar, W.R.	Ph.D.	Assoc. Prof. – Soil and Water
Garthe, J.W.	M.Sc.	Instr. – Ext. – Power and Machinery
Grout, A.R.	M.Sc.	Prof. – Ext. – Structures and Environment
Hilton, J.W.	Ph.D.	Asst. Prof. – Agricultural Mechanics
Hoover, J.R.	Ph.D.	Adjunct Asst. Prof. – Soil and Water
Jarrett, A.R.	Ph.D.	Asst. Prof. – Soil and Water
Keppeler, R.A.	Ph.D.	Assoc. Prof. – Food Engineering
Kjelgaard, W.L.	M.Sc.	Assoc. Prof. – Power and Machinery
Manbeck, H.B.	Ph.D.	Prof. – Structures and Environment
McCurdy, J.A.	M.Sc.	Prof. – Ext. – Electric Power and Processing
Meyer, D.J.	Ph.D.	Asst. Prof. – Ext. – Structures and Environment
Morrow, C.T.	Ph.D.	Assoc. Prof. – Horticultural Engineering
Murphy, D.J.	Ph.D.	Asst. Prof. – Ext. – Safety
Persson, S.P.E.	Ph.D.	Prof. – Power and Machinery
Sastry, S.K.	Ph.D.	Asst. Prof. – Food Engineering
Schroeder, M.E.	Ph.D.	Prof. – Structures and Environment
Shaw, M.D.	M.Sc.	Assoc. Prof. – Soil and Water
Stephenson, K.Q.	M.Sc.	Prof. – Electric Power and Processing
Wooding, N.H.	M.Sc.	Prof. – Ext. – Soil and Water

Status: Department established in 1921, state supported.

Teaching activities: Programs in Agricultural Engineering at the B.Sc., M.Sc. and Ph.D. levels – Average: 100 students. Program in Agricultural Mechanization leading to the B.Sc. degree – Average: 60 students. Research: Plant production area – reducing soil air entrapment, frost protection for fruit, anhydrous ammonia application, apple harvest mechanization, specialized machinery development. Animal production area – sheep facilities, feeding devices, hay and silage storage. Energy in agriculture – mushroom composting, solar energy, alternate liquid fuel. Properties of agricultural materials – failure criteria, cutting forces, damage evaluation. By-product management – anaerobic digestion of biomass, wastewater application to land, on-lot sewage disposal. Land and water resources – moisture distribution prediction, conservation practices, infiltration. Electronics and automation – microcomputer applications, energy conservation.

Publications: Methane production from animal manures, Energy conservation in greenhouses, Properties of agricultural materials, Mushroom production, Anhydrous ammonia application to soil and forages, Frost protection in orchards, Energy conservation in agricultural production, Soil air entrapment and infiltration.

Correspondence: English.

South Carolina

Department of Agricultural Engineering, Clemson University
Clemson, S.C. 29631

Webb, B.K.	Ph.D.	Head of Department
Allen, W.H.	Ph.D.	Structures and environment
Alphin, J.G.	Ph.D.	Machinery and processing
Barth, C.L.	Ph.D.	Agricultural waste management
Bunn, J.M.	Ph.D.	Electric power and processing
Christenbury, G.D.	Ph.D.	Field crop mechanization

Collier, J.A.	Ph.D.	Aquacultural engineering and machine design
Craig, J.T.	MS	Power and machinery
Davis, J.B.	Ph.D.	Crop production systems
Dodd, R.B.	MS	Power and machinery
Garner, T.H.	Ph.D.	Field crop mechanization
Garrett, T.R.	ME	Fruit and vegetable mechanization
Griffin, B.J.	MA	Housing
Hedden, F.H.	MS	Structures and Housing
Hegg, R.O.	Ph.D.	Waste management
Hood, C.E.	Ph.D.	Machinery and power
King, T.G.	MS	Waste management
Lambert, J.R.	Ph.D.	Computer simulation, soil and water
Ligon, J.T.	Ph.D.	Soil and water
Linvill, D.E.	Ph.D.	Meteorology
Miles, G.E.	Ph.D.	Systems analysis
Patton, F.C.	BS	Power and machinery
Payne, F.A.	Ph.D.	Energy
Privette, C.V.	MS	Irrigation, soil and water
Roberts, D.L.	Ph.D.	Energy safety
Spray, R.A.	Ph.D.	4-H, electric power and processing
Williamson, R.E.	Ph.D.	Fruit and vegetable mechanization
Wilson, T.V.	Ph.D.	Soil and water
Wolak, F.J.	Ph.D.	Power and machinery

<u>Branches at</u>: Edisto Experiment Station, Blackville, South Carolina - Pee Dee Experiment Station, Florence, South Carolina.

<u>Status</u>: The Agricultural Engineering Department was established in 1931 - Governmental Land Grant Institution, supported by the State of South Carolina.

<u>Activities</u>: Teaching: A four-year curriculum in Agricultural Engineering leading to the B.S.A.E. degree - Average number of students during past five years approximately 65. A four-year curriculum in Agricultural Mechanization and Business leading to the B.S. in Agricultural degree. Graduate programs leading to the M.S.A.E., Master of Engineering, Master of Agriculture, and Ph.D. degrees. - Research: Automatic controller for a grain combine - Animal utilization and treatment - Bulk handling systems for tree fruit crops - Irrigated crop production - Mechanical oyster harvesting - Hydrologic/water quality models - Energy reduction for on-farm processing - Flue cured tobacco bulk curing - Irrigation scheduling models - Mechanizing vegetable production - Non-point source pollution - Nutrient management of poultry wastes - Overland flow treatment of animal waste - Personnel computers in agriculture - Ambient air grain drying - Solar energy for home heating - Soybean production and management stimulation models - Trickle irrigation - Land application for animal waste - Viability of soybeans in storage - Control and reversible draining systems.

<u>Publications</u>: Crop responses to enhanced atmospheric CO_2 levels - Energy conservation in animal housing - On-farm digesters - Animal housing for confinement production - Corn and cotton root growth - Ambient air grain drying - Evaluation of herbicide incorporation uniformity - Alternative energy sources - Seepage from animal waste lagoons and Municipal lagoons - Methods of estimating soil moisture - Automatic data acquisition - Microcomputer control for grain combines - Engineering applications of microcomputers - A microprocessor-based soil plant atmosphere research system - Water flow in a controlled and reversible drainage system.

<u>Correspondence</u>: English.

Agricultural Engineering Department, South Dakota State University
Brookings, South Dakota 57007   Tel.: (605) 688-5141

Moe, D.L.	Head of Department and Director,	Institute of Irrigation Technology
Chisholm, T.S.	Assoc. Professor	Power and Machinery, Alcohol Fuels
Christianson, L.L.	Asst. Professor	Power and Machinery, Farm Structures, Ag. Energy
Chu, S-T.	Assoc. Professor	Soil and Water
Cluever, L-R.	Ext.Agr.Engr./Instructor	Soil and Water, Ag. Waste Mgt., Ag. Energy
DeBoer, D.W.	Professor	Soil and Water, Irrigation
Dittman, A.	Research Associate	Soil and Water, Ag. Waste Mgt.
Durland, G.R.	Ext.Ag.Engr./Assoc. Prof.	Power and Machinery, Alcohol Fuels
Heber, A.	Instructor	Farm Structures and Environment
Hellickson, M.A.	Professor	Farm Structures and Environment, Ag. Energy
Kelley, van C.	Instructor	Power and Machinery, Ag. Education
Klosterman, T.	Farm Supt.	Power and Machinery
Lubinus, L.	Ext.Agr.Engr./Assoc. Prof.	Farm Structures and Environment, Ag. Waste Mgt., Building Plans
Lush, J.	Ext.Agr.Engr./Asst.Prof.	Electric Power and Processing, Alcohol Fuels
Lytle, W.H.	Assoc. Professor	Weather Science, Climatology
Pahl, D.	Ext.Agr.Engr./Asst. Prof.	Soil and Water, Irrigation
Papendick, St.	Instructor	Soil and Water, Farm Structures
Ullery, C.H.	Water Res. Specialist/Assoc. Prof.	Soil and Water
Astleford, S.	Grad.Res.Asst.	Structures and Environment, Ag. Energy
Polak, Rita	Grad.Res.Asst.	" " " "
Resen, M.	Grad.Res.Asst.	Power and Machinery, Ag. Energy
Stampe, S.	Grad.Res.Asst.	" " " Alcohol Fuels

Branch: James Valley Research and Extension Center, Redfield, South Dakota

Status: The Agricultural Engineering Department was established in 1924. State supported Land-Grant University.

Activities: Study courses and curriculum leading to a Bachelor of Science degree in Agricultural Engineering - Master of Science degree granted in Agricultural Engineering - Options in AE: Structures and Environment, Power and Machinery, Electric Power and Processing, Water Resources Engineering, Environmental Management plus courses in Climatology and Meteorology - Curriculum in Mechanized Agriculture leading to a Bachelor of Science degree - Options in MA: Business, Science and Production, Irrigation, Equipment and Processing, and Vocational Agriculture Teacher - Research on: Use of Non-Anhydrous Ethanol in Internal Combustion Engines - Energy Utilization in South Dakota Agriculture - Non-Parametric Theory of Infiltration during Rainfall on Layered Soils for Water Resources Planning - Forage Production and Utilization Systems as a Base - Irrigation System Performance and Pumping Plant Capacity Evaluation - Wind Energy for Agricultural Applications - Evaluation and Development of Equipment for Reduced Tillage Systems - Energy Efficiency and Utilization in Agriculture - Livestock Confinement and Environmental Control Systems Relationship with Climate and Environment - Relationship of Daily and Climatological Weather Variables to Agricultural Production in South Dakota - Climatic Resources of the North Central Region - Soil and Water Management for Low pressure Irrigation Appliances.

Publications: Performance of a Solar Energy Intensifier with Thermal Energy Storage – Performance and Evaluation of Swine House Heating with a Solar Energy Intensifier-Thermal Energy Storage System – Evolution of a Solar Energy Intensifier-Thermal Energy Storage System for Agricultural Applications – Wind Energy Potential for Agricultural Application in South Dakota – Design and Performance of a Solar Energy Intensifier for Drying Shelled Corn – Wind and Solar Energy Combination for Agricultural Applications in South Dakota – A Solar Energy Intensifier-Thermal Energy Storage System for Agricultural Applications – Swine House Heating and Corn Drying with a Solar Energy Intensifier-Thermal Energy Storage System – Performance and Evaluation of Swine House Heating with a Solar Energy Intensifier-Thermal Energy Storage System – Design and Performance of a Solar Energy Intensifier System for Drying Shelled Corn – Design of a Vertical Axis Wind Turbine Variable Speed Heat Pump Alternative Energy System – Dry Matter and Nutrient Losses for Large Round Bales Stored Outside – Energy Efficiencies of Electric Powered Irrigation Pumping Plants – Pumps: Make Optimum Use – Dry Edible Beans – Water Management on Corn – Cropping Sequences Following Sunflowers – Feeding Value of Pro-Sil Treated High Moisture Ground Ear Corn with Two Groups of Crossbred Heifers – Effect of Length of Feeding Period on Performance of British and Exotic Crossbred Yearling Heifers – Thermal Resistance Measurement of Walls Using a Low-Cost Digital Electronic Device – Farmstead Facilities and Arrangements for 2025 – Farming in 2025 – Swine Performance Model for Summer Conditions – Center Pivot Irrigation Design – Pumping Energy Reduction by Modified Cost Analysis – Components of an Alcohol Plant – The Process of Making Alcohol – Farm Use of Alcohol Fuels – Vegetable Oil: One Alternative to Liquid Fuel – Energy Management in Dairy Facilities – Mechanical Ventilation for Swine Buildings.

Correspondence: English.

Tennessee

Agricultural Engineering Department, The University of Tennessee
P.O. Box 1071, Knoxville, Tennessee 37901   Tel.: 615/974-7237

Luttrell, D.H.	Ph.D.	Professor and Head
Baxter, D.O.	MS	Assistant Professor – Farm Structures
Bledsoe, B.L.	Ph.D.	Professor – Power and Machinery
Duckett, K.E.	Ph.D.	Associate Professor – Cotton Fiber Properties
Henry, Z.A.	Ph.D.	Associate Professor – Elec. Power and Processing
Safley, L.M., Jr.	Ph.D.	Assistant Professor – Waste Management
Shelton, C.H.	MS	Associate Professor – Soil and Water
Tomkins, F.D.	Ph.D.	Associate Professor – Power and Machinery
Wilhelm, L.R.	Ph.D.	Associate Professor – Elec. Power and Processing

Extension:

Luttrell, D.H.	Ph.D.	Professor and Head
DeBusk, K.E.	MS	Associate Professor – Rural Electrification
Grandle, G.F.	MS	Instructor – Farm Structures
Vaigneur, H.O.	Ph.D.	Professor – Soil and Water
Wills, J.B., Jr.	MS	Assistant Professor – Safety

Status: The Agricultural Engineering Department was established in 1927; accredited by ECPD in 1964.

Agricultural Engineering activities: Teaching in study courses leading to the following degrees: Agricultural Engineering – BS, MS, Ph.D. Agricultural Mechanization – BS, MS.

Research Experiments and Testing in: Alternative Systems of Swine Production – Minimizing Energy and Labour in Harvesting and Curing Air-Cured Tobacco – Hydrologic/Water Quality Models for Agriculture and Forestry – Evaluation of Selected Physical Properties Affecting Handling of Tennessee Vegetables – The Effect of Cotton Fiber Physical Properties on Dust Levels and on Yarn Characteristics – Abrasive Band Cutting Device for Severance of Soybean Plants – Management Practices for Erosion and Sedimentation Control – Engineering Systems and Energy Needs for Cotton Production – Animal Waste Utilization and Treatment Systems – Curing, Storing and Feeding High Value Hay in Large Packages – Use of the Earth as an Energy Source or Sink for Heating and Cooling Buildings – Energy Reduction for On-Farm Processing of Agricultural Products – Optimize Efficiency of Energy Utilization in Agricultural Housing Systems – Application of Titanium Dioxide on Agricultural Land – A modular Dryer for Large Hay Packages Using Solar Heated Air.

Publications: Roofed vs. No-Roof Silos: In Unloader Performance, Coating Durability, and Silage Consumption by Dairy Cows – Azimuthal Intensity Profiles of Scattered Light from the Suface of Convoluted Cotton Fibers – A Continuim Mechanics Approach to Twisted Yarns – Cotton Fiber Instrumentation Research in the Institute of Agriculture – Humidity and Heat Effects on the Coefficient of Energy Dissipation – Use of Hydrostatic Pressure in the Development of Stress-Strain Information on Tomato Skins – Date from Fiber and Spinning Tests – Effect of Sodium Hydroxide Concentrations on Selected Cotton Fiber Properties – Incorporation of Treflan for Controlling Johnsongrass in Soybeans – Quality of Mechanically Harvested Cotton – Response of Soybeans and Johnsongrass to Glyphosate Applications – Data from Fiber and Spinning Tests – The Qualities of Cotton Produced in the United States – A System Analysis Approach to Establishing Objectives for Control of Aflatoxin Production in Cotton Seed – Effects of Narrow Rows on Cotton Fiber Properties – Visible and Infrared Remote Sensing in Soil Moisture Determination – The Structural Condition of Rural Housing in Tennessee – Big Package Haymaking in Tennessee – The Application of Weak-Link Theory to the Determination of a True Zero-Gage Tensile-Test Length of Single Cotton Fibers – The Effects of Compression on Airflow Measurements of Cotton Fiber Immaturity – Energy Dissipation within Sheared Fiber Assemblies – Progress in Air Flow Measurements for Determining Cotton Fiber Specific Surface Area and Maturity – Some Observations on Single-Fiber and Flat-Bundle Tests – Surface Properties of Cotton Fibers – An Opener-Picker for Miniature Spinning Test – Electrical Charge Characteristics of Single Cotton Fibers – Instrumentation and Measurement for Environmental Sciences – A 5-Year Comparison of Seedbed Preparation Systems for Cotton on Memphis and Collins Silt Loams – Systems for Making, Handling, Storing and Feeding Large Hay Packages – A Discussion of the Cross-Point Theories of van Wyk – Quality Characteristics of Agricultural and Waste Disposal Runoff Water – Cooling Bulk Shipments of Snap Beans – Evaluation of Home-Size, Solar Dryers – Energy Conservation Waste Management – Animal Waste Lagoons in Tennessee – Soil Management Essential if Man is to Survive – The Energy Problem – Rural Housing in Tennessee – Optical Measurements of Flat-Bundle Cotton Masses Used in Standard Fiber Tenacity Tests – Moisture Diffusion in the Cured Barley Tobacco Leaf – A Laboratory Seed Separator-Grader – Methods of Applying Postemergence Herbicides to Cotton and Soybeans for Cocklebur Control – Solar Drying of Hay: Open Chamber Dryer Design – How Density Affects Keeping Qualities of Big Package Hay – Solar Drying of Hay in Big Packages – Reduced Tillage Vegetable Production: Many Questions Remain to be Answered – 1976 Minimum Tillage Field Tests.

Correspondence: English.

Texas

Department of Agricultural Engineering, Texas A&M University
College Station, Texas 77843

Edward, A.H.	Ph.D.	Professor and Head
Aldred, W.H.	M.Sc.	Associate Professor – Power and Machinery
Beach, W.E.	M.Sc.	Associate Professor – Mechanized Agriculture
Childers, R E.	M.Sc.	Extension Agricultural Engineer – Cotton Ginning and Mechanization
Coble, C.G.	Ph.D.	Associate Professor – Power and Machinery
Darcey, C.L.	Ed.D.	Assistant Professor – Mechanized Agriculture
Diehl, K.C.	Ph.D.	Assistant Professor – Food Engineering
Hawkins, G.W.	Ph.D.	Assistant Professor – Power and Machinery
Keese, C.W.	M.Sc.	Extension Agricultural Engineer – Irrigation
Kunze, O.R.	Ph.D.	Professor
LePori, W.A.	Ph.D.	Associate Professor – Energy
McCune, W.E.	M.Sc.	Professor – Power and Machinery, Farm Electrification
McFarland, M.J.	Ph.D.	Associate Professor – Soil and Water
Mounce, C.E.	M.Sc.	Lecturer – Mechanized Agriculture
Nelson, G.L.	Ph.D.	Extension Agricultural Engineer – Safety
Nieber, J.L.	Ph.D.	Assistant Professor – Soil and Water
O'Neal, H.P.	M.Sc.	Extension Agricultural Engineer – Energy
Parnell, C.B.	Ph.D.	Associate Professor – Air Quality
Reddell, D.L.	Ph.D.	Professor – Soil and Water
Searcy, S.W.	Ph.D.	Assistant Professor – Power and Machinery
Stewart, B.R.	Ph.D.	Extension Agricultural Engineer – Environmental Control
Stout, B.A.	Ph.D.	Professor – Energy in Agriculture
Sweat, V.E.	Ph.D.	Associate Professor – Food Engineering
Sweeten, J.M.	Ph.D.	Extension Agricultural Engineer – Waste Management
Suter, D.A.	Ph.D.	Associate Professor – Power and Machinery
Wilkes, L.H.	M.Sc.	Professor – Mechanized Agriculture
Withers, R.E.	Ph.D.	Extension Agricultural Engineer – Processing
Zingery, W.L.	MS	Lecturer – Electronic Measurement

Status: State of Texas financed institution, established in 1920.

Activities: Courses offered Fall and Spring semesters leading to B.Sc., M.Sc. and Ph.D. degree in agricultural engineering – Agricultural mechanization programmes – Research and short courses offered in summer – Research projects dealing with: Fruit, vegetable, nut and cotton mechanization – Drying, storage and processing of agricultural products – Preservation of product quality – Structural design – Drainage and irrigation – Rural electrification – Effective and efficient methods for the control of rainfall and water under various crop, soil, and climatic conditions – Environment control – Soil and water – Engineering and water pollution studies – Waste disposal – Mechanized feed processing and handling – Agricultural energy conservation and alternative energy sources – Irrigation scheduling techniques and application systems for increased water use efficiency – Air quality and systems for prevention of grain dust explosions – Food engineering and physical properties of perishable agricultural products – Sugarcane and sweet sorghum mechanization.

Publications: Energy, A challenge and an opportunity for Agriculture – Impact of increasing energy costs on irrigation and agricultural production – What's new in seed cotton storage? – Cotton ginning systems and operational trends – Analysis of cyclone separator collection performance for grain sorghum dust – Cotton harvesting, handling and ginning cost using HARVSUM-modified - Role of agricultural engineering in reducing cost of cotton production in the 1980s - Systems for drying rice - Interrelationships of equipment and procedures for rice harvesting and processing - Mechanical onion top removal and related pre-harvest practices - Grain sorghum response to inundation at three growth stages - Trickle and sprinkler irrigation of grain sorghum - Grain sorghum response to inundation duration at the early reproductive growth stage - Friction coefficients of sorghum grains on steel, teflon and concrete surfaces - Systems for harvesting nuts - Various publications on research in energy, waste management, cotton, sorghum grains, irrigation and drainage.

Correspondence: English.

Texas A&M University Agricultural Research and Extension Centre
Route 7, Box 999, Beaumont, Tex. 77706

Calderwood, D.L.   M.Sc.   Rice drying engineering

Status: Research at this Centre is financed by the state government of Texas, Texas Rice Improvement Association, and the U.S. Department of Agriculture.

Agricultural Engineering activities: Research on drying and storage of rough rice.

Publications: Use of aeration to aid rice drying - Drying and handling rough rice at commercial driers - Effect of the method of drier operation on performance and on the milling and cooking characteristics of rice - Analysis of pilot plant and commercial rice drying operations - Breakage of processed rice due to falling impact - Cereal chemistry today - Rough rice drying - Rice: chemistry and technology - Rough rice storage - Resistance to airflow of rough, brown and milled rice - Chemical preservatives for maintaining moist rice in storage - Drying, storing and handling - Rough rice (paddy) drying methods in the United States - Rice drying with solar heat.

Correspondence: English.

Agricultural Engineering Department, Texas Tech University
Lubbock, TX 79409

Dillingham, M.J.	MS	Assistant Professor - Agricultural Mechanics
Dvoracek, M.J.	MS	Associate Professor/Chairperson - Soil and Water; Hydrology
Eggenberger, L.	Ph.D.	Professor - Agricultural Mechanics
Foerster, E.P.	Ph.D.	Associate Professor - Soil and Water; Irrigation
Grub, W.	MS	Professor - Electrification; Design of Ag. Structures
Lewis, R.B.	Ph.D.	Visiting Assistant Professor - Soil and Water
Schacht, O.B.	Ph.D.	Assistant Professor - Power and Machinery
Ulich, W.L.	Ph.D.	Professor - Power and Machinery

Status: College established in 1923. Agricultural Engineering Department established in 1951; department jointly administered by College of Engineering and College of Agricultural Sciences. Awards MS degree in Agricultural Engineering; BS in Agricultural Engineering; BS in Mechanized Agriculture; Ph.D degree in Engineering (interdisciplinary), average student enrollment - 110. Research activities: Equipment for control of noxious weeds and brush; surface hydrology; irrigation; soil and water conservation; air pollution control.

Publications: Various publications related to irrigation, artificial recharge, brush control, ag. processing products.

Correspondence: English.

Utah

Department of Agricultural and Irrigation Engineering, Utah State University
Logan, Utah 84322

Anderson, B.H.	Ph.D.	Prof. - Water Resources Planning and Irrigation
Bishop, A.A.	Ph.D.	Prof.(Emeritus) - Water Resources Planning, Irrigation and Drainage
Christiansen, J.E.	M.Sc.	Prof.(Emeritus) - Water Resources Planning, Hydrology, Irrigation
Daines, D.R.	JD	Assoc.Prof. - International Water Law
Griffin, R.E.	M.Sc.	Assoc.Prof. - Irrigation, Water Use
Hargreaves, G.H.		Res. Engineer
Hill, R.W.	Ph.D.	Assist.Prof. - Hydro Water Res.
Jarrett, W.H.		Assoc.Prof. - Extension Farm Machinery Specialist
Keller, J.	Ph.D.	Prof. and Head - Water Resources Planning, Hydraulic Engineering, Irrigation and Drainage
Olsen, E.C.	Ph.D.	Assoc.Prof. - Irrigation Engineering
Peterson, H.B.	Ph.D.	Prof.(Emeritus) - Water Chemistry, Irrigation
Stringham, G.E.	Ph.D.	Assoc.Prof. - Water Resources Planning, Hydraulic Engineering
Stutler, R.K.	MS	Res.Asst.Prof. - Irrigation Engineer
Walker, W.R.	Ph.D.	Prof. - Surface Irrigation
Willardson, L.S.	Ph.D.	Prof. - Irrigation and Drainage

Status: Land grant institution with government status, supervised by Board of Trustees.

Activities: Approximately sixteen courses in the agricultural engineering curriculum, each having a duration of approximately 3 months - Average attendance is approximately 15 students - These courses are supplemented by courses in basic engineering sciences taught in the Civil, Mechanical, and Electrical Engineering Department - Courses lead to the following degrees: B.Sc.A.E. (soil-water field) - Ph.D. in agricultural engineering (soil-water field) - M.Sc. and Ph.D. in irrigation and drainage engineering - Research includes irrigation and drainage and water resources engineering - The University has a 120 acre drainage farm, a hydraulic research laboratory, and a new engineering center to support the research and instructional program - The work is closely integrated with the Engineering and Agricultural Experiment Stations of the University.

Publications on: Irrigation, drainage, water research, hydrology, arid lands, soils, water resources planning, crop production.

Correspondence: English, Spanish.

Vermont

Vocational Education and Technology Department, University of Vermont
Burlington, Vermont 05405

Fuller, G.R.	Ph.D.	Prof. Chairman of Department – Teaching
Moore, M.J.	M.Sc.	Extension Assoc. Prof. – Housing and Utilities Engineering
Wells, G.	Ph.D.	Extension Asst. Prof. – Extension, Teaching and Research
Zimmerman, A.	M.Sc.	Lecturer – Teaching

Status: Land grant institution with Government status, Department of Agriculture Engineering established in 1946.

Teaching activities: Study courses on farm power, machinery and electricity, farm structures and utilities, farm shop, soil and water management, water supply, sewage disposal leading to a B.Sc. degree in agriculture, with an average number of 10 students. Extension teaching activities in energy, structures, and utilities.

Research activities: Investigation of recovery of heat from the ventilation systems in livestock and poultry buildings – energy alternatives for agriculture.

Publications: Controlling dairy barn moisture with heat exchangers – Farm water supplies – Ventilation for your dairy barn – Guidelines for free stall loose housing systems – Preventing ice dams on house eaves – Sugar house design – Extension plan service – Specific solar adaptions to agriculture – Harvesting hay for quality and safe storage – Hay packing systems.

Correspondence: English.

Virginia

Agricultural Engineering Department, Virginia Polytechnic Institute and State University
Blacksburg, Virginia 24061

Haugh, C.G.	Ph.D.	Prof. and Head of Department
Baker, J.L.	Ph.D.	Asst.Prof. – Biological Engineering, Crop Processing
Bell, E.S.	M.Sc.	Assoc.Prof. – Electric Power and Processing
Carr, J.C.		Systems Analyst – Computers, Hydrology
Collins, E.R.	Ph.D.	Asst.Prof. – Swine, Agric. Wastes, Pesticide Wastes
Collins, W.H.	M.Sc.	Asst.Prof. – Dairy, Beef, Sheep
Cundiff, J.S.	Ph.D.	Assoc.Prof. – Energy Systems
Hagee, G.L.	Ph.D.	Asst.Prof. – Agricultural Mechanics
Hale, E.B.	M.Sc.	Assoc.Prof. – Water: Supply, Quality, Use, Irrigation
Hetzel, G.H.	M.Sc.	Asst.Prof. – Safety, Agricultural Mechanics
Hughes, H.A.	Ph.D.	Assoc.Prof. – Swine, Horses, Chicken, Housing, Energy
Lambert, A.J.	M.Sc.	Assoc.Prof. – Processing-soybeans, peanuts, tobacco
Lovingood, M.	Ph.D.	Asst.Prof. – Power and Machinery, Power Mechanics
Magette, W.	Ph.D.	Asst.Prof. – Soil and Water
Mason, J.P.	Ph.D.	Prof. – Structures and Environment
Parsons, B.L.	M.Sc.	Assoc.Prof. – Soil and Water Conservation
Perumpral, J.V.	Ph.D.	Assoc. Prof. – Power and Machinery, Soil-Machine Interact

Ross, B.B.	Ph.D.	Research Assoc. – Hydrology, Computers
Shanholtz, V.O.	Ph.D.	Assoc.Prof. – Water Management, Hydrology, Watersheds
Smith, E.S.	M.Sc.	Assoc.Prof. – Farm Machinery
Smith, J.R.	M.Sc.	Assoc.Prof. – Domestic Housing
Smolen, M.D.	Ph.D.	Asst.Prof. – Water Quality
Trice, R.H.	B.Sc.	Asst.Prof. – Electricity, Insulation, Microwaves
Vaughan, D.H.	Ph.D.	Assoc.Prof. – Solar Energy, Plant Modelling
Wilson, J.H.	Ph.D.	Asst.Prof. – Agricultural Mechanics, Metallurgy
Woeste, F.E.	Ph.D.	Asst.Prof. – Wood, Structures, Probability
Younos, T.M.	Ph.D.	Research Assoc. – Surveying, Waste Management

Status: Established 1920; State-supported Land-Grant University; directed by Board of Visitors

Agricultural Engineering activities: Teaching: B.S. and M.S. degrees awarded in Agricultural Engineering by the College of Engineering. Ph.D. degree program expected to be approved soon. Approximately 25 B.S. and 6 M.S. degrees awarded annually. Research: Agricultural hydrology, watershed modelling, no-tillage machinery, peanut drying, turkey housing and environment, tobacco handling and curing, harvest mechanization, plastic greenhouses, leafy vegetable washing, animal waste management, alternate energy sources, energy systems, wood engineering, farm structures, forest engineering, physical properties of agricultural products and soil machine interaction. Extension: Educational programs in agricultural structures including the housing environment, agricultural waste management, rural family housing forest, mechanization, tobacco mechanization, farm and home electrification, water supply, irrigation, conservation of natural resources, power and machinery, on-farm processing, vocational agriculture, youth programme, and alternate energy sources.

Publications: Watershed modelling, no-tillage, waste management, plastic greenhouses, tobacco curing, grain storage and drying, turkey housing environment, farm safety, house structures and equipment, flood damage prevention, natural resources (film), confinement housing for swine, bulk fertilizer distributors, machinery performance and cost, physical properties of apple skin and flesh, sweet potato harvester, youth program guides in agricultural engineering subjects, wood engineering, farm structures, forest engineering, soil-machine interaction.

Correspondence: English.

Washington

Agricultural Engineering Department, State College of Washington
Pullman, Wash. 99163

Activities: Study courses leading to a degree in agricultural engineering – Research and experiment work.

Correspondence: English.

West Virginia

Agricultural Engineering Department, West Virginian University
Morgantown, W.Va. 26506

Activities: Courses leading to a B.Sc. degree in Agricultural Engineering at the end of four years plus a minimum of one year to complete the M.Sc. in Agricultural Engineering. Forest Engineering is an option in the Agricultural Engineering Curriculum. Also a graduate program in Forest Engineering leading to a Master of Science in Engineering Degree (M.S.E.). Research activities.

Correspondence: English.

Wisconsin

Agricultural Engineering Department, College of Agricultural and Life Sciences, University of Wisconsin-Madison
460 Henry Mall, Madison 53706

Buelow, F.H.	Ph.D.	Prof. Chairman - Electric power and processing
Barquest, G.D.	Ph.D.	Professor - Farm mechanics
Barrington, G.P.	M.Sc.	Prof. - Power and machinery
Brevik, T.J.	M.Sc.	Prof. - Structures
Brooks, L.A.	M.Sc.	Prof. - Farm electricity
Bubenzer, G.D.	Ph.D.	Prof. - Soil and water conservation
Converse, J.C.	Ph.D.	Prof. - Structures
Cramer, C.O.	Ph.D.	Prof. - Structures
Detroy, B.F.	M.Sc.	USDA Collaborator - Electrification
Finner, M.F.	M.Sc.	Prof. - Power and machinery
Holmes, B.J.	Ph.D.	Asst.Prof. - Farmstead engineering
Jensen, D.V.	M.Sc.	Assoc.Prof. - Farm safety
Koegel, R.G.	Ph.D.	Assoc.Prof. - Power and machinery
Massie, L.R.	Ph.D.	Assoc.Prof. - Soil and water conservation
Petersen, J.B.	Ph.D.	Asst.Prof. Farm waste engineering
Stith, D.J.	M.Arch.	Prof. - Structures
Straub, R.J.	Ph.D.	Asst.Prof. - Power and machinery

Status: The Department was established in 1904 and is financed by State Government funds.

Teaching activities: Four year B.Sc. degree programmes offered in agricultural engineering (80 students), in agricultural mechanization and management (30 students), and in construction technology (124 students). Also M.Sc. degree programme in agricultural engineering (5 students) and Ph.D. programme in agricultural engineering (9 students).

Correspondence: English.

Wyoming

Agricultural Engineering Division, University of Wyoming
Laramie, Wyoming 82071

Becker, C.F.	Ph.D.	P.E.	Prof. Head of Division	- Agricultural Engineering
Borrelli, J.	Ph.D.	P.E.	Assoc. Prof.	- Irrigation and Drainage
Burman, R.D.	Ph.D.	P.E.	Prof.	- Irrigation
Brosz, D.J.	M.Sc.	P.E.	Prof. Ext. Agr. Eng.	- Irrigation

Fornstrom, K.J.	Ph.D.	P.E. Prof.	– Farm Mechanization
McNamee, M.A.	M.Sc.	Assoc. Prof. Ext. Agr. Eng.	– Structures and Energy Conservation
Pochop, L.O.	Ph.D.	P.E. Prof.	– Farm structures, Climatology and Plant Environment.

Status: Governmental institution, supervised by Board of Trustees.

Activities: Study courses leading to B.Sc. and M.Sc. in agricultural engineering and B.Sc. in agricultural mechanization – Research in Climatology – Microclimate – Evapotranspiration – Dry and irrigated land tillage methods – Irrigated crop mechanization.

Publications: Equipment for seeding hay meadows – Equipment for applying herbicides – Evapotranspiration – Sugar beet mechanization – Climatology – Land application of waste – Water requirement of lawns – Efficiency of stoves and fireplaces.

Correspondence: English.

UPPER VOLTA

Ouagadougou

Direction de l'Hydraulique et de l'Equipement Rural (H.E.R.) (Directorate of Hydraulics and Agricultural Equipment)
B.P. 330, Ouagadougou, Haute Volta

Service de l'Hydraulique Agricole et de l'Equipement Rural (H.A.E.R.)

Service de l'Inventaire des Recherches Hydrauliques (I.R.H.)

Service de l'Hydraulique Urbaine et Industrielle (H.U.I.)

Annexes:

Circonscription Centre Est, B.P. 330, Ouagadougou

Circonscription Ouest, B.P. 179, Bobo-Dioulasso

Structure: Institution gouvernementale, fondée en 1954 et contrôlée par la Direction des Services du Gouvernement.

Activités: Formation des Ingénieurs de l'Equipement Rural à l'Ecole des Ingénieurs de l'Equipement Rural de Ouagadougou (E.I.E.R.) B.P. 139. – Formation des Adjoints Techniques du Génie Rural à l'Ecole Inter-Etats de formation des A.T.G.R., B.P. 594, Ouagadougou. – Recherches Hydro-Agricoles auprès de la Station Expérimentale Hydro-Agricole de Mogtedo, B.P. 596, Ouagadougou.

Publications: Rapports annuels sur les activités du Service de l'H.E.R.

Correspondance: français.

Ecole Inter-Etats d'Ingénieurs de l'Equipement Rural (E.I.E.R.)
B.P. 7023, Ouagadougou, Haute Volta

Département Physique appliquée et Froid

Département Mathématiques appliquées

Département Hydraulique appliquée

Département Génie Rural

Département Génie Sanitaire

Structure: Créée en décembre 1968 par décision Conseil des Ministres des Etats devant constituer le Conseil d'Administration de l'Ecole.

Activités concernant le Génie Rural: Formation: Le diplôme d'Ingénieur de l'Equipement Rural est accordé à l'issu d'un cycle de formation de cinq ans se décomposant: Deux ans d'enseignement supérieur en Faculté (Diplôme Universitaire d'Etudes Scientifiques D.U.E.S. Option Mathématiques - Physique ou Agronomie) - Trois ans de formation à l'E.I.E.R. avec sélection en fin de première année. - Recherches: Programme de recherches appliquées menées par les élèves-ingénieurs de troisième année a débuté cette année.

Correspondance: Français.

URUGUAY

Cerro Largo

Escuela de Agronomía "Banado de Medina" (School of Agronomy Banado de Medina)
Cerro Largo

Terra, J.   Ing.Agr.   Maquinaria agrícola.

Investigación: Unidad Experimental - sistemas de producción - pruebas de comportamiento de cultivos.

Libertad

Escuela de Mecánica Agrícola, Universidad del Trabajo del Uruguay (School of Agricultural Mechanization, University of Work of Uruguay)
Libertad, San José

Gomez, A.   Ing.Agr.   Director

Carácter y estructura: Institución oficial de enseñanza dependiente de la Universidad del Trabajo, establecida desde el año 1956.

Actividades: Cursos de dos años de duración, con capacidad de 60 alumnos internos, expidiéndose certificados de estudio de Idóneos en Maquinaria Agrícola.

Correspondencia: Español, Inglés.

Montevideo

Universidad Mayor de la República Oriental del Uruguay, Facultad de Agronomía
(University of the Republic)
Av. Garzón 780, Montevideo

Cuñetti, E.   Ing.Agrón.   Curriculum: Investigación Ayudante de la
                              Cátedra de Maquinaria Agrícola
Molinari, J.   Ing.Agr.   Curriculum: Investigación Ayudante de la
Borges, M.   Ing.Agr.   Cátedra de Maquinaria Agrícola

Carácter y estructura: La Cátedra de Maquinaria agrícola comenzó sus actividades en 1909, año en que fue fundada la Facultad de Agronomía perteneciente a la Universidad Mayor de la República Oriental del Uruguay.

Actividades en ingeniería rural: Ensenanza: Se dicta un curso de seis meses de duración que corresponden al primer semestre del tercer año, correspondiente al Ciclo Básico. No se concede título. Promedio de alumnos: 250.

Investigación: Se realizan cursillos sobre "laboreo de suelos" para productores. Se efectúan ensayos sobre laboreo cero.

Correspondencia: Español.

Instituto Nacional de Colonización (I.N.C.) (National Institute of Colonization)
Casa Central: Cerrito 488 - Casilla de correo 505 - Montevideo, Uruguay

Cussac, C.M.   Ing.Agrón.
Gelos, A.   "   "
Irrisarri, E.   "   "
Jorge, H.   "   "
Jorge Hiriart, J.C.(Ms) Ing.Agrón.
Sere, M. (Ms.)   Ing.Agrón.
Cheveste de Glisenti, S. Arquitecto
Galceran, H.   "
Negrin de Marichal, S.   "

Dependencias en otras localidades: Oficinas Regionales del I.N.C.: Departamento de Artigas: en Bella Unión - Departamento de Salto: en Salto (Ciudad) - Departamento de Paysandú: en Paysandú (Ciudad) y en Guichón - Departamento de Rio Negro: en Colonia "Tomás Berreta" del INC, ubicada a 5 kms de Fray Bentos - Departamento de Soriano: en José E. Rodó - Departamento de Colonia: en Tarariras - Departamento de San José: en San José (Ciudad) - Departamento de Canelones: en San Jacinto - Departamento de Cerro Largo: en Melo - Departamento de Tacuarembó: en Tacuarembó (Ciudad) - Departamento de Florida: en Sarandí Grande, El Núcleo Colónico "Treinta y Tres" con Oficina en San Ramón (Depto Canelones) - Departamento de Lavalleja: en Minas.

Caractér y estructura: Organismo oficial con carácter de Ente Autónomo del Estado. Se financia con recursos propios y del Gobierno Nacional. Lo dirige un Directorio designado por el Poder Ejecutivo, actualmente así integrado: Presidente Interventor Ing.Agr. Walter Arías; Vicepresidente: Coronel(R) Gonzalo Madeiro; Segundo Vicepresidente: Ing.Agr. Omar Aguirre; Director Carlos J. Zitta. Gerente General: Octavio Y. Martinez; Secretario de Directorio: Dr. Arturo Gomeza.

Actividades de Ingeniería Rural: Asistencia Técnica a los colonos - Programas de Crédito supervisado - Asistencia a grupos de colonos a través de Cooperativas, Sociedades de Fomento, etc. - Viveros y Forestación - Recuperación de suelos con estímulos para el colono a traves de reducción de rentas de áreas controladas - Administración Técnica de Obras de Regadío.

Correspondencia: Español, Inglés, Francés.

VENEZUELA

Barinas

Universidad Nacional Experimental de los Llanos Ezequiel Zamora
Barinas, Estado Barinas

Caracas

Ministerio del Ambiente y de los Recursos Naturales Renovables (MARNR)

Actividades: todos los aspectos inherentes al manejo y conservación de los recursos renovables.

Departamento de Ingeniería, Instituto Agrario Nacional (Engineering Department, National Agricultural Institute)
Caracas, Distrito Federal

Actividades: Investigaciones y experiencias relacionadas con ingeniería rural.

Jusepín

Escuela de Ingeniería Agronómica, Universidad de Oriente (School of Agronomic Engineering, University of the East)
Jusepín - Estado Monagas

Rondon C., F.F.	Ing. M.S.	Riego y Drenaje
Zunico, H.L.	Ing.	Riego y Drenaje
Guzman, A.J.	Ing.	Riego y Drenaje
Verde J., H.	Ing.	Funcionamiento y Operación de Sistemas de Riego
Chirinos, I., J.A.	Ing.	Hidráulica e Hidrología
Martinez H., L.R.	Ing. M.S.	Agroclimatología - Jefe Depto. Ingeniería Agrícola
Boada S., C.A.	Ing.	Mecanización Agrícola
Lopez R., R.T.	Ing.	Maquinaria Agrícola
De La Riva F., R.	Ing.	Maquinaria Agrícola
Call, G.	Ing.	Topografía y Vialidad Rural
Febres G., E.	Arq.	Construcciones Rurales
Hurtado, R.	Topografía	Topografía

Carácter y Estructura: La Escuela de Ingeniería Agronómica fue fundada en el año 1962 y es una Universidad del Gobierno Venezolano.

Actividades de Ingeniería Rural: Enseñanza: Cursos dictados Riego y Drenaje I y II - Maquinaria Agrícola I y II - Topografía - Vialidad Rural - Construcciones Rurales. Investigación: Consumo de agua por las plantas - Cálculo de E.T.P. - Estudio de Drenaje - Estudio de variabilidad de precipitación - Estudio de infiltración - Retención de agua por los suelos.

Correspondencia: Español, Inglés.

Maracaibo

Facultad de Agronomía, Universidad Nacional del Zulia (Faculty of Agriculture, National University)
Maracaibo, Estado del Zulia

Actividades de Ingeniería Rural: La Universidad tiene cátedras de ingeniería agrícola.

Maracay

Departamento e Instituto de Ingeniería Agrícola, Facultad de Agronomía, Universidad Central de Venezuela (Department and Institute of Agricultural Engineering, School of Agronomy, Central University of Venezuela)
El Limón, Maracay, Estado Aragua

Estaciones Experimentales: Samán Mocho, al S.E. del Lago de Valencia, Estado Carabobo - El Laurel, 20 Km al S.E. de Caracas - Bajo Seco, 40 Km al N.O.E. de Caracas - San Nicolás, 30 Km al S.S.E. de Guanare, Estado Portuguesa.

Carácter y estructura: El Departamento e Instituto de Ingeniería Agrícola son dependencias de la Universidad Central de Venezuela (U.C.V.), creadas en 1959 y financiadas, como toda la Universidad Central, en más del 90% por asignaciones provenientes del presupuesto general de la Nación. El resto es financiado por ingresos propios y algunas contribuciones de entidades nacionales e internacionales.

Actividades de Enseñanza: Cursos obligatorios para todos los estudiantes de la Facultad de Agronomía en los seis primeros semestres. En los semestres siguientes (7° a 10°) se dan las materias de orientación en Ingeniería Agrícola. - Actividades de Investigación.

Correspondencia: Español, Inglés, francés.

Mérida

Facultad de Ciencias Forestales - Universidad de Los Andes (Faculty of Forest Science, Andean University)
Mérida, Edo Mérida, Venezuela

Bracho, J.	Ing.Forestal	Jefe del Departamento - Cursos de postgrado a nivel de Maestría en Fotogrametría y Mantenimiento Vial
Slezinger, E.	Prof.a Asist.	Mantenimiento Vial
Gonzáles, E.	M.S. Prof.Agreg.	Carreteras Rurales
Ponce, J. Espinoza, A.	Profesores Asistentes	Topografía
Pernia, E.	M.S. Prof.Asist.	Fotogrametría, Sensores Remotos y Cartografía
Franco, W.	Ing.For.Dr.(Agreg.)	Suelos

Gonzales, R.	Ing.For.,Asoc.	Suelos
Garcia Jarpa	Geologo, Asoc.	"
Reves De, M-Soledad	Geologo, Instruct.	"
Rojas S., D.	Ing.Forestal,Asist.	"
Hernandez B., E.	Ing.For.MSc Agr.	Manejo de Cuencas Hidrograficas, Hidrología
Leonet, R.	Ing.For. Asist.	" " "
Lopez A., J.M.	Ing.For.MSc.	" " " "
Barboza M., A.	Ing.For. Asist.	Hidrología
Vitela R., E.	Ing.Civil MSc	Construcciones Forestales

Dependencias en otras localidades: Escuela de Ingeniería Forestal Mérida Edo Mérida.

Carácter y estructura: Fundada en 1948. Los recursos financieros son aportados principalmente por el Gobierno. Su funcionamiento es autónomo.

Actividades en Ingeniería Rural: a) La Escuela de Ing. Forestal ofrece las siguientes opciones: Manejo de Bosques, Conservación de Cuencas, Conservación de Recursos Naturales Renovables y Tecnología de Productos Forestales. En estas opciones se ofrecen algunos cursos (no carrera u opción) relacionados con Ing. Rural. En todo caso el título concedido es: Ing. Forestal. En general la escuela tiene actualmente cerca de 2000 estudiantes, y egresan anualmente un total aproximado de 30 Ingenieros Forestales. b) Gran parte de los esfuerzos y el tiempo estan actualmente concentrados en docencia. Se realizan algunas investigaciones principalmente en forma individual.

Publicaciones: Algunas publicaciones de circulación interna: Carreteras Rurales Tablas Actualizadas - Aplicaciones practicas de la Altimetría - Analisis practico de Planimetría.

Correspondencia: Castellano o en su defecto inglés.

Centro de Estudios Forestales de Postgrado (Forest Science Study Centre)
Carretera Chorros de Milla, Mérida, Venezuela   Tel.:35555

Arroyo, J.	Ing.Forestal	Propiedades Madera
Barroeta, G.	Ing.Forestal	Planificación
Conejos, J.	M.S.	Preservación
Durán, J.	Ph.D.	Química de la Madera
Luna, A.	Ing.Forestal	Cartografía
Ninin, L.	Doctor	Aserraderos
Rivera, A.	M.S.	Tableros
Rodríguez, L.	M.S.	Ecología
Salinas, P.	Ph.D.	Ecología
Silva, R.	Ing.Forestal	Plantaciones
Vincent, L.	Ph.D.	Silvicultura

Coordinadores de Cursos:

García C., J.R.	Manejo de Bosques
Hernández, E.	Manejo de Cuencas
Ninin, L.	Tecnología Productos For.
Chaves, L.F.	Análisis Uso de la Tierra

Dependencias en otras localidades: Estación Experimental de Caparo (Reserva Forestal de Caparo, Estado Barinas).

Carácter y estructura: El Centro se creó en 1968. Es una dependencia de la Facultad de Ciencias Forestales de la Universidad de Los Andes, para la enseñanza al 4° nivel (postgrado) en Ciencias Forestales y Geográficas. La Universidad de Los Andes es Instituto Nacional Autónomo, financiado por el Estado Venezolano a través del Ministerio de Educación.

Actividades en Ingeniería Rural: a) Enseñanza: Cursos Largos de Magister Scientiae en los opciones de Manejo de Bosques Tropicales, Manejo de Cuencas Hidrográficas, Tecnología de Productos Forestales y Análisis del Uso de la Tierra. Promedio General de Alumnos al año: 10. - Cursos cortos de Actualización o Perfeccionamiento en áreas materias específicas: Aprovechamiento Forestal, Correción de Torrentes, Pulpa y Papel, etc. - Seminarios y Talleres sobre Metodología de Investigación, Política Forestal, etc. b) Investigación: Se conducen investigaciones en las área de estudio. Además se apoya la investigación individual que realizan los Profesores del Centro, a través del Consejo de Desarrollo Científico y Humanístico de la Universidad (CDCH)(4 proyectos). c) Extensión: El Centro asesora a organismos nacionales y empresas privadas que formulen solicitudes en tal sentido.

Publicaciones: a) Tesis y Trabajos Especiales de los Alumnos son multigrafiados. b) Artículos para revistas científicas y divulgativas sobre tópicos forestales y conservacionistas son preparados por los Profesores y publicados en el país o fuera de él. c) Informes técnicos también son reproducidos y enviados a organismos interesados.

Correspondencia: Español, Inglés.

VIET-NAM

Hanoi

Hoc Vien Thuyloi (The Hydraulic Research Institute)
Đông Da - Hànôi, S.R. Vietnam

Dao Khuong	Dr.Science	Director - Soil Improvement and Irrigation Techniques
Vu Tat Uyên	"	Deputy Director - Hydrodynamics
Nguyên Thanh Ngà	Engineer	Deputy Director - Hydraulic Construction and responsible for the Branch of the Institute based in Ho Chi Minh city.

Annexes situated in other locations: The Central Institute is located in Hanoi and its regional annex in Ho Chi Minh city. In addition, there is a network of experimental stations for hydraulic agriculture situated in different climatical regions of the country.

Structure: The Institute of Hydraulic Research was established in 1959 and since it has been placed under the control of the Ministry of Water Resources. The Institute is a governmental agency, which is financed by the national budget.

Activities: a) Scientific researches on soil improvement by irrigation and drainage methods, by irrigation techniques, on hydrodynamics and hydraulic construction. Many results of scientific researches by the Institute have been applied to production. b) Participating in the training of hydraulic engineers and doctors of hydraulic science.

Publications: Technical norms on regimes and techniques of irrigation for different plants, on construction materials (concrete, gravel, sand etc.) and on hydrodynamics.

Correspondence: Vietnamese, Russian, French, English.

WESTERN SAMOA

Apia

Agricultural Engineering Department, School of Agriculture, University of the South Pacific
P.O. Box 890, Apia, Western Samoa

Win, M.	B.E.(Agr.), M.Eng.Sc.	Lecturer and acting head of department
Fau, E.	D.T.A.	Technician
–		Appointment pending for other staff

Status: Established along with the School of Agriculture, University of the South Pacific in 1977.

Agricultural Engineering activities: a) Teaching: Dip in Tropical Agriculture (3 yr course) – B.Agr. Degree (4 yr course) – 30 students per course. b) Research: Village workshop establishment – Wet land taro mechanization and water management – Rural energy.

Publications: A few internal publications on: Incubator management – Irrigation and drainage problems.

Correspondence: English.

YUGOSLAVIA

Beograd

Institut za mehanizaciju poljoprivrede (Institute for Agricultural Mechanization)
Batajnicki drum 12 km p.f. 41, Beograd-Zemun   Tel.: 608482

Status: Semi-governmental Institution.

Activities: Testing and development of: tractors, engines, vehicles, constructions of agricultural machinery and agricultural transporting vehicles, machines for tillage of arable land, for drilling and planting, for fertilizing, for melioration, for agrophysics of soil for vegetable growing, machines and other equipment for harvesting, drying and storage of agricultural products, machines for fruit growing and vineyards and for plant protection, machines and equipment in livestock production, agricultural objects, repair and maintenance of agricultural machines, planning of agricultural objects, economics of mechanization of agriculture.

Publications: "Poljoprivredna tehnika" (Agricultural Technique), a quarterly periodical containing articles on the results of research work by the Institute.

Correspondence: Serb, Croate, English, French, German.

Institut za Poljoprivrednu tehniku, mašinstvo i racionalizacija rada, Poljoprivredni
fakultet, Univerzitet u Beogradu (Institute for Agricultural Engineering, Technology and Rational Farm Management, Faculty of Agriculture, Belgrade University)
11000 Beograd

Status: Nov. 1970. Teaching is financed by the Council of Education, Government of
Serbian Republic. Research is financed on the basis of agreement with associate enterprises.

Activities: Teaching: Bachelor's degree leading to Dipl.Ing.Master's degree leading to
Mag.Ing. Doctorate degree leading to Dr.Ing. and Development. - Research: Testing of
Farm Machinery for Combines for hilly regions, for sugar beets, for maize, grain and
hay dryers, plant protection machinery, seed and fertilizer drills and conveyor
machinery, etc.

Publications on above research activities.

Correspondence: English.

Ljubljana

Kmetijski inštitut Slovenije (Agricultural Institute of Slovenia)
61109 Ljubljana, Hacquetova 2, Yugoslavia

Jenčič, R.	Dr.	Farm mechanization
Berčič, S.	Mech.Eng.	
Marinc, V.	Agr.Eng.	
Miklič, Z.	M.S.	
Mrhar, M.	Agr.Eng.	
Muri, J.	Agr.Eng.	
Novak, M.	Dr.	

Branches: Biotehniška fakulteta, Katedra za kmetijsko strojništvo (Biotechnical
Faculty, Agricultural Mechanization Section) Ljubljana, Yugoslavia - Višja agronomska
šola, Katedra za kmetijko strojništvo (Higher Agricultural School, Agricultural
Mechanization Section), Maribor, Yugoslavia.

Status: Established 1952. Research Council of Slovenia (governmental).

Activities: Teaching: regular study of agronomy - Engineer of Agronomy (30 students).
Research: experiment, testing or extension work on agricultural machinery and farm
mechanization.

Publications: on agriculture, farm mechanization, mountain-farms, liquid manure,
environment protection, roughage harvesting, hay drying, solar energy. Textbooks.

Maribor

Univerza v Mariboru, Višja agronomska šola Maribor (University of Maribor, Agricultural Junior College Maribor)
62000 Maribor, Vrbanska 30

Mikluš, I.   Prof., dipl.agr.eng.   Dean

Vrabl, S.   Prof., D.Sc.dipl.agr.eng.   S. Dean
Novak, M.   Prof., D.Sc.dipl.agr.eng.   Head, Agricultural Engineering
Kravos, A.   Prof. M.A.,dipl.agr.eng.   Head, Agricultural Technology

Branches: Agricultural Junior College. This Institution is financed by Educational Community of Slovenia. Established 1.9.1960.

Activities: Teaching: Agricultural engineering - Agricultural technology. - Research: Development basis of agriculture in North-Eastern Slovenia for the period from 1978 to 2000 - Entomofauna of vine in Slovenia - Introduction of complex mechanization for viticulture and fruit growing - Regulation of blossom induction - Research and introduction of cultivation methods.

Publications on above activities.

Correspondence: English, German, Italian and French.

Novi Sad

Institut za mehanizaciju Fakulteta tehničkih nauka (Institute of Mechanization, Faculty of Technical Sciences) University Novi Sad
YU-21000 Novi Sad, Yugoslavia - Veljka Vlahovića 3   Tel.:(021) 55622

Tesić, M.   Dr.Mr.Dipl.Eng.Mech.   Asst.Prof. - Director, Head of Dept.
Križnar, M.   Dr.Dipl.Eng.Mech.   Prof., Head of Department
Babin, N.   Dr.Mr.Dipl.Eng.Mech.   Assoc.Prof., Head of Department
Plavšić, M.   Dr.Mr.Dipl.Eng.Mech.   Asst.Prof.

Status: Semi Governmental institution, established 1962. Teaching is financed by the Community of Interest in Education. Research, projecting, testing and certifying is financed through agreements with forms and institutions.

Activities: Teaching: Bachelor's degree leading to Dipl.Eng.Mech. Postgraduate degree leading to Mr. of technical Sc., Doctorate degree leading to Dr. of Techn.Sc. Research: Fundamental, Applied and new Product Research, Surveys, Projects, Studies and Reports, Engineering Design and Specifications, Constructions and Supervision, Testing and Certifying, all in the field of Agricultural Machinery, Transport Machinery, Engines and Vehicles.

Publications since 1979: Research reports on: Optimization of mechanization in agriculture for socially owned and individual farms - Improving nutrient value of by-products of plant origin by modern physico-chemical processing methods - Energy saving and improvement of economic effect of mechanized operations in field crop production under application of improved system of traction machine using new and safer energy sources - Harvesting, transport and handling of agricultural by-products, mainly straw, maize husks and sunflower tops - Load analysis of transmission elements in agricultural machinery during operation - Project: Transport of agricultural products. Topics: Dynamic and energetic interaction between engine and working machine; Investigation of some working parameters of transport means in connection with Vojvodina soils; Concept development of a new agricultural vehicle - A concept of maintenance and repair workshops on agricultural farms - Working out of a plan for a refuel dredger - Working out of Plan a cable dredger - Project of cable way for cement factory of Beočin - Testing and issuing certificates for transport, building and agricultural machines - Determining optimal characteristics and fixing points of vibrators for Prepare, Transport and Earth Loading Machines - Main project of equipment for apartment prefabrication - Main project cable way reconstruction; Ledinci - Application of hydrotransmission to transport, and agricultural machinery - Project and survey of river fleet for sand transport - Harvesting of Camomile.

Correspondence: English, German, French, Serbo-Croat.

Sarajevo

Institut za pedologiju, agrohemiju i melioracije Poljoprivrednog fakulteta Univerziteta Sarajevo (Institute for Soil Science, Agrochemistry and Land Reclamation, Faculty of Agriculture of University of Sarajevo)
Zagrebačka 18

Vlahinic, M.I.	Eng.Agr., Dr.Agr.Sci.	Professor of Land Reclamation (drainage, irrigation)
Hakl, Z.A.	Eng.Agr., M.S.	Professor of Land Reclamation

Status: Established 1947.

Activities: Teaching about 100 students attending the courses of second degree and about 10 students of master of science degree - Research: through the experimental field, different problems of drainage, irrigation and erosion control; Extension work and design in subject of land reclamation.

Publications: Different articles about land drainage issued in national "Vodoprivreda" (Water Economy) Beograd, Yugoslav Society of Soil Science and international publications.

Correspondence: English, French, Italian.

Skopje

Institut za lozarstvo i vinarstvo, Zemjodelski Fakultet na Univerzitetot "Kiril i Metodij" vo Skopje, Ul. "Natanail Kuceviski" (Institute for Viticulture and Enology, Faculty of Agriculture at the University "Kiril and Metodij" - Skopje "Natanail Kuceviski"
bb, Skopje, Yugoslavia

Removski, D.I.	Dr.Agr.Sci.	Specialist in viticulture, scientific adviser, Director of the Institute for Viticulture and enology
Viticulture Department:		
Boskov, S.	Dr.Agr.Sci.	Scientific adviser, specialist in viticulture
Bozinovic, Z.	M.Agr.Sci.	Specialist in viticulture, assistant
Cimburovski, B.	Grad.Agr.Eng.	" " "
Donevski, D.	Grad.Agr.Eng.	Professional adviser
Hristov, P.	Grad.Agr.Eng.	Specialist in viticulture, assistant
Jovanovski, D.	Grad.Agr.Eng.	Specialist in viticulture, assistant
Petrovski, G.	Dr.Agr.Sci.	Part-time professor in viticulture at Agricultural Faculty of Skopje.

Enology Department:

Jarev, T.	M.Ag.Sci.	Docent for enology at Agric.Faculty of Skopje
Naumova, C.	Grad.Agr.Eng.	Specialist in enology, professional adviser
Nockovska, E.	M.Ag.Sci.	Assistant
Vojnoski, B.	Grad.Agr.Eng.	Specialist in enology, assistant

Status: The Institute for Viticulture and Enology is founded in 1951 by the Government of People's Republic of Macedonia. At present it is financed by Governmental bodies and self-financing bodies.

Agricultural Engineering activities: a) Teaching: Professors for the subjects of viticulture and enology hold lectures at the viticulture and enology departments at the Agricultural Faculty, which in 1980 had 470 students in all years of studying. The students who finish their studies receive the title of Graduated Agricultural Engineer, specialist in viticulture and enology. The Agricultural Faculty of Skopje can hold specialistic courses, master and doctor courses in agricultural sciences.
b) Research, experiment and applied work: The Institute works on the following scientific problems: Hibridization and founding of new vine and dessert kinds - Specific selection in the kinds of vine - It studies the agro-biological and technological characteristics of the new-formed and perspective dessert and wine kinds - Application of contemporary agrotechnical measures for getting higher harvest and better quality in the production of grapes - Modernization of the production of young plant material - Technological solutions for grapes keeping (storage) - Modern working of the wine grapes - Determination of the elements in sweet and other wines - Application of thermovinificators in the production of black wines - Determination of the acidity in wines - Forming of sorts and kinds of wines, etc. The collaborators of the Institute take active part in the application of this scientific knowledge in the field of viticulture and enology, together with the organizations of joint labour and individual vine and grape workers.

Publications: mainly in the following issues: Works of the Agricultural Faculty of Skopje - Works of the Symposium for Viticulture at the Institute for Viticulture and Enology and Agrocombinate "Tikvesh" from Kavadarci - 'Viticulture' scientific issue of the Jugoslav scientific association in Belgrade.

Correspondence: French, English and Russian.

Zagreb

Institut for Farm Mechanization Technology and Buildings, Zagreb University, Agricultural Faculty
Ferenčica 104, 41000 Zagreb

Todorić, I.	Dr.Agr.	Director - complex corn drying
Brčić, J.	Prof.Dr.Agr.	Dep. Director - education, dairy and farm mech.
Gašparac, J.	Dr.Agr.	Dep. Director - farm electrification, forage machinery
Antonić, B.	Eng.Agr.	Work protection, horticultural mech.
Antončić, I.	Dr.Agr.	Complex. farm mech.
Barčić, J.	Mr.Agr.	Work on precipitous terrains, plow researching
Bedeković, J.	Mr.Agr.	Agricultural technology

Beštak, T.	Dr.Agr.	Education, programming of mechanization on large farms
Dujmović, M.	Prof.Dr.Agr.	Education, horticultural mech.
Gospodarić, Z.	Eng.Agr.	Work protection, ergonomy
Katić, Z.	Prof.Dr.Mach.	Silage, drying
Košutić, S.	Eng.Agr.	Complex.mach.
Komunjer, D.	Prof.Dr.Agr.	Education, farm mach.
Lacković, L.	Prof.Dr.Agr.	Education, soil tillage and cult
Lončarević, J.	Eng.Agr.	Education
Piria, I.	Dr.Agr.	Tractors, agric. vehicles, transport, electronic equipment
Roje, G.	Eng.Agr.	Dairy mech.
Šalamon, J.	Eng.Agr.	Dairy mech.
Šikić, D.	Prof.Dr.Archit.	Farm buildings
Vešnik, F.	Dr.Agr.	Agricult. technology
Zelenko, F.	Dr.Agr.	Dairy mech.
Dobricević, J.	Eng.Mach.	Tractors, electronic equipment

Location: The Institute has two locations: Administration and Machinery examination department are located in Zagreb I, Ferencica 104; Education department, Agricultural Technology and buildings are located in Zagreb, Simunska c. 25.

Status: The Institute is established in 1952. It works independently in concatenation of Agricultural College, Zagreb University.

Activities: Education. Research work in farm, dairy, horticulture mech. technology and buildings, tractor testing.

Publications: on contemporary problems of mechanization in agriculture, mechanization of agricultural production. Books: Mechanization of industrial production of vegetables on large farms - Textbook: Mechanization of agriculture.

Correspondence: Croate, Serb, English, French, German, Italian, Russian.

ZAIRE

Kinshasa

Département de l'Agriculture, République du Zaïre
B.P. 8722, Kinshasa

Machinisme et Motorisation Agricoles
INERA, B.P. 2037 Kinshasa I.
UNAZA, Campus Universitaire de Kinshasa, B.P. 799, Kinshasa XI.
Centre de Mécanisation Agricole de Mikondo, Département de l'Agriculture, Kinshasa I.

Correspondance: Français.

ZAMBIA

Magoye

Farm Machinery Research Unit, Regional Research Station, Ministry of Agriculture and Water Development
P.O. Box 11, Magoye, Zambia

Cullen, J.A.	Farm Machinery Engineer
Nalumino, I.	Engineering Assistant

Status: Established in 1971. Funded by the Government of the Republic of Zambia. Directed by Government, through a Technical Committee comprising members from Government Departments, Farmers Bureau and Trade Representatives.

Agriculture Engineering activities: Research: Testing of equipment to determine suitability for use in Zambia - Hand hoes and axes - ox-drawn ploughs, harrows, ridgers, cultivators and multi purpose toolbars - knapsack sprayers - groundnut lifters and shellers - small tractors under 30 HP - Development of simple machinery - modifications and improvement of existing designs - Liaison with local suppliers and manufacturers.

Publications: Test reports and recommendations on previously listed equipment.

Correspondence: English.

University of Zambia - Lusaka Campus - Faculty of Engineering
P.O.B. 32379 Lusaka

Dean: Whittaker, D.   Ph.D., CEng., FIFE

Natural Resources Development College
P.O.B. CH99, Chelston, Lusaka

Activities: 3-year dipl. course in agriculture, agricultural education and engineering.

Principal: Chungu, R.K.

Zambia College of Agriculture
P.O.B. 53. Monze

ZIMBABWE

Harare

Ministry of Natural Resources and Water Development, Division of Water Development
Private Bag 7712, Causeway, Harare

Head Office in Harare:

Shaw, D.N.	B.Sc.(Eng.) M.Zwi.I.E., M.I.C.E.	Secretary for Natural Resources and Water Development
Grizic, P.M.	B.Sc.(Eng.), M.Zwi.I.E., C.Eng., M.I.C.E., M.A.S.C.E.	Deputy Secretary, Division of Water Development
McDonald, N.G.	M.Zwi.M.	Under Secretary (Administration)
Elliott, K.D.	B.Sc.(Eng.), F.Zwi.I.E., C.Eng., M.I.C.E.	Management Engineer (Development)
Dell, D.S.	B.SC.(Eng.) M.Zwi.I.E., C.Eng., M.I.C.E.	Management Engineer (Planning)
Burke, N.A.	B.Sc.(Eng.), M.Zwi.I.E.	Designs Engineer I
Lawson, J.D.	B.Sc.(Eng.), M.Zwi.I.E.	Dam Safety and Betterment Engineer
Paron, J.M.M.	B.SC.(Eng.), M.Zwi.I.E., C.Eng. M.I.C.E.	Designs Engineer II
Purnell, D.G.	B.Sc.(Eng.), M.Zwi.I.E., C.Eng., M.I.C.E.	Chief Planning Engineer
Rowe, D.B.	B.Sc.(Eng.), M.Sc., C.Eng., M.Zwi.I.E. M.I.C.E.	Deputy Chief Planning Engineer I
Wells, R.	B.Sc.(Eng.)	Deputy Chief Planning Engineer II
Mitchell, T.B.	B.Sc.(Eng.), M.Zwi.I.E., C.Eng., M.I.C.E.	Chief, Hydrological Engineer
Landing, K.D.	B.Sc.(Eng.), M.Zwi.I.E., M.A.S.A.I.C.E., Pr.Eng.	Deputy Chief, Hydrological Engineer

Provincial Branch, Mashonaland (Harare):

Kieck, D.F.	B.Sc.(Eng.) M.Zwi.I.E., M.A.S.A.I.C.E.	Acting Provincial Water Engineer
Shurmer, R.	B.Sc.(Eng.), M.Zwi.I.E.	Deputy Provincial Water Engineer

Provincial Branch, Matabeleland (Bulawayo):

Nurton, J.B.	M.A.(Cantab.), M.Zwi.I.E., C.Eng., M.I.C.E.	Provincial Water Engineer
Tate, A.B.	M.Zwi.I.E., C.Eng., M.I.C.E.	Deputy Provincial Water Engineer

Provincial Branch, Midlands (Gwelo):

Timm, M.N.G.	B.Sc.(Eng.), M.Zwi.I.E.	Provincial Water Engineer
Varndell, G.J.	B.Sc.(Eng.)	Deputy Provincial Water Engineer

Provincial Branch, Victoria (Fort Victoria):
Lotter, M.G.   B.Sc.(Eng.),   Provincial Water Engineer
                      M.Zwi.I.E., C.Eng.,
                      M.I.C.E.
Martiz, P.J.   B.Sc.(Eng.),   Deputy Provincial Engineer
                      M.Zwi.I.E., M.S.A.I.C.E.

Provincial Branch, Manicaland (Umtali)
Morton, J.P.   B.Sc.(Eng.),   Provincial Water Engineer
                      M.Zwi.I.E., C.Eng.,
                      M.I.C.E.
Dayton, J.   B.Sc.(Eng.)   Deputy Provincial Water Engineer

<u>Status</u>: Government institution under the responsibility of the Minister of Natural Resources and Water Development.

<u>Agricultural Engineering activities</u>: Advisory services to farmers, land owners, other bodies and Ministries in regard to irrigation, water supply and water conservation. - Construction and maintenance of water conservation works, water supplies for Government Townships, for mines and other industries. - Hydrology: Irrigation and irrigation engineering. - Planning of water development and co-ordination of requirements and resources. - Surveys and investigations for water development. - Water boring for Government purposes and for private applicants. - Water pollution control. - Dam safety.

<u>Publications since 1973</u>: Hydrological year books - Annual Report of the Division of Water Development - A review of Limnology in Zimbabwe - A guide to the Design and Construction of Medium Sized Dams in Zimbabwe - Water Resources of Zimbabwe - Water Pollution Control - An Appraisal - Earth Rock Dam Construction in Zimbabwe - The yield of an average dam in Zimbabwe - The Philosophy of Reservoirs Yield and Operational System for the Maximum Utilization of Water Resources - Feasibility of the drop inlet spillway of Siya Dam - Reservoir Yield using the transition probability matrix method - The estimating of the yield of multiple dams - Palawan Dam Progress Report on the Constructions - Seke Effluent Disposal Scheme.

<u>Correspondence</u>: English.

The Institute of Agricultural Engineering, (Research and Specialist Services)
P.O. Box BW.330, Borrowdale, Harare

Spear, A.J.   Fl.Agr.E.   Chief Engineer, Head of Branch
Elwell, H.A.   M.Sc., C.Eng., M.I.C.E.   Conservation Engineering
Meikle, G.J.   AMI Agr.E. Ass.N.C.Ag.E.   Tillage/Vegetable Fuel Oil Research
Oliver, G.J.   MI Agr.E.   Oilseeds Mechanization Specialist
Radajewski, W.C. M.Sc.Eng.   Grain Drying
Smith, R.D.   B.Sc., Mech.Eng.   Tillage/Cotton Mechanization
Spence, S.M.   B.Sc., Mech.   Appropriate Technology
                  M.Sc., Agr.Eng.

<u>Branches</u>: Conex Extension Section - Conservation Branch.

<u>Status</u>: The Institute was established in 1968 and is within the Ministry of Agriculture although certain posts are funded by Farmers Organizations and the research programme is co-ordinated by the Agricultural Research Council of Zimbabwe.

Activities: The main activities of the Institute are research and development.

Publications: Most of the publications consist of scientific articles published in various journals, the main publications now being done by the Extension section based on the unpublished articles from the Research section.

University of Zimbabwe (Department of Land Management), Faculty of Agriculture
P.O. Box MP 167, Harare   Tel.: 303211

Ascough, W.J.   B.Sc.(Agric.), M.I.Agr.Eng.   Lecturer

Status: University institution since 1955. Statutory body. Government grant.

Activities: (a) Teaching: Courses in agricultural engineering and mechanization are offered in the Bachelor of Science Agriculture Degree with Honours; also short courses in appropriate technology. Average 30 students. (b) Research: Appropriate technology and small scale industry studies, associated with a Technology Promotion Centre.

Publications: Intermediate technology as a potential aid to a more productive subsistence agriculture. Zimbabwe agric.J., vol. 76(3).

Correspondence: English.

Gwebi College of Agriculture
Private Bag 376 B, Harare   Tel.: 32936/39

Gilling, F.W.   A.M.I.M.I., Rhod.Tech. I.E.   Agricultural Machinery

Status: Governmental Institution, established in 1949.

Activities in agricultural engineering: Two-year diploma course in general agriculture consisting of one-quarter agricultural engineering – 48 students per year, total 89 – Instruction (50 per cent of time on practical) and instructional projects – Small educational projects related to local conditions.

Correspondence: English.

Department of Agricultural Engineering, Tobacco Research Board
P.O. Box 1909, Harare   Tel.: 50411

Head of Department – Vacant at present
Tourle, L.   F.I. Agr.E.

Branches: Two other research stations at Banket and Fort Victoria.

Status: Parent Body, Tobacco Research Board, established in 1950. Agricultural Engineering Department in March 1955.

Activities: Lecturing at University of Zimbabwe. (B.Sc.(Agric.) Course). Development and testing of curing systems and mechanical aids for Virginia and air-cured tobaccos. Testing of proprietary apparatus for use in tobacco curing systems. Service and equipment design and fabrication for internal departments.

INTERNATIONAL INSTITUTIONS AND ORGANIZATIONS
INSTITUTIONS ET ORGANISATIONS INTERNATIONALES
INSTITUCIONES Y ORGANIZACIONES INTERNACIONALES

Ecole Inter-Etats d'Ingénieurs de l'Equipement Rural (E.I.E.R.)
B.P. 7023, Ouagadougou, République de Haute-Volta

Cette école, entreprise commune des Etats Membres de l'O.C.A.M. est un Etablissement d'Enseignement Supérieur à vocation interétatique. Formation et perfectionnement des ingénieurs concernés par la mise en valeur des ressources hydrauliques, les techniques sanitaires et l'équipement rural en Afrique.

Le Conseil d'Administration comprend des représentants des Etats suivants: Cameroun, Centrafrique, Congo, Côte d'Ivoire, Bénin, Gabon, Haute-Volta, Mali, Mauritanie, Niger, Sénégal, Tchad et Togo.

Le recrutement se fait par concours ou sur titres au niveau du diplôme universitaire d'enseignement scientifique - physique, chimie.

La durée de l'enseignement est de 30 mois avec une formation scientifique supérieure et une formation specialisée en hydraulique générale et urbaine, traitement des eaux et assainissement, hydraulique agricole, hydrologie, hydrogéologie, génie sanitaire, industries agricoles et alimentaires, techniques frigorifiques.

Le diplôme attribué est celui d'ingénieur de l'Equipement Rural.

Economic Commission for Europe (ECE), FAO/ECE Working Party on Mechanization of Agriculture (Commission Economique pour l'Europe, Groupe de travail mixte FAO/ECE de la mécanisation de l'agriculture)
Palais des Nations, CH-1211 Genève 10

Activities: Study of various specific aspects of mechanization of agriculture with the emphasis to energy problems, protection of the environment and some other wide problems in the context of farm mechanization. Caryying out reports by rapporteurs from different countries nominated for each particular study. Publication of the reports based on the contributions of a number of ECE countries in the AGRI/MECH series. Exchange of scientific and technical information through study tours, films and bibliographical material.

Publications in the AGRI/MECH series since 1970: Accelerated Testing of Agricultural Machinery, AGRI/MECH/43 - The Mechanization of Rice Cultivation, AGRI/MECH/44 - Methods and Equipment for Growing and Harvesting of Tobacco, AGRI/MECH/45 - Technical and Economic Problems of High-Speed Ploughing, AGRI/MECH/46 - Mechanical Equipment for Field Drainage and Ditching, AGRI/MECH/18/Rev.1 - Technical and Working Methods for Cowshed and Pigsty Cleaning, AGRI/MECH/47 - Automatic and Semi-automatic Feeding Systems for Livestock, AGRI/MECH/48 - Mechanized Cultivation and Harvesting of Vegetables in Open Ground, AGRI/MECH/49 - Present and Foreseeable Trends in the Development of Tractors, Especially High-Capacity Tractors, AGRI/MECH/50 - Present and Foreseeable Trends in Mechanization and Their Impact on European Agriculture (Horizon 1980), AGRI/GE.2/1, AGRI/MECH Report N 51, Volume I, Volume II in two parts - Country Reports, 1973 - Vehicles for the Transportation of Loads in Agriculture, AGRI/GE.2/2, AGRI/MECH Report N 52 - Technical Aspects in the Prevention of Noise from Agricultural Tractors, AGRI/GE.2/3, AGRI/MECH Report N 53 - Mechanization of Potato Harvesting, Cleaning and Grading, AGRI/GE.2/4, AGRI/MECH Report N54 - Mechanization of the Cultivation and Harvesting of Sugar Beet, AGRI/GE.2/5, AGRI/MECH Report N55 -

Methods of Application of Liquid Fertilizers, AGRI/GE.2/6, AGRI/MECH Report N.56 - Mechanization of Vegetable Cultivation and Harvesting for Protected Cropping, AGRI/GE.2/7, AGRI/MECH Report N 57 - Methods and Machines for Harvesting and Threshing Herbage Seed Crops, AGRI/GE.2/8, AGRI/MECH Report N 58 - Examples of Technical and Economic Analysis of Mechanized Processes in Various Agro-Technical Conditions, AGRI/MECH/32/Add.1 - Methods and Equipment for Producing Mixed Feed, AGRI/GE.2/13, AGRI/MECH Report N 59 - Methods of Manure Treatment with Special Regards to the Protection of the Environment, AGRI/GE.2/15, AGRI/MECH Report N 60 - Methods and Equipment for Minimum Tillage, AGRI/Ge.2/16, AGRI/MECH Report N 61 - Preparation and Conservation of Hay, AGRI/GE.2/19, AGRI/MECH Report N 62 - Farm Milk Cooling and Milk Collection, AGRI/GE.2/20, AGRI/MECH Report N 63 - Milking Methods and Milking Machines, FAO/ECE/AGRI/WP.2/1, AGRI/MECH Report N 64 - Methods for Fertilizer Placement and Band Sowing, FAO/ECE/AGRI/WP.2/2, AGRI/MECH Report N 65 - High-Temperature Drying of Green Forage, FAO/ECE/AGRI/WP.2/3, AGRI/MECH Report N 66 - Mechanization of Potato Production on Slopes and Mountains, FAO/ECE/AGRI/WP.2/4, AGRI/MECH Report N 67 - Use of Hydraulic Transmission Systems in Tractors and Self-Propelled Machines, FAO/ECE/AGRI/WP.2/9, AGRI/MECH Report N 68 - Equipment and Techniques for Manure Treatment, FAO/ECE/AGRI/WP.2/10, AGRI/MECH Report N 69 - Automation of the Control of Technological Processes in Mobile Agricultural Machines, FAO/ECE/AGRI/WP.2/13, AGRI/MECH Report N 70 - Ergonomic Aspects of the Design of Tractors, FAO/ECE/AGRI/WP.2/15, AGRI/MECH Report N 71 - Mechanical Production, Nursing and Handling of Seedlings, FAO/ECE/AGRI/WP.2/16, AGRI/MECH Report N 72 - Air Conditioning of Livestock Buildings: Modern Equipment and Development Trends, FAO/ECE/AGRI/WP.2/17, AGRI/MECH Report N 73 - Methodology Used by ECE Countries in Forecasting Mechanization Developments, FAO/ECE/AGRI/WP.2/18, AGRI/MECH Report N 74 - Means of Reducing Energy Consumption in the Heating of Greenhouses, FAO/ECE/AGRI/WP.2/21, AGRI/MECH Report N 75 - Forage Handling Systems from Field to Storage, FAO/ECE/AGRI/WP.2/22, AGRI/MECH Report N 76 - Feeding Methods and Equipment for Dairy and Beef Cattle, FAO/ECE/AGRI/WP.2/24, AGRI/MECH Report N 77 - Harvesting, Processing and Storage of Grain, FAO/ECE/AGRI/WP.2/25, AGRI/MECH Report N 77 - Harvesting, Processing and Storage of Grain, FAO/ECE/AGRI/WP.2/25, AGRI/MECH Report N 78 - Mechanization of Grape Harvesting, FAO/ECE/AGRI/WP.2/26, AGRI/MECH Report N 79 - High-Powered Tractors and Their Implements, Including Aspects of Their Impact on the Soil, FAO/ECE/AGRI/WP.2/27, AGRI/MECH Report N 80 - Means of Reducing Energy Consumption in Drying Equipment, FAO/ECE/AGRI/WP.2/28, AGRI/MECH Report N 81 - Methods and Equipment of Hail Prevention, FAO/ECE/AGRI/WP.2/30, AGRI/MECH Report N 82 - The Role of Agricultural Mechanization in Preserving and Improving the Environment (under Conditions of Cropping in Flat Areas), FAO/ECE/AGRI/WP.2/32, AGRI/MECH Report N 83 - Methods and Equipment for Placing Chemical Products in Agriculture, FAO/ECE/AGRI/WP2/34, AGRI/MECH Report N 84 - Methods and Equipment of Protection from Frost Damage, FAO/ECE/AGRI/WP.2/37, AGRI/MECH Report N 85 - Foreseeable Developments in Self-Propelled Crop Harvesters, FAO/ECE/AGRI/WP.2/38, AGRI/MECH Report N 86 - Technology of Storage and Mechanization of Work in Vegetable Store-Houses, FAO/ECE/AGRI/WP.2/39, AGRI/MECH Report N 87 - Review of Existing Technological Processes in the Mechanization of Crop Production in Order to Reduce Energy Consumption, FAO/ECE/AGRI/WP. 2/42, AGRI/MECH Report N 88 - Methods and Mechanical Equipment for Increasing the Nutritive Value of Straw, FAO/ECE/AGRI/WP.2/43, AGRI/MECH Report N 89 - Methods and Equipment for Gathering and Storage of Agricultural By-Products, FAO/ECE/AGRI/WP.2/44, AGRI/MECH Report N 90 - Methods and Equipment for the Mechanization of Agricultural Work on Sandy Soils, FAO/ECE/AGRI/WP.2/45, AGRI/MECH Report N 91.

<u>Correspondence</u>: English, French, Russian.

Instituto Interamericano de Ciencias Agrícolas de la OEA (IICA)
(Inter-American Institute of Agricultural Sciences of OEA - IICA)
Apartado 55, Coronado, Costa Rica

Merea, A.   Ing.   Coordinador del Comité de Tierras y Aguas

Especialistas en Conservación y Manejo de Tierras y Aguas
Arrunátegui, H.   Ing.Agr.
Aquize, J.   M.Sc.
Barrios, J.   M.Sc.
Chavez, O.   Ing.Agr.
Forsythe, W.   Ph.D.
González, N.   Ing.Agr.
Matute, E.   Ing.Agrícola
Millar, A.   Ph.D.
Mojica, I.   Ph.D.
Novello, F.   Ing.Agr.
Paulet, M.   Ph.D.

Dependencias en otras localidades: El IICA cuenta con Oficinas distribuidas en 28 países del Area Latinoamericana y el Caribe.

Carácter y Estructura: El Instituto es el Organismo especializado de la OEA para el sector agropecuario. Fue establecido en 1942 por los países americanos.

Actividades en Ingeniería Rural: El IICA tiene establecido y dirige el Programa de Conservación y Manejo de Tierras y Aguas, a través del cual, desde 1971, viene realizando muchas y varias actividades en la mayor parte de los países latinoamericanos y del Caribe, donde tiene establecidas sus oficinas nacionales. Básicamente, se está trabajando en el fortalecimiento de las instituciones nacionales responsables en sus países de la conservación y manejo de tierras y aguas; en la actualización y complementación de leyes, reglamentos y normas en materia de aguas de riego; sistematización de tierras de riego; diseno de infraestructura a nivel predial; inventario de recursos de los sistemas de riego; drenaje; métodos de riego; normas para la operación y mantenimiento de los proyectos de riego; sectorización de los proyectos de riego y organización de sus usuarios; cobros de los servicios de operación y mantenimiento y de recuperación de las inversiones de fondos públicos y en la infraestructura de los proyectos de riego; control y evaluación de la distribución de aguas de riego; planes nacionales de irrigación; optimización del uso y conservación de los recursos de tierras y aguas disponibles; estudios en cuencas pilotos experimentales; pequena irrigación; estudios socio-económicos de los productores regentes y en actividades de capacitación y adestramiento en muchos de los aspectos indicados.

Publicaciones: El IICA tiene dos Series de Publicaciones Técnico-Científicas, a saber: Series Periódicas: Revista Turrialba, especializada en agricultura tropical, consignando artículos científicos, resenas, notas y comentarios. - Series de Libros: Libros de texto para apoyo de los cursos de capacitación que se dictan sobre diversos aspectos relacionados con la problematica de tierras y aguas - Adicionalmente, se publican materiales didácticos para apoyo de los cursos cortos de capacitación que se dictan en materia de operación y mantenimiento de proyectos de riego; sistematización de tierras; relación agua-suelo-planta; drenaje; planes de cultivo y riego; etc.; informes finales de Seminarios y Reuniones Técnicas nacionales, multinacionales y latinoamericanas; documentos técnicos referentes a estudios e investigaciones sobre aspectos varios referentes a tierras y aguas.

Correspondencia: Espanol.

Commission Internationale du Génie Rural (C.I.G.R.) (International Commission of Agricultural Engineering)
17-21 rue de Javel, 75015 Paris tél : 577-75-78 et 577-60-66

Lehoczky, L. Président Director of the Farm Machinery Institute, University of Agricultural Sciences, Agricultural Engineering Faculty, Pater Karoly u. 1, H - 2103 Gödöllö (Hungary)

Carlier, M. Secrétaire Général - Ingénieur Général du Génie Rural, Professeur à l'Institut National Agronomique, 17-21 rue de Javel, 75015 Paris

1ère Section - Les Sciences du sol et des eaux dans leurs applications aux travaux de Génie Rural - Techniques de la défense et conservation des sols, de l'aménagement agricole des eaux et des aménagements fonciers

Baquero de la Cruz, G. Secrétaire Général de l'Institut National des réformes et du développement agraires (IRYDA), Avenida del Generalissimo 2, Madrid 16, Espagne

2ème Section: Constructions rurales et équipements connexes

Henriksson, R. Président Department of Farm Buildings L.T.B., Box 624, S-220 06 Lund 6 (Suède)

3ème Section: Machinisme agricole

Manfredi, E. Président Directeur de l'Institut de machinisme agricole, via Filippo Re 4, 40126 Bologne (Italie)

4ème Section: Distribution de l'électricité dans les zones rurales et ses applications agricoles dans le contexte général énergétique

Wakeford, P. Président The Electricity Council, Farm-electric Centre, National Agricultural Centre Stoneleigh, Kenilworth Warwickshire CV 8 2LS (Royaume-Uni)

5ème Section: Organisation scientifique du travail en agriculture

Maton, A. Président Directeur de la Station de Génie Rural, Van Gangsberghelaan 61, 9220 Merelbeke (Belgique)

Annexes situées dans d'autres localités: La CIGR rassemble 33 pays de tous les continents dont les 26 pays suivants sont représentés par des Associations Nationales groupant, dans chaque pays, les Professeurs, chercheurs, ingénieurs, techniciens et maîtres d'oeuvre de Génie Rural: Afrique du Sud, Allemagne (RFA), Autriche, Belgique, Brésil, Danemark, Espagne, Finlande, France, Hongrie, Israel, Irlande, Italie, Japon, Maroc, Nigéria, Norvège, Pays-Bas, Pologne, Portugal, Royaume-Uni, Suède, Suisse, Tchécoslovaquie, U.S.A., Yougoslavie. Les 7 pays suivants sont, provisoirement, représentés par des membres individuels: Allemagne (R.D.A.), Egypte, Grèce, Islande, Luxembourg, Nouvelle-Zélande, Turquie.

Structure: La CIGR est une organisation internationale non gouvernementale fondée en 1930. Elle bénéficie du statut consultatif spécial de la F.A.O. et du statut de Consultant de l'U.N.E.S.C.O. et du Conseil de l'Europe. L'administration et la gestion de la CIGR sont assumées par le Secrétariat Général, sous contrôle et suivant les directives d'un Comité Directeur qui comprend le Président de la C.I.G.R., le Premier Vice-Président, les 5 Vice-Présidents, le Secrétaire Général et un représentant de chaque pays membre.

Activités: La CIGR établit et entretient des relations de caractère scientifique et technique entre ses membres au moyen de réunions de travail des Sections techniques (en moyenne chaque Section tient une telle réunion de travail tous les deux ou trois ans) et de Congrès internationaux de Génie Rural groupant toutes les Sections (un tel Congrès a lieu, en moyenne, tous les cinq ans) - Elle encourage et effectue des études techniques de Génie Rural d'intérêt international, notamment à la demande de la F.A.O. et de l'U.N.E.S.C.O.

Publications: Liste disponible sur demande. Il s'agit essentiellement des comptes rendus des Journées d'études des Sections techniques ainsi que des Congrès internationaux de Génie Rural.

Correspondance: Français, anglais et allemand.

International Commission on Irrigation and Drainage (ICID)
48 Nyaya Marg, Chanakyapuri, New Delhi, 110021, India

Darves-Bornoz, R.   President   (France)
Framji, K.K.   Secretary-General
Garg, B.C.   Secretary

Affiliated Branches: 76 National Committees in the following 76 countries: Algeria, Angola, Arab Republic of Egypt, Argentina, Australia, Austria, Bangladesh, Brazil, Bulgaria, Burma, Canada, Chile, Colombia, Cuba, Cyprus, Czechoslovakia, Dominican Republic, Ecuador, Ethiopia, Federal Republic of Germany, France, German Democratic Republic, Ghana, Great Britain, Greece, Guyana, Honduras, Hungary, India, Indonesia, Iran, Iraq, Ireland, Israel, Italy, Ivory Coast, Japan, Jordan, Kenya, Lebanon, Malawi, Malaysia, Mexico, Morocco, Nepal, The Netherlands, New Zealand, Nigeria, Pakistan, Peru, Philippines, Poland, Portugal, Syria, Republic of China, Republic of Korea, Mozambique, Zimbabwe, Romania, Saudi Arabia, Senegal, Spain, Sri Lanka, Sudan, Surinam, Switzerland, Thailand, Tunisia, Turkey, Uganda, U.S.A., U.S.S.R., Venezuela, Socialist Republic of Vietnam, Yugoslavia, Zambia.

Status: Non-governmental international organization, established in 1950, governed by the International Executive Council comprising one representative from each National Committee and Office-Bearers (one President, nine Vice-Presidents and one Secretary-General) - Financed by the National Committees.

Activities: Stimulation and promotion of the development and application of the science and technique of irrigation, drainage, flood control and river training in the engineering, economic and social aspects; interchange of information; periodical meetings; organization of studies and experiments; publication of reports, proceedings and documents, special publications, bulletins, etc.

Publications: Multilingual Technical Dictionary on Irrigation and Drainage (English-French) 1967 ed. - Controlling seepage Losses from Irrigation Canals - A World-wide Survey (1967 ed.) - Contrôle des Pertes par Infiltration des Canaux d'Irrigation - Etude mondiale 1967 - World-wide Survey of Experiments and Results on the Prevention of Evaporation Losses from Reservoirs (Rev.ed. 1967) - International Co-operation in the Development of Water Resources for Agriculture - Co-opération Internationale dans

la Mise en Valeur des Ressources Hydrauliques pour l'Agriculture - Irrigated Wheat - A World-wide Survey (1972) - Design Practices of Irrigation Canals in the World (1972) - ICID Technical Memoirs No.1 (1972) - Irrigated Cotton - A World-wide Survey (1973) - ICID Technical Memoirs No.2 (1974) - Irrigation Efficiency in Small Farm Areas (1974) - ICID Silver Jubilee Commemorative Volume, July 1975 - Flood Control in the World - A Global Review, Vol.I (1976), vol.II(1977) - Irrigation and Salinity - A World-wide Survey (1976) - Irrigated Rice - A World-wide Survey 1977 - State-of-the-Art Irrigation, Drainage and Flood Control - Drainage Construction Techniques for vertical/tubewell drainage - ICID Technical Memoirs No.3 (1979) - Canal Construction . Open Channels Construction - Machinery and Techniques.
Transactions of Congresses, Proceedings of Sessions, Symposia and Seminars; Annual and bi-annual ICID bulletins.

International Rice Research Institute
P.O. Box 933, Manila, Philippines   Tel.: 88-48-69

Bockhop, C.W.	Ph.D.	Agricultural Engineer - Head, Agricultural Engineering Department
Duff, J.B.	MS	Associate Agricultural Economist
Wicks, J.A.	Ph.D.	Associate Agricultural Economist
Fischer, R.C., Outreach Staff	MS	Associate Agricultural Engineer - IRRI-Thai Machinery Development Programme, Bangkok
Khan, A.U. "	Ph.D.	Agricultural Engineer - IRRI-PAK Agricultural Machinery Programme, Islamabad, Pak.
Peterson, G. "	MS	Agricultural Engineer - Rama, Dacca-2, Bangladesh
Reddy, V.R. "	MS	Agricultural Engineer - IRRI-DITPROD Industrial Extension Project Pasar Monggu, Jakarta Selatan, Indonesia
Townsend, J.S. "	Ph.D.	Project Engineer - CANADA-IRRI-BURMA Project Rangoon, Burma

Status: Date of establishment: 1965 - Government bodies which direct and finance the institution: Consultative on International Agricultural Research, Washington, USA; U.S. Agency for International Development, USA

Correspondence: English.

The World Ploughing Organization (WPO)
Foulsyke, Loweswater, Cockermouth, Cumberland

Stehouwer, A.C.	Chairman of International Governing Board
Geiger, F.	Vice-Chairman
Hall, A.	General Secretary

Branches: WPO has at present 21 affiliated national bodies.

Status: International non-governmental organization.

Activities: Organization of world ploughing contests held in a different country each year - Promotion of better understanding of the value of soil cultivation practices - Determination of techniques and types of equipment; practical demonstrations and instruction on the principle of learning by doing; advice upon promotion of ploughing competitions to achieve high standards of workmanship - Arrangement of conferences on soil tillage problems - The 19th World Contest was held at Vernon Center, Mankato, Minnesota, USA, in September 1972, the 20th World Contest near Wexford, Republic of Ireland in October, 1973 - The 21st World Contest near Helsinki, Finland, in August, 1974 - A conference on "Soil fertility, tillage and conservation" was also held in conjunction with the 19th Contest in the USA - A soil tillage conference was held in conjunction with the 20th Contest in Ireland in October, 1973 at the Johnstown Agricultural Research Institute. Other World Contests: 1975 Shawa, Ontario, Canada; 1976 Bjertorp, Vara, Sweden; 1977 Flevohof, Netherlands; 1978 Wickstadt, near Friedberg, Germany; 1979 Limavady, Northern Ireland; 1980 Christchurch, New Zealand; 1981 Wexford, Ireland, 1982 Tasmania, Australia.

Publications: Yearly handbook of each world contest printed in English and the language of the host country - Bulletins of news and information at irregular intervals - Descriptive brochure.

Correspondence: English and German.

Centro Internacional de Mejoramiento de Maiz y Trigo - International Maiz and Wheat Improvement Center (CIMMYT)
Londres 40, Aptdo. Postal 6-641, México 6, D.F., México

International Center for Agricultural Research in the Dry Areas (ICARDA)
P.O.B. 114/5055, Beirut, Lebanon   Telex: 22509 LE   Cable: ICARDA BEIRUT, Tel.303860

International Crops Research Institute for the Semi-Arid Tropics (ICRISAT)
1-11-256, Begumpet, Hyderabad-500016 (A.P.), India

International Institute of Tropical Agriculture (IITA)
Oyo Road, P.M.B. 5320, Ibadan, Nigeria

Centro Internacional de Agricultura Tropical, CIAT
Apartado Aéreo 67-13, Cali, Colombia   Cables: CINATROP

Food and Agriculture Organization of the United Nations
via Terme di Caracalla, 00100 Rome, Italy   Tel.: 57971   Telegr.: FOODAGRI ROME
                                                              Telex : 610181 FAO I

Agricultural Services Division
Nicholas, M.S.O.   Director

Agricultural Engineering Service
von Hülst, H.J.   Chief
Rånnfelt, C.A.   Senior Officer   Agricultural Engineering
Caro-Greiffenstein, A.   Post-Harvest Technology
Corbett, G.G.   Storage of Food Crops and Inputs
Gifford, R.C.   Farm Mechanization
Lester, T.   Mechanization of Irrigated Crop
                                            Production
Lundström, B.   Rural Construction
Oodally, G.M.   Farm Power and Machinery
Vacant post   Systems Engineering

Food and Agricultural Industries Service
Asselbergs, E.A.   Chief
Barreveld, W.H.   Senior Officer   Agricultural Industries
Barat, S.K.   Agricultural Industries (Hides and Skins
                                            and Animal Byproducts)
Karam, R.   Agricultural Industries (Natural Fibres)
Vacant post   Senior Officer   Food Industries
Faure, J.C.   Food Industries
Vacant post   Food Industries
Vacant post   Agricultural Industries (Near East)

Land and Water Development Division
Dudal, R.J.   Director
Vivekananthan, T.   Senior Officer

Water Resources, Development and Management Service
Horning, H.M.   Chief
Dieleman, P.J.   Senior Officer   Drainage and Land Reclamation
Florin, R.   Land and Water Development

Water Resources Group
Thomas, R.G.   Senior Officer   Water Resources Planning
Vacant post   Water Use Planning
Sagardoy, J.A.   Water Resources
Underhill, H.W.   Hydrology
Radelet, O.   Water Planning

Water Development Group
Mather, T.H.   Senior Officer   Water Resources Development
Vermeiren, L.   Water Development
Siegert, K.   Hydraulic Eng.

Water Management Group
Vacant post   Senior Officer   Water Management
Smith, M.   Water Management

Vacant post   Irrigation Eng.
Vacant post   Water Quality

Soil Resources, Management and Conservation Service
Hauck, F.W.   Chief
Higgins, G.M.   Coordinator   Agro-ecological Zones Proj.

Soil Resources Group
Pecrot, A.J.   Senior Officer

Soil Management and Conservation Group
Arens, P.   Senior Officer
Sanders, D.W.   Senior Officer   Soil Conservation

Remote Sensing Centre
Howard, J.A.   Chief
Kalensky, Z.D.   Senior Officer   Technical Support Group

Ref S 673 .I68 1983

International directory of
agricultural engineering